Protein NMR Techniques

METHODS IN MOLECULAR BIOLOGY™

John M. Walker, SERIES EDITOR

METHODS IN MOLECULAR BIOLOGY™

Protein NMR Techniques

Second Edition

Edited by

A. Kristina Downing

Department of Biochemistry, University of Oxford,

Oxford, UK

HUMANA PRESS ✳ TOTOWA, NEW JERSEY

© 2004 Humana Press Inc.
999 Riverview Drive, Suite 208
Totowa, New Jersey 07512

www.humanapress.com

This publication is printed on acid-free paper. ∞
ANSI Z39.48-1984 (American Standards Institute)

Permanence of Paper for Printed Library Materials.

Production Editor: Angela L. Burkey
Cover design by Patricia F. Cleary.

Cover illustration: Background: Figure 3A from Chapter 19, "Membrane Protein Structure Determination Using Solid-State NMR," by Anthony Watts, Suzana K. Straus, Stephan L. Grage, Miya Kamihira, Yuen Han Lam, and Xin Zhao. Inset: Figure 5 from Chapter 15, "NMR Studies of Protein–Nucleic Acid Interactions," by Gabriele Varani, Yu Chen, and Thomas C. Leeper.

For additional copies, pricing for bulk purchases, and/or information about other Humana titles, contact Humana at the above address or at any of the following numbers: Tel.: 973-256-1699; Fax: 973-256-8341; E-mail: humana@humanapr.com; or visit our Website: www.humanapress.com

Photocopy Authorization Policy:

Printed in the United States of America. 10 9 8 7 6 5 4 3 2 1

e-ISBN: 1-59259-809-9

ISSN:1064-3745

Library of Congress Cataloging-in-Publication Data

Protein NMR techniques / edited by A. Kristina Downing. -- 2nd ed.
    p. ; cm. -- (Methods in molecular biology ; 278)
    Includes bibliographical references and index.
    ISBN 1-58829-246-0 (alk. paper)
    1. Proteins--Analysis. 2. Nuclear magnetic resonance spectroscopy.   I. Downing, A. Kristina (Anna Kristina) II. Series: Methods in molecular biology (Clifton, N.J.) ; v. 278.
    [DNLM: 1. Proteins--analysis. 2. Nuclear Magnetic Resonance, Biomolecular--methods. W1 ME9616J v.278 2004 / QU 55
    P9672 2004]
    QP551.P3976 2004
    572'.6--dc22

                                                                                    2003027535

Preface

When I was asked to edit the second edition of *Protein NMR Techniques*, my first thought was that the time was ripe for a new edition. The past several years have seen a surge in the development of novel methods that are truly revolutionizing our ability to characterize biological macromolecules in terms of speed, accuracy, and size limitations. I was particularly excited at the prospect of making these techniques accessible to all NMR labs and for the opportunity to ask the experts to divulge their hints and tips and to write, practically, about the methods.

I commissioned 19 chapters with wide scope for *Protein NMR Techniques*, and the volume has been organized with numerous themes in mind. Chapters 1 and 2 deal with recombinant protein expression using two organisms, *E. coli* and *P. pastoris,* that can produce high yields of isotopically labeled protein at a reasonable cost. Staying with the idea of isotopic labeling, Chapter 3 describes methods for perdeuteration and site-specific protonation and is the first of several chapters in the book that is relevant to studies of higher molecular weight systems. A different, but equally powerful, method that uses molecular biology to "edit" the spectrum of a large molecule using segmental labeling is presented in Chapter 4. Having successfully produced a high molecular weight target for study, the next logical step is data acquisition. Hence, the final chapter on this theme, Chapter 5, describes TROSY methods for structural studies.

In Chapters 6–12 of *Protein NMR Techniques*, the focus shifts to studies of aligned molecules, beginning with Chapter 6, which describes different options for the preparation of an aligned sample. So many labs have contributed to the development of new media that I believe it will be particularly useful to have this information summarized in an easily digestible format. Residual dipolar coupling (RDC) data acquisition and incorporation into structure calculations are presented in an equally straightforward manner. Chapters 6 and 7 make it clear that RDCs, a powerful source of structural data that are complimentary to NOE-derived distance restraints, can be measured and used routinely. An exciting chapter on the use of RDCs to study protein dynamics highlights the range of information accessible using these methods and, also, their particular potential to inform on motions on a timescale that previously has not been accessible. Having opened the discussion on RDCs and relaxation, there is an elegant description of how the methods can be combined to yield insight into the properties of multimodule constructs.

Three chapters round out this section: Chapter 10 discusses the correct interpretation of relaxation data and the dangers of ignoring anisotropy; Chapter 11 introduces TROSY methods for the study of dynamics; and Chapter 12 provides a detailed and fascinating account of new methods for the characterization of intermediate timescale motions.

NMR studies often share common methodological features, and, in other features, they necessarily vary. Chapters 13–15, discussing studies of partially folded states and of protein–protein and protein–nucleic acid complexes, are designed to highlight three special areas of interest. A computational section focusing on the automation of both assignment and structure calculation methods follows. Having personally spent much time on these aspects of structure determination, I find these methods particularly exciting because of their potential to significantly expedite the process. Finally, Chapter 19 comprehensively reviews the application of solid state methods to the study of membrane proteins, a particularly important but difficult class of targets.

Editing *Protein NMR Techniques* has been more work than I had expected, but it was also even more rewarding than I anticipated. I hope that you will find it to be a valuable resource in your research.

A. Kristina Downing

Contents

Contributors

JENNIFER BLAKE-HALL • *Department of Chemistry and Biochemistry, University of Maryland, College Park, MD*

RAFAEL BRÜSCHWEILER • *Gustaf H. Carlson School of Chemistry and Biochemistry, Clark University, Worcester, MA*

YU CHEN • *Departments of Biochemistry and Chemistry, University of Washington, Seattle, WA*

DAVID COWBURN • *New York Structural Biology Center, New York, NY*

EVA DE ALBA • *Laboratory of Biophysical Chemistry, National Heart, Lung, and Blood Institute, National Institutes of Health, Bethesda, MD*

A. KRISTINA DOWNING • *Department of Biochemistry, University of Oxford, Oxford, UK*

KIERAN FLEMING • *Department of Biological Sciences, Imperial College of Science, Technology and Medicine, London, UK*

DAVID FUSHMAN • *Department of Chemistry and Biochemistry, University of Maryland, College Park, MD*

XIAOLIAN GAO • *Department of Chemistry, University of Houston, TX*

STEPHAN L. GRAGE • *Biomembrane Structure Unit, Department of Biochemistry, University of Oxford, Oxford, UK*

CHRISTIAN GRIESINGER • *Max Planck Institute for Biophysical Chemistry, Göttingen, Germany*

PETER GÜNTERT • *RIKEN Genomic Sciences Center, Yokohama, Japan*

MICHAEL HABECK • *Unité de Bio-Informatique Structurale, Institut Pasteur, Paris, France*

LORRAINE HEWITT • *Laboratory of Molecular Biophysics, Department of Biochemistry, University of Oxford, Oxford, UK*

BRUCE A. JOHNSON • *Molecular Systems, Merck Research Labs, Rahway, NJ*

MIYA KAMIHIRA • *Biomembrane Structure Unit, Department of Biochemistry, University of Oxford, Oxford, UK*

JAMES G. KEMPF • *Department of Chemistry, Yale University, New Haven, CT*

YUEN HAN LAM • *Biomembrane Structure Unit, Department of Biochemistry, University of Oxford, Oxford, UK*

ERNEST D. LAUE • *Department of Biochemistry, University of Cambridge, Cambridge, UK*

THOMAS C. LEEPER • *Department of Biochemistry and Department of Chemistry, University of Washington, Seattle, WA*

DONGHAI LIN • *Department of Biochemistry, The Hong Kong University of Science and Technology, Hong Kong, SAR, People's Republic of China*

JENS P. LINGE • *Unité de Bio-Informatique Structurale, Institut Pasteur, Paris, France*

J. PATRICK LORIA • *Department of Chemistry, Yale University, New Haven, CT*

STEPHEN MATTHEWS • *Department of Biological Sciences, Imperial College of Science, Technology and Medicine, London, UK*

JAMES M. MCDONNELL • *Laboratory of Molecular Biophysics, Department of Biochemistry, University of Oxford, Oxford, UK*

JENS MEILER • *Department of Molecular Biochemistry, University of Washington, Seattle, WA*

HELEN R. MOTT • *Department of Biochemistry, University of Cambridge, Cambridge, UK*

TOM W. MUIR • *Laboratory of Synthetic Protein Chemistry, The Rockefeller University, New York, NY*

PETER R. NIELSEN • *Department of Biochemistry, University of Cambridge, Cambridge, UK*

DANIEL NIETLISPACH • *Department of Biochemistry, University of Cambridge, Cambridge, UK*

MICHAEL NILGES • *Unité de Bio-informatique Structurale, Institut Pasteur, Paris, France*

JOANNE M. O'LEARY • *Department of Biochemistry, University of Oxford, Oxford, UK*

JENNIFER J. OTTESEN • *Laboratory of Synthetic Protein Chemistry, The Rockefeller University, New York, NY*

WOLFGANG PETI • *Department of Molecular Biology, The Scripps Research Institute, La Jolla, CA*

ANDREW R. PICKFORD • *Department of Biochemistry, University of Oxford, Oxford, UK*

CHRISTINA REDFIELD • *Oxford Centre for Molecular Sciences, University of Oxford, Oxford, UK*

WOLFGANG RIEPING • *Unité de Bio-Informatique Structurale, Institut Pasteur, Paris, France*

LUKE M. ROONEY • *Department of Biochemistry, University of Oxford, Oxford, UK*

SACHCHIDANAND • *Department of Biochemistry, University of Oxford, Oxford, UK*

ALEXANDER SHEKHTMAN • *New York Structural Biology Center, New York, NY*

KATHERINE M. STOTT • *Department of Biochemistry, University of Cambridge, Cambridge, UK*

SUZANA K. STRAUS • *Department of Chemistry, University of British Columbia, Vancouver, British Columbia, Canada*

ABARNA THIRU • *Department of Biochemistry, University of Cambridge, Cambridge, UK*

NICO TJANDRA • *Laboratory of Biophysical Chemistry, National Heart, Lung and Blood Institute, National Institutes of Health, Bethesda, MD*

GABRIELE VARANI • *Department of Biochemistry and Department of Chemistry, University of Washington, Seattle, WA*

OLIVIER WALKER • *Department of Chemistry and Biochemistry, University of Maryland, College Park, MD*

ANTHONY WATTS • *Biomembrane Structure Unit, Department of Biochemistry, University of Oxford, Oxford, UK*

JÖRN M. WERNER • *Department of Biochemistry, University of Oxford, Oxford, UK*

YOULIN XIA • *Department of Chemistry, University of Houston, TX*

RONG XU • *New York Structural Biology Center, New York, NY*

XIN ZHAO • *Biomembrane Structure Unit, Department of Biochemistry, University of Oxford, Oxford, UK*

GUANG ZHU • *Department of Biochemistry, The Hong Kong University of Science and Technology, Hong Kong, SAR, People's Republic of China*

1

Screening and Optimizing Protein Production in *E. coli*

Lorraine Hewitt and James M. McDonnell

Summary

Significant improvements in the technologies used for protein production have been driven by impending genome-scale proteomics projects. These initiatives have favored *Escherichia coli*-based expression systems, which allow rapid cloning and expression of proteins at low cost. The range of commercially available molecular biology kits, vectors, affinity tags, and host cell lines have increased dramatically in recent years. For the structural biology community, where protein production is often a rate-limiting step, these developments have made the process of producing and purifying large amounts of protein for structural studies simpler and faster. The large-scale automated screening approaches for optimizing protein production employed by structural genomics initiatives can be adapted to a more practical targeted approach appropriate for individual structural biology groups. This chapter describes simple, rapid screening methods for testing optimal vector/host combinations using a 96-well format.

Key Words: Structural biology; protein expression; vectors; fusion proteins; affinity tags; restriction endonucleases.

1. Introduction

With the recent successes in whole genome sequencing, the next great challenge for the biochemistry community will be efforts toward describing the structural genome, the characterization of the interaction proteome, and studies of functional proteomics. Nuclear magnetic resonance (NMR) spectroscopy could potentially play a significant role in aspects of all of these projects. Structural and functional analysis of gene products often requires substantial quantities of proteins, and production of this material represents the bottleneck for high-throughput structural studies. Although many functional genomics approaches, such as "proteome chip"-based research, will require much smaller amounts of any individual protein, production and purification of this

From: *Methods in Molecular Biology, vol. 278: Protein NMR Techniques*
Edited by: A. K. Downing © Humana Press Inc., Totowa, NJ

protein will still be the rate-limiting step in these studies. Consequently, there have been significant efforts in recent years to improve the efficiency of protein production.

In practice, there are only a few structural biology laboratories that actually apply true high-throughput approaches to solving structures, but clearly many laboratories benefit from recent improvements in the technologies involved in protein production. The requirements for rapid cloning, expression, and purification of gene products has resulted in development of new vectors, cell lines, and purification protocols that are of general benefit to the structural biology community. In a structural genomics initiative, the process of choosing targets may be rather different than for a structural biology group. Whereas the "low-hanging fruit" represents the primary targets for structural genomics *(1)*, the targets of most structural biology laboratories generally are dictated by the functional activity of the proteins. But the overall aims of maximizing the efficiency of the approach and minimizing the personnel time involved in the process remain the same for both disciplines. As an NMR group with a focus on protein structure and interactions, the production of protein is very often a limiting step, and any improvements in this area are important for our research productivity.

Traditional approaches to protein production offer many choices to the experimenter. The first decision is which expression system to use. Because of its well-characterized genetics, ease of manipulation, and low cost, *Escherichia coli*-based expression systems are the most commonly used. If the protein requires posttranslational modification (glycosylation, phosphorylation, etc.) for biological activity, then it may be impossible to produce a functional product in *E. coli*, but there are other systems, including yeast, insect, mammalian, or cell-free in vitro translation methods, that are capable of these modifications. The requirement for metabolic labeling (^2H, ^{13}C, ^{15}N) of proteins for NMR studies makes the microbial systems far more practical and affordable; nevertheless, it is possible, albeit expensive, to produce labeled protein in baculovirus and mammalian expression systems *(2,3)*. The cell-free in vitro translation systems that produce proteins using bacterial extracts continue to show improvements in yields *(4,5)* and successes in labeling *(6)*, but these methods are not in general use, and their successful application for high-level protein production has been limited to a few groups. In recent years, the methylotrophic yeast *Pichia pastoris* has become a powerful host for the heterologous expression of proteins; however, at present, laborious cloning protocols limit it as an expression system for rapid screening. A chapter on protein production and isotopic labeling in the yeast system *Pichia pastoris* is found elsewhere in this book (*see* **Chapter 2**). This chapter focuses on *E. coli*-based protein expression.

1.1. From Genes to Proteins

The first step in this process is obtaining the complementary DNA (cDNA) of the gene of interest. Deriving a cDNA from a gene without introns is a straightforward application of recombinant DNAmethods. Deriving cDNAs from a gene with introns has historically been made more challenging because of the lack of availability of a full-length cDNA within an existing library, or a more laborious preparation of full-length cDNA from messenger RNA (mRNA) *(7)*. However, ongoing efforts to produce nearly complete cDNA libraries (such as the Mammalian Gene Collection *(8)* should soon simplify this process. An alternative, but more costly, approach is the commercially available whole gene synthesis *(9)*. It is common practice to establish the cDNA in a cloning vector, which is then used as the source for subcloning into expression vectors. The subcloning process ideally should be "cut and paste" in character. For the sake of improved efficiency and reliability this is routinely achieved by establishing a universal set of restriction endonuclease sites for the cloning vector and different expression vectors. Several new commercial systems are moving away from the use of restriction enzymes in the subcloning process altogether—instead employing a universal entry clone and then a series of compatible recombination vectors. The commercial systems Gateway™ from Invitrogen (Carlsbad, CA) and Creator™ from BD Biosciences (Palo Alto, CA) are examples of this approach *(10)*.

There is a vast array of available bacterial expression vectors. The standard approach is to use an inducible T7 RNA polymerase promoter to drive protein production. It is often convenient to produce the gene product as a fusion protein, using the fusion partner to increase expression levels, improve solubility, target to defined subcellular locations, or merely to act as affinity tags to facilitate protein purification. Fusion proteins with cleavable glutathione *S*-transferase (GST) and polyhistidine, at either N- or C-termini, have been the most commonly employed, with the polyhistidine tag being the choice of many large-scale automated structural genomics efforts. This may be particularly attractive to structural studies by NMR, in which the presence of a short flexible terminus generally is innocuous for structural analysis.

The vectors can then be transformed into numerous available *E. coli* host strains. Some of the host strains are good generic lines for protein production, whereas others have more specialized function. For example, the Rosetta™ (Novagen, Madison, WI) and Codon Plus™ (Stratagene, Cedar Creek, TX) lines encode genes for transfer RNA (tRNA) for rare codons, allowing for more efficient production of eukaryotic proteins that contain rarely used *E. coli* codons. Certain vector/host combinations are optimized for specialized function. Novagen uses a pET-32 vector to make a thioredoxin fusion protein in the host strain Origami™, which provides mutations in thioredoxin reductase (*trxB*) and

Table 1
Standard Expression Vectors

Vector	Supplier	Characteristics
pET vector series	Novagen	IPTG-inducible T7 promoter *(11)*. Broad range of vectors giving wide choice of multiple restriction sites and affinity tags. For example: pET32 series has a Trx fusion and hexahistidine tag. pET41 series is a GST fusion system with both hexahistidine and S-tag options. pET43.1 series has a hexahistidine affinity tag and a NusA fusion for solubility.
pCAL vectors	Stratagene	CBP (calmodulin-binding peptide) affinity tag.
pMAL vectors	New England Biolabs	Maltose Binding Protein affinity tag with protease cleavage site; pMAL-p2x has a signal sequence to direct protein to periplasm, which improves solubility.

Table 2
Standard Expression Hosts

Cell line	Supplier	Characteristics
BL21 (DE3) B834	Various	Use with T7 promoter systems.
C41 (BL21 derived)	Not commercially available *(12)*	As BL21 but adapted for increased expression of "toxic" proteins.
BL21 AI	Invitrogen	Arabinose induced tighter regulation to improve expression of toxic proteins.
BL21 Codon Plus™	Stratagene	Codon usage bias.
AD494	Novagen	*TrxB* mutants to facilitate disulfide bond formation.
Origami™ OrigamB™	Novagen	*Trx/gor* mutants enhance disulfide bond formation giving high yields of properly folded, active protein.
Rosetta-gami™	Novagen	Have both the codon usage plasmid to adjust for codon bias as well as the *trxB* and *gor* mutations.

glutathione reductase (*gro*) genes, to enhance disulfide formation in the bacterial cytoplasm. A list of standard expression vectors and hosts can be found in **Tables 1** and **2**.

Unfortunately no single expression vector or host line works well for all protein constructs; lack of success is typically because of either insufficient protein production or solubility. To date, it has not been possible to reliably predict the expression characteristics of an untested protein. Therefore, it is advisable to screen a matrix of expression vector/host combinations and culture conditions.

It is clear that most of the steps in protein production, as simple liquid handling protocols, have the potential to be handled robotically in an automated process. Steps that do not lend themselves to robotic handling can be replaced, such as using magnetic bead-based separations instead of centrifugation steps. However, liquid-handling robots have yet to replace centrifuges in most laboratories. Thus, in the interest of practical applicability, this chapter focuses on more targeted screening approaches for testing protein expression using equipment routinely available in most structural biology laboratories.

2. Materials

2.1. Polymerase Chain Reaction (PCR) and Agarose Gels

The percentage of agarose gel is dependent on the size of the DNA to be analyzed (*see* **Table 3**).

DNA fragments <500 bp may be better analyzed using acrylamide gel electrophoresis *(13)*.

Agarose gels can be run in either TAE or TBE buffer.

TAE/L of 50X	TBE/L of 10X
242 g Tris base	109 g Tris base
57.1 mL glacial acetic acid	55.6 g boric acid
100 mL 0.5 *M* ethylenediaminetetraacetic acid (EDTA), pH 8.0 (*see* **Note 1**)	9.3 g EDTA

These buffers are stored at room temperature.

Table 3
Agarose Gel Percentages

Size range (kb)	% of Gel
0.2–3	2.0
0.3–5	1.5
0.4–12	1.0
0.5–20	0.8

Table 4
Antibiotic Stock Solutions

Antibiotic	Concentration	Solvent	Final concentration/dilution
Ampicillin	100 mg/mL	Water	100 µg/mL (1/1000)
Carbenicillin	100 mg/mL	Water	100 µg/mL (1/1000)
Chloramphenicol	4 mg/mL	Ethanol	170 µg/mL (1/200)
Kanamycin	10 mg/mL	Water	50 µg/mL (1/200)
Streptomycin	10 mg/mL	Water	50 µg/mL (1/200)
Tetracyclin	5 mg/mL	Ethanol	50 µg/mL (1/200)
(*see* **Note 4**)			

2.2. Acrylamide Gels

1. Acrylamide, *N,N'*-methylene-*bis*-acrylamide (bis-acrylamide), 10% ammonium persulfate (APS), and *N,N,N',N'*-tetramethylethylenediamine (TEMED) are all available from Sigma-Aldrich (Gillingham, UK).
2. 30% acrylamide/0.8% bisacrylamide: Mix 30 g acrylamide and 0.8 g *N,N'*-methylene-*bis*-acrylamide in a total volume of 100 mL water.
3. 4X Tris-HCl/sodium dodecyl sulfate (SDS), pH 8.8: Dissolve 182 g Tris base and 4 g SDS in 600 mL water. Adjust pH to 8.8 with HCl. Add water to total volume of 1000 mL.
4. Both of these stock solutions should be filtered through a 0.45-µ*M* filter and stored at 4°C in the dark.

2.3. Ethidium Bromide Solution for Visualization of DNA in Agarose Gels

Ethidium bromide (Sigma-Aldrich) stock solution of 10 mg/mL in water, TAE, or TBE (*see* **Note 2**). Store at 4°C. Dilute stock solution to 0.5 mg/mL in water, TAE, or TBE.

2.4. LB Plates/Liter

Add 15 g Agar to the above Luria Bertani (LB) reagents; make up to 1 liter and autoclave. Allow to cool to <40°C before adding antibiotics (*see* **Table 4**) and/or nutrients before pouring into culture plates. Allow approx 25 mL per plate. The mix can be separated into smaller volumes providing that the agar is evenly distributed through the solution prior to autoclaving. Plates can be stored at 4°C for up to 2 wk.

2.5. Stock Solutions for Antibiotics

Stock solutions should be filter-sterilized and stored in 500 µL or 1-mL aliquots at −20°C (*see* **Note 3**). Thaw and refreeze cycles should be kept to a minimum to avoid loss of activity.

3. Methods

3.1. Standard PCR

Cycles of PCR are comprised of three steps: melting of the template DNA so that it is denatured into two strands usually at 95°C for 1–2 min followed by an annealing step so that primers can hybridize to the template DNA. The temperature of the annealing step can be roughly calculated from the composition of the primers according to the Wallace rule (*see* **[i]** below). The third step is extension by a polymerase, and this is carried out at the optimal temperature for the polymerase according to the manufacturer's instructions. On the last cycle, a final extension period of 15–20 min is carried out to ensure that all synthesis is completed. All PCR reactions should be carried out carefully to avoid cross-contamination.

[i] Wallace Rule: Tm (in °C) = [(G+C) × 4 + (A+T) × 2]

There are numerous factors that can affect the success of PCR reactions: annealing temperature—further useful information can be found online (http://www.sigma-genosys.com/oligo_meltingtemp.asp)—template concentration, primer concentration, $MgCl_2$ concentration (normally in supplied buffer), extension time, and number of cycles. Many suppliers have developed kits, for example, FailSafe PCR™ (Epicentre, Madison, WI) or *Taq* PCR Master Mix kit (Qiagen, Valencia), that provide various combinations of the PCR reagents to ensure a successful outcome (*see* **Table 5**). Some suppliers publish helpful handbooks, supplied free of charge, including *Critical Factors for Successful PCR* from Qiagen.

Table 5
PCR Reagent Combinations

Reagent	Volume
10X Buffer (supplied)	5 µL
dNTP mix (*see* **Note 6**)	1 µL
5′ primer (100 pmol) (*see* **Note 7**)	1 µL
3′ primer (100 pmol) (*see* **Note 7**)	1 µL
Template DNA (0–1 µg)	x µL
DMSO (*see* **Note 8**)	2.5 µL
DNA polymerase (*see* **Note 9**)	y µL
Sterile H_2O (*see* **Note 10**)	z µL
TOTAL VOLUME	50 µL

dNTP, deoxynucleotide-triphosphate; DMSO, dimethyl sulfoxide.

In the reactions here, the variable volumes are shown as "x," "y," and "z" and are to be adjusted in accordance with the concentration of the reagents being used to give the final total volume shown (*see* **Note 5**).

3.2. Vector and Insert Digest (See Table 6)

Incubate at 37°C for at least 1 h (*see* **Note 14**).

3.3. Ligation

Ligation reaction is incubated overnight at 14°C (*see* **Table 7**). Alternatively, commercial kits, such as Quick Ligation Kit™ (New England Biolabs, Beverly, MA), are now available to carry out the ligation in 5 min at room temperature.

The insert:vector ratio is a principal consideration in the ligation reaction. The 1:1 ratio can be calculated with this equation:

$$\frac{[(\text{DNA of Vector}) \times \text{size of insert}]}{[\text{size of vector}]}$$

It is useful to set up a range of reactions using 1:1, 3:1, and 5:1 ratios.

3.4. Preparation of Competent Cells: Calcium Chloride Method (see Notes 15 and 16)

1. Prepare overnight culture by picking a single colony from stock plate (or a glycerol stock) into 5 mL of LB.
2. Next morning, inoculate 10 mL of LB with 100 mL from overnight and grow to an optical density $(\text{OD})_{600}$ of approx 0.5.
3. Spin down 1.5-mL aliquots for 5 min at 1000g and remove supernatant.

Table 6
PCR Vector and Insert Digest

Reagent	Volume
10X Buffer (supplied)	5 µL
Phosphatase	1 µL
(*see* **Note 11**)	
Enzyme A	1 µL
Enzyme B	1 µL
BSA (*see* **Note 12**)	x µL
Vector or insert DNA	y µL
(*see* **Note 13**)	
Sterile H$_2$O	z µL
TOTAL VOLUME	50 µL

BSA, bovine serum albumin.

Table 7
Ligation Combinations

Reagent	Volume
10X Buffer	5 μL
Vector DNA	x μL
Insert DNA	y μL
T4 DNA Ligase	1 μL
Sterile H$_2$O	z μL
TOTAL VOLUME	10 μL

4. Add 500 μL of ice-cold 100 m*M* CaCl$_2$ (*see* **Note 17**), carefully resuspend pellet, and place tubes on ice for 1 h.
5. Spin down tubes again for 5 min at 1000*g*, remove supernatant, resuspend in 200 μL of ice-cold 100 m*M* CaCl$_2$, and place on ice for 2 h or more (up to overnight if required).

3.5. Transformation of Competent Cells

1. Add 2 μL of plasmid DNA (10–100 ng), mix thoroughly by flicking tube, and put on ice for 30 min to 1.5 h.
2. Heat shock tubes by placing in water bath at 42°C; then immediately place back on ice for 2 min.
3. Add 1 mL of sterile LB to each transformation, and incubate for 30 min to 1 h before plating out onto LB plates containing the appropriate antibiotic for the system (*see* **Note 18**). Place overnight in 37°C incubator.

3.6. Trial Growths to Assess Protein Expression

Conditions for optimal growth should be determined on a small scale before growing larger cultures. Typically, we will test the expression profile characteristics of four or five different expression vectors, transformed into roughly the same number of host cell lines. Each matrix of vector/host pairs is then grown at several different temperatures. If nothing is known about expression characteristics of the protein, then we will screen a diverse selection of vectors and hosts. If we know more about the protein, or a closely related homolog, then the matrix might explore a smaller set of more closely related conditions. The small-scale test growths are performed in deep-well microtiter plates (Biomek 96, BD Biosciences).

1. Pick a single colony from the plate and inoculate into 2 mL media (*see* **Note 19**) containing the appropriate antibiotic for the system used. Grow overnight at 37°C.
2. Inoculate 50 μL of the overnight culture into 2 mL LB containing the appropriate antibiotic, and grow (at 18°C, 28°C, and 37°C) until OD$_{600}$ = 0.6–0.8. Remove

0.5-mL aliquot to prepare an uninduced sample. The induction method is determined by the cell line and plasmid used and may be a chemical additive (e.g., IPTG).

3. Protein expression can be monitored by removing aliquots at various time-points, for example, at hourly intervals, and analyzed with SDS-polyacrylamide gel electrophoresis (SDS-PAGE) *(15)*. It is recommended that both soluble- and whole-cell pellets be analyzed. Samples are prepared by pelleting cells and decanting the supernatant. Cell pellets are then resuspended in 200 µL buffer (50 mM Tris-HCl, 100 mM NaCl, 1 mM EDTA, pH 7.4) and sonicated. For the SDS-PAGE sample, add 50 µL SDS sample buffer to 50 µL cell lysate.

For the whole-cell sample, centrifuge the remaining lysate and resuspend in 50 µL sample buffer. Whole-cell samples should be heated to 100°C for 5 min and the soluble fraction for 2 min.

A higher volume of sample buffer and longer boiling time helps to reduce the viscosity of the cell lysate sample to facilitate sample loading.

4. Once growth conditions have been determined larger volumes can be tried, such as 1 or 2 L in baffled flasks or even a fermenter in which aeration and pH during growth can be controlled.

For larger volumes, cells can be disrupted by various methods, including sonication (with or without the addition of lysozyme) or french press. There are commercially available additives for protein extraction, for example, BugBuster™, PopCulture™ (Novagen), or CellLytic™ reagents (Sigma-Aldrich). Nucleic acid contamination can be removed by various methods including deoxyribonuclease I, benzonase (supplied with the BugBuster kit™), or polymin P precipitation *(16)*.

5. When a transformant has been isolated that expresses the protein of interest, 500-mL aliquots of glycerol stock should be prepared by adding sterile LB/glycerol to a final concentration of 25%. Aliquots should be rapidly frozen in liquid nitrogen or dry ice/ethanol and stored at –80°C. In addition, plasmid DNA from the transformant should be prepared using either a commercial kit (Promega, Madison, WI) or the alkaline lysis method *(13)*.

3.7. Approaches to Improving Protein Solubility

When substantial yields of protein can be obtained, but the protein is insoluble, various approaches can be tried (*see* **Note 20**). Probably the simplest approach to solubilize the protein can be made by adjusting growth conditions (*see* **Note 21**).

1. Cell cultures can be grown at 37°C until they reach the point of induction, and then the temperature should be reduced to between 18 and 28°C. It should be noted that the reduction in temperature will slow down cell growth as well as protein synthesis so that the length of induction time should be adjusted accordingly.

2. Subclone into a vector that has a weaker promoter than the T7 (e.g., trc) or use a plasmid that has a lower copy number.
3. Reduce the amount of the inducer.
4. Different growth media can be tried including sorbitol-betaine *(17)* (*see* **Note 19**). However, if ultimately the protein is to be expressed in minimal media, then changing the growth media may not be very useful, although there are some additives such as yeast nitrogen base that can be added to enrich minimal media without affecting isotopic labeling.
5. Other additives that can help are ethanol, NaCl, and low molecular weight thiols and disulfides or cofactors required by the protein *(18)*.
6. Use an *E. coli* system that coexpresses a molecular chaperone such as DNAK-DNAJ-GrpE. There are also *E. coli* systems that coexpress foldases, which can facilitate the pathway of protein folding, such as DsbA and DsbC *(19)*.
7. There are vectors that contain a signal sequence to direct the protein of expression to the periplasm such as the pMAL vector that produces a fusion protein of MBP and the protein of interest. This fusion protein can also be directed to the oxidizing environment of the periplasm where it can be isolated by osmotic shock. There are other fusion proteins available that can enhance solubility including GST, pDEST™ (Invitrogen), or NusA (Nus.Tag™ pET43.1 series, Novagen) *(20)*. Fusion protein tags are also useful when they can be used as affinity tags for protein purification.
8. The choice of host cell lines, such as AD494 and Origami™ (Novagen), can improve solubility of the expressed protein.
9. It has been shown that thioredoxin and glutathione have a role in the correct formation of disulfide bonds, and cell lines are available that have mutations in the thioredoxin reductase (*trx*B) and glutathione reductase (*gor*) genes (**Table 2**), which can improve disulfide bond formation in the bacterial cytoplasm.
10. If a biological assay is available, try targeted proteolysis to isolate active fragments of the protein. Often, constructs of a smaller fragment prove to be more soluble than the full-length construct.
11. If the protein can only be expressed as inclusion bodies, denaturing reagents, such as guanidine hydrochloride or urea, can be used under reducing conditions to unfold and refold the protein *(21)*. An alternative is the "sparse matrix" described by Lindwall and coworkers *(22)*.

3.8. SDS-PAGE Electrophoresis

1. Gel percentage should be chosen on the basis of molecular weight of the protein of interest (*see* **Tables 8** and **9**).
2. Prepare gel kit glass plates by cleaning with ethanol and drying off with tissue; assemble kit according to the manufacturer's instructions.
3. A gel mix of resolving gel buffer, water, and acrylamide (*see* **Note 22**) can be prepared up to 1 d in advance, *but* the APS and TEMED should be added just before pouring the gel.
4. Prepare resolving gel mix, add 10% APS and TEMED, and pour between glass plates. Overlay poured gel gently with either deionized H_2O or water-saturated

Table 8
SDS-PAGE Gel Percentages

Range	Molecular weight range (Daltons)	Gel concentration range
High	215,000–40,000	6–8%
Mid	97,500–14,400	8–15%
Low	31,000–3500	16–20%

Table 9
Acrylamide Percentages and Volumes

Component	Volume for different percentages of acrylamide				
	8%	10%	12%	15%	20%
4X Tris-HCl/ SDS, pH 8.8	2.5 mL	2.5 mL	2.5 mL	2.5 mL	2.5 mL
Water	4.8 mL	4.2 mL	3.5 mL	2.5 mL	2.5 mL
30% acrylamide/ 0.8% bis-acrylamide	2.7 mL	3.3 mL	4.0 mL	5 mL	6.7 mL
10% APS	50 μL	50 μL	50 μL	50 μL	50 μL
TEMED	8 μL	8 μL	8 μL	8 μL	8 μL

isopropanol. Leave to polymerize. Prepare stacking gel mix, remove overlay with filter paper, add APS and TEMED to stacker, pour, and insert combs. Standard recipes can be found in *The Protein Protocols Handbook (23)* or online (http://www.bio-rad.com/).

A range of ready-prepared acrylamide mixes can be purchased from various suppliers, including National Diagnostics (Atlanta, GA), Sigma-Aldrich, and Fisher Scientific (Loughborough, UK) (*see* **Note 23**).

4. Notes

1. EDTA is poorly soluble at low pH; therefore, when making stock solutions of 0.5 *M* or 1 *M*, it is advisable to add sodium hydroxide pellets and adjust to pH 8.0.
2. Ethidium bromide is a mutagen and carcinogen and should be handled with great care.
3. Some suppliers recommend alternative concentrations for antibiotics used with their cell lines (e.g., Novagen Origami™ cell lines), and users should refer to the manual supplied.
4. Tetracyclin is light sensitive so container should be foil wrapped.
5. There are several important factors regarding the success of all DNA reactions outlined previously:
   a. Enzymes should also be kept in the freezer until required and then kept on ice.

    b.   Always add the enzyme to the reaction last, and then mix all the components by flicking the tube. Do not vortex. Adequate mixing of the reaction mixtures may seem trivial but is an important factor for a successful outcome.

6.   dNTPs can be prepared by mixing equimolar amounts of dATP, dCTP, dGTP, and dTTP to give a 10 mM solution of each. Many suppliers provide ready-mixed dNTPs with each nucleotide at 10 mM concentration.

7.   Oligonucleotide primers can be supplied by many companies, including MWG Biotech (Ebersberg, Germany), Sigma-Aldrich, and Thermo Electron (Ulm, Germany). They are often supplied in lyophilized form and can be made up to 100 pmol/µL.

8.   The addition of DMSO is optional.

9.   There are a range of polymerases, including the basic *Taq* polymerase (New England Biolabs) as well as others that have additional proofreading capabilities, such as Pwo (Roche Diagnostics, Basel, Switzerland) and KOD (Novagen), which has a faster elongation rate.

10.   It is helpful to use control reactions, including a negative control, where DNA template is replaced with water; others can include vector-only and no-insert controls; template DNA, but no primers; with and without DMSO.

11.   The phosphatase catalyzes the hydrolysis of nucleotide-5′ phosphate groups, thus preventing religation and recircularization of a linearized vector. There are two commonly used reagents for this process: calf intestinal phosphatase (CIP) and shrimp alkaline phosphatase (SAP). In contrast to CIP, SAP can be irreversibly inactivated by heating at 65°C for 15 min. Digest products should be cleaned to remove enzymes, especially the phosphatase; this prevents dephosphorylation of the insert DNA, which would result in an unsuccessful ligation reaction.

12.   BSA: Addition of BSA is dependent on the restriction enzyme used and should be added according to the supplier's advice.

13.   Volume of DNA used should contain approx 1 µg of purified DNA.

14.   A complete digest of the vector is essential because undigested vector can efficiently transform competent bacterial cells and result in a high background of transformants. The digest can be monitored by agarose gel electrophoresis.

15.   An alternative method using rubidium chloride giving higher transformation efficiency can be found on-line (http://www.nwfsc.noaa.gov/protocols/rbc).

16.   Although in nature DNA can be taken up by bacterial cells, it is a very inefficient process. Even with $CaCl_2$ treatment, transformation of bacterial cells is not particularly efficient and can vary markedly on the cell line and/or plasmid used *(13)*.

17.   Calcium chloride solution should be prepared fresh each time from an autoclaved 1 M stock diluted with sterile H_2O.

18.   Unless you know your system very well, it is a good idea to take 100 µL from the tube and plate out, then spin down remainder, take off 800 µL LB, resuspend pellet in the remaining 100 µL, and spread onto plate to give a 10X concentration of cells.

19.   Place reagents in large baffled flask for increased aeration of cultures—for example, 2-L flask for 500 mL or 1 L growths and seal with a stopper that has been wrapped with foil. After autoclaving, caps on bottles should be fastened, and

flasks and bottles can be stored at room temperature indefinitely. One of the more routinely used media for bacterial expression is LB. Alternative media can be tried to optimize protein expression broth including terrific broth (TB), although for isotopic labeling of proteins for NMR minimal media is used *(24)*. There are commercially available kits that provide different compositions of media to enable screening for optimal protein production (Molecular Dimensions, Soham, UK).

a. LB/L: 10 g tryptone, 5 g yeast extract, 10 g NaCl.
b. 5X M9 minimal media/L: 30 g Na_2HPO, 15 g KH_2PO_4, 5 g NH_4Cl, 2.5 g NaCl, 15 mg $CaCl_2$ (optional).
c. Sorbitol-betaine media/L: 10 g tryptone, 5 g yeast extract, 1 g NaCl, 182 g sorbitol, 383 mg glycine-betaine.
Autoclave at 121°C for 20 min; allow to cool to <40°C before adding nutrients or antibiotics.

20. It is advisable to follow the suppliers' instructions for cell lines and plasmid vectors, initially, and use this as a starting point for empirical determination of the optimal expression protocol. If using cDNA, the codon usage may not be optimal for the bacterial cell line chosen for expression, but a range of cell lines are now available from numerous suppliers that address this issue This bias of codons rarely used by *E. coli,* being employed generally by other organisms, can result in low levels of expression, early termination, or misincorporation of amino acids.
21. There are optimization screens available to maximize production of the protein of interest such as Media Optimization Kit™ (Molecular Dimensions).
22. Acrylamide is a potent accumulative neurotoxin; gloves should always be worn when handling solutions and gels.
23. If using a ready-made acrylamide mix, the percentage of the gel mix should be taken into account when making the gel mixes. Gel mixes are available from Sigma-Aldrich (40% mix, which can also be used for DNA gel electrophoresis) or Protogel™ (National Diagnostics), a 30% mix economical for everyday gel running.

Appendix: Suppliers' Websites

Many suppliers have very useful technical resources online that provide useful practical information on PCR, restriction enzymes, many molecular biology protocols, protein expression, and purification. Useful Web addresses are listed as follows:

http://www.epicentre.com
http://www.thermohybaid.de
http://www.invitrogen.com
http://www.moleculardimensions.com
http://www.bio-rad.com/
http://www.mwg-biotech.com/
http://www.neb.com

http://www.novagen.com
http://www.promega.com
http://www.qiagen.com
http://www.sigma-aldrich.com
http://www.stratagene.com
http://www.thermohybaid.de/

References

1. Edwards, A. M., Arrowsmith, C. H., Christendat, D., Dharamsi, A., Friesen, J. D., Greenblatt, J. F., et al. (2000) Protein production: feeding the crystallographers and NMR spectroscopists. *Nat. Struct. Biol.* **7**, 970–972.
2. Coughlin, P. E., Anderson, F. E., Oliver, E. J., Brown, J. M., Homans, S. W., Pollak, S., et al. (1999) Improved resolution and sensitivity of triple-resonance NMR methods for the structural analysis of proteins by use of backbone-labeling strategy. *J. Am. Chem. Soc.* **121**, 11,871–11,874.
3. Creemers, A. F. L., Klaassen, C. H. W., Bovee-Geurts, P. H. M., Kelle, R., Kragl, U., Raap, J., et al. (1999) Solid state ^{15}N NMR evidence for a complex Schiff base counterion in the visual G-protein coupled receptor rhodopsin. *Biochemistry* **38**, 7195–7199.
4. Jermutus, L., Ryabova, L. A., and Pluckthun, A. (1998) Recent advances in producing and selecting functional proteins by using cell-free translation. *Curr. Opin. Biotechnol.* **9**, 534–548.
5. Kigawa, T., Yabuki, T., Yoshida, Y., Tsutsui, M., Ito, Y., Shibata, T., et al. (1999) Cell-free production and stable-isotope labeling of milligram quantities of proteins. *FEBS Lett.* **442**, 15–19.
6. Guignard, L., Ozawa, K., Pursglove, S. E., Otting, G., and Dixon, N. E. (2002) NMR analysis of in vitro-synthesized proteins without purification: a high throughput approach. *FEBS Lett.* **524**, 159–162.
7. Harwood, A. J. (1996) Basic DNA and RNA protocols. In *Methods in Molecular Biology,* vol. 58 (Walker, J. M., ed.). Humana, Totowa, NJ, pp.
8. Strausberg, R. L., Feingold, E. A., Klausner, R. D., and Collins, F. S. (1999) The mammalian gene collection. *Science* **286**, 455–457.
9. Stewart, L., Clark, R., and Behnke, C. (2002) High-throughput crystallisation and structure determination in drug discovery. *Drug Discov. Today* **7**, 187–196.
10. Liu, Q., Li, M. Z., Leibham, D., Cortez, D., and Elledge, S. J. (1998) The univector plasmid-fusion system, a method for rapid construction of recombinant DNA without restriction enzymes. *Curr. Biol.* **8**, 1300–1309.
11. Studier, F. W. (1991) Use of T7 RNA polymerase to direct expression of cloned genes. *Meth. Enzymol.* **185**, 286–299.
12. Miroux, B. and Walker, J. E. (1996) Overproduction of proteins in Escherichia coli: mutant hosts that allow synthesis of some membrane proteins and globular proteins at high levels. *J. Mol. Biol.* **260**, 289–298.
13. Sambrook, J. (1989) *Molecular Cloning: A Laboratory Manual,* 2nd ed. Cold Spring Harbor Laboratory Press, Woodbury, NY.

14. Wallace, R. B., Shaffer, J., Murphy, R. F., Bonner, J., Hirose, T., and Itakura, K. (1979) Hybridization of synthetic oligodeoxyribonucleotides to phi chi 174 DNA: the effect of a single base pair mismatch. *Nucleic Acids Res.* **6,** 3543–3557.

15. Laemmli, U. K. (1970) Cleavage of structural proteins during the assembly of the head of the bacteriophage T4. *Nature* **227,** 680–685.

16. Burgess, R. R. (1991) Use of polyethylenimine in purification of DNA-binding proteins. *Meth. Enzymol.* **208,** 3–11.

17. Blackwell, J. R. and Horgan, R. (1991) A novel strategy for production of a highly expressed recombinant protein in an active form. *FEBS Lett.* **295,** 10–12.

18. Georgiou, G. and Valax, P. (1996) Expression of correctly folded proteins in *Escherichia coli. Curr. Opin. Biotechnol.* **7,** 190–197.

19. Battistoni, A., Mazzetti, A. P., and Rotilio, G. (1999) In vivo formation of Cu, Zn superoxide dismutase disulfide bond in *Escherichia coli. FEBS Lett.* **443,** 313–316.

20. Davis, G. D., Elisee, C., Newham, D. M., and Harrison, R. G. (1999) New fusion protein systems designed to give soluble expression in *Escherichia coli. Biotechnol. Bioeng.* **65,** 382–388.

21. Marston, F. A. O. and Hartley, D. L. (1990) Solubilization of protein aggregates. *Meth. Enzymol.* **182,** 264–276.

22. Lindwall, G., Chau, M., Gardner, S. R., and Kohlstaedt, L. A. (2000) A sparse matrix approach to the solubilization of overexpressed proteins. *Protein Eng.* **13,** 67–71.

23. Walker, J. M. (1996) *The Protein Protocols Handbook,* 1st. ed. Humana, Totowa, NJ.

24. Ausubel, F., Brent, R., E., K. R., Moore, a. D., Seidman, J. G., Smith, J. A., and Struhl, K. (1999) Short protocols in molecular biology. *Current Protocols in Molecular Biology.* 4th ed. Wiley, Hoboken, NJ.

2

Isotopic Labeling of Recombinant Proteins From the Methylotrophic Yeast *Pichia pastoris*

Andrew R. Pickford and Joanne M. O'Leary

Summary

The methylotrophic yeast *Pichia pastoris* is now an established expression system for the production of recombinant protein for nuclear magnetic resonance (NMR) studies. It is capable of expressing high levels of heterologous proteins and possesses the ability to perform many of the posttranslational modifications of higher eukaryotes. Here, we describe efficient methods for the production of uniformly ^{13}C,^{15}N-labeled proteins in shake flasks and of uniformly ^{13}C,^{15}N-labeled and ^{2}H,^{13}C,^{15}N-labeled proteins in fermenters. We also provide details of two chromatographic procedures, cation exchange and concanavalin A lectin affinity, that have proved useful in purifying *P. pastoris*-expressed proteins for NMR studies.

Key Words: *Pichia pastoris*; isotopic labeling; fermentation; nuclear magnetic resonance; recombinant protein.

1. Introduction

In recent years, the methylotrophic yeast *Pichia pastoris* has succeeded the baker's yeast *Saccharomyces cerevisiae* as the most-favored lower eukaryotic host for the recombinant expression of heterologous proteins (*1*). Like *S. cerevisiae*, *P. pastoris* can easily be manipulated at the molecular genetic level (*2*) and is capable of many higher eukaryotic protein modifications, such as glycosylation, disulfide-bond formation, and proteolytic processing (*1*). However, *P. pastoris* has numerous significant advantages over *S. cerevisiae*. First, *P. pastoris* prefers a respiratory rather than a fermentative mode of growth and, therefore, does not accumulate the toxic levels of ethanol and acetic acid that restrain cultures of *S. cerevisiae* (*3*). Second, unlike *S. cerevisiae*, which tends to hyperglycosylate *N*-linked sugars, *P. pastoris* glycosylation is more similar to that of higher eukaryotes with 8–14 mannose residues per chain (*4*). Finally,

From: *Methods in Molecular Biology, vol. 278: Protein NMR Techniques*
Edited by: A. K. Downing © Humana Press Inc., Totowa, NJ

in *P. pastoris*, the unusually efficient and tightly regulated promoter from the alcohol oxidase 1 gene (*AOX1*) can be used to drive high-level transcription of the foreign gene, resulting in protein-expression levels many fold higher than those achievable in *S. cerevisiae* ***(1)***.

The early reports of the extraordinary protein-expression levels achievable with *P. pastoris* ***(5)***, together with the simplicity of the cell culture media, which easily allows for isotopic labeling, soon made the system highly attractive to the nuclear magnetic resonance (NMR) community ***(6)***. Its popularity grew rapidly to the point where it is now the second most-used expression host for NMR studies behind *Escherichia coli*.

This chapter describes methods for the production of uniformly ^{13}C,^{15}N- and ^{2}H,^{13}C,^{15}N-labeled proteins for NMR studies using shake-flask and fermenter cultures of *P. pastoris*. Particular attention is paid to the differences in methodology from non-NMR applications that are imposed by the limited availability of isotopically labeled carbon and nitrogen sources and by the specific requirements of deuteration. It is beyond the scope of this chapter to discuss features of the *P. pastoris* expression system in detail. Comprehensive guides and detailed methodologies for the construction and analysis of recombinant strains of *P. pastoris* can be found elsewhere ***(7–10)***. However, given the number of strains and vectors available (and their differing compatibility with the isotopic labeling procedures described here), we begin this chapter with a survey of the *P. pastoris* expression system options.

1.1. The P. pastoris *Expression System*

A wide variety of expression vectors are available from Invitrogen (Carlsbad, CA) offering varying promoters, secretion signal sequences, affinity tags, and antibiotic/antifungal resistance. However, not all of these are compatible with the methods described here for efficient production of isotopically labeled proteins. We do not recommend using the vectors pPIC9, pHIL-D2, or pHIL-S1 supplied in the original *Pichia* expression kit, as these do not allow for the selection of multicopy transformants that is necessary for the highest levels of recombinant expression ***(8)***. Instead, we recommend using either pPIC9K, pPICZα, or pPIC6α, each of which contains a dose-dependent antibiotic resistance gene allowing for multicopy transformant selection through resistance to the antifungals G418 (for pPIC9K), Zeocin® (for pPICZα), or blasticidin (for pPIC6α). Each of these vectors contains the *AOX1* promoter for high-level transcription of the foreign gene and incorporates the α-factor pre-pro signal sequence from *S. cerevisiae* for secretion into the culture medium. In addition, pPICZα and pPIC6α contain polyhistidine tags C-terminal to their multiple cloning sites for convenient purification using metal chelating resins.

A wide variety of *P. pastoris* host strains are available, not all of which are compatible with all vectors. For pPIC9K constructs, we recommend using GS115 (*his4*) or the protease-deficient SMD1168 (*his4 pep4*) as the host strain. Completed pPIC9K constructs should *not* be linearized with the restriction enzyme *Bgl*II prior to transformation, as this may result in replacement of the endogenous *AOX1* gene *(7)*. The resulting methanol utilization slow phenotype (*mut*^{S}) transformants are incompatible with the methods described here. Instead, the restriction enzymes *Sac*I or *Sal*I should be used for linearization, each of which will produce the required wildtype methanol utilization (*mut*$^+$) phenotype *(7)*. For pPICZα and pPIC6α constructs, the wildtype X-33 or the protease-deficient SMD1168H (*pep4*) should be used as the host strain. With these vectors and host strains, all transformants will have the required *mut*$^+$ phenotype regardless of the choice of restriction enzyme for linearization.

1.2. Shake-Flask Cultures of **P. pastoris**

The most convenient method of producing recombinant protein from *P. pastoris* transformants is in shake-flask culture because no specialized equipment is required. This chapter describes a method for ^{13}C,^{15}N-labeling of protein from *mut*$^+$ transformants, which uses ^{13}C-glucose and ^{13}C-methanol as carbon sources and ^{15}N-ammonium sulfate as the nitrogen source.

1.3. Fermentation of **P. pastoris**

Although recombinant expression of heterologous proteins from *P. pastoris* can be done in shake-flask culture, the expression levels achievable in fermenters are typically much higher *(3)*. For most applications, a three-phase fermentation scheme is employed *(11,12)*. In the first phase, the engineered strain is batch-cultured (in a simple, defined medium) with the repressing carbon source, glycerol, to accumulate biomass. The second phase is a fed-batch transition stage in which the glycerol is fed to the culture at a growth-limiting rate to further increase the biomass. Finally, in the induction phase, methanol is added to the culture at a slow rate to induce expression of the recombinant protein. The methanol feed rate is then adjusted upward periodically until the desired growth rate and expression level are reached. Throughout the fermentation, ammonium hydroxide is used as a nitrogen source and pH control solution.

The methods for producing ^{13}C,^{15}N- and ^{2}H,^{13}C,^{15}N-labeled protein by fermentation described in this chapter differ from the previously mentioned scheme in three important ways. First, because there is little to be gained in accumulating isotopically labeled biomass, the batch phase of the fermentation is kept short (typically 24 h), and there is no fed-batch phase. Thus, in this application, the role of the fermenter is not to produce ultra-high cell densities but simply to allow the pH, aeration, and methanol feeding rate of the culture to be tightly controlled and so optimize the protein-expression level. Second, because of the greater expense of ^{13}C-glycerol compared to ^{13}C-glucose, ^{13}C-glucose is used as the repressing carbon source during the batch phase. Finally, given the impracticality and expense of using ^{15}N-ammonium hydroxide in cell culture, ^{15}N-ammonium sulfate is used as the nitrogen source with potassium hydroxide as the pH control solution.

High-density fermentation of *P. pastoris* does have one significant disadvantage—not only does the concentration of recombinant protein increase with cell density, so does that of the debris from cell lysis, in particular vacuolar proteases *(1)*. Three strategies have proved effective in minimizing the proteolytic instability of foreign proteins secreted into the *P. pastoris* culture medium. One is the addition of amino acid-rich supplements, such as peptone or Casamino acids, to the culture medium. These supplements appear to reduce product degradation by acting as excess substrates for one or more problem proteases *(13)*. However, this strategy is inappropriate for isotopic labeling as the additives themselves would need to be labeled. A second is changing the culture medium pH. *P. pastoris* is capable of growing across a relatively broad pH range from 3.0 to 7.0, which allows considerable flexibility in adjusting the pH to one that is not optimal for a problem protease. A third is the use of a protease-deficient *P. pastoris* host strain, such as SMD1168 (*his4 PEP4*) or SMD1163 (*his4 PEP4 PRB1*) *(14)*. The *PEP4* gene encodes proteinase A, a vacuolar aspartyl protease required for the activation of other vacuolar proteases, such as carboxypeptidase Y and proteinase B. Proteinase B, prior to processing and activation by proteinase A, has about half the activity of the processed enzyme. The *PRB1* gene codes for proteinase B. Therefore, *PEP4* mutants display a substantial decrease or elimination in proteinase A and carboxypeptidase Y activities and a partial reduction in proteinase B activity. The *PEP4 PRB1* double mutant shows a substantial reduction or elimination in all three of these protease activities.

1.4. Purification of Recombinant Protein

Because *P. pastoris* secretes only low levels of endogenous proteins and because its culture medium contains no added proteins, a secreted heterologous protein makes up the vast majority of the total protein in the medium. Thus, in separating the foreign protein from the bulk of cellular proteins, secretion serves as a major first step in purification. Subsequent purification steps must be tailored to the characteristics of the foreign protein. However, here we have included two purification methods that we have found to be particularly well suited to recombinant proteins expressed from *P. pastoris*, cation exchange and concanavalin A (Con A) affinity chromatography.

1.5. Disclaimer

The shake-flask and fermentation methods described in this chapter are adapted from those published previously *(6,10,12,15)*. They have been developed to make sufficient quantities of recombinant proteins for NMR structural and functional studies, using cost-effective amounts of labeled substrates. However, they remain "first-shot" approaches, which should be tried with unlabeled reagents to assess the performance of a given clone; the scale of the culture and/or the feed times can be adjusted to compensate for clones that perform poorly. As with all other protein-expression systems, the results are protein specific. Fermentation of *P. pastoris* is not difficult, but we strongly advise anyone attempting it to consult a fermentation scientist for general information about safe handling and disposal; also, there are variations in fermentation systems that may necessitate small adaptations to our protocols.

2. Materials

All unlabeled reagents were standard laboratory reagent grade; all solutions were prepared using Milli-Q water (Millipore, Billerica, MA). D_2O (99.9 At %), ^{15}N-ammonium sulfate (>98 At %), U-$[^{13}C]$-glucose (>99 At %), ^{13}C-methanol (>99 At %) and $^2H,^{13}C$-methanol (>99 At %) were obtained from CK Gas Products (Finchampstead, UK).

2.1. Seed Culture for Uniform $^{13}C,^{15}N$-Labeling in Shake Flasks or a Fermenter

1. Potassium phosphate solution, pH 6.0 (per liter): Dissolve 23 g potassium dihydrogen phosphate and 118 g dipotassium phosphate in water. If necessary, adjust the pH to 6.0 ± 0.1 with potassium hydroxide (corrosive) or phosphoric acid (85% wt; corrosive) (*see* **Note 1**). Autoclave and store at room temperature for up to 1 yr.
2. Yeast nitrogen base (YNB) solution (per liter): Dissolve 34 g YNB without amino acids or ammonium sulfate (Difco, UK) in water. Filter-sterilize and store at 4°C for up to 1 yr.
3. Biotin solution (per 100 mL): Dissolve 20 mg D-biotin in 10 mM sodium hydroxide. Filter-sterilize and store at 4°C for up to 3 mo.
4. ^{13}C-glucose solution (per 25 mL): Dissolve 1.25 g U-$[^{13}C]$-D-glucose in water. Filter-sterilize and store at 4°C for up to 1 yr (*see* **Notes 2** and **3**).
5. ^{15}N-ammonium sulfate solution (per 50 mL): Dissolve 5 g ^{15}N-ammonium sulfate in water. Filter-sterilize and store at 4°C for up to 1 yr (*see* **Notes 2** and **3**).
6. $^{13}C,^{15}N$-labeled buffered minimal glucose ($^{13}C,^{15}N$-BMD), pH 6.0 (per 50 mL): Autoclave 30 mL of water and allow to cool to room temperature. Aseptically, add 5 mL of potassium phosphate, pH 6.0, solution, 5 mL of YNB solution, 0.1 mL of

biotin solution, 5 mL of ^{13}C-glucose solution, and 5 mL of ^{15}N-ammonium sulfate solution. Store at 4°C for up to 1 mo.

2.2. Uniform ^{13}C,^{15}N-Labeling in Shake Flasks

1. ^{13}C,^{15}N-BMD, pH 6.0 (per 200 mL): Scale-up from the 50-mL recipe in **Subheading 2.1., step 6.**
2. ^{13}C-methanol feed solution (per 100 mL): Add 10 mL ^{13}C-methanol (toxic) to 90 mL water. Filter-sterilize and store at 4°C for up to 1 mo (*see* **Notes 2** and **3**).
3. ^{13}C,^{15}N-labeled buffered minimal methanol (^{13}C,^{15}N-BMM), pH 6.0 (per liter): Autoclave 600 mL of water and allow to cool to room temperature. Aseptically, add 100 mL of potassium phosphate, pH 6.0, solution, 100 mL of YNB solution, 2 mL of biotin solution, 100 mL of ^{13}C-methanol solution, and 100 mL of ^{15}N-ammonium sulfate solution. Store at 4°C for up to 1 mo.

2.3. Fermenter Vessel Preparation for Uniform ^{13}C,^{15}N-Labeling

A 1-L glass fermentation vessel was obtained from Electrolab (Tewkesbury, UK) with fittings for variable-speed impeller, temperature control via "cold finger" and heater mat, adjustable sterile air supply introduced through a sparger, and an air outlet fitted with a condenser. The vessel head plate has ports for temperature, pH, dissolved oxygen (dO$_2$), and foam probes, for acid-, base-, antifoam-, and nutrient-feed lines, and for inoculation and sampling. The fermenter was operated using a controller (from Electrolab) with feedback regulation of temperature, pH (via a base pump), aeration (via the impeller), and foaming (via an antifoam pump). A manually controlled, variable-speed, nutrient-feed pump (from Electrolab) was used to deliver methanol.

1. Basal salts medium (per 400 mL): Dissolve 0.47 g calcium sulfate, 7.5 g magnesium sulfate heptahydrate, and 13.4 mL orthophosphoric acid (85% wt; corrosive) in water. Adjust the pH to 3.0 with solid potassium hydroxide (corrosive) (*see* **Note 4**). Sterilize *in situ* by autoclaving (*see* **Subheading 3.3., step 3**).
2. *Pichia* trace metals (PTM$_1$) (per 100 mL): Dissolve 0.6 g copper (II) sulfate, 8 mg sodium iodide, 0.3 g manganese sulfate monohydrate, 20 mg sodium molybdate dehydrate, 2 mg boric acid, 50 mg cobalt chloride, 2 g zinc chloride, 6.5 g iron (II) sulfate, 0.5 mL sulfuric acid (98%; corrosive) in water. Store at room temperature for up to 1 yr. Filter-sterilize before use.

2.4. Uniform ^{13}C,^{15}N-Labeling in a Fermenter

1. ^{13}C-methanol feed solution (per 100 mL): Prepare as in **Subheading 2.2., step 2**.
2. Antifoam solution: Autoclave polypropylene glycol 1025, and store at room temperature indefinitely (*see* **Note 5**).

3. Base control solution: Prepare and autoclave 2 *M* potassium hydroxide (corrosive), and store at room temperature indefinitely (*see* **Note 6**).

2.5. Seed Culture for Uniform $^2H,^{13}C,^{15}N$-Labeling in a Fermenter

1. Deuterated potassium phosphate solution, pH 6.0 (per 10 mL): Prepare potassium phosphate solution (*see* **Subheading 2.1.**, **step 1**). Lyophilize 10 mL of the solution, and resuspend in 10 mL D$_2$O. Filter-sterilize and store at room temperature for up to 1 yr (*see* **Note 2**).
2. Deuterated YNB solution (per 10 mL): Prepare YNB solution (*see* **Subheading 2.1.**, **step 2**). Lyophilize 10 mL of the solution, and resuspend in 10 mL D$_2$O. Filter-sterilize and store at room temperature for up to 1 yr (*see* **Note 2**).
3. Deuterated biotin solution (per 10 mL): Prepare biotin solution (*see* **Subheading 2.1.**, **step 3**). Lyophilize 10 mL of the solution, and resuspend in 10 mL D$_2$O. Filter-sterilize and store at room temperature for up to 1 yr (*see* **Note 2**).
4. Deuterated ^{13}C-glucose solution (per 30 mL): Dissolve 1.5 g U-[^{13}C]-glucose in 30 mL D$_2$O. Lyophilize the solution, and resuspend in 30 mL D$_2$O. Filter-sterilize and store at 4°C for up to 1 yr (*see* **Notes 2** and **3**).
5. Deuterated ^{15}N-ammonium sulfate solution (per 55 mL): Dissolve 5.5 g ^{15}N-ammonium sulfate in 55 mL D$_2$O. Lyophilize the solution, and resuspend in 55 mL D$_2$O. Filter-sterilize and store at 4°C for up to 1 yr (*see* **Notes 2** and **3**).
6. 70%-deuterated ^{13}C,^{15}N-BMD (70%-^{2}H,^{13}C,^{15}N-BMD), pH 6.0 (per 50 mL): Filter-sterilize 30 mL D$_2$O. Aseptically, add 5 mL of deuterated potassium phosphate, pH 6.0, solution, 5 mL of deuterated YNB solution, 0.1 mL of deuterated biotin solution, 5 mL of deuterated ^{13}C-glucose solution, and 5 mL of deuterated ^{15}N-ammonium sulfate solution. Store at 4°C for up to 1 mo.
7. 100%-deuterated ^{13}C, ^{15}N-BMD (100%-^{2}H,^{13}C,^{15}N-BMD), pH 6.0 (per 50 mL): Filter-sterilize 30 mL D$_2$O. Aseptically, add 5 mL of deuterated potassium phosphate, pH 6.0, solution, 5 mL of deuterated YNB solution, 0.1 mL of deuterated biotin solution, 5 mL of deuterated ^{13}C-glucose solution, and 5 mL of deuterated ^{15}N-ammonium sulfate solution. Store at 4°C for up to 1 mo.

2.6. Fermenter Vessel Preparation for Uniform $^2H,^{13}C,^{15}N$-Labeling

1. Deuterated BSM (per 390 mL): Dissolve 27.2 g potassium dihydrogen phosphate and 12.4 g magnesium sulfate heptahydrate in 100 mL D$_2$O. Lyophilize the solution, and resuspend in 390 mL D$_2$O (*see* **Note 4**). Filter-sterilize and store at room temperature for up to 1 yr (*see* **Note 2**).
2. Deuterated calcium chloride solution (per 10 mL): Dissolve 1.47 g calcium chloride dihydrate in 10 mL D$_2$O. Lyophilize the solution and resuspend in 10 mL D$_2$O. Filter-sterilize and store at room temperature for up to 1 yr (*see* **Note 2**).
3. Deuterated PTM$_1$ (per 10 mL): Prepare PTM$_1$ solution (*see* **Subheading 2.3.**, **step 2**). Lyophilize 10 mL of the solution, and resuspend in 10 mL D$_2$O. Filter-sterilize and store at room temperature for up to 1 yr (*see* **Note 2**).

2.7. Uniform $^2H,^{13}C,^{15}N$-Labeling in a Fermenter

1. Deuterated $^2H,^{13}C$-methanol feed solution (per 100 mL): Add 10 mL $^2H,^{13}C$-methanol (toxic) to 90 mL D_2O. Filter-sterilize and store at 4°C for up to 1 mo (*see* **Notes 2** and **3**).
2. Antifoam solution: Prepare as in **Subheading 2.4., step 2**.
3. Deuterated base control solution (100 mL): Dissolve 21.2 g tri-potassium phosphate (corrosive) in 100 mL D_2O. Filter-sterilize and store at room temperature for up to 1 yr (*see* **Notes 2** and **6**).

2.8. Protein Purification by Cation-Exchange Chromatography

SP-Sepharose Fast Flow was supplied by Amersham Biosciences (Little Chalfont, UK).

1. Cation-exchange equilibration buffer (per liter): Prepare 20 mM sodium citrate, 0.02% (w/v) sodium azide (highly toxic), pH 3.0. Store at room temperature for up to 1 yr.
2. Cation-exchange elution buffer (per liter): Prepare 20 mM sodium citrate, 1 M sodium chloride, 0.02% (w/v) sodium azide (highly toxic), pH 3.0. Store at room temperature for up to 1 yr.

2.9. Glycoprotein Purification by Lectin Affinity Chromatography

Con A Sepharose 4B was supplied by Amersham Biosciences (Little Chalfont, UK).

1. Con A regeneration buffer: Prepare 20 mM Tris, 0.02% (w/v) sodium azide (highly toxic), pH 8.5. Store at room temperature for up to 1 yr.
2. Con A equilibration buffer: Prepare 20 mM Tris, 0.5 M sodium chloride, 1 mM manganese (II) chloride, 1 mM calcium chloride, 0.02% (w/v) sodium azide (highly toxic), pH 7.4. Store at room temperature for up to 1 yr.
3. Con A elution buffer: Prepare 0.5 M methyl-α-D-glucopyranoside, 20 mM Tris, 0.5 M sodium chloride, 0.02% sodium azide (w/v) (highly toxic), pH 7.4. Store at room temperature for up to 1 yr.

3. Methods

3.1. Seed Culture for Uniform $^{13}C,^{15}N$-Labeling in Shake Flasks or a Fermenter

1. Thaw an aliquot of a *P. pastoris* strain that has been stored as a 20% (v/v) glycerol stock at −80°C. Pellet the cells by centrifugation at 3000g for 5 min at room temperature. Resuspend the cell pellet in 1 mL of $^{13}C,^{15}N$-BMD.
2. Inoculate a sterile 250-mL baffled flask containing 49 mL of $^{13}C,^{15}N$-BMD with the cell suspension and grow, shaking at 30°C for 18–24 h, until the optical density $(OD)_{600}$ >2 (*see* **Note 7**). This will constitute the seed culture for a 1-L shake-flask growth (*see* **Subheading 3.2., step 1**) or 0.5-L fermentation (*see* **Subheading 3.4., step 2** and **Note 8**).

3.2. Uniform $^{13}C,^{15}N$-Labeling in Shake Flasks

1. Inoculate a sterile 2-L baffled flask containing 200 mL of $^{13}C,^{15}N$-BMD with the 50 mL seed culture (*see* **Subheading 3.1., step 2**). Incubate with shaking at 30°C for 48–60 h, until the OD_{600} = 10–20 (*see* **Note 7**).
2. Pellet the cells in sterile containers at 3000g for 5 min at room temperature. Discard the supernatant.
3. Gently resuspend the cells in 1 L of $^{13}C,^{15}N$-BMM. Equally divide the cell suspension into four 2-L baffled flasks (*see* **Note 9**). Incubate with shaking at 30°C, until the optimal induction time is reached (*see* **Note 10**). Take samples regularly (every 8–12 h) for analysis of cell density (OD_{600}) and of expression level by sodium dodecyl sulfate-polyacrylamide gel electrophoresis (SDS-PAGE) (*see* **Note 11**).
4. Every 24 h during the induction phase, feed the cultures by adding 12.5 mL ^{13}C-methanol feed solution.
5. At the end of the induction phase, pellet the cells by centrifugation at 10,000g for 10 min at 4°C (*see* **Note 10**). Remove the yellow-green supernatant, measure its volume, and store it at 4°C (*see* **Note 12**). Resuspend the cell pellet in a volume of 100 mM HCl equivalent to the original supernatant volume. Pellet the cells by centrifugation, as above; combine this supernatant with the original supernatant, and clarify this solution by passing through a 0.2 μM filter (*see* **Note 13**). Start purification immediately to minimize proteolytic degradation (*see* **Subheading 3.8.**, and *see* **Note 14**).

3.3. Fermenter Vessel Preparation for Uniform $^{13}C,^{15}N$-Labeling

1. Calibrate the fermenter pH probe with pH 4.0 and pH 7.00 buffers, rinse it with deionized H_2O, and introduce it into one of the vessel's ports (*see* **Note 15**).
2. Test the fermenter dO_2 probe with zeroing gel (available from Mettler-Toledo, Leicester, UK), rinse it with deionized H_2O, and introduce it into one of the vessel's ports (*see* **Note 16**). If available, connect a polarization module to the electrode (*see* **Note 17**).
3. Mask the external ends of all ports and lines of the fermenter vessel with aluminum foil and autoclave tape, clamp any lines that will be in contact with the media (e.g., the air inlet and sampling line), and sterilize the vessel containing 400 mL of BSM. Also sterilize three empty aspirators for base control, antifoam, and methanol feed (*see* **Note 5**).
4. Assemble the vessel on the fermentation system, insert the temperature probe, and connect the facilities for temperature control (e.g., heating mat and cold finger, or vessel jacket with circulating water bath). Allow the media to equilibrate to 30°C.
5. Connect the impeller and the dO_2 probe to the controller. If a polarization module has not been used, allow 6 h for probe polarization before inoculating the vessel with the starter culture (*see* **Note 17**). The dO_2 level can now be controlled by feedback regulation of the impeller speed (*see* **Note 18**).
6. Aseptically, fill the base control aspirator with base control solution, and connect it to the vessel via the base control pump. Connect the pH probe to the controller.

The system can now regulate the culture pH (*see* **Note 6**). Allow the media to equilibrate to the desired pH for fermentation (*see* **Notes 1, 4,** and **19**).

7. Aseptically, fill the antifoam aspirator with antifoam solution, and connect it to the vessel via the antifoam pump. Connect the foam probe to the controller to allow the system to manage excessive frothing (*see* **Note 5**).

8. Set the impeller speed to 250 rpm to aid temperature equilibration. Aseptically, introduce 20 mL ^{13}C-glucose solution, 45 mL ^{15}N-ammonium sulfate solution, 2.5 mL biotin solution, and 2.5 mL PTM$_1$ via the inoculation port (*see* **Note 19**).

9. Connect the sterile air-line, open the air outlet, and increase the impeller speed and air-flow rates to the maximum value to be used during the fermentation, typically 1000 rpm and 5 vol of air per volume of culture per min (vvm). After 10 min, calibrate the dO$_2$ probe at 100% saturation. Set the air-flow rate to 2 vvm.

3.4. Uniform ^{13}C,^{15}N-Labeling in a Fermenter

1. Activate the aeration feedback control with a target dO$_2$ of 20%, and maximum and minimum impeller speeds of 1000 rpm and 200 rpm, respectively (*see* **Note 18**). The saturated media should cause the impeller speed to drop to 200 rpm.

2. Inoculate with the 50-mL seed culture (*see* **Subheading 3.1., step 2**), and start the controller to monitor dO$_2$, impeller speed, temperature, and pH. Take a 10-mL sample via the sample port for assay and determination of wet cell weight and/or OD$_{600}$ (*see* **Note 11**).

3. Aseptically, fill the methanol feed aspirator with the ^{13}C-methanol feed solution, and connect the aspirator to the vessel via the nutrient-feed pump.

4. When the dO$_2$ "spikes" (typically 18–24 h after inoculation), take a 10-mL sample for analysis, and add 50 mL ^{13}C-methanol feed solution (toxic) via the inoculation port (*see* **Note 20**). Start the methanol feed at 2.5 mL/h, increasing to 5 mL/h after 6 h. Take samples regularly (every 8–12 h) for analysis of cell density (OD$_{600}$) and of expression level by SDS-PAGE (*see* **Note 11**). Continue the fermentation until it has consumed at least 250 mL methanol feed solution (*see* **Note 3**).

5. To prevent methanol accumulating in the vessel, but at the same time maintaining optimal cell growth, the feed rate of methanol can be increased with time, but it must remain limiting. Under conditions of limiting carbon source, the respiration rate rises and falls as methanol is constantly depleted, then repleni- shed, in the vessel; this is signaled by multiple dO$_2$ "spikes" (*see* **Note 21**).

6. Harvest the expressed protein by centrifugation as described in **Subheading 3.2., step 5**, and start purification immediately to minimize proteolytic degradation (*see* **Subheading 3.8.** and **Note 14**).

3.5. Seed Culture for Uniform ^{2}H,^{13}C,^{15}N-Labeling in a Fermenter

To produce significant quantities of ^{2}H,^{13}C,^{15}N-labeled protein, *P. pastoris* cells must be "conditioned" for growth in a deuterated environment. This is achieved by subculturing the cells into first 70-, then 100%-deuterated media.

1. Thaw an aliquot of a *P. pastoris* strain that has been stored as a 20% (v/v) glycerol stock at −80°C. Pellet the cells by centrifugation at 3000*g* for 5 min at room temperature. Resuspend the cell pellet in 1 mL of 70%-^{2}H,^{13}C,^{15}N-BMD.
2. Inoculate a sterile 250-mL baffled flask containing 49 mL of 70%-^{2}H,^{13}C,^{15}N-BMD with the cell suspension and grow, shaking at 30°C for 36–48 h, until the OD$_{600}$ >2 (*see* **Note 7**). Pellet a 10-mL aliquot of the cells by centrifugation at 3000*g* for 5 min at room temperature. Resuspend the cell pellet in 1 mL of 100%-^{2}H,^{13}C,^{15}N-BMD.
3. Inoculate a sterile 250-mL baffled flask containing 49 mL of 100%-^{2}H,^{13}C,^{15}N-BMD with the cell suspension and grow, shaking at 30°C for 36–48 h, until the OD$_{600}$ >2 (*see* **Note 7**). This will constitute the seed culture for a 0.5-L fermentation (*see* **Subheading 3.7.**, **step 2** and **Note 8**).

3.6. *Fermenter Vessel Preparation for Uniform ^{2}H,^{13}C,^{15}N-Labeling*

To achieve the highest levels of deuteration, every care should be taken not to introduce water into the fermenter vessel.

1. Calibrate the fermenter pH probe with pH 4.0 and pH 7.00 buffers, and immerse its electrode in a measuring cylinder of deionized H$_2$O in preparation for autoclaving (*see* **Note 15**).
2. Test the fermenter dO$_2$ probe with zeroing gel, rinse it with deionized H$_2$O, and immerse its electrode in a measuring cylinder of deionized H$_2$O in preparation for autoclaving (*see* **Note 16**). If available, connect a polarization module to the electrode (*see* **Note 17**).
3. Mask the external ends of all ports and lines of the fermenter vessel with aluminum foil and autoclave tape, and clamp any lines that will be in contact with the media (e.g., the air inlet and sampling line). Cover the ports to be used for the pH and dO$_2$ probes with gauze. Sterilize the vessel, the pH and dO$_2$ probes, and three empty aspirators for deuterated base control, antifoam, and deuterated methanol feed.
4. Dry the sterile fermenter vessel in an oven at approx 80°C for 24–48 h to remove any traces of water. Once cooled, aseptically fill the dry vessel with 390 mL deuterated BSM and 10 mL of deuterated calcium chloride solution.
5. Aseptically, air-dry the sterile pH and dO$_2$ probes for 10 min, rinse their electrodes with filter-sterilized D$_2$O to remove any traces of water, and introduce them into the gauze-covered ports on the vessel.
6. Assemble the vessel on the fermentation system, and connect the facilities for temperature, dO$_2$, pH, and foam control as described in **Subheading 3.3.**, **steps 4–7**.
7. Set the impeller speed to 250 rpm to aid temperature equilibration. Aseptically, introduce 20 mL deuterated ^{13}C-glucose solution, 45 mL deuterated ^{15}N-ammonium sulfate solution, 2.5 mL deuterated biotin solution, and 2.5 mL deuterated PTM$_1$ via the inoculation port (*see* **Note 19**).
8. Calibrate the dO$_2$ probe as described in **Subheading 3.3.**, **step 9**.

3.7. Uniform $^2H,^{13}C,^{15}N$-Labeling in a Fermenter

1. Activate the aeration feedback control as described in **Subheading 3.4., step 1**.
2. Inoculate with the 50-mL seed culture (*see* **Subheading 3.5., step 3**) and start the controller to monitor dO_2, impeller speed, temperature, and pH. Take a 10-mL sample via the sample port for assay and determination of wet cell weight and/or OD_{600} (*see* **Note 11**).
3. Aseptically, fill the methanol feed aspirator with the $^2H,^{13}C$-methanol feed solution and connect the aspirator to the vessel via the nutrient-feed pump.
4. When the dO_2 "spikes" (typically 36–48 h after inoculation), take a 10-mL sample for analysis, and add 50 mL deuterated $^2H,^{13}C$-methanol feed solution (toxic) via the inoculation port (*see* **Note 20**). Start the methanol feed at 1.25 mL/h, increasing to 2.5 mL/h after 6 h. Take samples regularly (every 16–24 h) for analysis of cell density (OD_{600}) and of expression level by SDS-PAGE (*see* **Note 11**). Continue the fermentation until it has consumed at least 250 mL methanol feed solution (*see* **Note 3**).
5. *See* **Subheading 3.4., step 5**.
6. Harvest the expressed protein by centrifugation as described in **Subheading 3.2., step 5**, and start purification immediately to minimize proteolytic degradation (*see* **Subheading 3.8.** and **Note 14**).

3.8. Protein Purification by Cation-Exchange Chromatography

The following protocol is a convenient first-step procedure that serves not only to partially purify the recombinant protein but also to concentrate the sample volume for subsequent steps. However, if the heterologous protein is acid labile, insoluble at low pH, or is highly acidic (has a low isoelectric point) an alternative purification method should be used.

1. Dilute the crude supernatant twofold with Milli-Q water. Adjust the pH to 3.0 with 2 *M* hydrochloric acid (corrosive) or 2 *M* sodium hydroxide (corrosive).
2. Equilibrate the cation-exchange column of SP-Sepharose FF with 5 column volumes of cation-exchange equilibration buffer.
3. Apply the diluted protein sample onto the column at 10–20 mL/min. Collect the yellow-green nonbinding material that contains alcohol oxidase (AOX) and other acidic proteins (*see* **Note 12**).
4. Wash the column with 2 column volumes of cation-exchange equilibration buffer.
5. Elute the bound proteins by applying a gradient of 0–100% cation-exchange elution buffer and collecting fractions equivalent to 0.1 column volumes.
6. Analyze the fractions and the nonbinding material by SDS-PAGE, and pool those containing the recombinant protein.

3.9. Glycoprotein Purification by Lectin Affinity Chromatography

Separation of glycosylated from nonglycosylated proteins secreted from *P. pastoris* can be achieved using affinity chromatography with immobilized jack bean Con A lectin. Con A has specificity for branched mannoses and carbohydrates with terminal mannose or glucose, and therefore, it binds to both N- and O-linked oligosaccharides produced by *P. pastoris.*

1. Regenerate a column of Con A-Sepharose (to remove any bound glycoproteins or sugar analogs) by washing with 10 column volumes of Con A regeneration buffer, followed by 10 column volumes of Con A equilibration buffer.
2. Add $CaCl_2$, $MgCl_2$, and $MnCl_2$ to the protein sample, each to a final concentration of 1 m*M*. Adjust the sample pH to 7.4 with solid Tris base (*see* **Note 22**).
3. Apply the protein sample to the column at 0.1–0.5 mL/min, and wash the column with 5 column volumes of Con A equilibration buffer. Collect the nonbinding fraction that contains the nonglycosylated proteins.
4. To elute the glycoproteins and free oligosaccharides, wash the column with 5 column volumes of Con A elution buffer. Analyze the eluted glycoproteins and the nonbinding aglycosyl proteins by SDS-PAGE.

3.10. Enzymatic Deglycosylation of Glycoproteins

The large hydrodynamic radius of *N*-linked, high-mannose sugars on recombinant proteins greatly increases their rotational correlation time (τ_c) leading to extensive line-broadening in their NMR spectra. In the majority of cases, the sugar moiety is not essential for structural integrity or functional activity of the protein and can be trimmed using the enzyme endoglycosidase (Endo) H_f (*see* **Note 23**). Endo H_f cleaves the chitobiose core of high-mannose and hybrid type *N*-linked oligosaccharides to leave a single *N*-linked *N*-acetylglucosamine (GlcNAc). Although enzymatic removal of *O*-linked oligosaccharides is possible and has been described in detail elsewhere *(16)*, in our experience it is not economically viable to do so on the quantity of protein required for an NMR sample.

1. Adjust the protein sample to pH 5.5 with 1 *M* sodium hydroxide (corrosive).
2. Add 1 kU of Endo H_f (supplied by New England Biolabs) per milliliter of glycoprotein solution. Incubate at 37°C for 1–3 h.
3. Analyze the glycosylated and aglycosyl protein by SDS-PAGE. *N*-linked glycosylation results in an increase in apparent molecular weight of approx 7 kDa per high-mannose chain, and a significant smearing of the protein band.
4. The removed oligosaccharides may be separated from the deglycosylated protein by Con A affinity chromatography (*see* **Subheading 3.9.**).

4. Notes

1. Selection of the growth and induction pH of *P. pastoris* can help to overcome proteolysis and insolubility of the recombinant product. The optimum value should be determined in small-scale cultures prior to isotopic labeling. *P. pastoris* is capable of growing across a relatively broad pH range from 3.0 to 7.0, which allows considerable leeway in adjusting the pH to one that is not optimal for a problem protease.

2. Solutions of ammonium sulfate and D-glucose can be sterilized by autoclaving. Filter-sterilization is used instead as a precaution against accidental loss of the isotopically labeled compounds. D_2O solutions should never be sterilized by autoclaving.

3. The amounts of ^{15}N-,^{13}C-, and ^{2}H,^{13}C-labeled isotopes given here have resulted in good yields in our experience, but given their high cost, it is highly recommended that a "dry run" be performed using unlabeled ammonium sulfate, D-glucose, and methanol to establish the minimum quantities needed for the required expression level.

4. Final adjustment of the fermentation media pH is performed after autoclaving as inorganic salts may precipitate during the sterilization process if the BSM pH is >3.5. Brady et al. *(17)* have solved this problem by reducing the concentration of all salts in the BSM to one-quarter of those given here without any adverse effect on cell growth rate, biomass yield, or the level of their recombinant protein. The reduced ionic strength may be beneficial when using ion-exchange chromatography as a first step in protein purification.

5. Different molecular-weight preparations of polypropylene glycol are available; we routinely use 1025 obtained from Merck (Poole, UK). The disadvantage of polypropylene glycol as an antifoam reagent is that its presence, even at low concentrations, can significantly reduce flow rates through certain ultrafiltration membranes. The addition of antifoam is usually necessary, but when growing *P. pastoris* at relatively low cell densities only small quantities are required. If the methanol feed is continued for longer periods, more antifoam solution may be needed. If the culture foams excessively, it is a sign of carbon source limitation, low pH, or ill health of the culture. If automated antifoam addition is unavailable, add the antifoam manually, but use the minimum amount to control foaming.

6. In general, no acid is required for *Pichia* fermentation because a healthy culture always acidifies the media. If the pH of the culture is increasing, it is a sign of carbon source depletion or ill health of the culture.

7. An $OD_{600} = 1$ is equivalent to 5×10^7 *P. pastoris* cells.

8. The volume of seed culture can be scaled according to the volume of the fermentation. In our experience, good results are obtained from using a seed culture that is 10% of the fermentation volume.

9. Aeration, particularly during the induction phase, is the most important parameter for efficient expression in shake flasks. Baffled flasks should always be used, and the volume of the culture should not exceed 30% of the flask volume.

10. Typical induction times range from 36 to 72 h, but will depend on the desired expression level, the stability of the expressed product, and the amount of ^{13}C- or ^{2}H,^{13}C-methanol feed solution available. Optimal induction times should be determined from SDS-PAGE analysis of culture samples taken during the extended induction phase of a "dry run" using unlabeled D-glucose and methanol.

11. Regular sampling is important to monitor the growth of the fermentation, allows assay for any recombinant product, and permits microscopic inspection of the culture for contaminating organisms. Any precipitated inorganic salts in the sample

can interfere with these tests but can be dissolved by the addition of an equal volume of 100 mM HCl without leading to cell lysis. If the recombinant product is acid labile, then a sample of supernatant should be removed for assay before addition of the HCl.

12. The yellow-green color of the culture supernatant is the result of AOX that, although not actively secreted, accumulates in the media because of leakage from the cells. The AOX can be conveniently purified away from the recombinant material by cation-exchange chromatography (*see* **Subheading 3.8.**).

13. The purpose of washing the cells is threefold: (a) to increase the yield of recombinant protein; (b) to dilute the supernatant and, therefore, to reduce its ionic strength prior to ion-exchange chromatography; and (c) to reduce the pH to approx 3.0 for cation exchange. If the recombinant protein is acid sensitive, then the cells should be washed with sterile H_2O instead of 100 mM HCl.

14. To minimize proteolysis in the crude supernatant, add phenylmethylsulfonyl fluoride (PMSF) (toxic) and ethylenediaminetetraacetic acid to a concentration of 1 mM. The PMSF should be refreshed every few hours.

15. The performance of the gel-filled autoclavable pH probes will deteriorate with repeated sterilization but can be prolonged by periodic cleaning as recommended by the manufacturer. Probes should be stored in 3 M KCl.

16. We have only used polarographic oxygen probes; information on the use of galvanic oxygen probes should be obtained from the manufacturer. Polarographic oxygen probes require routine maintenance to prolong their working life, and these details should also be obtained from the manufacturer.

17. It takes approx 6 h to completely polarize a dO_2 electrode. The polarization module applies the same voltage to the dO_2 electrode as the controller. Autoclavable modules offer the speed advantage of performing electrode polarization during sterilization.

18. The fermenter impeller speed can be controlled manually, but this is not recommended given the extended duration of *P. pastoris* fermentations. During the induction phase, the dO_2 level must remain above 15% at all times to prevent the buildup of methanol in the vessel, but it must also stay below 50% to avoid methionine oxidation of the recombinant protein.

19. As the pH rises above 3.5, and particularly on addition of PTM$_1$ salts, some inorganic salts will precipitate. This does not affect cell growth.

20. When a culture is growing normally and consumes the entire carbon source, respiration will slow dramatically, causing the dO_2 level to rise sharply or "spike." Addition of more carbon source should lead to an almost instantaneous rise in respiration signaled by a concomitant fall in dO_2. Many fermenter controllers (including ours) have automated feedback control, which adjusts the impeller speed to maintain the dO_2 at a preset level. This is very convenient when the system is left unattended for long periods of time but can mask dO_2 fluctuations. In this case, test for carbon limitation by fixing the impeller speed at the current variable level and watch for dO_2 "spikes."

21. The dO_2 "spike" method of ensuring growth-limiting methanol concentration is labor intensive and repeatedly exposes the cells to noninducing levels of methanol.

This can be avoided through the use of a methanol probe that allows continuous measurement and control of the methanol concentration in the media.

22. Con A is a metalloprotein that contains four metal binding sites. To maintain the binding characteristics of Con A, both Mn^{2+} and Ca^{2+} must be present in the binding buffer and protein sample.

23. Endo H_f is a protein fusion of Endo H (cloned from *Streptomyces plicatus*) and maltose binding protein. It has identical activity to Endo H but is less expensive. PNGase F may be used as an alternative to Endo H_f. This enzyme cleaves between the innermost GlcNAc and asparagine residues of high mannose, hybrid, and complex oligosaccharides from N-linked glycoproteins. However, the activity of this enzyme is more susceptible to steric hindrance of the scissile bond.

Acknowledgments

The authors thank the Wellcome Trust for financial assistance and Prof. Iain D. Campbell and Dr. A. K. Downing for their support and direction.

References

1. Higgins, D. R. and Cregg, J. M. (1998) Introduction to *Pichia pastoris*. In *Methods in Molecular Biology*, vol. 103: *Pichia Protocols* (Higgins, D. R. and Cregg, J. M., eds.), Humana, Totowa, NJ, pp. 1–15.
2. Cregg, J. M., Shen, S., Johnson, M., and Waterham, H. R. (1998) Classical genetic manipulation. In *Methods in Molecular Biology,* vol. 103: *Pichia Protocols* (Higgins, D. R. and Cregg, J. M., eds.), Humana, Totowa, NJ, pp. 17–26.
3. Lin Cereghino, G. P., Lin Cereghino, J., Ilgen, C., and Cregg, J. M. (2002) Production of recombinant proteins in fermenter cultures of the yeast *Pichia pastoris*. *Curr. Opin. Biotechnol.* **13,** 1–4.
4. Grinna, L. S. and Tschopp, J. F. (1989) Size distribution and general structural features of N-linked oligosaccharides from the methylotrophic yeast, *Pichia pastoris*. *Yeast* **5,** 107–115.
5. Cregg, J. M., Vedvick, T. S., and Raschke, W. C. (1993) Recent advances in the expression of foreign genes in *Pichia pastoris*. *Biotechnology* **11,** 905–910.
6. Laroche, Y., Storme, V., De Meutter, J., Messens, J., and Lauwereys, M. (1994) High level secretion and very efficient isotopic labelling of tick anticoagulant peptide (TAP) expressed in the methylotrophic yeast *Pichia pastoris*. *Biotechnology* **12,** 1119–1124.
7. Cregg, J. M., Barringer, K. J., Hessler, A. Y., and Madden, K. R. (1985) *Pichia pastoris* as a host system for transformations. *Mol. Cell. Biol.* **5,** 3376–3385.
8. Scorer, C.A., Clare, J. J., McCombie, W. R., Romanos, M. A., and Sreekrishna, K. (1994) Rapid selection using G418 of high-copy number transformants of *Pichia pastoris* for high-level foreign gene expression. *Biotechnology* **12,** 181–184.
9. Romanos, M. A. (1995) Advances in the use of *Pichia pastoris* for high-level gene expression. *Curr. Opin. Biotechnol.* **6,** 527–533.

10. Invitrogen Corporation. *Pichia Expression System: Manual of Methods for Expression of Recombinant Proteins in Pichia pastoris*, Version G. Available at http://www.invitrogen.com/content/sfs/manuals/pich_man.pdf. Accessed 03/01/04.

11. Stratton, J., Chriuvolu, V., and Meagher, M. (1998) High cell-density fermentation. In *Methods in Molecular Biology*, vol. 103: *Pichia Protocols* (Higgins, D. R. and Cregg, J. M., eds.), Humana, Totowa, NJ, pp. 107–120.

12. Invitrogen Corporation. *Pichia Fermentation Process Guidelines.* Available at http://www.invitrogen.com/content/sfs/manuals/pichiaferm_prot.pdf. Accessed 03/01/04.

13. Clare, J. J., Romanos, M. A., Rayment, F. B., Rowedder, J. E., Smith, M. A., Payne, M. M., et al. (1991) Production of mouse epidermal growth factor in yeast: high-level secretion using *Pichia pastoris* strains containing multiple gene copies. *Gene* **105,** 205–212.

14. Gleeson, M. A. G., White, C. E., Meininger, D. P., and Komives, E. A. (1998) Generation of protease-deficient strains and their use in heterologous protein expression. In *Methods in Molecular Biology*, vol. 103: *Pichia Protocols* (Higgins, D. R. and Cregg, J. M., eds.), Humana, Totowa, NJ, pp. 81–94.

15. Bright, J. R., Pickford, A. R., Potts, J. R., and Campbell, I. D. (1999) Preparation of isotopically labelled recombinant fragments of fibronectin for functional and structural study by heteronuclear nuclear magnetic resonance spectroscopy. In *Methods in Molecular Biology*, vol. 139: *Extracellular Matrix Protocols* (Streuli, C. and Grant, M., eds.), Humana, Totowa, NJ, pp. 59–69.

16. Cremata, J. A., Montesino, R., Quintero, O., and Garcia, R. (1998) Glycosylation profiling of heterologous proteins. In *Methods in Molecular Biology*, vol. 103: *Pichia Protocols* (Higgins, D. R. and Cregg, J. M., eds.), Humana, Totowa, NJ, pp. 95–105.

17. Brady, C. P., Shimp, R. L., Miles, A. P., Whitmore, M., and Stowers, A. W. (2001) High-level production and purification of P30P2MSP1$_{19}$, an important vaccine antigen for malaria, expressed in the methylotrophic yeast *Pichia pastoris. Protein Expr. Purif.* **23,** 468–475.

3

Perdeuteration/Site-Specific Protonation Approaches for High-Molecular-Weight Proteins

Stephen Matthews

Summary

Among the factors that limit the application of nuclear magnetic resonance (NMR) to biological macromolecules are increasing resonance overlap and fast transverse relaxation. Multidimensional NMR combined with ^{13}C and ^{15}N labeling has alleviated these problems temporarily; however, they resurface at molecular weight (mol wt) in excess of 30 kDa. Combined perdeuteration/site-specific protonation together with segmental labeling (*see* Chapter 4), transverse relaxation-optimized spectroscopy (TROSY) (*see* Chapter 5), and residual dipolar couplings (*see* Chapter 7) have all helped to dramatically extend the mol wt limit. This article describes some of the practical aspects of the combined perdeuteration/site-specific protonation approach, which has proved so useful in the global fold determination of large proteins.

Key Words: Perdeuteration; site-specific protonation; large proteins; global folds; residual dipolar couplings.

1. Introduction

The deuteration of molecules for the simplification of nuclear magnetic resonance (NMR) spectra is not a new strategy, but its combination with triple-resonance spectroscopy and site-specific protonation has generously extended the molecular size limitation for NMR applications to macromolecules. The deuteron has a significantly smaller magnetogyric ratio (1/6.5 that of a proton), and consequently deuteration removes the dominant ^{1}H-^{13}C dipolar relaxation mechanism for ^{13}C nuclei. Typically, the transverse relaxation of the C$_{\alpha}$ is approximately fivefold slower in deuterated proteins. This, in combination with ^{2}H-decoupling during ^{13}C evolution, ensures that triple-resonance experiments, namely HNCA, HNCOCA, HNCACB, HNCOCACB, and HNCACO, work very efficiently on proteins in excess of 30 kDa *(1,2)*. In addition, the cross-peak

From: *Methods in Molecular Biology, vol. 278: Protein NMR Techniques*
Edited by: A. K. Downing © Humana Press Inc., Totowa, NJ

phase in the double-constant time d-HN(CA)CB and d-HN(COCA)CB experiments is particularly useful for rapid sequential assignment *(1)*. The $^{13}C_\beta$ nuclei that are coupled to an odd number of aliphatic carbons give one sign and $^{13}C_\beta$ nuclei that are coupled to an even number of aliphatic carbons give the opposite sign.

Transverse relaxation in deuterated proteins above 60 kDa becomes sufficiently fast that the standard triple-resonance approaches to assignment begin to fail. Pervushin and coworkers have developed a new class of NMR experiment that offers a further relaxation benefit in deuterated proteins *(3)*. Transverse relaxation-optimized spectroscopy (TROSY) selects transitions that experience mutual cancellation of dipole–dipole and chemical shift anisotropy relaxation mechanisms. Furthermore, any residual relaxation because of dipolar interactions with remote protons can almost be entirely removed by perdeuteration. The result is a dramatic decrease in resonance line-widths that enables NMR spectra to be obtained for proteins in excess of 100 kDa. Numerous improvements and derivatives have been published *(4,5)*, with perhaps the current highlight being the application of cross-correlated relaxation-enhanced polarization transfer and cross-correlated relaxation-induced polarization transfer to 470 and 800 kDa GroEL/GroES complexes *(6)*. These TROSY-based elements have also been incorporated into the repertoire of multidimensional assignment experiments and have further extended their applicability *(7)*. Notably, Tugarinov et al. have applied these techniques successfully to the 723-residue protein, malate synthase G *(8)*.

One of the limiting factors of perdeuteration is loss of protons between which nuclear Overhauser effect (NOE) measurements can be made and the subsequent impact this has on the quality of resulting structures. Mal et al. demonstrated on Fn3 that the longer distances measurable in deuterated macromolecules facilitate the characterization of the fold for an exclusively β-sheet protein using reprotonated amides alone *(9)*. α-helical proteins possess many fewer long-range amide–amide NOEs and, therefore, are not amenable to this approach. Gardner et al. established the limitations of such an approach using an SH2 domain from phospholipase C γ1 *(10)*. To increase the number of NOEs observed, several groups have assessed the feasibility of specific protonation of amino acids within a deuterium background *(11,12)*. Amino acids containing methyl groups were chosen for maximum impact because they provide improved line-widths, fall in well-resolved regions of the spectrum, and are common within protein hydrophobic cores. Rosen et al. described a novel route to achieving methyl protonated samples that used $^{13}C,^1H$-pyruvate as the carbon source rather than supplementing D_2O-based growth media with the protonated amino acids *(13)*. By far the most elegant method was that of Gardner et al., which involved the selective reintroduction of protons into the methyl group positions of valine, leucine, and isoleucine *(10,14–16)*. *Escherichia coli* will readily use the biosynthetic precursors of these amino acids, namely [3,3-^{2}H$_2$] ^{13}C-2-ketobuterate and ^{13}C-α-ketoisovalerate, in preference to complete biosynthesis. The final product is a deuterated protein that is specifically protonated at the terminal methyl groups of valine, leucine, and isoleucine. Simulations suggest that the numerous NOE correlations between methyl groups within the hydrophobic core are sufficient to confidently characterize the global fold of many, if not all, proteins to within a few angstrom root-mean-square deviation (RMSD) *(10)*. In cases where higher precision is required, one strategy is to introduce additional protons, with perhaps aromatic sidechains being the most likely to provide substantial benefits *(12)*. The strong ^{13}C–^{13}C J-coupling and poor chemical shift dispersion can greatly hinder the assignment of NOEs involving the aromatic rings, particularly phenylalanine. Wang and coworkers proposed a method of selectively labeling the ε-proton in phenylalanine sidechains, which, when combined with selective methyl group labeling, would improve the precision of structures considerably *(17)*.

The introduction of residual dipolar couplings (RDCs) can afford further improvements in the quality of structures derived from partially deuterated material. Moreover, RDCs are highly applicable to deuterated molecules as they can be measured between nonproton nuclei, that is, ^{15}N-^{13}C and ^{13}C-^{13}C. The practical aspects of RDC applications are covered in Chapter 7 of this book and are not addressed in this chapter, although they represent a natural extension to this work.

2. Materials

2.1. Deuterium and ^{13}C/^{15}N Minimal Media

1. M9 salts: 6.0 g Na$_2$HPO$_4$, 3.0 g KH$_2$PO$_4$, 0.5 g NaCl, 0.7 g ^{15}NH$_4$Cl, 1 L D$_2$O.
2. 20% ^{13}C/^{1}H glucose.
3. 1 M MgSO$_4$.
4. 0.1 M CaCl$_2$.
5. 1 M thiamine.
6. *E. coli* trace elements, pH 7.0; 5 g ethylenediaminetetraacetic acid, 800 mL D$_2$O, 0.5 g FeCl$_3$, 0.05 g ZnCl$_2$, 0.01 g CuCl$_2$, 0.01 g CoCl$_2$.6H$_2$O, 0.01 g H$_3$BO$_3$.
7. 100 mg/mL antibiotic solution.

2.2. Specific Protonation

1. 50 mg [3,3-^{2}H$_2$] α-ketobutyrate (Cambridge Isotope Laboratory, Andover, MA, cat. no. CDLM4611).
2. 100 mg ^{15}N, ^{13}C-[2,3-^{2}H$_2$]-valine (Cambridge Isotope Laboratory, Andover, MA, cat. no. CDNLM4281) or 75 mg ^{13}C α-ketoisovalerate.
3. 30 mg ε-^{13}C-phenylalanine.

2.3. NMR Spectroscopy and Structure Calculation

1. Four-channel pulsed-field gradient NMR spectrometer equipped with deuterium decoupling capabilities.
2. NMRview *(18)* or other suitable NMR assignment software.
3. XPLOR/CNS program *(19–22)*.

3. Methods

The methods described in this section outline preparation, NMR spectroscopy, and structure calculation of deuterated proteins with specific protonation.

3.1. Preparation of Deuterated Proteins With Specific Protonation

DNA encoding Int280 (EPEC strain E2348/69) was cloned into the expression vector pET3d and freshly transformed into a strain of BL21 pLysS *E. coli* *(23,24)*.

1. Prepare 1 L of D_2O minimal culture medium (*see* **Note 1**).
2. Add approx 1/10 of labeled amino acid (i.e., 5 mg [3,3-$^2H_2$] α-ketobutyrate, 10 mg ^{15}N, ^{13}C-[2,3-$^2H_2$]-valine and/or 3 mg ε-^{13}C-phenylalanine) (*see* **Note 2**).
3. Filter-sterilize culture medium.
4. Inoculate 1 L of D_2O minimal culture media and grow at 37°C.
5. At 1 h prior to induction add the rest of the labeled amino acid(s) (i.e., 45 mg [3,3-$^2H_2$] α-ketobutyrate, 90 mg ^{15}N, ^{13}C-[2,3-$^2H_2$]-valine and/or 27 mg ε-^{13}C-phenylalanine) (*see* **Note 2**).
6. Induce at optical density $(OD)_{600}$ approx 0.6 by adding isopropyl-β-D-thiogalactopyranoside to 120 mg/L, and grow for at least 5 h before harvesting.
7. Disrupt cells, purify protein, and prepare for NMR spectroscopy.

3.2. NMR Spectroscopy

Pure protein was extensively dialyzed against 20 m*M* sodium acetate buffer at pH 5.2, concentrated to 1 m*M* in 300 µL, and placed in a 5-mm Shigemi tube (Shigemi, Allison Park, PA). All NMR spectra were recorded at 500 MHz proton frequency on a four-channel Bruker DRX500 equipped with a z-shielded gradient, triple-resonance probe. NMR data were processed using XWINNMR and NMRPipe *(25)* and analyzed using NMRview *(18)*. The temperature was maintained at 310 K throughout the experiments *(23,24)*. One-dimensional and two-dimensional (2D) NMR spectra of site-specifically protonated, deuterated Int280 are shown in **Figs. 1** and **2**.

1. Determine sequence-specific backbone ^1H_N, ^{15}N, $^{13}C_\alpha$, $^{13}C_\beta$, and C′ assignments using the standard triple-resonance approach *(1,2)* (*see* **Note 3**).
2. Assign valine, leucine, and isoleucine methyl group resonances using (H)C(CO)NH-total correlation spectroscopy (TOCSY) *(1)* (*see* **Note 4**).

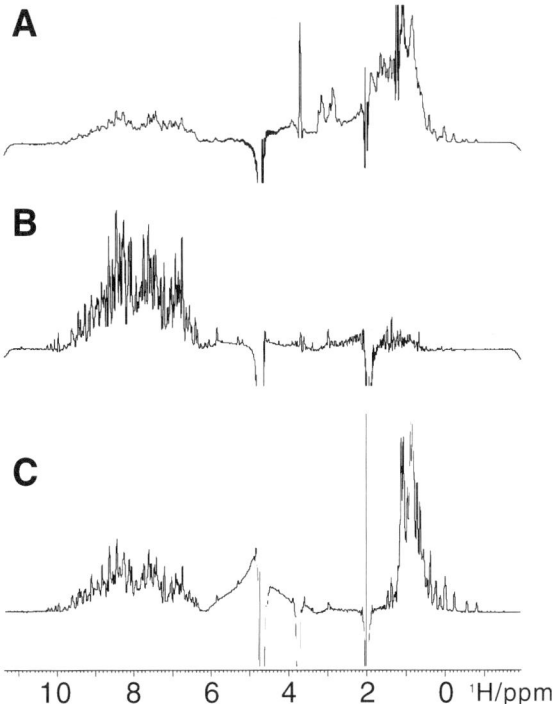

Fig. 1. 1D ^{1}H NMR spectra of **(A)** ^{1}H,^{13}C,^{15}N Int280, **(B)** ^{2}H,^{13}C,^{15}N Int280 showing the dramatic improvement in amide proton line-widths on deuteration, and **(C)** ^{13}CH$_3$–Val, Leu,^{2}H,^{13}C,^{15}N Int280, illustrating the selectivity of methyl group labeling.

3. Assign NOEs for use in structure calculation from ^{1}H/^{15}N- and ^{1}H/^{13}C-edited NOE heteronuclear single-quantum coherence (NOESY–HSQC) spectroscopy experiments *(26,27)*, as well as 3D/4D ^{15}N/^{15}N-, ^{15}N/^{13}C- and ^{13}C/^{13}C-edited HMQC–NOESY–HSQC experiments *(28)* (*see* **Note 5**). Two mixing times of 100 and 200 ms were chosen for this protein, which has a correlation time of 16 ns.
4. Identify amide protons involved in hydrogen bonds by their resistance to exchange in the presence of D$_2$O. NH resonances observed in HSQC spectra recorded after dissolving or buffer exchanging into 100% D$_2$O are assigned as hydrogen bond donors.

3.3. Structure Calculation

1. Measure NOE cross-peak intensities at both 100 and 200 ms mixing times (*see* **Note 6**).
2. Calibrate NOE-based distance restraints using known sequential distances, that is, NH–NH$_{(i,i+1)}$ in the center of a helix.

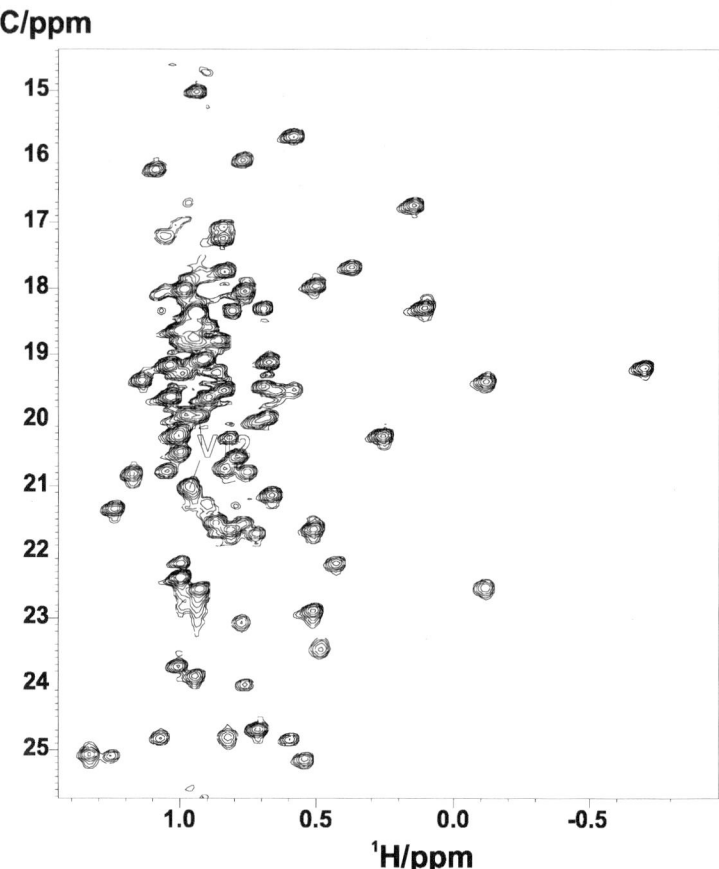

Fig. 2. Methyl group region of the 2D ^{1}H-^{13}C CT-HSQC NMR spectrum of ^{13}CH$_3$–Val, Leu,^{2}H,^{13}C,^{15}N Int280 illustrating the high quality of data and selectivity of methyl group labeling.

3. Place NOEs observed at a mixing time of 100 ms into two categories on the basis of estimated NOE cross-peak intensity: strong (<3.0 Å) and medium (<5.0 Å).

4. Set NOEs only appearing in the 200 ms spectrum to weak (<6.8 Å).

5. Add a distance of 0.5 Å to all NOEs involving methyl groups (i.e., 0.5 Å for NH–Me and 1.0 Å for Me–Me NOEs).

6. Introduce hydrogen bonds as two distance restraints, d_{HN-O}=1.8–2.6 Å and d_{N-O}=1.8–3.5 Å.

7. Estimate dihedral angle restraints within regions of secondary structure from chemical shifts using TALOS *(29)* (*see* **Note 7**).

Structure calculations were performed starting from extended structures and using a hybrid algorithm including torsion angle dynamics (TAD) and Cartesian dynamics executed within the program XPLOR or CNS *(19–22)*. An initial TAD phase was performed at a temperature of 50,000 K consisting of 2000 molecular dynamics steps each of 15 fs. A TAD cooling phase to 2000 K followed, which included 2000 steps each of 15 fs. Finally, a Cartesian dynamics cooling phase to 300 K was executed using 3000 steps each with a 0.3 fs time step. All parameters were scaled using default CNS values. One hundred structures were calculated and were ranked in overall energy terms, the best 10% were selected as representative and are presented in **Fig. 3**.

4. Notes

1. One liter of D_2O minimal media contains the following: 1 L M9 salts, 10 mL glucose, 10 mL *E. coli* trace elements, 1 mL $MgSO_4$, 1 mL $CaCl_2$, 1 mL thiamine, and 1 mL antibiotic solution. An approx 87% deuterated protein can be prepared by growth of the bacteria on minimal media prepared in D_2O minimal media, and for >97% deuteration $^{13}C_6$,^{1}H-glucose is replaced with $^{13}C_6$,^{2}H-glucose (**Fig. 1B**). Once optimized, the yields are not severely affected by this method and by reclaiming the D_2O the cost is only marginally more expensive than double-labeled preparations.

2. Either ^{15}N, ^{13}C-[2,3-$^{2}H_2$]-valine or ^{13}C α-ketoisovalerate can be used for the methyl group labeling of valine and leucine residues. Various combinations of the aforementioned amino acids can be labeled depending on the sequence of the target protein. In our example, specific protonation of the methyl groups of valine and leucine was chosen as the strategy. Amino acid labeling in all cases was highly specific and >85% efficient, as adjudged by NMR (*see* **Figs. 1C** and **2**). A detailed study of the relative efficiencies of different specific labeling strategies has been carried out by Goto et al. *(16,30)*.

3. During purification, the protein sample experienced a total of 3 wk exchange in H_2O prior to the recording of NMR data. Reprotonation of the amide positions was judged to be essentially complete after a further 3 wk at 310 K by comparison with nondeuterated spectra. After this incubation period, experiments were repeated to identify missing resonances. The re-exchange can be accelerated by dialysis against buffer containing 1–2 *M* urea for a few days, at which time it is removed. Improved TROSY-based triple-resonance methodology has also been described *(8,31)*.

4. Methyl group assignments are unambiguously completed using a modified version of the (H)C(CO)NH-TOCSY pulse sequence for deuterated proteins *(1)*. In cases where complete assignment is not possible by this method, $^{13}C/^{15}N$-edited NOESY information can be used once the fold has been characterized.

5. Zwahlen et al. have also introduced NOESY experiments with significantly improved resolution for valine and isoleucine methyl groups *(15)*.

6. The choice of NOESY mixing time is particularly important when measuring distances longer than 5′Å, and careful calibration is required. Other mixing times chosen for perdeuterated/site-specifically protonated proteins include 175 ms for

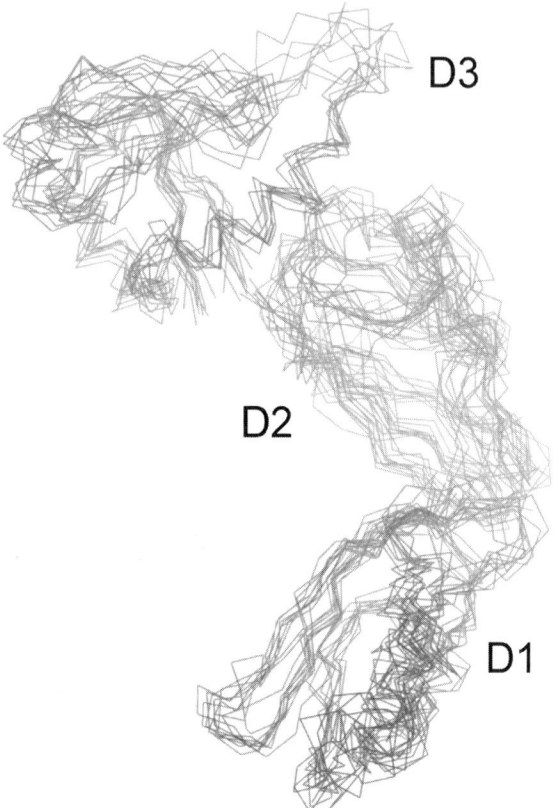

Fig. 3. Backbone traces for the 10 lowest energy structures of Int280. **D1** (residues 1–91), **D2** (residues 93–182), and **D3** (residues 183–280). Calculations were based on 1084 NOE and 78 H-bond distance restraints. The NOE restraints were composed of 576 short-range (residue i to residues $1 < i \leq 4$) and 508 long-range (residue i to residues $i > 4$) connectivities, of which a total of 502 involved methyl groups. One hundred and ten dihedral angle restraints were used for initial calculations of the fold, but were subsequently removed during refinement. The orientation of **D1** with respect to the rest of the structure is ill defined. To illustrate precision of the **D1** family, it has been superimposed separately on a representative structure. RMSDs are 1.5, 1.8, and 1.8 Å for backbone atoms position in **D1**, **D2** and **D3**, respectively.

the 370-residue maltodextrin-binding protein *(10)*, 300 ms for the α-helical, antiapoptotic protein Bcl-xL *(32)*, 175 ms for Dbl homology domain *(33)*, and 200 ms deuterated OmpA in micelles *(34)*.

7. The $^{13}C_{\alpha}$, $^{13}C_{\beta}$ and $^{13}C'$ shifts were corrected for deuterium isotope effects using the method described by Venter et al. *(35)*. In cases where angles found by TALOS do not satisfy the acceptance criteria, chemical shift index predictions can be used *(36)*.

Acknowledgments

S. Matthews would like to acknowledge the support of the Wellcome Trust and the BBSRC. The atomic coordinates for Int280 have been deposited in the Brookhaven PDB with accession number 1INM and assignments in BioMag–ResBank under the accession number 4111.

References

1. Shan, X., Gardner, K. H., Muhandiram, D. R., Rao, N. S., Arrowsmith, C. H., and Kay, L. E. (1996) Assignment of ^{15}N, ^{13}Cα, ^{13}Cβ and HN resonances in an ^{15}N, ^{13}C, ^{2}H labeled 64 kDa trp repressor–operator complex using triple resonance NMR spectroscopy and ^{2}H-decoupling. *J. Am. Chem. Soc.* **118,** 6570–6579.

2. Yamazaki, T., Lee, W., Arrowsmith, C. H., Muhandiram, D. R., and Kay, L. (1994) A suite of triple resonance NMR experiments for the backbone assignment of ^{15}N, ^{13}C, ^{2}H labeled proteins with high sensitivity. *J. Am. Chem. Soc.* **116,** 11,655–11,666.

3. Pervushin, K., Riek, R., Wider, G., and Wuthrich, K. (1997) Attenuated T-2 relaxation by mutual cancellation of dipole–dipole coupling and chemical shift anisotropy indicates an avenue to NMR structures of very large biological macromolecules in solution. *Proc. Natl. Acad. Sci. USA* **94,** 12,366–12,371.

4. Riek, R., Pervushin, K., and Wuthrich, K. (2000) TROSY and CRINEPT: NMR with large molecular and supramolecular structures in solution. *Trends Biochem. Sci.* **25,** 462–468.

5. Riek, R., Fiaux, J., Bertelsen, E. B., Horwich, A. L., and Wuthrich, K. (2002) Solution NMR techniques for large molecular and supramolecular structures. *J. Am. Chem. Soc.* **124,** 12,144–12,153.

6. Fiaux, J., Bertelsen, E. B., Horwich, A. L., and Wuthrich, K. (2002) NMR analysis of a 900K GroEL–GroES complex. *Nature* **418,** 207–211.

7. Salzmann, M., Pervushin, K., Wider, G., Senn, H., and Wuthrich, K. (2000) NMR assignment and secondary structure determination of an octameric 110 kDa protein using TROSY in triple resonance experiments. *J. Am. Chem. Soc.* **122,** 7543–7548.

8. Tugarinov, V., Muhandiram, R., Ayed, A., and Kay, L. E. (2002) Four-dimensional NMR spectroscopy of a 723-residue protein: chemical shift assignments and secondary structure of malate synthase G. *J. Am. Chem. Soc.* **124,** 10,025–10,035.

9. Mal, T. K., Matthews, S. J., Kovacs, H., Campbell, I. D., and Boyd, J. (1998) Some NMR experiments and a structure determination employing a {N-15,H-2} enriched protein. *J. Biomol. NMR* **12,** 259–276.

10. Gardner, K. H., Rosen, M. K., and Kay, L. E. (1997) Global folds of highly deuterated, methyl-protonated proteins by multidimensional NMR. *Biochemistry* **36,** 1389–1401.

11. Metzler, W. J., Wittekind, M., Goldfarb, V., Mueller, L., and Farmer, B. T. (1996) Incorporation of ^{1}H/^{13}C/^{15}N-(Ile, Leu, Val) into a perdeuterated, ^{15}N-labeled protein: potential in structure determination of large proteins by NMR. *J. Am. Chem. Soc.* **118,** 6800, 6801.

12. Smith, B. O., Ito, Y., Raine, A., Teichmann, S., Ben-Tovim, L., Nietlispach, D., et al. (1996) An approach to global fold determination using limited NMR data from larger proteins selectively protonated at specific residues. *J. Biomol. NMR* **8**, 360–368.

13. Rosen, M. K., Gardner, K. H., Willis, R. C., Parris, W. E., Pawson, T., and Kay, L. E. (1996) Selective methyl group protonation of perdeuterated proteins. *J. Mol. Biol.* **263**, 627–636.

14. Gardner, K. H. and Kay, L. E. (1997) Production and incorporation of N-15, C-13, H-2 (H-1-delta 1 methyl) isoleucine into proteins for multidimensional NMR studies. *J. Am. Chem. Soc.* **119**, 7599, 7600.

15. Zwahlen, C., Vincent, S. J. F., Gardner, K. H., and Kay, L. E. (1998) Significantly improved resolution for NOE correlations from valine and isoleucine (C-gamma 2) methyl groups in N-15,C-13- and N-15,C-13,H-2-labeled proteins. *J. Am. Chem. Soc.* **120**, 4825–4831.

16. Goto, N. K., Gardner, K. H., Mueller, G. A., Willis, R. C., and Kay, L. E. (1999) A robust and cost-effective method for the production of Val, Leu, Ile (delta 1) methyl-protonated N-15-, C-13-, H-2-labeled proteins. *J. Biomol. NMR* **13**, 369–374.

17. Wang, H., Janowick, D. A., Schkeryantz, J. M., Liu, X., and Fesik, S. W. (1999) A method for assigning phenylalanines in proteins. *J. Am. Chem. Soc.* **121**, 1611, 1612.

18. Johnson, B. A. and Blevins, R. A. (1994) NMRView: a computer program for the visualization and analysis of NMR data. *J. Biomol. NMR* **4**, 603–614.

19. Stein, E., G., Rice, L. M., and Brünger, A. T. (1997) Torsion-angle molecular dynamics as a new efficient tool for NMR structure calculation. *J. Magn. Reson. B* **124**, 154–164.

20. Nilges, M., Gronenborn, A. M., and Clore, G. M. (1988) Determination of 3-dimensional structures of proteins by simulated annealing with interproton distance restraints—application to crambin, potato carboxypeptidase inhibitor and barley serine proteinase inhibitor-2. *Protein Eng.* **2**, 27–38.

21. Brünger, A. T. (1992) *XPLOR Manual Ver. 3.1.* Yale University, New Haven, CT.

22. Brünger, A. T., Adams, P. D., Clore, G. M., DeLanod, W. L., Grose, P., Grosse-Kunstleve, R. W., et al. (1998) Crystallography and NMR system: a new software suite for macromolecular structure determination. *Acta. Crystallogr. D Biol. Crystallogr.* **D54**, 905–921.

23. Kelly, G., Prasannan, S., Daniell, S., Frankel, G., Dougan, G., Connerton, I., et al. (1998) Sequential assignment of the triple labeled 30.1 kDa cell-adhesion domain of intimin from enteropathogenic *E-coli. J. Biomol. NMR* **12**, 189–191.

24. Kelly, G., Prasannan, S., Daniell, S., Fleming, K., Frankel, G., Dougan, G., et al. (1999) Structure of the cell-adhesion fragment of intimin from enteropathogenic *Escherichia coli. Nat. Struct. Biol.* **6**, 313–318.

25. Delaglio, F., Grzesiek, S., Vuister, G. W., Zhu, G., Pfeifer, J., and Bax, A. (1995) NMRPipe: a multidimensional spectral processing system based on Unix pipes. *J. Biomol. NMR* **6**, 277–293.

26. Norwood, T. J., Boyd, J., Heritage, J. E., Soffe, N., and Campbell, I. D. (1990) Comparison of techniques for ¹H-detected heteronuclear 1H-15N spectroscopy. *J. Magn. Reson. B* **87,** 488–501.
27. Marion, D., Driscoll, P. C., Kay, L. E., Wingfield, P. T., Bax, A., Gronenborn, A., et al. (1989) Overcoming the overlap problem in the assignment of larger proteins by the use of three-dimensional heteronuclear ¹H-¹⁵N Hartmann-Hahn-multiple quantum coherence and nuclear Overhauser multiple quantum coherence spectroscopy: application to interleukin 1. *Biochemistry* **28,** 6150–6156.
28. Vuister, G. W., Clore, G. M., Gronenborn, A. M., Powers, R., Garrett, D. S., Tschudin, R., et al. (1993) Increased resolution and improved spectral quality in 4-dimensional ¹³C/¹³C-separated HMQC-NOESY-HMQC spectra using pulsed field gradients. *J. Magn. Reson. B* **101,** 210–213.
29. Cornilescu, G., Delaglio, F., and Bax, A. (1999) Protein backbone angle restraints from searching a database for chemical shift and sequence homology. *J. Biomol. NMR* **13,** 289–302.
30. Goto, N. K. and Kay, L. E. (2000) New developments in isotope labeling strategies for protein solution NMR spectroscopy. *Curr. Opin. Struct. Biol.* **10,** 585–592.
31. Mulder, F. A. A., Ayed, A., Yang, D. W., Arrowsmith, C. H., and Kay, L. E. (2000) Assignment of H-1(N), N-15, C-13(alpha), (CO)-C-13 and C- 13(beta) resonances in a 67 kDa p53 dimer using 4D-TROSY NMR spectroscopy. *J. Biomol. NMR* **18,** 173–176.
32. Medek, A., Olejniczak, E. T., Meadows, R. P., and Fesik, S. W. (2000) An approach for high-throughput structure determination of proteins by NMR spectroscopy. *J. Biomol. NMR* **18,** 229–238.
33. Aghazadeh, B., Zhu, K., Kubiseski, T. J., Liu, G. A., Pawson, T. P., Zheng, Y., et al. (1998) Structure and mutagenesis of the Dbl homology domain. *Nat. Struct. Biol.* **5,** 1098–1107.
34. Arora, A., Abildgaard, F., Bushweller, J. H., and Tamm, L. K. (2001) Structure of outer membrane protein A transmembrane domain by NMR spectroscopy. *Nat. Struct. Biol.* **8,** 334–338.
35. Venter, R. A., Farmer, B. T., Fierke, C. A., and Spicer, L. D. (1996) Characterizing the use of perdeuteration in NMR studies of large proteins: C-13, N-15 and H-1 assignments of human carbonic anhydrase II. *J. Mol. Biol.* **264,** 1101–1116.
36. Wishart, D. S. and Sykes, B. D. (1994) The ¹³C chemical shift index: a simple method for the identification of protein secondary structure using 13C chemical shift data. *J. Biomol. NMR* **4,** 171–180.

4

Segmental Isotopic Labeling
for Structural Biological Applications of NMR

David Cowburn, Alexander Shekhtman, Rong Xu, Jennifer J. Ottesen, and Tom W. Muir

Summary

This chapter describes the preparation of precursor domains for the formation of multidomain segmentally labeled proteins by protein ligation.

Key Words: Expressed protein ligation; inteins; isotopic labeling; reduced proton density labeling (REDPRO).

1. Introduction

It is widely recognized that the composition of the human proteome has resulted from extensive domain duplication, with a significant proportion of the most duplicated of domain superfamilies being specific to multicellular function *(1)*. Many of these domains within a particular protein may be independently folded and have a specific recognition function. The composite function of such a protein is nonetheless dependent on the organization of these domains and is rarely simply the sum of the parts. Although X-ray crystallography remains the gold standard for atomic resolution structures, the orientation of domains in crystal forms may be influenced by the crystal packing, and the interconversion between different ligated forms may be hard to study in the solid state. Thus the individual domain–domain interfaces require additional investigation by other methods. Nuclear magnetic resonance (NMR) can significantly play this role, especially with recent advances in improved sensitivity and extension of molecular weight limits *(2)*. Even with the powerful transverse relaxation-optimized spectroscopy methods *(3)*, there still remains the "too many notes" problem of multiple signals *(4)* and excess signal complexity, especially in nonsymmetric

From: *Methods in Molecular Biology, vol. 278: Protein NMR Techniques*
Edited by: A. K. Downing © Humana Press Inc., Totowa, NJ

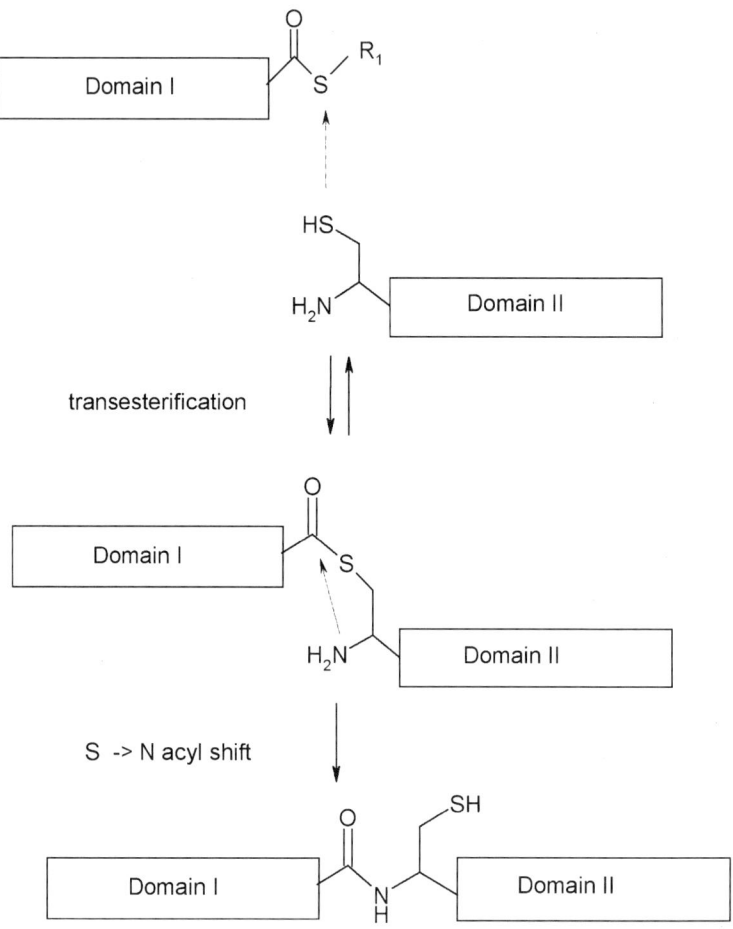

Fig. 1. Thioester/N-terminal cysteine chemical ligation.

higher molecular weight proteins. Isotopic labeling of a segment of a protein permits a wide range of structural and mapping studies to be conducted by NMR (for a review, *see* **ref. 5**).

The total synthesis of peptides, predominantly by solid state methods, is well developed and provides essentially complete control of the placement of isotopic labels, regio- and stereospecifically, but the economies and time-scales of such syntheses currently preclude their use for routine NMR analysis of products larger than about 3 kDa. The stitching together of segments of a protein labeled separately in bacterial or other expression systems provides a more general and effective route to segmental labeling. How is such stitching together to be done? The junction needs to resemble the natural peptidic linkage, and this presents a major challenge to any synthetic method compatible with native expressed proteins. With synthetic products, the most effective chemical ligation method generally is accepted to be that of the reaction of a C-terminal thioester with an N-terminal cysteine residue *(6)*. An overview of this reaction is in **Fig. 1**.

This chemical reaction was subsequently discovered to be very similar to that used by nature in intein fusion of proteins known as *protein splicing (7,8)*. The fascinating details of protein splicing are beyond the scope of this chapter and may be reviewed elsewhere *(9,10)*. Subsequent molecular biology developments identified methods for the production of C-terminal α-thioesters and for the production of split inteins. These provide the base for the two current methods of segmental labeling—expressed protein ligation (EPL) and *trans*-splicing. The two methods are contrasted in **Fig. 2**.

This chapter describes the use of EPL. Considerable success with the alternate *trans*-splicing method has been described *(11–13)*. Our preference for the use of EPL is based on its ability to couple together native expressed fragments, without refolding, and its much less restrictive requirements on the linker composition. (*see* **ref. 14**).

The major procedures involved in these protocols are continuously evolving— the general steps are illustrated in **Fig. 3**.

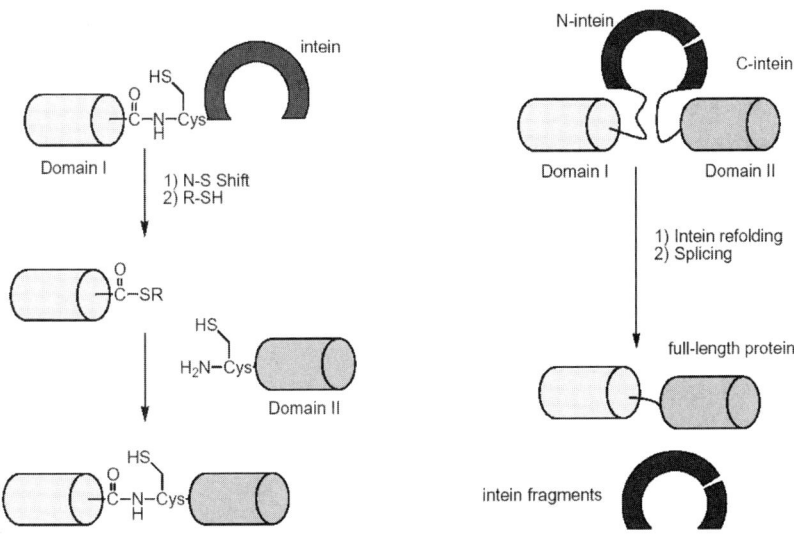

Expressed Protein Ligation   *trans*-splicing

Fig. 2. Comparison of segmental labeling by EPL, and by *trans*-splicing.

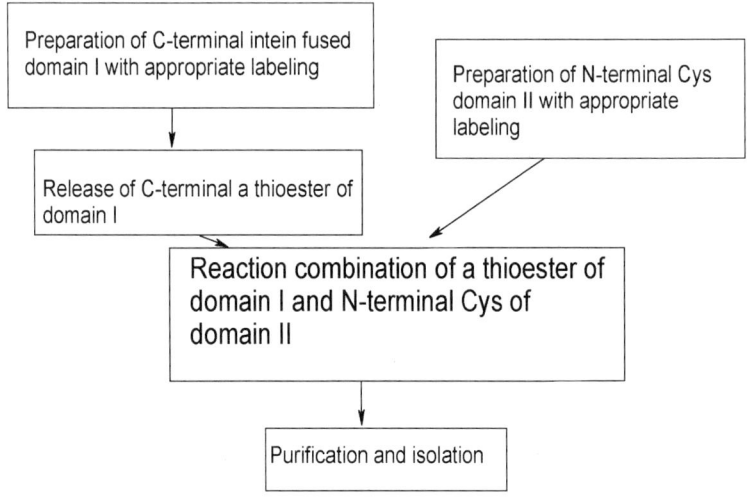

2. Materials

Because the exact reagents required depend on some of the strategies for the above steps, we list only the key material requirements, especially for the reaction combination.

There are multiple strategies for producing the two segments described in **Subheading 1.** Many projects will involve joining domains that have already been expressed, and these are likely to provide useful starting points with appropriate cloning sites and substituted codons for efficient expression.

1. C-Terminal Intein Fusion: Multiple vectors are commercially available from New England Biolabs (Beverly, MA; www.neb.com). Many are described in Table 1 of **ref. *14***. The pTXB1 vector has been generally useful (e.g., **ref. *15***) in BL21(DE3) or derivatives of *Escherichia coli.*
2. N-Terminal Cys-Containing Protein: Commercial primers for N-terminal mutation/insertion are available from a large number of businesses and academic resource facilities. Examples include Integrated DNA Technologies (Coralville, IA; www.idtdna.com) and Invitrogen (Carlsbad, CA; www.invitrogen.com). For cells, it is our experience that frequent replenishment from a commercial source is desirable (e.g., Novagen, Madison, WI; www.novagen.com). Factor Xa for cleavage of precursor proteins available from Amersham Biosciences, Piscataway, NJ.
3. Equipment: The experimental approach outlined will require resources generally available in an NMR lab currently using genetic engineering and protein expression. Without being totally exhaustive, this list includes

analytical high-pressure (performance) liquid chromatography, preparative liquid chromatography, polymerase chain reaction (PCR) equipment, cell-growth equipment (incubators, etc.), preparative centrifuge, concentrators, items required for the target's functional assay, and gel electrophoresis. Required analytical services, possibly in a central resource, include protein mass spectrometry and DNA sequencing. This chapter does not deal with the NMR applications to the targets, described in other chapters, and the spectrometer equipment so required.

3. Methods

3.1. Overall Strategic Issues

One of the major advantages of using EPL is that it can avoid refolding steps, but to do so does require that (a) one or both precursor segments are folded (as in **ref. 16**), (b) one of the precursor segments is folded and the other folds spontaneously when ligated (as in **ref. 15**), or (c) the precursor segments are not fully folded, but ligation induces their spontaneous fold. Conditions (a) and (b) are most readily identified with systems where significant "domainology" is already available, as has been the case in the work to date *(15,16)*.

3.1.1. Position of the Ligation Point

Comparison of domain behaviors should permit identification of likely linker regions between folded subdomains. It is desirable to maintain the same length of a linker and to substitute two or three residues in the final fusion at the junction site. When a linker is believed to be entirely flexible, it may be possible to insert additional residues at the junction site. Preexisting structural data, or homology modeling is likely to identify suitable areas, especially those containing hydrophilic and small side chain residues, which will be changed to –Gly-Cys-Gly- or –Gly-Cys- in the final fusion product. In other words, the "Domain-I" will have a C-terminal Gly preceding the α-thioester, and the "Domain-II" will have an N-terminal Cys, possibly followed by a Gly. The purpose of the first -Gly-, preceding the α-thioester, is to minimize steric hindrance of the transesterification of **Fig. 2**. The second Gly in the "Domain-II" facilitates the preparation of the N-terminal Cys Domain-II by cleavage using Factor Xa, and possibly other proteases. It is not obligatory even in these cases. Of course, one danger of these short peptide substitutions is that they may result in structural and/or functional modification related to the question under study. A conservative approach is then to prepare the mutant whole protein by conventional methods to test its relevant properties in advance of the vector preparation of the precursor sequences and isotopic labeling (*see* **Note 1**).

3.1.2. Composition of the Ligation Surrounds

The rate of ligation in **Fig. 1** will be partly limited by the conformational flexibility of ligating ends of the sequences. To reduce this, it is wise to avoid inserting the linker point in any region likely to have secondary structure. In addition, very large clustering of similarly charged residues or of hydrophobic amino acids may lead to restricted conformational flexibility and reduce the rate of ligation.

3.2. Preparation of α-Thioester Protein

3.2.1. Expression and Purification of Protein–Intein Fusion Precursor

For the preparation of the α-thioester protein, the vendor's recommended cloning procedures and expression methods with chitin-binding domain (CBD)-mutated inteins are generally appropriate, as is the affinity chromatography of the expressed fusion protein on chitin-agarose beads. As a starting estimate, about 300 μg protein are likely to be absorbed per milliliter of chitin beads, and the volume of beads can be scaled to the estimated protein yield. The lysate from cell cleavage may be resuspended in a suitable column buffer (e.g., 20 mM sodium HEPES, 350 mM sodium chloride, 1 mM ethylenediaminetetraacetic acid, 0.1% Triton X-100, pH 7.0), and applied to the chitin bead column. The column is then washed with the same buffer and stored at 4°C. Loading of this column may be determined by treating 100 μL of beads for 12 h with 200 mM sodium phosphate, 200 mM sodium chloride, 100 mM dithiothreitol, pH 7.2, washing the beads with 1:1 acetonitrile:water, and loading the solution and washes onto a high-pressure liquid chromatograph (HPLC) for quantitation.

3.2.2. α-Thioester Protein Production

Fusion can be achieved either with or without intermediate isolation of the α-thioester protein. In either case, it is probably worthwhile to ensure that coupling with a test peptide can be achieved with released α-thioester protein. A model peptide such as NH$_2$-CGRGRGRK[fluorescein]-CONH$_2$ *(16)* can be used to readily ensure that the construct designed can be released and coupled, prior to reaction with the domain-II product. The production of the fluorescein conjugated domain-I can then be monitored by the fluorescent band on gel electrophoresis.

3.2.3. Release of Thioester Protein Using Ethanethiol *(Caution: See Note 2)*

Domain-I-intein-CBD fusion bound to chitin beads is equilibrated and suspended in 0.2 M sodium phosphate, pH 6.0, 0.2 M NaCl, and 3% (v/v) ethanethiol. The suspension is gently agitated overnight, the supernatant removed, and the beads washed with a releasing buffer, such as 1:1 (v/v) acetonitrile (HPLC grade)/water. All washes and the supernatant are combined and loaded onto a retaining chromatographic system, such as C_{18} preparative HPLC, and purified. α-thioesters are highly reactive with aldehydes, and ketones—all glassware or plasticware should be prerinsed in buffers, and under no circumstances should glassware be dried with acetone (*see* **Note 3**).

3.2.4. Alternative Release Using 2-Mercaptoethanesulfonic Acid (MENSA)

Production of stable α-thoiesters has also been reported using MENSA *(17)*.

3.3. N-Terminal Cysteine-Containing Protein

Several methods have been developed that include mutated inteins releasing an N-terminal cysteine protein *(14)*. The locally preferred method uses Factor Xa to release the final product from a specific fusion system *(15,16,18)*. The N-terminal insertion sequence –IEGRC- may be used. This contains the previously well-recognized –IEGR- motif for Factor Xa, with a C-terminal glycyl residue improving product yield *(15)*. Alternative methods are available *(19,20)*.

3.4. Final Product Preparation

3.4.1. Optimizing the Ligation Reaction

The general limitation of the synthesis in **Fig. 1** is that the speed of the second order reaction depends on the concentration of both precursors. This concentration may be limited by solubility or by the amount of material on hand. A large excess of one of the sequence reactants is a possible approach to this issue, but it will commonly be limited by lack of availability, especially for stable isotope-labeled materials. Some effort should be devoted to surveying a range of conditions of concentration, temperature, pH, and so forth, using analytical polyacrylamide electrophoresis of the products to assess the degree of product formed. As starting conditions for assessing ligation, the limiting component sequence should be in excess of 200 μ*M* with buffer conditions of 200 m*M* sodium phosphate, pH 7.2, 200 m*M* sodium chloride, thiophenol (1.5% w/v), and benzyl mercaptan (1.5% w/v).

3.4.2. Isotopic Labeling Issues

The general issues of isotopic labeling for NMR are well known (e.g., **ref. 21**). The most obvious use of segmental labeling is to characterize only one of the subsequent domains using labeling and leaving the other "cryptic." More elaborate schemes are likely to be developed.

3.4.2.1. THE REDUCED PROTON (REDPRO) APPROACH

The REDPRO procedure has the potential for applications to structure determination, for mapping of chemical shift changes, and for application to higher molecular weight cases *(22)*. Detailed descriptions of related procedures may also be consulted *(23,24)*. It was previously observed *(23)* that *E. coli* BL21 can be grown without prior adaptation on minimal medium containing 100% ^2D_2O supplemented with [U-^{13}C,^{1}H] glucose and [U-^{15}N,^{1}H] ammonium chloride. *E. coli* BL21 (DE3) is freshly transformed before each expression with the appropriate construct. Cells are cultured on unlabeled M9 medium in ^1H_2O, typically overnight to an optical density $(OD)_{600}$ of less than 0.2, collected by centrifugation, washed in sterile phosphate-buffered saline, and resuspended in labeled minimal media with ^2H_2O, typically 2 g/L glucose and 1 g/L ammonium chloride. After a short 2- to 3-h adaptation process, cells reach an OD_{600} of 0.5–0.8. Induction is started with 1 mM isopropyl-β-D-thiogalactopyranoside, and cells are aerated at 37°C for up to 20 h. Some systems may express better at lower temperatures.

4. Notes

1. If the linker is believed to be entirely uninvolved in the function, then the direct insertion and testing of the construct [Optional Tag 1]-[Domain I with C-terminal (3′) restriction site 1 (RS1)]-IEGRC-[Domain II with N-terminal (5′) restriction site 2 (RS2)]-[Optional Tag 2] for function will then permit a simple PCR process to the intermediary [Optional Tag 1]-[Domain I/RS1]-G-Intein and [Optional Tag III]-IEGR-C-[Domain II/RS2]-[Optional Tag 2].
2. Ethanethiol presents little hazard to experienced chemists working in well-ventilated hoods. However, users should consult their material data safety sheets, or the local equivalent, before using this flammable, volatile, and odiferous material.
3. It cannot be sufficiently emphasized that the conditions of reaction require highly purified materials of exactly the sequences designed. Therefore, it is essential to use DNA sequencing on all new vectors, and to use mass spectrometry to identify rigorously the unlabeled precursors, before runs with isotope labeling. Measurements of protein concentrations in mixtures can be estimated for gels, or directly measured against standards using analytical HPLC.

References

1. Muller, A., MacCallum, R. M., and Sternberg, M. J. (2002) Structural characterization of the human proteome. *Genome Res.* **12,** 1625–1641.
2. Riek, R., Fiaux, J., Bertelsen, E. B., Horwich, A. L., and Wuthrich, K. (2002) Solution NMR techniques for large molecular and supramolecular structures. *J. Am. Chem. Soc.* **124,** 12,144–12,153.
3. Pervushin, K., Riek, R., Wider, G., and Wuthrich, K. (1997) Attenuated T_2 relaxation by mutual cancellation of dipole–dipole coupling and chemical shift anisotropy indicates an avenue to NMR structures of very large biological macromolecules in solution. *Proc. Natl. Acad. Sci. USA* **94,** 12,366–12,371.
4. *The Grove Concise Dictionary of Music* (1994) (Sadie, S., ed.), Macmillan, New York, p. 541.

5. Ottesen, J. J., Blashke, U. K., Cowburn, D., and Muir, T. W. (2003) Segmental isotopic labeling: prospects for a new tool to study the structure–function relationships in multidomain proteins. In *Biological Magnetic Resonance*, vol. 20 (Krishna, N. R. and Berliner, L. J., eds.), Kluwer Academic, New York, pp. 35–51.

6. Dawson, P. E., Muir, T. W., Clark-Lewis, I., and Kent, S. B. H. (1994) Synthesis of proteins by native chemical ligation. *Science* **266**, 776–779.

7. Xu, M. Q., Comb, D. G., Paulus, H., Noren, C. J., Shao, Y., and Perler, F. B. (1994) cinimide formation. *EMBO J.* **13**, 5517–5522.

8. Chong, S., Shao, Y., Paulus, H., Benner, J., Perler, F. B., and Xu, M. Q. (1996) Protein splicing involving the *Saccharomyces cerevisiae* VMA intein: the steps in the splicing pathway, side reactions leading to protein cleavage, and establishment of an in vitro splicing system. *J. Biol. Chem.* **271**, 22,159–22,168.

9. Paulus, H. (2000) Protein splicing and related forms of protein autoprocessing. *Annu. Rev. Biochem.* **69**, 447–496.

10. Liu, X. Q. (2000) Protein-splicing intein: genetic mobility, origin, and evolution. *Annu. Rev. Genet.* **34**, 61–76.

11. Yamazaki, T., Otomo, T., Oda, N., Kyogoku, Y., Uegaki, K., and Ito, N. (1998) Segmental isotope labeling for protein NMR using peptide splicing. *J. Am. Chem. Soc.* 120 (**22**), 5591,5592.

12. Otomo, T., Ito, N., Kyogoku, Y., and Yamazaki, T. (1999) NMR observation of selected segments in a larger protein: central-segment isotope labeling through intein-mediated ligation. *Biochemistry* **38**, 16,040–16,044.

13. Otomo, T., Teruya, K., Uegaki, K., Yamazaki, T., and Kyogoku, Y. (1999) Improved segmental isotope labeling of proteins and application to a larger protein. *J. Biomol. NMR* **14**, 105–114.

14. Xu, M. Q. and Evans, T. C., Jr. (2001) Intein-mediated ligation and cyclization of expressed proteins. *Methods* **24**, 257–277.

15. Camarero, J. A., Shekhtman, A., Campbell, E. A., Chlenov, M., Gruber, T. M., Bryant, D. A., et al. (2002) Autoregulation of a bacterial σ factor explored by using segmental isotopic labeling and NMR. *Proc. Natl. Acad. Sci. USA* **99**, 8536–8541.

16. Xu, R., Ayers, B., Cowburn, D., and Muir, T. W. (1999) Chemical ligation of folded recombinant proteins: segmental isotopic labeling of domains for NMR studies. *Proc. Natl. Acad. Sci. USA* **96**, 388–393.

17. Evans, T. C., Jr., Benner, J., and Xu, M. Q. (1998) Semisynthesis of cytotoxic proteins using a modified protein splicing element. *Protein Sci.* **7**, 2256–2264.

18. Erlanson, D. A., Chytil, M., and Verdine, G. L. (1996) The leucine zipper domain controls the orientation of AP-1 in the NFAT.AP-1.DNA complex. *Chem. Biol.* **3**, 981–991.

19. Cowburn, D. and Muir, T. W. (2001) Segmental isotopic labeling using expressed protein ligation. *Meth. Enzymol.* **339**, 41–54.

20. Tolbert, T. J. and Wong, C.-H. (2002) New methods for proteomic research: preparation of proteins with N-terminal cysteines for labeling and conjugation. *Angew. Chem. Int. Ed. Engl.* **114**, 2275–2278.

21. Gardner, K. H. and Kay, L. E. (1998) The use of ^{2}H, ^{13}C, ^{15}N multidimensional NMR to study the structure and dynamics of proteins. *Annu. Rev. Biophys. Biomol. Struct.* **27,** 357–406.
22. Shekhtman, A., Ghose, R., Goger, M., and Cowburn, D. (2002) NMR structure determination and investigation using a reduced proton (REDPRO) labeling strategy for proteins. *FEBS Lett.* **524,** 177–182.
23. Leiting, B., Marsilio, F., and O'Connell, J. F. (1998) Predictable deuteration of recombinant proteins expressed in *Escherichia coli*. *Anal. Biochem.* **265,** 351–355.
24. Marley, J., Lu, M., and Bracken, C. (2001) A method for efficient isotopic labeling of recombinant proteins. *J. Biomol. NMR* **20,** 71–75.

5

TROSY-Based Correlation and NOE Spectroscopy for NMR Structural Studies of Large Proteins

Guang Zhu, Youlin Xia, Donghai Lin, and Xiaolian Gao

Summary

Transverse relaxation-optimized spectroscopy (TROSY) is based on the fact that cross-correlation relaxation rates associated with the interferences between chemical shift anisotropy and dipole–dipole interactions can be dramatically reduced. TROSY selects these slowly relaxing components of $^{15}N-^{1}H^{N}$ or $^{13}C-^{1}H$ antiphase coherences to significantly enhanced signal sensitivity and spectral resolution for large proteins (>30 kD). The basic principles and applications of three- and four-dimensional TROSY-based triple-resonance experiments and NOESY experiments for structure-function studies of proteins are discussed in this chapter. To make applications of these experiments easier, some of the experimental setups are also described.

Key Words: TROSY; TROSY–HN(CO)CACB; TROSY–HNCACB; TROSY–HNCA; TROSY–HNCO; NOESY–TROSY.

1. Introduction

Conventional multidimensional nuclear magnetic resonance (NMR) spectroscopy for structure–function studies of biological macromolecules has been limited to molecules smaller than 50 kDa *(1,2)*. Yet, NMR is one of the principle experimental techniques in structural biology. In recent years, the introduction of transverse relaxation-optimized spectroscopy (TROSY)-based NMR methods *(3–8)* has opened a new avenue for the structure determination of larger biological macromolecules, particularly for proteins and nucleic acids. As has been demonstrated recently, the cross-correlated relaxation-induced polarization transfer (CRIPT)–TROSY-type experiments *(9)*, modified versions of the first-introduced insensitive nucleus enhancement by polarization transfer (INEPT)–TROSY-type experiments, enable scientists to study proteins up to 900 kDa *(10)*.

From: *Methods in Molecular Biology, vol. 278: Protein NMR Techniques*
Edited by: A. K. Downing © Humana Press Inc., Totowa, NJ

TROSY *(3)* is based on the fact that cross-correlated relaxation introduced by the interference of the dipole–dipole interaction (DD) and chemical shift anisotropy (CSA) gives rise to much smaller transverse relaxation rates in a high field for a system of two coupled spins (1/2), I and S, such as the ^{15}N-^{1}H moiety in protein backbone *(3)* and ^{13}C-^{1}H moiety in the aromatic group of amino acids *(4)*, allowing NMR studies of much larger proteins and nucleic acids.

For structures of proteins in the molecular weight range up to about 150 kDa, the TROSY scheme is sufficient and effective for obtaining workable correlation spectra *(5–8,11,12)*, triple-resonance spectra for sequential assignment *(8,13–16)*, and nuclear Overhauser effect spectroscopy (NOESY) spectra for resonance assignment and the collection of conformational constraints *(11,17–20)*. It has also been demonstrated that TROSY-based triple-resonance experiments are more sensitive than the conventional triple-resonance experiments for a 23-kDa doubly or triply labeled protein *(8)*.

1.1. TROSY-Based Correlation Spectroscopy

In a high magnetic field, both protein backbone ^{15}N-^{1}H and aromatic side chain ^{13}C-^{1}H moieties have a TROSY effect, which is more pronounced if proteins are deuterated *(8)*.The TROSY effect will yield the narrowest line width for ^{1}H^N at 1 GHz and for ^{15}N at 900 MHz *(21)*. This property can be readily used in many types of correlation spectroscopy applied to isotopically labeled proteins or nucleic acids.

There are some fundamental differences between two-dimensional (2D) ^{15}N-^{1}H TROSY and ^{15}N-^{1}H heteronuclear single-quantum coherence (HSQC). 2D TROSY spectra of a deuterated protein manifest much narrower line widths, the peak positions are shifted by ^{1}J$_{HN}$/2 $\cong$ 45 Hz in the ^{15}N and ^{1}H^N dimensions, and the NH$_2$ proton resonances do not appear as in 2D HSQC spectra. Other characteristics of 2D TROSY are similar to that of 2D HSQC, as are their applications and experimental setups.

It is well known that the single transition to single transition polarization transfer TROSY can only select 50% of the spin population to have a TROSY effect. The advantage of TROSY-type experiments can only be effectively realized when the transverse relaxation time, T$_2$, of a protein or nucleic acid is sufficiently short (**Table 1**), such that the spectral line-narrowing effect of TROSY can compensate for the intrinsic lower sensitivity of TROSY. To achieve this experimentally, a longer evolution time is required.

1.2. 3D TROSY-Based Triple-Resonance Experiments

TROSY was originally introduced for the study of large deuterated proteins and protein complexes. Therefore, the backbone assignment strategy is slightly different from that of conventional multidimensional NMR methods. For protein

Table 1
Transverse $^1H^N$ and ^{15}N Relaxation Times Predicted for a 23-kDa Protein (τ_c = 15 ns) at 750 MHz in TROSY-HNCA and HSQC-HNCA[a]

	TROSY-HNCA		HSQC-HNCA		Sensitivity gain
	$R_2(^{15}N)[s^{-1}]$	$R_2(^1H^N)[s^{-1}]$	$R_2(^{15}N)[s^{-1}]$	$R_2(^1H^N)[s^{-1}]$	
Isolated ^{15}N-^1H group	3.0	3.2	20.9	20.3	8.0
β-Sheet $^{13}C/^{15}N$-labeled	10.6	41.1	28.5	58.2	1.8
α-helix $^{13}C/^{15}N$-labeled	8.7	31.5	26.6	48.6	2.0
β-Sheet $^2H/^{13}C/^{15}N$-labeled	3.7	6.3	21.6	23.5	4.7
α-helix $^2H/^{13}C/^{15}N$-labeled	5.0	13.2	22.6	30.3	2.9

[a](Reproduced with permission from **ref. 8**.)

backbone assignments, a series of three-dimensional (3D) and four-dimensional (4D) triple-resonance experiments **(Table 2)** have been developed such as 3D TROSY–HNCA *(8,13,14,16,22,23)*, TROSY–HN(CO)CA *(13,16)*, TROSY–HNCO *(8,13,16,24)*, TROSY–HN(CA)CO *(13,16)*, TROSY–HNCACB *(13,16,25)*, TROSY–HN(CO)CACB *(13,16)*, 4D-TROSY–HNCACO *(26)*, and TROSY–HNCOCA *(26)*. Other types of correlation spectroscopy, such as $[^{13}C, {}^1H]$-TROSY–HNC *(4,27–29)*, have also been proposed.

The basic design in TROSY-based triple-resonance experiments is the same as for conventional ones with the only exception being that in the TROSY experiments, one has to ensure that the slowly relaxing magnetization components are retained during evolution of the magnetization throughout the whole pulse sequence and during detection. Similar to the conventional HSQC-based experiments, 3D TROSY–HNCA (H^N_i, N_i, C^α_i, {C^α_{i-1}}) *(8,13,14,16,22,23)*, 3D TROSY–HN(CA)CO (H^N_i, N_i, C'_i) *(13,16)*, and TROSY–HNCACB (H^N_i, N_i, C^α_i, {C^α_{i-1}}, C^β_i, {C^β_{i-1}}) *(13,16,25)* provide intraresidue correlations **(Table 2)**. The smaller $^2J_{NC\alpha}$ coupling can sometimes provides weak interresidue peaks in TROSY–HNCA and TROSY–HNCACB spectra. These experiments rely on the $^1J_{HN} \cong 90$ Hz, $^1J_{NC\alpha} = 8$–12 Hz and $^1J_{CC'} = 55$ Hz couplings for the magnetization transfer.

For obtaining interresidue correlations, TROSY–HN(CO)CACB (H^N_i, N_i, C^α_{i-1}, C^β_{i-1}) *(13,16)*, TROSY–HN(CO)CA (H^N_i, N_i, C^α_{i-1}) *(13,16)*, and TROSY–HNCO (H^N_i, N_i, C'_{i-1}) *(8,13,16,24)* **(Table 2)** have been designed with

Table 2
TROSY based Triple Resonance Experiments Used for Backbone Assignment

Experiment	Correlation observed	Magnetization transfer	J couplings	Ref.
3D TROSY-HNCA	$^1H_i^N$ — $^{15}N_i$ — $^{13}C_i^\alpha$ $^1H_i^N$ — $^{15}N_i$ — $^{13}C_{i-1}^\alpha$		$^1J_{NH}$ $^1J_{NC^\alpha}$ $^2J_{NC^\alpha}$	8,13,14,16, 22,23
3D TROSY-HN(CO)CA	$^1H_i^N$ — $^{15}N_i$ — $^{13}C_{i-1}^\alpha$		$^1J_{NH}$ $^1J_{NCO}$ $^1J_{C^\alpha CO}$	13,16
3D TROSY-HNCO	$^1H_i^N$ — $^{15}N_i$ — $^{13}CO_{i-1}$		$^1J_{NH}$ $^1J_{NCO}$	8,13,16,24
3D TROSY-HN(CA)CO	$^1H_i^N$ — $^{15}N_i$ — $^{13}CO_i$ $^1H_i^N$ — $^{15}N_i$ — $^{13}CO_{i-1}$		$^1J_{NH}$ $^1J_{NC^\alpha}$ $^2J_{NC^\alpha}$ $^1J_{C'\,CO}$	13,16

(continued)

Table 2 (continued)

Experiment	Transfer pathway		Couplings	Ref.
3D TROSY-HNCACB	$^1H_i^N - ^{15}N_i - ^{13}C_i^\alpha / ^{13}C_i^\beta$ $^1H_i^N - ^{15}N_i - ^{13}C_{i-1}^\alpha / ^{13}C_{i-1}^\beta$		$^1J_{NH}$ $^1J_{NC^\alpha}$ $^1J_{C^\alpha C^\beta}$ $^2J_{NC^\alpha}$	13,16,25
3D TROSY-HN(CO)CACB	$^1H_i^N - ^{15}N_i - ^{13}C_{i-1}^\alpha / ^{13}C_{i-1}^\beta$		$^1J_{NH}$ $^1J_{NCO}$ $^1J_{C^\alpha CO}$ $^1J_{C^\alpha C^\beta}$	13,16
4D TROSY-HNCACO	$^1H_i^N - ^{15}N_i - ^{13}C_i^\alpha - ^{13}CO_i$ $^1H_i^N - ^{15}N_i - ^{13}C_{i-1}^\alpha - ^{13}CO_{i-1}$		$^1J_{NH}$ $^1J_{NC^\alpha}$ $^1J_{C^\alpha CO}$ $^2J_{NC^\alpha}$	26
4D TROSY-HNCOCA	$^1H_i^N - ^{15}N_i - ^{13}CO_{i-1} - ^{13}C_{i-1}^\alpha$		$^1J_{NH}$ $^1J_{NCO}$ $^1J_{C^\alpha CO}$	26
4D TROSY-HNCO$_{i-1}$CA$_i$	$^1H_i^N - ^{15}N_i - ^{13}CO_{i-1}, ^{13}C_i^\alpha$		$^1J_{NH}$ $^1J_{NCO}$ $^1J_{NC^\alpha}$	31

61

the use of the $^1J_{HN}$, $^1J_{NC'}$ ($\cong$ 15 Hz) and $^1J_{CC'}$ couplings for the magnetization transfer.

1.3. 4D TROSY-Based Triple-Resonance Experiments

Because the T_2 relaxation time in TROSY is much longer compared with that in HSQC, it becomes ideal to design 4D TROSY-based triple-resonance experiments for the backbone assignment of larger $^{13}C/^{15}N/^2H$-labeled proteins *(30)*. 4D TROSY–HNCACO (H^N_i, N_i, C^α_i, $\{C^\alpha_{i-1}\}$, CO'_i, $\{CO'_{i-1}\}$), TROSY–HNCO-CA (H^N_i, N_i, C^α_{i-1}, CO'_{i-1}), *(26)* and TROSY–HNCO$_{i-1}$CA$_i$ (H^N_i, N_i, C^α_i, C'_{i-1}) *(31)* provide both inter- and intraresidue correlations based on $^1J_{HN}$, $^1J_{NC\alpha}$, $^1J_{NC'}$, and $^1J_{CC'}$ for the magnetization transfer. A list of 3D and 4D triple-resonance experiments used for backbone assignment is shown in **Table 2**.

1.4. Application of TROSY-Based Experiments for Protein Backbone Assignment (30)

The TROSY-based triple-resonance experiments described in the preceding sections have been used for the backbone assignment of 67-kDa $^{13}C/^{15}N/^2H$-labeled p53 in a manner similar to that used in HSQC-based multidimensional NMR *(30)*. Spectra of the 3D and 4D experiments were processed with the use of linear prediction methods *(32,33)* to extend the truncated free induction decays (FIDs) in the indirectly detected dimensions before the use of conventional Fourier transformation (FT) procedures to enhance resolution and reduce truncation artifacts.

Protein sample used:

0.4 m*M* $^{15}N/^{13}C/^2H$-labeled p53 (393 a.a., dimer, 67 kDa) in 25 m*M* phosphate, 250 m*M* NaCl, 10 m*M* dithiothreitol (DTT), 5 m*M* MgCl$_2$, pH 6.5, 90% H$_2$O/10% D$_2$O.

Equipment: 600 MHz NMR spectrometer with three channels.

Temperature: 25°C.

Assignment strategy: If ($^{13}C^\alpha$, $^{13}C'$) and ($^1H^N$, ^{15}N) spin pairs are unique, then it is possible to link ($^1H^N$, ^{15}N) of the *i*th residue to that of the *i*+1th residue through the common ($^{13}C^\alpha$, $^{13}C'$) chemical shifts with the use of the 4D TROSY–HNCACO (H^N, N_i, C^α_i, CO'_i), and TROSY–HNCOCA (H^N_i, N_i, C^α_{i-1}, CO'_{i-1}) experiments.

In the case of chemical-shift degeneracy of ($^{13}C^\alpha$, $^{13}C'$) or ($^1H^N$, ^{15}N) spin pairs, additional experiments must be analyzed, such as 4D $^{15}N/^{15}N$–NOESY–HSQC or 4D $^{15}N/^{15}N$–NOESY–TROSY *(20)* (*see* **Subheading 2.3.**), 4D TROSY–HNCO$_{i-1}$CA$_i$ (H^N_i, N_i, C^α_i, C'_{i-1}) *(31)*, 3D TROSY–HN(CA)CB ($^1H^N_i$, $^{15}N_i$, $^{13}C^\beta_i$) and 3D TROSY–HN(COCA)CB ($^1H^N_{i-1}$, $^{15}N_{i-1}$, $^{13}C^\beta_i$) *(13)*.

In addition to the above backbone assignment strategy using 4D TROSY-based experiments, 3D TROSY-based experiments are widely employed to obtain backbone assignments of a protein. The inter- and intraresidue connections can be readily obtained using the 3D TROSY-based experiments listed in **Table 2**. Normally, C^β chemical shifts have much less degeneracy. Hence, the 3D TROSY–HNCACB ($^1H^N_i$, $^{15}N_i$, $^{13}C^\alpha_i$ {$^{13}C^\alpha_{i-1}$}, $^{13}C^\beta_i$ {$^{13}C^\beta_{i-1}$}) and 3D TROSY–HN(CO)CACB ($^1H^N_{i-1}$, $^{15}N_{i-1}$, $^{13}C^\alpha_i$, $^{13}C^\beta_i$) *(13,16,25)* and 3D NOESY–TROSY *(11)* are useful in resolving ambiguous assignments.

2. TROSY-Based NOESY Experiments

NMR-derived protein structures are based on interproton distance restraints obtained from ^{15}N- and ^{13}C-edited 2D, 3D, and 4D NOESY data sets, torsion angle restraints generated from various heteronuclear and homonuclear scalar coupling correlation experiments *(34,35)*, and residual dipolar couplings obtained using samples dissolved in liquid crystal media *(36–38)*. Among these, NOE restraints are essential for defining the structure to high resolution. Specifically, NOE restraints are obtained from 3D ^{15}N-edited *(39)*, ^{13}C-edited *(40)*, or 3D or 4D ^{15}N- and ^{13}C-edited NOESY experiments *(41,42)*. TROSY-based NOESY experiments *(11,17–20,43–46)* have been shown to be superior in signal enhancement for large $^{15}N/^{13}C/^2H$-labeled proteins. In the following sections, these ^{15}N-separated TROSY-based NOESY experiments *(11,17,18,20)* are discussed in detail.

2.1. 3D Single-Quantum NOESY–TROSY Experiment

3D single-quantum NOESY–TROSY *(11)* was the earliest experiment used to introduce TROSY into 3D ^{15}N separated NOESY experiments. In the 3D single-quantum NOESY–TROSY, water flip-back en-TROSY is used so that the sensitivity is enhanced by a factor of $\sqrt{2}$ compared with the original version of TROSY *(3)*, and the signal attenuation resulting from the chemical exchange between water protons and amide protons is reduced. The pulse sequence of this experiment is shown in **Fig. 1**. The advantage of TROSY-type experiments can only be effectively realized when the transverse relaxation time, T_2, of a protein or nucleic acid is small enough such that TROSY can have better sensitivity than HSQC. To achieve this experimentally, a longer evolution time is required. With this in mind, single-quantum magnetization is used during the t_1 evolution time to obtain higher sensitivity as depicted in the **Fig. 1**.

Application to proteins:

Protein sample used:

1 m*M* ^{15}N-labeled calmodulin (147 a.a.) in 90% H_2O/10% D_2O at pH 6.8.

Equipment: 500-MHz NMR spectrometer with three channels.

Experimental parameters and results:

Temperature: 15°C.

Experimental parameters: WATERGATE uses a 1.7-ms pulse with a power of 110 Hz. $\tau_1 = 2.7$ ms and $\tau_{mix} = 100$ ms (**Fig. 1**). The quadrature component

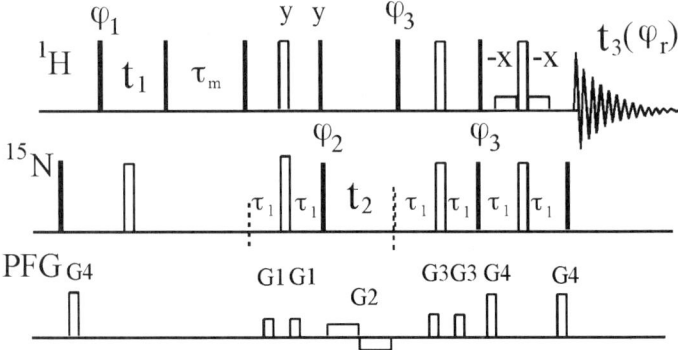

Fig. 1. Pulse sequence of the 3D single quantum NOESY-TROSY experiment. Filled bars and open bars represent 90° and 180° pulses, respectively. Default phases are x. $\varphi = 4(x), 4(-x)$; $\varphi_2 = (y, x, -y, -x)$, $\varphi = (y)$; $(x, -y, -x, y, -x, y, x, -y)$.

in the t_1 dimension is acquired by altering the phase of φ_1 in the States-time proportional phase increment (TPPI) manner. Echo/antiecho selection during t_2 is done by reversing φ_3 and the even-number phases of the receiver. The durations and strengths of the gradients are of conventional values and the bipolar gradient G2 is 0.5 G/cm.

Processing parameters: A 3D NOESY–TROSY data matrix of 64×32×512 complex points was acquired. The spectrum (**Fig. 2**) was processed with the use of linear prediction in the t_1 and t_2 time domains to extend the size of the FIDs by 50% *(33,47)*. A cosine-square bell window function was applied in t_1 and t_2, and the data are zero-filled one time before FT.

The 3D single quantum NOESY–TROSY experiment is widely used to obtain distance restraints and to verify assignments for doubly and triply labeled large proteins because of its simplicity and sensitivity. This experiment can be applied to triply labeled proteins above approx 40 kDa to provide the following NOEs: dNN(i, i+1), (Strong [α-helix], Weak [β-sheet]), dNN (medium [cross-β strands]) and other long-range NOEs used for obtaining spectral assignments and distance constrains.

2.2. 2D and 3D Transverse Relaxation-Optimized NOESY Experiments

When the T_2 relaxation time of a protein or nucleic acid is short enough and the t_1 evolution time is sufficiently long, the sensitivity of the single-quantum NOESY–TROSY experiment described in the preceding section is no longer optimized. In these cases, it is necessary to enhance the 3D NOESY experiment by TROSY in all three dimensions *(17,19,46)*. Here, two sensitivity-enhanced NOESY–TROSY experiments, namely, the 2D spin-state selective excitation method (S3E)–NOESY–S3E and 3D TROSY–NOESY–S3E experiments *(17)*, with the use of the S3E *(48,49)* are described. These experiments have the TROSY effect in all of the indirectly and directly detected dimensions, and so provide optimal resolution of amide protons. Furthermore, these two experiments provide an additional useful feature in that the diagonal peaks of the amide proton region are canceled or greatly reduced. This enables clear identification of NOESY cross peaks close to the diagonal peaks.

The pulse sequences of the 2D S3E–NOESY–S3E and 3D TROSY–NOESY–S3E experiments are depicted in **Fig. 3A** and **B**, respectively. The 2D S3E–NOESY–S3E experiment is a combination of the S3E method and the ordinary NOESY experiment. This experiment not only gives H_N–H_N and H_N–H_C (for partially deuterated proteins) NOE cross peaks with the characteristic narrow TROSY H_N line widths and greatly reduced diagonal peaks, but it also provides normal NOE cross-peaks, although with reduced intensities. In the 3D TROSY–NOESY–S3E experiment, the selection of the slowly relaxing components of ^1H and ^{15}N magnetization before the NOE mixing time is achieved by the use of the ^{15}N-^1H TROSY scheme, and the selection of the TROSY ^1H component before detection is accomplished by the use of the S3E

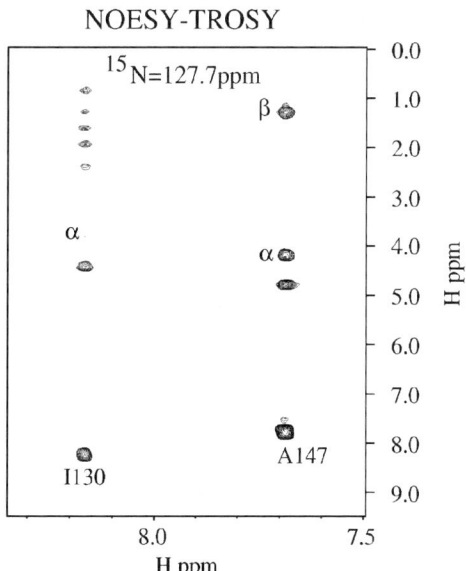

Fig. 2. 2D cross-section taken form 3D NOESY-TROSY spectrum (at ^{15}N = 127.7 ppm). The peak at ^1H = 4.8 ppm is the exchange peak between water and NH of A 147 at the terminus of the protein. The NOE mixing time is 100 ms. (Reproduced from **ref. *11*** with kind permission of Kluwer Academic Publishers.)

A   **B**

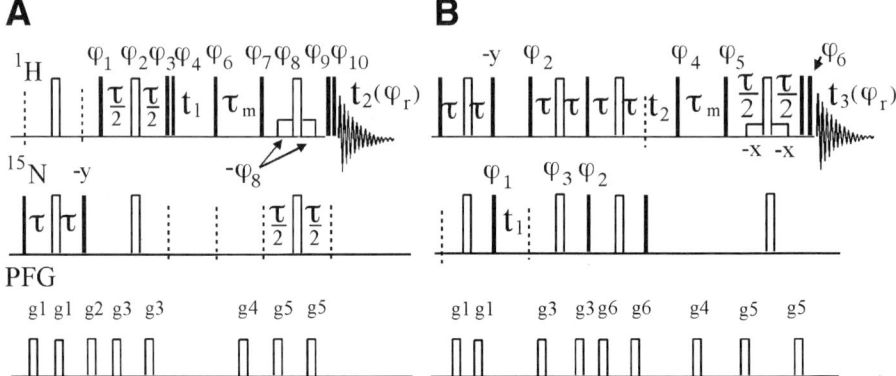

Fig. 3. 2D and 3D TROSY enhanced NOESY pulse sequences. Filled bars and open bars represent 90° and 180° pulses, respectively. Default phases are in the x direction. (A) S3E-NOESY-S3E pulse sequence: $\varphi_1 = (x, -x, -x, x, y, -y, -y, y) + 45°$; $\varphi_2 = 4(y)$, $4(-x)$; $\varphi_3 = 4(x)$, $4(y)$; $\varphi_4 = (x, -x, x, -x, y, -y, y, -y)$; $\varphi_6 = 4(x)$, $4(y)$; $\varphi_7 = [16(x)$, $16(y), 16(-x), 16(-y)] + 45°$; $\varphi_8 = 16(y)$, $16(-x)$; $\varphi_9 = 16(x)$, $16(y)$; $\varphi_{10} = 8(x)$, $8(-x)$, $8(y)$, $8(-y)$; $\varphi_r = 4(x, x, -x, x)$, $4(y, -y, -y, y)$, $4(-x, x, x, -x)$, $4(-y, y, y, -y)$. **(B)** TROSY-NOESY-S3E pulse sequence: $\varphi_1 = (x, y, -x, -y)$; $\varphi_2 = (y)$; $\varphi_3 = (x)$; $\varphi_4 = (x, y)$; $\varphi_5 = [4(x), 4(-x)] + 45°$; $\varphi_6 = 8(x)$, $8(-x)$; $\varphi_{10} = (-x, x)$; $\varphi_r = (x, x, -x, -x, -x, -x, x, x)$. (Reproduced from **ref. *17*** with kind permission of Kluwer Academic Publishers.)

sequence. Thus, the characteristic TROSY spectral line widths are obtained in all three dimensions of the 3D spectrum.

The diagonal peaks in the H_N–H_N region in the two 2D S3E–NOESY–S3E and 3D TROSY–NOESY–S3E spectra are canceled or greatly reduced, as discussed in detail in the literature *(17,19,46)*. The drawback of these experiments is that they are less sensitive than the corresponding normal 2D NOESY and 3D NOESY–HSQC experiments. However, this sensitivity loss is compensated by the narrower spectral line widths afforded by TROSY for large proteins *(3)*. Partial deuteration of proteins can largely reduce the contributions of ^1H–^1H and ^1H–^{15}N dipolar interactions to the relaxation of the H^N and N, so that the TROSY effect can be enhanced *(8)*.

Application to Proteins:

Protein sample used: 1 mM ^{15}N-labeled calmodulin (147 a.a.) in 90% H_2O/10% D_2O at pH 6.8.

Equipment: 750-MHz NMR spectrometer with three channels.

Experimental parameters and results:

Temperature: 15°C.

Experimental parameters: The experimental recovery delay is 1 s, $\tau = 1/(4^1J_{NH})$ $\approx$ 2.7 ms, $\tau_m = 80$ ms. (A) S3E–NOESY–S3E pulse sequence: The quadrature component in the t_1 dimension is acquired by altering the phases φ_1 to φ_4 in the States-TPPI manner. (B) TROSY–NOESY–S3E pulse sequence: Four transients, $S_{\text{phase1,phase2}}$, are acquired through altering φ_2 and φ_4 (if phase1 = 2, $\varphi_2 = \varphi_2 + 180°$ and $\varphi_4 = \varphi_4 + \varphi_{10}$, otherwise φ_2 and φ_4 are unchanged; if phase2 = 2, $\varphi_4 = \varphi_4 - 90°$, otherwise φ_4 is unchanged) for every pair of t_1 and t_2 values. The proton transmitter offset before the NOE mixing period is set at 8.5 ppm. The durations and strengths of the gradients are of conventional values.

Processing parameters: In the 2D S3E–NOESY–S3E, spectral widths in both dimensions were 10,500 Hz; 600 FIDs were acquired each with 64 transients. In the 3D TROSY–NOESY–S3E, 64(^{15}N) × 128(^{1}H) FIDs were recorded each with 16 scans. A cosine-square bell window function was applied in t_1 and t_2, and the data were zero-filled one time before FT.

Results: **Fig. 4A** and **B** show the amide proton regions of a conventional 2D NOESY spectrum and the 2D S3E–NOESY–S3E spectrum, respectively. It is clear that the diagonal peaks from **Fig. 4B** are canceled or greatly reduced in the H_N region of the spectrum, revealing more cross-peaks with enhanced

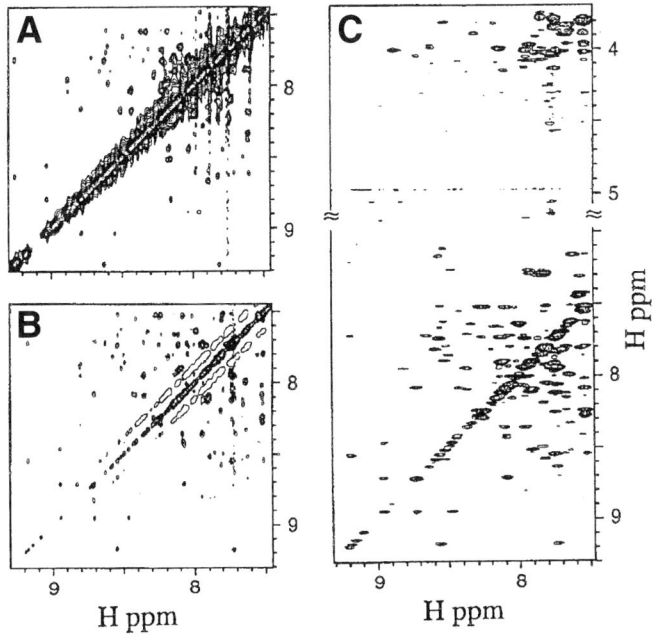

Fig. 4. 2D and 3D transverse relaxation optimized NOESY spectra. (**A**) and (**B**) show the amide proton region of a normal 2D NOESY spectrum and the S3E-NOESY-S3E 2D spectrum, respectively, and (**C**) the 2D spectrum obtained by the projection of 2D ^{1}H-^{1}H cross-section of the 3D TROSY-NOESY-S3E spectrum along the ^{15}N dimension. (Reproduced from **ref. *17*** with kind permission of Kluwer Academic Publishers.)

resolution when compared with the normal NOESY spectrum (**Fig. 4A**). The line-widths in **Fig. 4B** are about 4 Hz narrower on average than those in **Fig. 4A**. For deuterated proteins, the spectral line-widths will be further reduced. The open contours near the diagonal peaks in **Fig. 4B** are the negative $^1J_{HN}$ splittings of residual diagonal peaks.

Figure 4C displays the 2D NOESY spectrum obtained by projecting 2D ^1H–^1H cross-sections of the 3D TROSY–NOESY–S3E spectrum along the ^{15}N dimension. As seen in **Fig. 4C**, the 3D TROSY–NOESY–S3E experiment can produce a spectrum with canceled or greatly reduced diagonal peaks in the H_N region, but with some loss of cross-peaks in the H^N–H^α region resulting from the water suppression scheme used in the S3E (*17*).

2.3. 3D/4D $^{15}N/^{15}N$ Separated HSQC–NOESY–TROSY Experiment

3D single quantum NOESY–TROSY (*11*) and 3D TROSY–NOESY experiments (*17,19,46*) have been proposed to measure NOEs between protons. Nevertheless, the 3D NOESY–TROSY experiments may still be insufficient to provide unambiguous measurements of NOEs among the overlapping proton signals. This is particularly acute in proteins with a high level of helical content. To further alleviate the spectral overlap problem, four NOESY experiments, namely, 3D and 4D $^{15}N/^{15}N$ separated HSQC–NOESY–TROSY and TROSY–NOESY–TROSY have been proposed (*20*). Perdeuteration of the protein sample not only increases the amide proton T_2 and TROSY effect dramatically, but it also reduces the occurrence of spin diffusion, permitting detection of NOEs between protons separated more than 5Å, by using longer NOE mixing times (*50,51*). Hence, the TROSY-based 3D and 4D $^{15}N/^{15}N$ separated NOESY experiments could be very useful for structure determination of large proteins.

The two pulse sequences are shown in **Fig. 5A** and **B**, respectively. In these two pulse sequences, the TROSY sequences replace the latter gradient- and sensitivity-enhanced HSQC sections in the 3D and 4D $^{15}N/^{15}N$ separated HSQC–NOESY–HSQC pulse sequences. The ideas underlying use of HSQC or TROSY before the mixing time are the same as those discussed in **Subheading 2.1.**

Application to Proteins:

Protein sample used: 1 m*M* 100% ^{15}N/ 70% ^2H-labeled trichosanthin (256 a.a.) in 90% H_2O/10% D_2O at pH 7.0.

Equipment: 750-MHz NMR spectrometer with three channels.

Experimental parameters and results:

Temperature: 5°C.

Experimental parameters: Experimental recovery delay is 1 s, τ = 2.5 ms, τ_m = 100 ms. The durations and strengths of the gradients are of conventional values and gb are smaller bipolar gradients used to suppress the effect of water

A

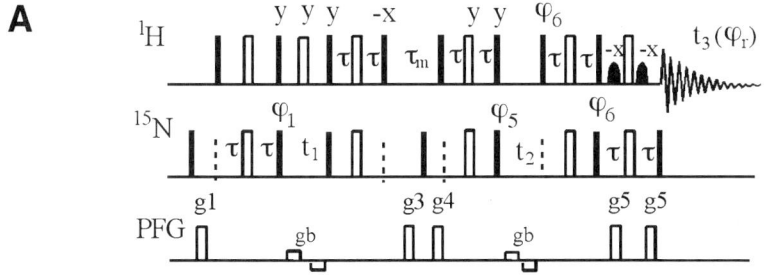

B

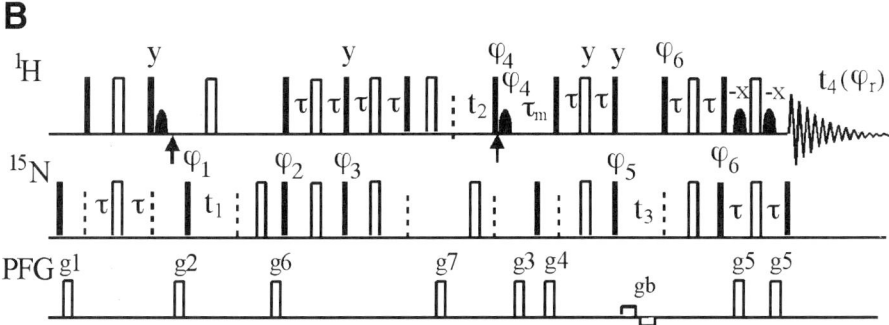

Fig. 5. HSQC-NOESY-TROSY pulse sequences: (**A**) 3D version and (**B**) 4D version. Filled bars and open bars represent 90° and 180° pulses, respectively. Fileed shaped pulses are 1.1 ms sinc-modulated rectangular 90° pulses to selectively excite water resonance. (phase1 = 1, 2; phase2 = 1, 2; phase3 = 1, 2), were recorded. Default phases are x. (**A**): $\varphi_1 = (x, x, -x, -x)$; $\varphi_5 = (y, x, y, x, -y, -x, -y, -x)$; $\varphi_6 = (y)$; $\varphi_r(x, -y, -x, y, -x, y, x, -y)$. If phase1 = 2, $\varphi_1 = \varphi_1 + 90°$, else φ_1 is not valid ; if phase2 = 2, $\varphi_6 = \varphi + 180°$ and the sign of φ_r is reversed for the even number step phases, otherwise φ_6 and φ_r are not varied. (**B**): $\varphi_1 = (y, -y)$; $\varphi_2 = (-x)$; $\varphi_3 = (y)$; φ_4 (y); $\varphi_5 = (y, y, x, x)$; $\varphi_6 = (y)$; $\varphi_r = (x, -x, -y, y)$. If phase1 = 2, $\varphi_4 = \varphi_4 - 90°$, otherwise φ_4 is not varied; if phase2 = 2. $\varphi_2 = \varphi_2 + 180°$ and the sign of g6 is inverted, otherwise φ_2 and g6 are not varied; if phase3 = 2, $\varphi_6 = \varphi_6 + 180°$ and the sign of φ_r is reversed for the third and fourth step phases otherwise φ_6 and φ_r are not varied. (Reproduced from **ref. *20*** with kind permission of Kluwer Academic Publishers.)

radiation damping. For the 4D experiment, the proton carrier position between the two arrows is set at 7.7 ppm and elsewhere returned to 4.7 ppm. For the 3D experiment, 4 transients, $S_{phase1,phase2}$ (phase1 = 1, 2; phase2 = 1, 2 as explained below), are recorded; for 4D experiment, 8 transients, $S_{phase1,phase2,phase3}$, are recorded.

Processing parameters: The 3D data matrix was composed of 54×108×1024 complex points with spectral widths of 1825×1825×10,500 Hz. The number of scans for each transient was eight. Cosine-bell window functions and zero-filling were used in all dimensions.

Results: 2D [^{15}N–^{15}N] (F$_1$–F$_2$) slices of the 3D ^{15}N/^{15}N separated HSQC–NOESY–TROSY spectrum of TCS and its corresponding HSQC–NOESY–HSQC spectrum taken at a ^{1}H chemical shift of F$_3$ = 7.92 ppm are shown in **Fig. 6A** and **B**, respectively. The number beside each peak of **Fig. 6A** is the ratio of peak intensity in **Fig. 6A** over its corresponding peak intensity in **Fig. 6B**. For all well-isolated peaks examined, the signal sensitivities in 3D HSQC–NOESY–TROSY experiment are enhanced by 18 to 178% compared with that of 3D HSQC–NOESY–HSQC. On average, a sensitivity enhancement of 62% is obtained. The line widths in the H$_N$ (F$_3$) and N (F$_2$) dimensions are reduced by 3–55% and 2–45%, respectively. On average, the line widths are decreased by 20 and 18% in the H$_N$ (F$_3$) and N (F$_2$) dimensions, respectively.

Two 2D [^{15}N–^{15}N] (F$_1$–F$_2$) slices of the 3D ^{15}N/^{15}N separated HSQC–NOESY–TROSY spectrum taken at ^{1}H chemical shift F$_3$ = 7.66 ppm and F$_3$ = 8.13 ppm are shown in **Fig. 7A** and **B**, respectively. For those protons with degenerate ^{1}H chemical shift, their NOE cross-peaks cannot be observed in a normal 2D [^{1}H–^{1}H]–NOESY spectrum, or a 3D ^{15}N separated NOESY spectrum. However, their cross-peaks are clearly seen in **Fig. 7**. It should be noted that the cross-peaks on the corners of dashed boxes, corresponding to different coherence transfer paths, respectively, are present on the same plane owing to the degeneracy of their proton chemical shifts.

The experimental data show that signal sensitivity in 4D HSQC–NOESY–TROSY is enhanced by 8% compared with that of 4D HSQC–NOESY–HSQC experiment. If the number of data points in the t$_3$ dimension is doubled (from 32 to 64 complex points), the sensitivity enhancement would be 18%. For a perdeuterated sample, NOEs between amide protons are essential distance constraints, and the 3D and 4D ^{15}N/^{15}N separated HSQC–NOESY–TROSY experiments proposed here are important to obtain NOEs for large proteins. For even larger proteins (e.g., >80 kD), TROSY–NOESY–TROSY experiment is expected to be more useful *(20)*.

2.4. 3D Zero-Quantum NOESY–TROSY and Other NOESY–TROSY Experiments

Transverse relaxation of single-quantum coherences of ^{15}N and ^{1}H can be optimized by exploiting the interference between ^{15}N CSA and the ^{15}N–^{1}H DD interaction (SQ–TROSY) *(3)*. In addition, transverse relaxation of zero-quantum coherence of ^{15}N and ^{1}H can also be optimized by exploiting the interference between ^{15}N and ^{1}H CSAs (ZQ–TROSY) *(18)*. Zero-quantum TROSY has been introduced into the 3D NOESY experiment *(18)*. Here, 2D

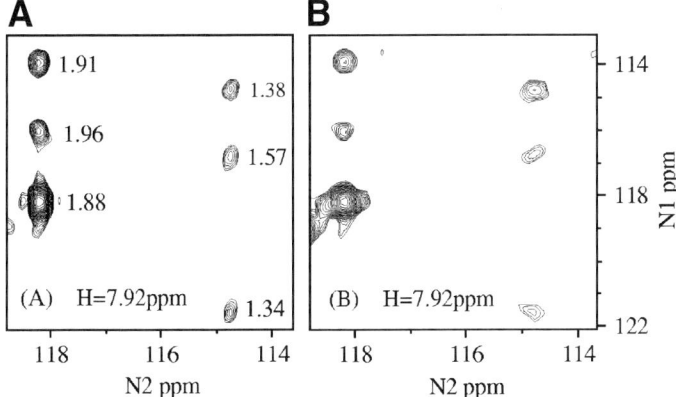

Fig. 6. 2D [^{15}N–^{15}N]-slices of a ^{15}N/^{15}N separated 3D HSQC-NOESY-TROSY spectrum (**A**) and its corresponding HSQC-NOESY-HSQC spectrum (**B**) of trichosanthin taken at ^{1}H chemical shift F_3 = 7.92 ppm. The chemical shifts of the N2 and H_N dimension in **A** are shifted by 45 Hz compared to those in **B**. The numbers beside peaks of **A** are the intensity ratios of peaks in **A** and their corresponding peaks in **B**. **A** and **B** have the identical lowest contour level and a contour factor of 1.2. (Reproduced from **ref. 20** with kind permission of Kluwer Academic Publishers.)

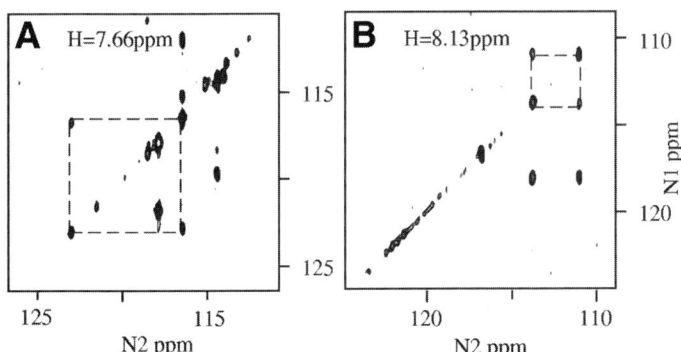

Fig. 7. 2D [^{15}N–^{15}N] (F_1–F_2) slices of the 3D ^{15}N/^{15}N separated HSQC-NOESY-TROSY spectrum taken at ^{1}H chemical shifts of F_3 = 7.66 ppm (**A**) and F_3 = 8.13 ppm (**B**), respectively. (Reproduced from **ref. 20** with kind permission of Kluwer Academic Publishers.)

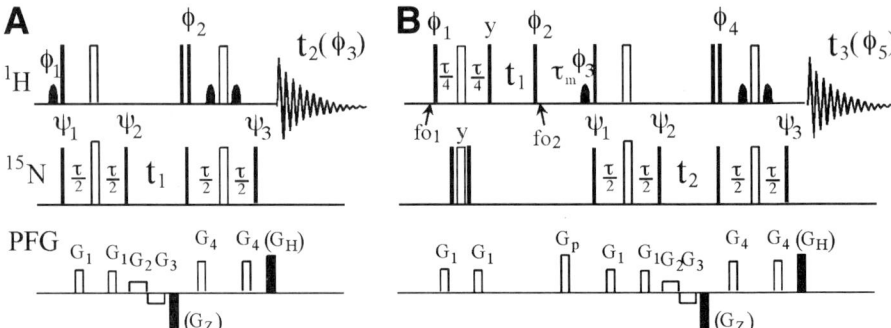

Fig. 8. Schemes for 2D ZQ-TROSY (**A**) and 3D NOESY-ZQ-TROSY (**B**). Filled bars and open bars represent 90° and 180° pulses, respectively, and curved shapes represent water-selective 90° rf pulses. (**A**) The phases of the rf pulses are: $\phi_1 = (-x)$; $\phi_2 = (x)$; $\phi_3 = (x, -x, -y, y)$; $\psi_1 = (-x, x, -y, y)$; $\psi_2 = (y, -, x, -x,)$; $\psi_3 = (y)$; x on all other pulses. To obtain a complex interferogram, a second FID is recorded for each t_1 delay, with the following phases: $\phi_1 = (x)$; $\phi_2 = (-x)$; $\phi_3 = (x, -x, y, -y)$; $\psi_3 = (-y)$. (**B**) The phases of the rf pulses are: $\phi_1 = (4(45°), (4(225°))$; $\phi_2 = (X)$; $\phi_3 = (-x)$; $\phi_4 = (x)$; $\phi_5 = (x, -x, -y, y, -x, x, y, -y)$; $\psi_1 = (-x, x, -y, y)$; $\psi_2 = (y, -y, x, -x)$; $\psi_3 = (y)$; x on all other pulses. Quadrature detection in t_1 dimension is achieved by the States-TPPI method applied with phase ϕ_2. For each t_2 increment, a second FID is recorded with the following different phases: $\phi_3 = (X)$; $\phi_4 = (-x)$; $\phi_5 = (x, -x, y, -y, -x, x, -y, y)$; $\psi_3 = (-y)$. Details of data processing may be found in the literature (*18*). (Reproduced with permission from **ref.** *18* copyright [1999] National Academy of Sciences, U.S.A.)

zero-quantum TROSY and 3D zero-quantum NOESY–TROSY experiments are described.

The pulse sequences of 2D ZQ–TROSY and 3D NOESY–ZQ–TROSY are shown in **Fig. 8**. In the 2D ZQ–TROSY experiment, two different types of the TROSY cross-correlation effects are involved. The cross-correlation effect of CSAs of ^{15}N and ^{1}H is used in t_1 period, whereas the cross-correlation effect of ^{1}H CSA and ^{15}N–^{1}H DD coupling is used in t_2 period. In the 3D NOESY–ZQ–TROSY experiment, the selection of the slowly relaxing magnetization component in the t_1 domain is done by S3E, and after the mixing time, ZQ–TROSY is used for detection. Therefore, transverse relaxation is optimized in all three dimensions of 3D NOESY–ZQ–TROSY.

Application to Proteins:

Protein sample used: 100% ^{15}N- and 70% ^{2}H-labeled protein 7,8-dihydro-neopterin aldolase (110 kDa).

Equipment: 750-MHz NMR spectrometer with three channels.

Experimental parameters and results:

Temperature: 30°C.

Experimental parameters: $\tau = 5.4$ ms. (A) In 2D ZQ–TROSY, the complex interferogram is multiplied by $\exp(-i\Omega_H t_1)$ after FT in the ω_2 dimension, where Ω_H is the offset in the ω_2 dimension relative to the ^1H carrier frequency in $rad.s^{-1}$. (B) In 3D NOESY–ZQ–TROSY, the ^1H carrier frequency offsets are set to 8.7 ppm at time-point fo_1 and to 4.7 ppm at fo_2.

Results: The ZQ–TROSY experiment is evaluated in **Fig. 9**. by comparison with the single-quantum TROSY and 2D HSQC. The figure shows cross-sections along both the ^{15}N and ^1H chemical shift axes of a well separated cross-peak. The 2D ZQ–TROSY (**Fig. 9A**) provides better sensitivity and resolution than the conventional 2D HSQC (**Fig. 9C**), but the single-quantum TROSY experiment clearly gives the best result (**Fig. 9B**). **Figure 10A** and **B** present two sets of strips taken from 3D NOESY–ZQ–TROSY and conventional 3D NOESY–HSQC. Cross peaks in **Fig. 10A** are sharper, and some of them are stronger than that in **Fig. 10B**. Furthermore, the absence of strong diagonal peaks in 3D NOESY–ZQ–TROSY helps to resolve NOE peaks close to the diagonal. In addition, the resonances of the $^{15}N-^1H_2$ groups of Gln and Asn residues are not suppressed by the ZQ–TROSY scheme. The suppression of the strong diagonal peaks may be useful in practice.

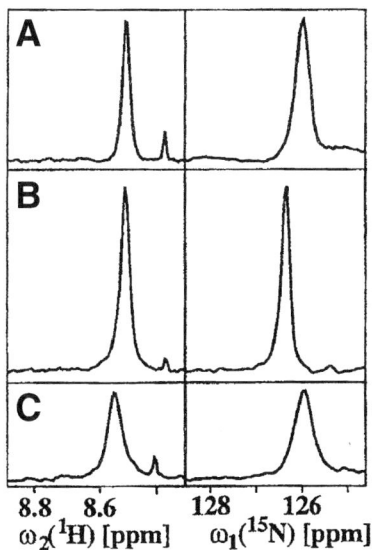

Fig. 9. Comparison of 2D ZQ-Trosy (**A**), SQ-TROSY (**B**), and HSQC (**C**). The measurement time and the experiment setup were identical for all three spectra. (Reproduced with permission from **ref.** *18* copyright [1999] National Academy of Sciences, U.S.A.)

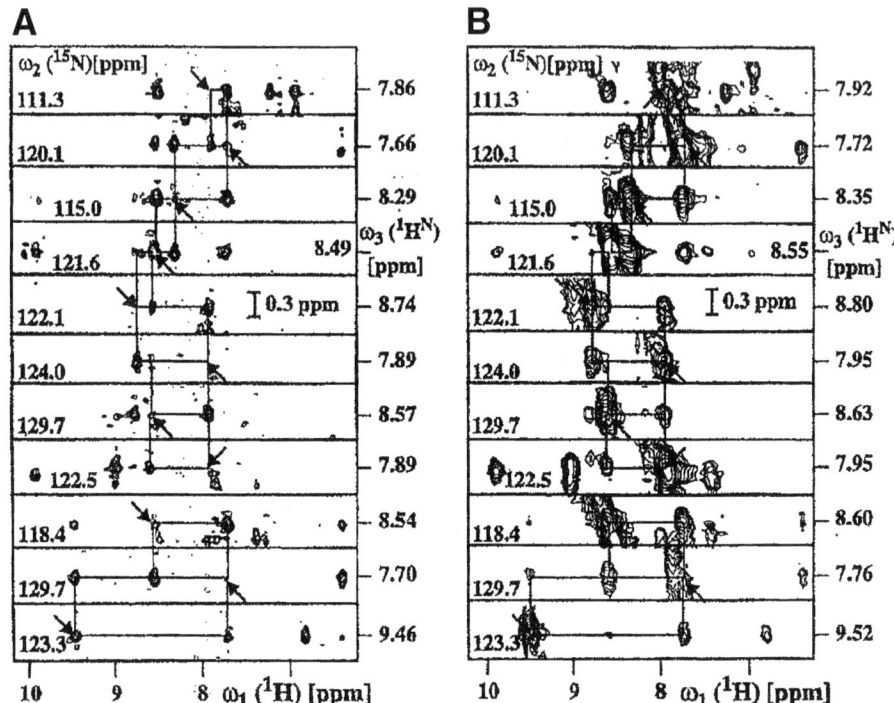

Fig. 10. Comparison of corresponding spectral region in a 3D NOESY-ZQ-TROSY **(A)** and a conventional 3D NOESY-HSQC **(B)**. Both experiments were carried out under the same experimental conditions. NOE connectivities are shown with thin horizontal and vertical lines. (Reproduced with permission from **ref. *18*** copyright [1999] National Academy of Sciences, U.S.A.)

In addition to the NOESY–TROSY experiment described above, a TROSY–NOESY experiment has been developed for the measurement NOEs using the TROSY effect on the aromatic ^{13}C–^1H moiety *(43)*. A NOESY–[^{15}N,^1H]/[^{13}C,^1H]–TROSY, which can simultaneously detect NOEs between HN and aromatic protons by taking advantage of TROSY effects of ^{15}N–^1H and ^{13}C–^1H moieties on the backbone and aromatic sidechains of a protein, has also been designed *(44)*.

3. Conclusions

The TROSY-based methods represent a major advance in NMR studies of large proteins, nucleic acids, and their complexes. 2D TROSY provides a ^{15}N–^1H fingerprint that is highly sensitive to the changes in the protein and provides an NMR probe for studies of intermolecular interaction of proteins or protein complexes up to 100 kD. If the CRIPT–TROSY is used, this limit can go as high as 900 kD *(10)*. TROSY-based experiments can be readily applied to small membrane proteins reconstituted in lipid micelles or water-soluble detergents *(52)*. The most suitable magnetic field for the TROSY effect for ^{15}N is 900 MHz and for $^{1}H^{N}$ is 1 GHz *(21)*. As the number of high field magnets increases, many more large proteins and protein complexes will be studied by TROSY-based NMR experiments.

References

1. Bax, A. (1994) Multidimensional nuclear magnetic resonance methods for protein studies, *Curr. Opin. Struct. Biol.* **4,** 738–744.
2. Gardner, K. H., Zhang, X. C., Gehring, K, and Kay, L. E. (1998) Solution NMR studies of a 42 KDa *Escherichia coli* maltose binding protein beta-cyclodextrin complex: chemical shift assignments and analysis. *J. Am. Chem. Soc.* **120,** 11,738–11,748.
3. Pervushin, K., Riek, R., Wider, G., and Wüthrich, K. (1997) Attenuated T-2 relaxation by mutual cancellation of DD coupling and chemical shift anisotropy indicates an avenue to NMR structures of very large biological macromolecules in solution. *Proc. Natl. Acad. Sci. USA* **94,** 12,366–12,371.
4. Pervushin, K., Riek, R., Wider, G., and Wüthrich, K. (1998) Transverse relaxation-optimized spectroscopy (TROSY) for NMR studies of aromatic spin systems in C-13-labeled proteins. *J. Am. Chem. Soc.* **120,** 6394–6400.
5. Pervushin, K., Wider, G., and Wüthrich, K. (1998) Single transition-to-single transition polarization transfer (ST2-PT) in [N-15,H-1]-TROSY. *J. Biomol. NMR* **12,** 345–348.
6. Zhu, G., Kong, X., Yan, X., and Sze, K. (1998) Sensitivity enhancement in transverse relaxation optimized NMR spectroscopy. *Angew. Chem. Int. Ed. Engl.* **37,** 2859–2861.
7. Weigelt, J. (1998) Single scan, sensitivity- and gradient-enhanced TROSY for multidimensional NMR experiments. *J. Am. Chem. Soc.* **120,** 10,778, 10,779.
8. Salzmann, M., Pervushin, K., Wider, G., Senn, H., and Wüthrich, K. (1998) TROSY in triple-resonance experiments: new perspectives for sequential NMR assignment of large proteins. *Proc. Natl. Acad. Sci. USA* **95,** 13,585–13,590.
9. Riek, R., Wider, G., Pervushin, K., and Wüthrich, K. (1999) Polarization transfer by cross-correlated relaxation in solution NMR with very large molecules. *Proc. Natl. Acad. Sci. USA* **96,** 4918–4923.
10. Fiaux, J, Bertelsen, E. B., Horwich, A. L., and Wüthrich, K. (2002) NMR analysis of a 900K GroEL-GroES complex. *Nature* **418,** 207–211.
11. Zhu, G., Kong, X. M., and Sze, K. H. (1999) Gradient and sensitivity enhancement of 2D TROSY with water flip-back, 3D NOESY-TROSY and TOCSY-TROSY experiments. *J. Biomol. NMR* **13,** 77–81.
12. Rance, M., Loria, J. P., and Palmer, A. G. (1999) Sensitivity improvement of transverse relaxation-optimized spectroscopy. *J. Magn. Reson.* **136,** 92–101.
13. Salzmann, M., Wider, G., Pervushin, K., Senn, H., and Wüthrich, K. (1999) TROSY-type triple-resonance experiments for sequential NMR assignments of large proteins. *J. Am. Chem. Soc.* **121,** 844–848.

14. Salzmann, M., Pervushin, K., Wider, G., Senn, H., and Wüthrich, K. (1999) [C-13]-constant-time [N-15,H-1]-TROSY–HNCA for sequential assignments of large proteins. *J. Biomol. NMR* **14**, 85–88.

15. Yang, D. W. and Kay, L. E. (1999) Improved lineshape and sensitivity in the HNCO-family of triple resonance experiments. *J. Biomol. NMR* **14**, 273–276.

16. Loria, J. P., Rance, M., and Palmer, A. G. (1999) Transverse-relaxation-optimized (TROSY) gradient-enhanced triple-resonance NMR spectroscopy. *J. Magn. Reson.* **141**, 180–184.

17. Zhu, G., Xia, Y., Sze, K., and Yan, X. (1999) 2D and 3D TROSY–enhanced NOESY of N-15 labeled proteins. *J. Biomol. NMR* **14**, 377–381.

18. Pervushin, K. V., Wider, G., Riek, R., and Wüthrich, K. (1999) The 3D NOESY-[H-1,N-15,H-1]-ZQ-TROSY NMR experiment with diagonal peak suppression. *Proc. Natl. Acad. Sci. USA* **96**, 9607–9612.

19. Meissner, A. and Sørensen, O. W. (1999) Suppression of diagonal peaks in TROSY-type H-1 NMR NOESY spectra of N-15-labeled proteins. *J. Magn. Reson.* **140**, 499–503.

20. Xia, Y., Sze, K., and Zhu, G. (2000) Transverse relaxation optimized 3D and 4D N-15/N-15 separated NOESY experiments of N-15 labeled proteins. *J. Biomol. NMR* **18**, 261–268.

21. Wüthrich, K. (1998) The second decade-into the third millennium. *Nat. Struct. Biol.* **5(Suppl. S)**, 492–495.

22. Pervushin, K., Gallius, V., and Ritter, C. (2001) Improved TROSY–HNCA experiment with suppression of conformational exchange induced relaxation. *J. Biomol. NMR* **21**, 161–166.

23. Permi, P. and Annila, A. (2001) A new approach for obtaining sequential assignment of large proteins. *J. Biomol. NMR* **20**, 127–133.

24. Xia, Y., Sze, K. H., Li, N., Shaw, P. C., and Zhu, G. (2002) Protein dynamics measurements by 3D HNCO based NMR experiments. *Spectrosc-Int J.* **16**, 1–13.

25. Meissner, A. and Sørensen, O. W. (2001) Sequential HNCACB and CBCANH protein NMR pulse sequences. *J. Magn. Reson.* **151**, 328–331.

26. Yang, D. W. and Kay, L. E. (1999) TROSY triple-resonance four-dimensional NMR spectroscopy of a 46 ns tumbling protein. *J. Am. Chem. Soc.* **121**, 2571–2575.

27. Riek, R., Pervushin, K., Fernandez, C., Kainosho, M., and Wüthrich, K. (2001) [C-13,C-13]- and [C-13,H-1]-TROSY in a triple resonance experiment for ribose-base and intrabase correlations in nucleic acids. *J. Am. Chem. Soc.* **123**, 658–664.

28. Meissner, A. and Sørensen, O. W. (1999) Optimization of three-dimensional TROSY-type HCCHNMR correlation of aromatic H-1-C-13 groups in proteins. *J. Magn. Reson.* **140**, 447–450.

29. Simon, B., Zanier, K., and Sattler, M. (2001) A TROSY relayed HCCH-COSY experiment for correlating adenine H2/H8 resoances in uniformly ^{13}C-labeled RNA molecules. *J. Biomol. NMR* **20**, 173–176.

30. Mulder, F. A. A., Ayed, A., Yang, D. W., Arrowsmith, C. H., and Kay, L. E. (2000) Assignment of H-1(N), N-15, C-13(alpha), (CO)-C-13 and C-13(beta) resonances

in a 67 kDa p53 dimer using 4D-TROSY NMR spectroscopy. *J. Biomol. NMR* **18**, 173–176.

31. Konrat, R., Yang, D. W., and Kay, L. E. (1999) A 4D TROSY-based pulse scheme for correlating (HNi)-H-1,Ni-15,C-13(i)alpha,C-13'(i-1) chemical shifts in high molecular weight, N-15,C-13, H-2 labeled proteins. *J. Biomol. NMR* **15**, 309–313.

32. Zhu, G. and Bax, A. (1990) Improved linear prediction for truncated signals of known phase. *J. Magn. Reson.* **90**, 405–410.

33. Zhu, G. and Bax, A. (1992) Improved linear prediction of truncated damped sinusoids using modified backward-forward linear prediction. *J. Magn. Reson.* **100**, 202–207.

34. Wagner, G. (1989) Heteronuclear nuclear magnetic-resonance experiments for studies of protein conformation. *Methods Enzymol.* **176**, 93–113.

35. Clore, G. M. and Gronenborn, A. M. (1991) Structures of larger proteins in solution—3-dimensional and 4-dimensional heteronuclear NMR-spectroscopy. *Science* **252**, 1390–1399.

36. Clore, G. M., Gronenborn, A. M., and Bax, A. (1998) A robust method for determining the magnitude of the fully asymmetric alignment tensor of oriented macromolecules in the absence of structural information. *J. Magn. Reson.* **133**, 216–221.

37. Tjandra, N., Grzesiek, S., and Bax, A. (1996) Magnetic field dependence of nitrogen-proton J splittings in N-15-enriched human ubiquitin resulting from relaxation interference and residual dipolar coupling. *J. Am. Chem. Soc.* **118**, 6264–6272.

38. Tjandra, N. and Bax, A. (1997) Direct measurement of distances and angles in biomolecules by NMR in a dilute liquid crystalline medium. *Science* **278**, 1111–1114.

39. Zuiderweg, E. R. P. and Fesik, S. W. (1989) Heteronuclear 3-dimensional NMR-spectroscopy of the inflammatory protein c5a. *Biochemistry* **28**, 2387–2391.

40. Muhandiram, D. R., Farrow, N., Xu, G. Y., Smallcombe, S. H., and Kay, L. E. (1993) A gradient C-13 NOESY-HSQC experiment for recording NOESY spectra of C-13-labeled proteins dissolved in H_2O. *J. Magn. Reson.* **B102**, 317–321.

41. Kay, L. E., Clore, G. M., Bax, A., and Gronenborn, A. M. (1990) 4-dimensional heteronuclear triple-resonance NMR-spectroscopy of interleukin-1-beta in solution. *Science* **249**, 411–414.

42. Farmer, B. T., II. (1991) Simultaneous [C-13, N-15]-HMQC, a pseudo-triple-resonance experiment. *J. Magn. Reson.* **93**, 635–641.

43. Brutscher, B., Boisbouvier, J., Pardi, A., Marion, D., and Simorre, J. P. (1998) Improved sensitivity and resolution in H-1-C-13 NMR experiments of RNA. *J. Am. Chem. Soc.* **120**, 11,845–13,590.

44. Pervushin, K., Braun, D., Fernández, C., and Wüthrich, K. (2000) [N-15,H-1]/[C-13,H-1]-TROSY for simultaneous detection of backbone N-15-H-1, aromatic C-13-H-1 and side-chain N-15-H-1(2) correlations in large proteins. *J. Biomol. NMR* **17**, 195–202.

45. Brutscher, B., Boisbouvier, J., Kupce, E., Tisne, C., Dardel, F., Marion, D., Simorre, J.P. (2001) Base-type-selective high-resolution C-13 edited NOESY for sequential assignment of large RNAs. *J. Biomol. NMR.* **19**, 141–151.

46. Meissner, A. and Sørensen, O. W. (2002) Enhanced diagonal peak suppression in three-dimensional TROSY-type N-15-resolved H-1(N)-H-1(N) NOESY spectra. *Concepts Magn. Reson.* **14,** 1–8.

47. Delaglio, F., Grzesiek, S., Vuister, G. W., Zhu, G, Pfeifer, J., and Bax, A. NMRPIPE —a multidimensional spectral processing system based on Unix pipes. *J. Biomol. NMR* **6,** 277–293.

48. Meissner, A. Duus, J. Ø., and Sørensen, O. W. (1997) Integration of spin-state-selective excitation into 2D NMR correlation experiments with heteronuclear ZQ/2Q pi rotations for (1)J(XH)-resolved E.COSY-type measurement of heteronuclear coupling constants in proteins. *J. Biomol. NMR* **10,** 89–94.

49. Sørensen, M. D., Meissner, A., and Sørensen, O. W. (1997) Spin-state-selective coherence transfer via intermediate states of two-spin coherence in IS spin systems: application to E.COSY-type measurement of J coupling constants. *J. Biomol. NMR* **10,** 181–186.

50. Torchia, D. A., Sparks, S. W., and Bax, A. (1988) Delineation of alpha-helical domains in deuteriated staphylococcal nuclease by 2D NOE NMR-spectroscopy. *J. Am. Chem. Soc.* **110,** 2320–2321.

51. Venters, R. A., Metzler, W. J., Spicer, L. D., Mueller, L., and Farmer, B. T., II. (1995) Use of H1(N)-H1(N) NOEs to determine protein global folds in perdeuterated proteins. *J. Am. Chem. Soc.* **117,** 9592–9595.

52. Fernández, C., Adeishvili, K., and Wüthrich, K. (2001) Transverse relaxation-optimized NMR spectroscopy with the outer membrane protein Ompx in dihezanoyl phosphatidylcholine micelles. *Proc. Natl. Acad. Sci. USA* **98,** 2358–2363.

6

Media for Studies of Partially Aligned States

Kieran Fleming and Stephen Matthews

Summary

Measurement of residual dipolar couplings for proteins in nuclear magnetic resonance (NMR) requires a degree of molecular alignment. This may be achieved through the use of liquid crystals or compressed hydrated gels. Several media have been described in the literature, and this chapter describes five of the most commonly used systems. For two of these systems, bicelles and filamentous bacteriophage, the media can be purchased in a form ready for experiments. The remainder must be made by the investigator, yet they are generally straightforward to synthesize in the laboratory. Poly(ethylene glycol)/alcohol mixtures and Helfrich phases are made by simply mixing the ingredients in the correct manner, and strained gels use techniques familiar to all molecular biologists.

Key Words: Liquid crystal; NMR; residual dipolar couplings; bacteriophage; strain-induced alignment; bicelles.

1. Introduction

Residual dipolar couplings (RDCs) may be used as restraints in macromolecular structure determinations by nuclear magnetic resonance (NMR). The ability to measure RDC values accurately has resulted in a significant improvement in the accuracy of protein solution structures over the past 2–3 yr, particularly in cases where the molecule under study is multidomain. To measure RDCs, a small degree of alignment of the protein of interest in solution is required. As such, various liquid crystal technologies have been described that confer the necessary partial alignment and retain many of the advantages of solution NMR. The various liquid crystalline media are quite diverse *(1–6)*, and each has its own set of characteristics that makes it suitable for particular applications.

From: *Methods in Molecular Biology, vol. 278: Protein NMR Techniques*
Edited by: A. K. Downing © Humana Press Inc., Totowa, NJ

The suitability of a particular liquid crystal system for use in NMR studies is based on several criteria. Essentially, they should:

1. Form a homogeneous liquid crystalline phase at a concentration commensurate with potential NMR studies, that is, at a suitable pH and temperature.
2. Show no deleterious interaction with the molecule under study, that is, they may align the molecule under study via steric and/or electrostatic interactions but should not bind strongly to it.
3. Produce a tunable alignment in the magnetic field, that is, the degree of alignment of a macromolecule can be adjusted by altering the liquid crystal concentration or in the case of gels by the amount of applied strain.
4. Remain stable over a long period of time.

As a result of the rather unpredictable behavior of proteins in anisotropic phases, it is generally desirable to test as many systems as possible to maximize the likelihood of success and to provide more than one alignment tensor. Despite this, the choice of system can be refined based on properties of the protein under study. Notably, in cases of highly charged macromolecules (i.e., proteins with extreme pIs), a neutral system is more likely to prove successful— for example, poly(ethylene glycol)/alcohol or strained gels. Furthermore, for weakly charged proteins, it can be advantageous to choose a system that is oppositely charged (e.g., dimyristoylphosphatidylcholine [DMPC]:dihexanoylphosphatidylcholine [DHPC]:cetyltrimethylammonium bromide [CTAB] for negatively charge proteins). If the target protein is particularly valuable, then a liquid crystal system that is readily removable is preferable—for example, phage or strained gels. Several of the more commonly used media are now commercially available in a form ready to use for NMR experiments; therefore, attempting several systems is not time-consuming. In cases where the medium is not commercially available, details of its manufacture will be provided. However, the preparation of protein-containing samples for NMR is the main focus of this text.

2. Materials

One particular point of note when manufacturing liquid crystalline phases is that use of the highest purity materials often results in more stable phases and increases the chances of first-time success.

2.1. Bicelles

1. Premixed lipids for bicelle formation at the desired DMPC:DHPC ratio (Avanti Polar Lipids, Alabaster, AL). These should be stored at −18°C (*see* **Note 1**).
2. Buffer solution: 10 mM sodium phosphate, pH 6.6, 0.15 mM sodium azide, 90% H_2O (HPLC-grade), 10% D_2O (99.9%).
3. Tetradecyltrimethylammonium bromide (TTAB), 99% (Sigma, Gillingham, UK).

2.2. Filamentous Bacteriophage

1. Pf1 filamentous bacteriophage in 10 m*M* potassium phosphate buffer, pH 7.6, 2 m*M* MgCl$_2$, and 0.05% NaN$_3$ (ASLA Biotech, Riga, Latvia).

2.3. Poly(ethylene glycol)/Alcohol Mixtures

1. C12E5—*N*-dodecyl-penta(ethylene glycol), 98% (Fluka, Buchs, Switzerland).
2. *N*-hexanol.

2.4. Helfrich Phases

1. Cetylpyridinium bromide (CPBr) monohydrate, 98% (Sigma, Gillingham, UK).
2. Sodium bromide (NaBr), 99.99+% (Sigma, Gillingham, UK).
3. Hexanol.
4. Relevant NMR buffer, for example, 50 m*M* (to be diluted to 10 m*M*) sodium phosphate, pH 5.7, 90% H$_2$O:10% D$_2$O.

2.5. Strained Gels

1. Acrylamide.
2. *N,N'*-methylenebisacrylamide.
3. Ammonium persulphate.
4. *N,N,N'*-tetramethylethylenediamine.
5. Glass mold made from an NMR tube, internal diameter 3.4 mm (524PP, Wilmad Glass).
6. Shigemi microcell NMR tube, internal diameter 4.2 mm (BMS-005B, Shigemi, Allison Park, PA).
7. Buffer solution: 10 m*M* sodium phosphate, pH 5.7, 90% H$_2$O:10% D$_2$O.

3. Methods

3.1. Bicelles

The most commonly used bicelle medium is composed of a mixture of DMPC and DHPC *(1,7–12)*. Mixing of the individual lipids with one another is not recommended, especially as premixed lipid powders for bicelle formation are available for purchase. The example given in this section is for a basic 3:1 DMPC:DHPC suspension. An ideal final concentration for the bicelle suspension for use in protein NMR is around the 3.5% (w/v) to 5% (w/v) mark.

1. Make up a 15% (w/v) stock solution of DMPC/DHPC containing 2.4 m*M* TTAB (which increases bicelle stability) by adding 50 mg of the premixed lipids to 280 μg of buffer, for example (*see* **Note 2**).
2. Vortex for 10 min at 4°C and leave at this temperature for an additional 15 min. A milky solution should result, where the lipids are not completely dissolved (*see* **Note 3**).
3. Vortex again at room temperature for 1 min, and place in a water bath at 38°C for 30 min.

4. Store at 4°C for 15 min to allow the mixture to cool, and then vortex at 4°C for 30 min.
5. Allow to warm to room temperature for 15 min, vortex for 1 min, and then place in a water bath at 38°C for 30 min.

   Repeat **steps 1–5** until the solution becomes clear and is fluid at low temperatures. At about 25°C, it should become white and milky and then return to a clear and transparent fluid at 38°C. The viscosity should also increase markedly at elevated temperatures. The solution may appear to have a slight blue taint (*see* **Note 4**).
6. Dilute liquid crystalline NMR samples can be obtained by adding buffer to the 15% (w/v) stock solution followed by vortexing and slow-speed centrifugation. A concentration of 3.5% (w/v) produces a HDO quadrupolar splitting of about 6 Hz at 25°C (*see* **Note 5**).
7. Addition of the diluted solution to a minimal sample of protein, for example, 50 µL, in the same buffer, followed by slow-speed centrifugation completes the preparation (*see* **Note 6**).

3.2. Phage

Several different bacteriophages have been used for RDC measurement, including Pf1 and *fd (2,3,9–11,13–15)*. Production of this phage can be a tricky process and is not recommended in laboratories where protein expression is regularly undertaken because the risk of cross-contamination is high. Fortunately, Pf1 may be purchased in various buffers ready for use. The following assumes a sample of Pf1 phage in 10 mM potassium phosphate buffer, pH 7.6, 2 mM MgCl$_2$, and 0.05% NaN$_3$ has been purchased (*see* **Note 7**).

1. Reduce the phage concentration from 50 mg/mL, as supplied by Asla, to the desired concentration (for example, 15 mg/mL) by weighing out the correct mass of stock solution and adding to buffer. Verification of the concentration can be made by measuring A$_{270}$ and assuming an extinction coefficient, ε = 2.25 cm^{-1} mg^{-1} mL (*see* **Note 8**).
2. Centrifuge at 10,000g for 2 min to remove air bubbles resulting from the pipetting, and check the HDO quadrupolar deuterium splitting in the NMR magnet. A concentration of 15 mg/mL should give a splitting of approx 20 Hz at 25°C.
3. For preparation of protein samples, pipet the phage to the desired concentration into a small volume (50 µL) of protein in the same buffer as the phage. To ensure complete mixing, continuous pipetting followed by brief centrifugation is required (*see* **Note 9**).
4. The phage buffer may be changed by spinning down the Pf1 at 370,000g in a Beckman TLA-100.3 rotor for 1 h at 5°C. Then resuspend the phage in the new buffer. The amount of phage recovered will be only approx 50% of the original yield (http://www.asla-biotech.com/asla-phage.htm).

3.3. Poly(ethylene glycol)/Alcohol Mixtures

Although this medium is not commercially available, it is very simple to produce with easily obtainable materials. It is more versatile than bicelles or phage in terms of its stability over a wide pH range. Numerous different poly(ethylene glycol) derivatives may be used, and a choice of alcohol is available in the form of hexanol or octanol *(4,9,11,12,15)*. As with bicelles, use of different proportions of the starting materials yields phases with slightly different properties. In this case, the temperature range over which the phase is stable is dictated by the proportions of the mixed ingredients. One of the basic mixtures, 3% C12E5/hexanol (molar ratio = 0.96), is described. This ordered phase is stable in the temperature range 26–39°C and, therefore, is suitable for most proteins.

1. Weigh out C12E5 to yield a final concentration of 3% (w/w) and make up to 5 mL with a 90% H_2O:10% D_2O solution.
2. The pH may be adjusted from the base level of pH 3.5, as desired, using sodium hydroxide.
3. Add the correct weight of hexanol dropwise, while vigorously shaking, to a final molar ratio C12E5:hexanol of 0.96 (*see* **Note 10**).
4. Remove air bubbles via centrifugation at 5000*g* for 2–3 min.
5. Check the HDO quadrupolar deuterium splitting in the NMR spectrometer. A splitting of approx 20 Hz at 38°C is observed.
6. Pipet 200 µL of the C12E5:hexanol stock solution into a small volume (50 µL) of protein in buffer.

3.4. Helfrich Phases

Helfrich phases, the common name for mixtures of CPBr and hexanol, are straightforward to prepare, although care must be taken at several stages during the mixing *(5,9,11,12,16)*. The preparation of a 6.5% (w/v) stock solution is described in the following steps.

1. Weigh out amounts of CPBr and hexanol to achieve a final ratio of 1:1.33 (w/w) in a 6.5% (w/v) solution, and add to the correct volume of a 90% H_2O:10% D_2O solution.
2. Seal the tube containing the solution, vortex, and heat to 70°C in a water bath until the solution becomes clear.
3. Leave to cool at room temperature and then centrifuge at 4000*g* for 2 min.
4. The degree of liquid crystal formation of this stock solution can be tested by leaving the sample to equilibrate in the NMR magnet for at least 6 h and then measuring the HDO quadrupolar deuterium splitting. A splitting of around 20 Hz should be observed at 25°C (*see* **Note 11**).
5. Prepare an NMR buffer in the normal manner, for example, a 50 mM sodium phosphate buffer, pH 5.7, 90% H_2O:10% D_2O. Also prepare a stock solution of 1 M NaBr (*see* **Note 12**).
6. Add the buffer and NaBr solution (to a final concentration of 30 mM) to a sample of protein. Make up this to the volume of an NMR sample—for example, 250 µL—using the stock Helfrich phase and NaBr to a final liquid crystal concentration of 5% (w/v); with the buffer at the desired concentration—for example, 10 mM sodium phosphate and an NaBr concentration of 30 mM (*see* **Note 13**).

7. The tube should then be sealed, gentle mixing applied, and centrifuged at 4000*g* for 2 min. If an excess of hexanol exists, observed as a thin layer on top of the liquid crystal phase, this may be removed carefully using a pipet (*see* **Note 14**).

3.5. Strained Gels

In addition to liquid crystal media, an alternative approach has recently been adopted whereby the application of compression or stretching forces to a hydrated gel induces the alignment required for residual dipolar coupling measurement *(6,9,11,17–20)*. Termed *strain-induced alignment in a gel*, this method is particularly attractive to the molecular biologist because it relies on the casting of polyacrylamide gels.

1. Prepare a polyacrylamide gel solution containing 7% (w/v) acrylamide and 0.4% (w/v) *N,N'*-methylenebisacrylamide.
2. Initiate polymerization via addition of 0.08% (w/v) ammonium persulphate and 0.08% (v/v) *N,N,N'*-tetramethylethylenediamine.
3. Pipet approx 2.5 mL into the glass mold immediately and leave the gel to set (*see* **Note 15**).
4. Once set, the gel should be gently removed from the mold and thoroughly washed with distilled H_2O. Submerge the gel in a large excess of distilled H_2O under gentle agitation for at least 3–4 h (*see* **Note 16**).
5. Cut the gel into sections of the desired length, for example, 25 mm.
6. Introduction of the protein into the gel may be performed in two ways. The success of either method will depend on the protein under study. The first approach is to soak the section of gel in a 250 μL buffered solution containing 2 m*M* protein and 20% D_2O at the desired pH overnight in the Shigemi microcell NMR tube. Typically, the final concentration in the tube can be expected to be around half the starting concentration, with a similar uptake of D_2O. The second method is to first soak the gel section in a buffer solution for several hours. Then soak in a small volume, approx 500 μL, of protein solution at a concentration of approx 750 μ*M* overnight. The final concentration in the gel may consequently be up to two-thirds of this value, that is, 500 μ*M*.
7. Once the protein has been added, the plunger of the microcell can be used to apply compression. Typically, a compression ratio (length of gel after compression divided by length of gel before compression) of 0.75 is a good starting point (*see* **Note 17**).
8. Use parafilm and tape to hold the plunger in place.
9. A HDO quadrupolar deuterium splitting of a few Hz should be observed at 25°C.

4. Notes

1. The diacyl phospholipids of regular bicelles can be replaced by their hydrolysis-resistant dialkyl analogs, in which the carbohydrate chains are connected to the glycerol part by ether instead of ester bonds—for example, ditetradecyl-PC

and dihexyl-PC instead of DMPC and DHPC, respectively. These form liquid crystals over a wide pH range, remain chemically stable for several weeks, and provide the same degree of alignment for macromolecules as the DMPC/DHPC system *(21)*.

2. Because of the instability of bicelle mixtures, it is recommended that solutions are made immediately prior to any experiments. Use of HPLC-grade water is very important in bicelle preparation because the phase is so delicate. DHPC is extremely hygroscopic; prepare solutions in a dry box or dilute with buffer immediately after opening. Use low salt concentration (<50 m*M*). The pH is a very important consideration with bicelles, so care should be taken to ensure that a pH range of 6.5–7.0 is maintained at all times to minimize ester bond hydrolysis. At the recommended pH, the solution is stable for weeks rather than hours. There will be a slow decrease in the transition temperature as the sample ages, as a result of more efficient hydrolysis of DHPC relative to DMPC (see also www.biochem.ucl.ac.uk/~diego/rdc/content.html).

3. Do *not* use ultrasound for mixing the lipids.

4. Samples may be stored frozen, refrigerated, or at room temperature with no noticeable effect on the reproducibility of the liquid crystalline phase once the sample is reheated at 38°C. However frozen samples (at –80°C) hydrolyze much more slowly.

5. Test the stability and alignment of the bicelles in the NMR magnet prior to addition of protein. Before using the bicelles for protein NMR, briefly vortex and heat until the phase regains a clear appearance to ensure that the lipids are properly mixed and that there are no small particles in the suspension.

6. Recently, many variations on this theme have been applied to increase thermal stability, to modulate the alignment tensor significantly, and to reduce the rate of hydrolysis of the fatty acid chains on the surfactants *(21–25)*. For example, addition of CTAB instead of TTAB has a similar stabilizing effect and, in addition, significantly modulates the alignment tensor. Further reading of these references is advised if a particular protein is compatible with the use of bicelles.

7. Do not freeze the phage; keep at 4°C.

8. The phage solutions are highly viscous so it is recommended that a modified pipetting apparatus be used. Attaching a small length (1 cm or so) of silicon tubing to a 1-mL pipet tip allows precise pipetting of the phage suspension.

9. The salt concentration in protein-containing samples may need to be increased, for example, up to 50 m*M*, to decrease electrostatic interactions between the highly charged Pf1 phage and protein.

10. Care should be taken when adding the hexanol. This step is crucial to the success of the phase and should be done slowly with very vigorous shaking. Once the liquid crystal phase is reached, its identification is straightforward, as the solution is biphasic at low hexanol concentrations but suddenly becomes clear and opalescent when the correct hexanol concentration is achieved. In addition, the viscosity increases markedly when the phase boundary is reached.

11. The stock solution may be kept at room temperature for a period of weeks and will remain stable.

12. Various buffers may be used, including Tris-HBr (pH 8.1), sodium acetate (pH 4.2), and glycine-HBr (pH 3.2).

13. A salt (NaBr) concentration of 25–30 mM is desirable for most proteins.

14. The concentration of the Helfrich phase for NMR can vary greatly depending on the protein under study. The degree of alignment induced by this medium is high. A concentration of 4–6.5% (w/w) is useful for small, spherical proteins, yet for larger proteins a reduction in the liquid crystal concentration down to values as low as 3% (w/w) may be required.

15. The glass mold can be made by cutting the end off an NMR tube with an internal diameter of 3.4 mm. The set gel can then be recovered by pushing gently from one end. Casting the gel so that it is approx 0.8 mm thinner than the Shigemi microcell that will hold it during NMR experiments is intended to allow for the inevitable expansion that occurs when the gel is soaked in protein solution, and when compression is applied.

16. Washing the gel is very important because unreacted acrylamide/initiators must be removed before addition of any protein to avoid denaturation.

17. Care must be taken at all times when handling the gels to prevent introduction of any heterogeneity into the gel structure as this will result in poor NMR spectra. For example, do not twist the plunger in the microcell because this introduces shearing forces and is likely to tear the gel.

Acknowledgments

S. Matthews would like to acknowledge the support of The Wellcome Trust and the BBSRC.

References

1. Tjandra, N. and Bax, A. (1997) Direct measurement of distances and angles in biomolecules by NMR in a dilute liquid crystalline medium. *Science* **278,** 1111–1114.

2. Hansen, M. R., Mueller, L., and Pardi, A. (1998) Tunable alignment of macromolecules by filamentous phage yields dipolar coupling interactions. *Nat. Struct. Biol.* **5,** 1065–1074.

3. Clore, G. M., Starich, M. R., and Gronenborn, A. M. (1998) Measurement of residual dipolar couplings of macromolecules aligned in the nematic phase of a colloidal suspension of rod-shaped viruses. *J. Am. Chem. Soc.* **120,** 10,571–10,572.

4. Ruckert, M. and Otting, G. (2000) Alignment of biological macromolecules in novel nonionic liquid crystalline media for NMR experiments. *J. Am. Chem. Soc.* **122,** 7793–7797.

5. Barrientos, L. G., Dolan, C., and Gronenborn, A. M. (2000) Characterization of surfactant liquid crystal phases suitable for molecular alignment and measurement of dipolar couplings. *J. Biomol. NMR* **16,** 329–337.

6. Tycko, R., Blanco, F., and Ishii, Y. (2000) Alignment of biopolymers in strained gels: a new way to create detectable dipole–dipole couplings in high-resolution biomolecular NMR. *J. Am. Chem. Soc.* **122,** 9340, 9341.

7. Tjandra, N., Omichinski, J. G., Gronenborn, A. M., Clore, G. M., and Bax, A. (1997) Use of dipolar ^{1}H-^{15}N and ^{1}H-^{13}C couplings in the structure determination of magnetically oriented macromolecules in solution. *Nat. Struct. Biol.* **4,** 732–738.

8. Ottiger, M. and Bax, A. (1998) Characterization of magnetically oriented phospholipid micelles for measurement of dipolar couplings in macromolecules. *J. Biomol. NMR* **12,** 361–372.

9. Peti, W., Meiler, J., Bruschweiler, R., and Griesinger, C. (2002) Model-free analysis of protein backbone motion from residual dipolar couplings. *J. Am. Chem. Soc.* **124,** 5822–5833.

10. Prestegard, J. H., Al-Hashimi, H. M., and Tolman, J. R. (2000) NMR structures of biomolecules using field oriented media and residual dipolar couplings. *Q. Rev. Biophys.* **33,** 371–424.

11. Bax, A., Kontaxis, G., and Tjandra, N. (2001) Dipolar couplings in macromolecular structure determination. *Method Enzymol.* **339,** 127–174.

12. Gaemers, S. and Bax, A. (2001) Morphology of three lyotropic liquid crystalline biological NMR media studied by translational diffusion anisotropy. *J. Am. Chem. Soc.* **123,** 12,343–12,352.

13. Hansen, M. R., Hanson, P., and Pardi, A. (2000) Filamentous bacteriophage for aligning RNA, DNA, and proteins for measurement of nuclear magnetic resonance dipolar coupling interactions. *Method Enzymol.* **317,** 220–240.

14. Zweckstetter, M. and Bax, A. (2001) Characterization of molecular alignment in aqueous suspensions of Pf1 bacteriophage. *J. Biomol. NMR* **20,** 365–377.

15. Trempe, J. F., Morin, F. G., Xia, Z., Marchessault, R. H., and Gehring, K. (2002) Characterization of polyacrylamide-stabilized Pf1 phage liquid crystals for protein NMR spectroscopy. *J. Biomol. NMR* **22,** 83–87.

16. Prosser, R. S., Losonczi, J. A., and Shiyanovskaya, I. V. (1998) Use of a novel aqueous liquid crystalline medium for high-resolution NMR of macromolecules in solution. *J. Am. Chem. Soc.* **120,** 11,010–11,011.

17. Sass, H. J., Musco, G., Stahl, S. J., Wingfield, P. T., and Grzesiek, S. (2000) Solution NMR of proteins within polyacrylamide gels: diffusional properties and residual alignment by mechanical stress or embedding of oriented purple membranes. *J. Biomol. NMR* **18,** 303–309.

18. Ishii, Y., Markus, M. A., and Tycko, R. (2001) Controlling residual dipolar couplings in high-resolution NMR of proteins by strain induced alignment in a gel. *J. Biomol. NMR* **21,** 141–151.

19. Chou, J. J., Kaufman, J. D., Stahl, S. J., Wingfield, P. T., and Bax, A. (2001) A simple apparatus for generating stretched polyacrylamide gels, yielding uniform alignment of proteins and detergent micelles. *J. Biomol. NMR* **21,** 377–382.

20. Chou, J. J., Kaufman, J. D., Stahl, S. J., Wingfield, P. T., and Bax, A. (2002) Micelle-induced curvature in a water-insoluble HIV-1 Env peptide revealed by NMR dipolar coupling measurement in stretched polyacrylamide gel. *J. Am. Chem. Soc.* **124,** 2450, 2451.

21. Ottiger, M. and Bax, A. (1999) Bicelle-based liquid crystals for NMR-measurement of dipolar couplings at acidic and basic pH values. *J. Biomol. NMR* **13,** 187–191.

22. Ramirez, B. E. and Bax, A. (1998) Modulation of the alignment tensor of macro-molecules dissolved in a dilute liquid crystalline medium. *J. Am. Chem. Soc.* **120,** 9106, 9107.

23. Tan, C., Fung, B. M., and Cho, G. (2002) Phospholipid bicelles that align with their normals parallel to the magnetic field. *J. Am. Chem. Soc.* **124,** 11,827–11,832.

24. Losonczi, J. A. and Prestegard, J. H. (1998) Improved dilute bicelle solutions for high-resolution NMR of biological macromolecules. *J. Biomol. NMR* **12,** 447–451.

25. Cavagnero, S., Dyson, H. J., and Wright, P. E. (1999) Improved low pH bicelle system for orienting macromolecules over a wide temperature range. *J. Biomol. NMR* **13,** 387–391.

7

Residual Dipolar Couplings in Protein Structure Determination

Eva de Alba and Nico Tjandra

Summary

Each magnetic nucleus behaves like a magnetic dipole able to create a local magnetic field at the position of nearby nuclei. In the presence of an external magnetic field, the local field modifies the original Larmor frequency of the affected nucleus. Such an interaction is called the *dipole–dipole interaction* or *dipolar coupling*. Its magnitude depends on, among other factors, the distance between the interacting nuclei and the angle that the internuclear vector forms with the magnetic field. Through this angular dependence it is possible to relate the position of the two interacting nuclei with respect to an arbitrary axis system of reference. Therefore, dipolar couplings can be used to obtain structural information.

In liquid samples, which usually provide high-resolution nuclear magnetic resonance (NMR) spectra, the internuclear vector moves isotropically and the dipolar coupling averages to zero. In the solid state, where this vector has a fixed orientation, the dipole–dipole interactions are numerous and strong, broadening NMR signals such that structural information at high resolution cannot be obtained. An intermediate situation is achieved by partially restricting molecular tumbling of liquid samples. The alignment of a fraction of molecules in the presence of the magnetic field allows the measurement of dipolar couplings. Because they are scaled down owing to partial alignment, we refer to them as residual dipolar couplings (RDCs). The structural information obtained from RDCs has impacted enormously traditional protein structure determination based on nuclear Overhauser effect-derived distance restraints. Methods to measure RDCs and their application to protein structure determination are illustrated.

Key Words: NMR; residual dipolar coupling; magnetic alignment; protein structure.

1. Introduction

The residual dipolar couplings (RDCs) most frequently measured involve nuclei that are chemically bonded; therefore, the internuclear distance is known *a priori*. Nevertheless, measurement and application of RDCs is not limited to directly bonded nuclei. At the same time, these nuclei are also scalar-coupled,

From: *Methods in Molecular Biology, vol. 278: Protein NMR Techniques*
Edited by: A. K. Downing © Humana Press Inc., Totowa, NJ

which means that the energy population of one of the nuclei will be split into two populations, having each a Larmor frequency modified in equal amount with the opposite sign. The difference between those two frequencies is called the *J coupling constant*. Under conditions of partial alignment, the RDC, which through the dipole–dipole interaction also modify the Larmor frequency, add or subtract from J. Therefore, the difference between the obtained J values, in the presence of isotropic conditions (i.e., nonaligned samples), and J + RDC under anisotropic conditions (i.e., partially aligned samples), provides the values of the RDC. The nuclear magnetic resonance (NMR) experiments designed to measure RDCs of chemically bonded nuclei are based on the measurement of J, and J + RDC, either from chemical shift differences or from magnetization intensity. Examples of these NMR experiments are discussed.

The relationship between the RDCs and the structural information they provide can be understood by the following equations:

In the principal axis frame of the alignment tensor, A, the RDC between two nuclei, P and Q, as a function of the polar coordinates, θ and φ is given by:

$$RDC_{P,Q(\theta,\varphi)} = D_a{}^{PQ}\left\{(3\cos^2\theta - 1) + 3/2\, R \sin^2\theta \cos2\varphi\right\} \tag{1}$$

where,

$$D_a{}^{PQ} = -\left(\mu_0\hbar / 16\pi^3\right) S\, \gamma_P\, \gamma_Q < r_{PQ}{}^{-3} > \tag{2}$$

A_a is the dimensionless axial component of the alignment tensor, that is, $A_{zz} - (A_{xx} + A_{yy})/2$, and $R = (A_{xx} - A_{yy})/A_{zz}$. R is the rhombicity of the tensor, which indicates the deviation from axial symmetry. S is the generalized order parameter, which accounts for the effect of internal dynamics (1). γ_P and γ_Q are the gyromagnetic ratios of nuclei P and Q. μ_o is the magnetic susceptibility. The polar coordinates θ and φ describe the orientation of the vector in the coordinate system of the alignment tensor (**Fig. 1.**). The brackets indicate time averaging of the inverse cube of the P–Q distance (r_{PQ}).

Equation 1 establishes the dependence of the dipolar coupling value on the orientation of the vector connecting the two atoms with respect to the alignment tensor. The orientation of each vector with respect to a reference frame can be derived once the orientation of the alignment tensor, the dipolar coupling, and the distance between the two atoms are known. The use of RDCs as structural restraints in protein structure determination is explained.

2. Methods

2.1. Commonly Used NMR Experiments to Measure RDC

2.1.1. Practical Issues

In the case of proteins, RDCs are commonly measured for nuclei belonging to the protein backbone (2). These are the amide N–H, as well as C_α–H_α, C_α–C′, and C′–N. The protein has to be enriched in magnetically active ^{15}N and ^{13}C. To

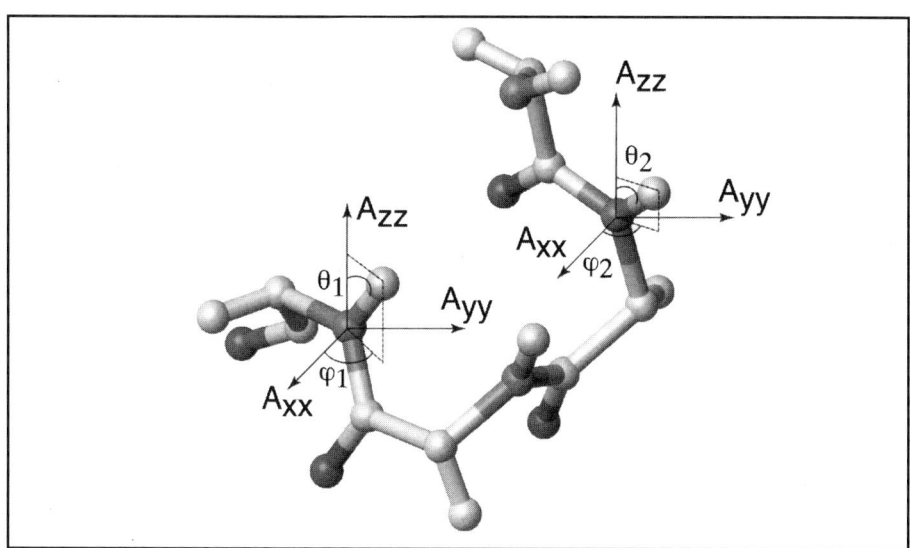

Fig. 1. Orientation of two dipolar coupling vectors in a protein segment. The vectors connect the amide $^1H^N$ and ^{15}N atoms and they coincide with the chemical bond. The axis system of the alignment tensor is designated as A_{xx}, A_{yy}, A_{zz}. The angles θ_1, φ_1, and θ_2, φ_2 define the orientation of both dipolar vectors with respect to the alignment tensor.

measure RDCs, NMR experiments have to be performed on the protein sample under isotropic and anisotropic conditions. There are numerous media for protein alignment published in the literature (*see* Chapter 6 of this volume). Some of them interact sterically with a protein *(3)*, whereas others are either negatively or positively charged, and therefore may give rise to electrostatic interactions. The protein structure and dynamic behavior can be modified by interactions with the alignment medium. The structural data obtained from RDCs will be biased under these circumstances. Biochemical information about the protein under study should be taken into account to select an appropriate orienting medium. For example, if the protein is known to bind to membranes, liquid crystals formed by detergent mixtures that arrange to form bicelles should not be used. Proteins that are known to bind DNA presumably contain positively charged regions in their electrostatic surface and are expected to interact with negatively charge phages, which are also a common media for macromolecular alignment *(4,5)*. Electrostatic interactions can be shielded by the addition of different types of salt to the medium. An easy way to detect structural or dynamic changes under anisotropic conditions is by recording a ^{15}N-heteronuclear single-quantum coherence (HSQC) experiment *(6,7)* in the presence and absence of the orienting particles. Only minor changes in the chemical shifts and line widths of the NMR signals are expected. When interactions between the protein and the alignment media are present, NMR line widths will be typically very broad because of the presence of long-range dipolar couplings (i.e., between atoms that are more that one chemical bond apart) and to a decrease in the protein tumbling rate.

Before performing NMR experiments to measure RDCs it is convenient to have a rough estimation of the magnitude of protein alignment. A two-dimensional (2D) ^{15}N-HSQC with ^{1}H-^{15}N scalar coupling in the ^{15}N dimension will provide the range of RDC. The chemical shift difference between the two components of the amide ^{1}H-^{15}N doublet is approx 94 Hz (^{1}H–^{15}N J-coupling constant) under isotropic conditions. In contrast, when the protein is weakly aligned these chemical shift differences should be in an approximate range of 80 to 110 Hz. This is an indication of an adequate magnitude of orientation. Each individual cross-peak should not be split into several peaks. Additional splitting indicates the presence of long-range dipolar couplings and is a result of excessive protein alignment, which should be avoided. The magnitude of alignment can be reduced, for example, by decreasing the concentration of the orienting particles, when bicelles *(8)* or phages are used, or by increasing the pore size of polyacrylamide gels *(9–12)*.

A different way to estimate the magnitude of alignment is by measuring the ^{2}H signal splitting of a ^{2}H spectrum because of the residual ^{2}H quadrupolar coupling constant under anisotropic conditions.

2.1.2. The In-Phase Antiphase (IPAP) Strategy for the Measurement of Amide N–H RDC

The J and J + RDC of amide N–H pairs can be obtained from the chemical-shift difference between the split signal of a ^{15}N-HSQC experiment with N–H coupling in the ^{15}N dimension. Signal splitting implies that the number of signals doubles with respect to a decoupled spectrum. For medium-size proteins, this will result in crowded spectra and extensive signal overlap. With the aim of alleviating this problem, two coupled spectra can be acquired. One will contain the antiphase component of the coupled magnetization and the other the in-phase component. Both spectra will be added and subtracted generating two spectra. One of them contains the upfield component and the other contains the downfield component of the doublet. With the IPAP strategy *(13)* NMR peaks are separated into two spectra, reducing signal overlap. In addition, the intensity loss owing to signal splitting is compensated by the addition and subtraction of the originally acquired spectra. The NMR pulse sequence of this experiment is illustrated in **Fig. 2A**.

2.1.3. The HNCO Experiment for the Measurement of C_α–C′

The HNCO is a three-dimensional NMR experiment *(14)* commonly used in the chemical shift assignment of the ^{1}H, ^{15}N and ^{13}C′ backbone nuclei. In the ^{13}C dimension, C_α–C′ coupling can be active to measure J and J + RDCs of

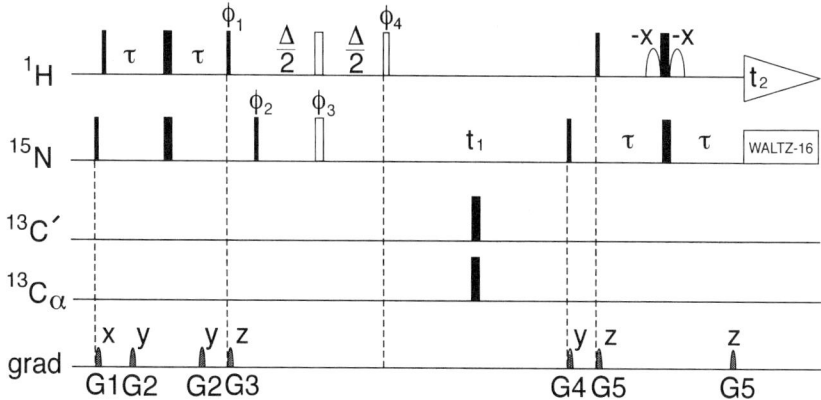

Fig. 2A. Schematic representation of the pulse program IPAP-^{15}N-HSQC. *Narrow* and *wide bars* represent 90° and 180° pulses, with phase x unless indicated. The *white bars* represent pulses that are applied only when the antiphase spectrum is acquired. The antiphase and in-phase spectra are recorded in an interleaved manner. Technical details may be found in the original publication (*13*).

C_α–C′ pairs. Signal overlap is not frequent in this experiment because protein amide pairs appear in different planes of the 3D experiment according to their ^{15}N chemical shift.

2.1.4. ^{15}N-HSQC–IPAP and HNCO for the Measurement of C′–N RDCs

The same experiment used for the measurement of amide N–H RDCs can be applied to measure RDCs of C′–N pairs by allowing coupling between the amide N and the carbonyl C′ in the ^{15}N dimension.

A modification of the HNCO experiment can also be used for the measurement of this dipolar coupling with higher accuracy (**Subheading 3.**).

2.1.5. Constant Time ^{13}C-HSQC J Modulated for the Measurement of C_α–H_α RDCs

In the presence of scalar and/or dipolar coupling, the magnetization intensity follows a sinusoidal-type modulation that depends on the value of J or J + RDC and the time during which the coupling is active. This dependence has been used to modify the constant time ^{13}C-HSQC (*15*) to measure RDC of C_α–H_α. Several constant time ^{13}C-HSQC spectra are acquired. In each of them the intensity of the 2D cross-peaks is modulated by the J or J + RDC value (*16*). A fitting of the measured intensity values to the appropriate equation provides the value of J and J + RDC. This experiment can be used to measure RDC of methylene and methyl sites in amino acid sidechains (**Subheading 3.**). For medium- to large-size proteins, the 2D ^{13}C-HSQC may have extensive overlap

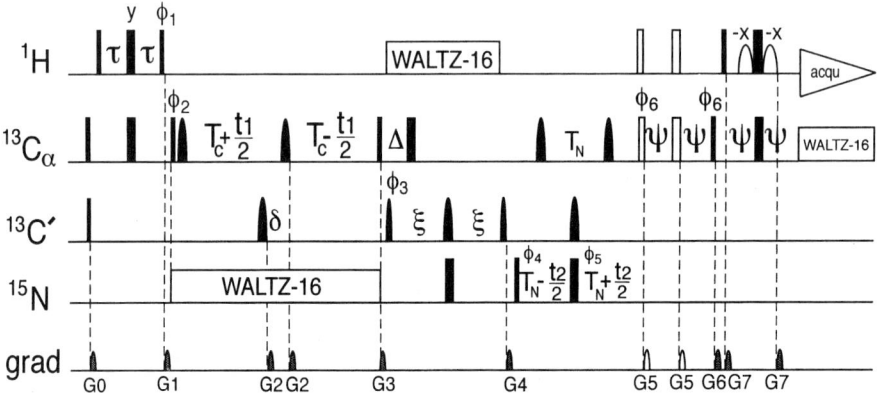

Fig. 2B. Schematic representation of the pulse programs used for the modified version of the (HA)CA(CO)NH experiment. The *white bars* represent the extra pulses that are applied in an interleaved manner to collect the in-phase ^{15}N magnetization. The antiphase ^{15}N magnetization is collected without the application of the white pulses. For further details see the original paper *(18)*.

in the chemical shift region where C_α–H_α signals appear. In this situation, 3D experiments, such as (HA)CA(CO)NH, *(17,18)* would probably be a better choice at a cost of signal-to-noise ratio (**Fig. 2B**).

2.2. Application of Residual Dipolar Couplings As Structural Restraints in Protein Structure Calculation

2.2.1. Information Related to the Alignment Tensor

When RDCs are the only available source of structural information, it is necessary to know *a priori* the magnitude and orientation of the alignment tensor A (**Eqs. 1** and **2**) to use them as structural restraints in a structure calculation protocol. In contrast, if nuclear Overhauser effect (NOE) data are available, too, only a rough estimation of tensor magnitude needs to be known in advance. In this circumstance, the tensor orientation can be determined during the calculation process by treating it as one of the variables. The distribution of probabilities of the different RDC values resembles that of a powder pattern, assuming that all possible orientations of internuclear vectors are represented. On the basis of this similarity, it is possible to obtain good estimations of magnitude (D_a) and rhombicity (R) of the alignment tensor using the following equations *(19)*:

$$A_{zz} = 2D_a \tag{3}$$

$$A_{yy} = -D_a (1 + 1.5R) \tag{4}$$

$$A_{xx} = -D_a (1 - 1.5R) \qquad (5)$$

$$|A_{zz}| \geq |A_{yy}| \geq |A_{xx}| \qquad (6)$$

Because of the presence of regular secondary structure elements in protein structure, different sets of dipolar couplings need to be used to ensure an isotropic distribution of bond–vector orientations. For example, if only N–H dipolar couplings are used, and the protein contains α-helices that run almost parallel to each other, then the orientation of the N–H vectors will be similar, and the distribution will not be isotropic. Therefore, dipolar couplings of C_α–H_α, C_α–C′, and C′–N pairs should be used as well. The powder pattern can be obtained once all dipolar couplings are normalized with respect to one particular set by taking into account the differences in the gyromagnetic ratios together with the different bond distances. **Figure 2C** shows the histogram distributions of different types of RDCs and the powder pattern shape obtained

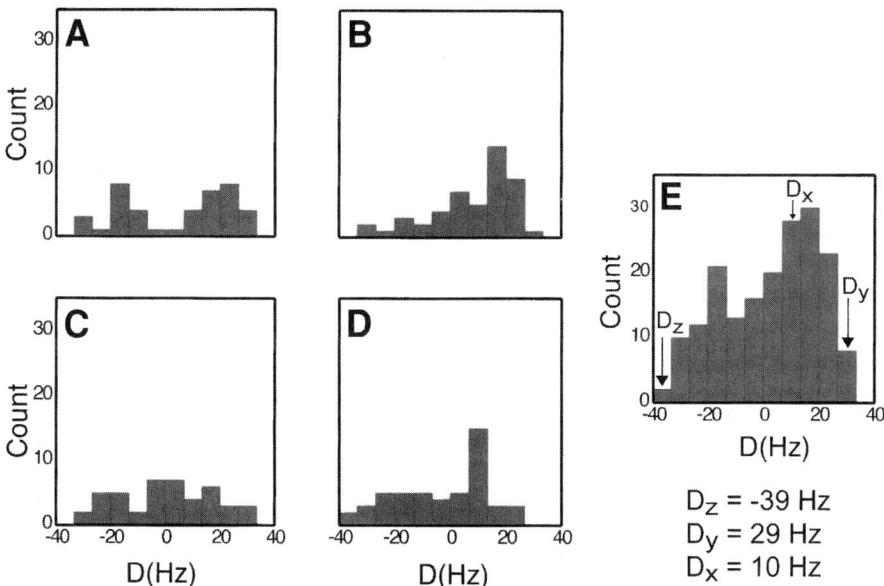

Fig. 2C. Histograms of residual dipolar couplings observed for the C-terminal KH domain of the ribonucleoprotein K in 3.2% w/v bicelle solution *(31)*. (A) RDC_{NH}, (B) $RDC_{C\alpha H\alpha}$, (C) $RDC_{C'N}$, (D) $RDC_{C'C\alpha}$. None of each type of dipolar couplings properly represents a powder pattern. In contrast, a histogram that contains the addition of all normalized dipolar couplings provides a good representation of a powder pattern distribution (E). CαHα, C′N and C′Cα dipolar couplings are normalized to those of NH using the appropriate factor $\gamma_N \gamma_H [r_{NH}^{-3}]/\gamma_A \gamma_B [r_{AB}^{-3}]$ where r_{AB} is the bond distance between A and B and γ_i is the gyromagnetic ratio of i.

once all normalized sets are included. D_a and R can be further optimized through a grid search by slight modifications of the initially estimated values that would lead to structures with lowest energies. It is also possible to optimize them automatically during the structure calculation process. *(20)*

2.2.2. Incorporation of RDCs into a Protein Structure Calculation Protocol

One of the most commonly used methods to incorporate RDCs into protein structure calculation was developed by Tjandra et al. *(21)* using the program Xplor-NIH *(22)*. RDCs are included as an empirical energy term that is added to the other conventional terms of the target function (torsion angles, covalent geometry, distance restraints derives form NOE data, etc). This energy terms has the form:

$$E_{dip} = k_{dip}\left(RDC_{cal} - RDC_{obs}\right)^2 \tag{7}$$

k_{dip} is a force constant and RDC_{cal} and RDC_{obs} are the calculated and observed values of the residual dipolar coupling, respectively.

The alignment tensor is considered as a reference axis that is represented by a pseudomolecule located far away from the biopolymer. This pseudo-molecule consists of four equidistant atoms (OXYZ). One represents the axis origin (O). The other three represent the three axes, OX, OY, and OZ, which are orthogonal to one another. E_{dip} is evaluated by calculating the angles that the dipolar vectors form with the reference axis. During the calculation process, the pseudo-molecule OXYZ is continually reoriented to yield the best fit between the calculated and the observed dipolar couplings. The force constant in **Eq. 7** is chosen such that the errors between the predicted dipolar couplings from the calculated structure and the observed ones reflect the experimental errors.

It is possible to incorporate in the structure calculation RDC information obtained in different alignment media. An alignment tensor corresponding to each orienting medium has to be represented by a different pseudomolecule, which should be located far away from the other (**Fig. 2D**).

The energy landscape of a simulating annealing procedure in which only dipolar couplings are included tends to have many local minima, therefore RDCs are usually introduced once a preliminary structure has been calculated from a small set of NOE data. It is important to increase the force constant of the E_{dip} term during the calculation process, in concert with other energy terms, to ensure an appropriate minimization. As a side advantage, the use of RDCs together with NOE-derived distance restraints can help to identify incorrectly assigned NOEs.

When using RDCs as structural restraints, it is useful to have some previous qualitative knowledge of the internal dynamics of the protein. As illustrated in **Eq. 2** the RDC depends on the dynamic behavior of the bond–vector that connects

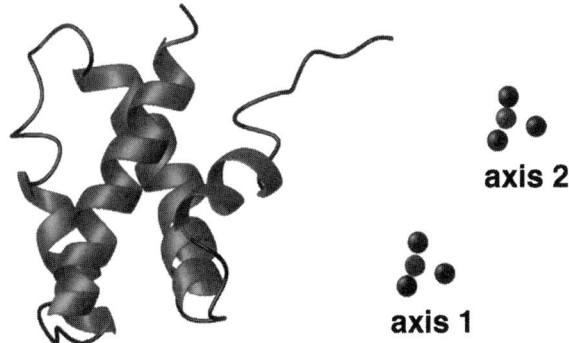

axis 2

axis 1

Fig. 2D. Ribbon diagram of a protein structure determined with RDC restraints obtained in two different alignment media. The pseudomolecules representing the two alignment tensors that were used during the calculation are also shown.

the interacting nuclei. If the vector has the degree of mobility characteristic of secondary structural elements, then the RDC will be scaled down by the generalized order parameter that accounts for fast internal motion (S), which is known to vary in the range of 0.85 to 0.95. In contrast, if the degree of mobility is higher, then the measured RDC values will be smaller than what is expected for less-flexible regions. RDC of this bond–vector can also be used in the calculation protocol, provided that the measured RDC is considered as a lower limit. In practice, the energy function related to RDCs of nonflexible regions is a harmonic potential, whereas for flexible regions a half-open square well potential is used in the calculation protocol.

Figure 2E shows a schematic representation of the protocol used in protein structure calculation using XPLOR and including RDCs as structural restraints. **Figure 2F** is an example of the script used as an input file for XPLOR containing the different commands and information necessary for the calculation process represented in **Fig. 2E**.

2.2.3. How Do We Know That the Structure Calculation With RDC Restraints Is Being Performed Correctly?

Incompatibilities between NOE data and RDC information or between RDC restraints obtained from different alignment media, are usually reflected in either NOE or RDC violations. Both types of data should be checked to identify incorrect assignments. In addition, the axes of the pseudomolecule, which represents the alignment tensor, should remain orthogonal at the end of the calculation. Deviations from this condition can be an indication of misinterpreted data or incorrect estimations of the magnitude and rhombicity of the alignment tensor. Alternatively, such deformation could result if the force constant for the

Restrained molecular dynamics protocol

1) **Equilibration**

   Temperature: 4000 K
   Number of steps: 1000
   Total time: 1ps
   Energy terms: bonds, angles, improper torsions, van der Waals (Lennard-Jones)

2) **Simulated annealing with RDC restraints**

   Initial temperature: 4000 K
   Number of cooling steps: 20,000
   Final Temperature: 100 K
   Number of molecular dynamics cycles: 128
   Total time: 256 fs
   Energy terms: bonds, angles, improper torsions, van der Waals (Lennard-Jones), dihedrals, dipolars

3) **Minimization**

   600 cycles of energy minimization
   Initial temperature: 4000 K
   Number of cooling steps: 20,000
   Final Temperature: 100 K
   Number of steps: 128
   Energy terms: bonds, angles, improper torsions, NOE, van der Waals (Lennard-Jones), dihedrals, dipolar

Fig. 2E. Scheme of a molecular dynamics protocol for protein structure determination with RDC restraints.

RDC is set too high relative to the force constant constraining the axis system as orthogonal.

RDCs improve structure precision and quality. A structure quality factor (Q) *(23)* that depends on RDC can be used to assess structure quality. The Q factor describes the agreement between the calculated structure and the experimental data, by comparing the predicted dipolar couplings of a determined structure and the experimentally measured dipolar couplings. The Q factor is defined as follows:

$$Q = \left[\sum_{i=1,\ldots,N} \left(RDC_{obs} - RDC_{cal} \right)^2 / N \right]^{1/2} / RDC_{rms} \tag{8}$$

RDC_{obs} and RDC_{cal} are the experimentally measured and the predicted dipolar couplings, respectively. The summation extends over the N residues for which dipolar couplings are known. RDC_{rms} is the root-mean-square value of dipolar couplings when they follow a random distribution. The alignment tensor

```
REMARKS Script for simulated annealing using RDC
restraints from two different alignment media
!--------------------------------------------------------------------
! read in the PSF file and initial structure
parameter
{===>}
  @parallhdg_new.pro
  @par_axis.pro
end

structure
{===>}
  @protein.psf @axis_in_medium1.psf @axis_in_medium2.psf
end
coordinates
           @protein_axis1_axis_2_coordinates.pdb
coordinates copy end
!--------------------------------------------------------------------
! initialize weights for energy terms
evaluate   ($knoe  = 25.0)  ! noes
evaluate   ($ksani = 0.3)    ! dipolar coupling force constant
evaluate   ($asym  = 0.2)    ! SOFT noe potential
evaluate   ($kcdi = 10.0 )   ! dihedral
evaluate   ($cool_steps = 20000)
evaluate   ($init_t   = 4000.00)
!--------------------------------------------------------------------
! Read experimental restraints
noe
  nrestraints=16000             ! allocate space for NOEs
  ceiling 100
  set echo on message  on end
  class dist
  @noes4D.tbl
  @noes3D.tbl
  averaging          dist sum
  potential          dist soft
  scale              dist $knoe
  sqconstant         dist 1.0
  sqexponent         dist 2
  soexponent         dist 1
  rswitch            dist 0.6
  sqoffset           dist 0.0
  asymptote          dist $asym
end

restraints dihedral
  scale $kcdi
  nass = 400
  @psi.tbl
  @phi.tbl
end

!Force constants for the different dipolar couplings
evaluate ($kdnh_axis1 =0.32*$ksani)
evaluate ($kdnh_axis2 = 0.15*$ksani)
evaluate ($kdch_axis1 = 0.16*$ksani)
evaluate ($kdcaco_axis1 =9.0*$ksani)

!Magnitudes of alignment scaled to NH and rhombicities
evaluate ($da_axis1=23.0)
evaluate ($da_axis2=16.6)
evaluate ($dach_axis1=$da_axis1*2.09)
evaluate ($dacaco_axis1=$da_axis1*0.19)
evaluate ($rhom_axis1=0.31)
evaluate ($rhom_axis2=0.36)

set echo on message  on end
```

```
dipo
  nres=10000

class dnh_axis1
    type fixd
    scale 1.0
    sign on
    average sum
    force $kdnh_axis1
    potential harmonic
    coef 0.0000 $da_axis1 $rhom_axis1
    @nh_axis1.tbl

class dnh_mob_axis1
    type fixd
    scale 1.0
    sign on
    average sum
    force $kdnh_axis1
    potential square
    coef 0.0000 $da_axis1 $rhom_axis1
    @nh_mob_axis1.tbl

class dnh_axis2
    type fixd
    scale 1.0
    sign on
    average sum
    force $kdnh_axis2
    potential harmonic
    coef 0.0000 $da_axis2 $rhom_axis2
    @nh_axis2.tbl

class dch_axis1
    type fixd
    scale 1.0
    sign on
    average sum
    force $kdch_axis1
    potential harmonic
    coef 0.0000 $dach_axis1 $rhom_axis1
    @ch_axis1.tbl

class dcaco_axis1
    type fixd
    scale 1.0
    sign on
    average sum
    force $kdcaco_axis1
    potential harmonic
    coef 0.0000 $dacaco_axis1 $rhom_axis1
    @caco_axis1.tbl

end

! Fixing the axis origin
vector do (refx=x) (all)
vector do (refy=y) (all)
vector do (refz=z) (all)
constraints fix(resid 500:600 and name OO) end

!--------------------------------------------------------------------

! Set the Flags
flags
  exclude * include bond angle impr vdw  noe cdih dipo
end
```

```
! Equilibration

vector do (mass = 100.0) (all)       ! uniform mass for all atoms
vector do (fbeta = 10.0 ) (all)      ! coupling to heat bath

evaluate ($count = 0)
evaluate ($endcount = 200)

while ($count < $endcount) loop main
evaluate ($count = $count + 1)
coor swap end
coor copy end

evaluate ($knoe = 2.0)        !slope of NOE potential
evaluate ($ksani = 0.0001)    ! dipolar coupling force constant
evaluate ($kcdi = 10.0)       ! torsion angles
evaluate ($rcon = 0.004)      ! force constant for vdW

noe
  scale   dist $knoe
end

evaluate ($kdnh_axis1 = 0.32*$ksani)
evaluate ($kdnh_axis2 = 0.15*$ksani)
evaluate ($kdch_axis1 = 0.16*$ksani)
evaluate ($kdcaco_axis1 = 9.0*$ksani)

dipo
  class dnh_axis1
    force $kdnh_axis1
  class dnh_mob_axis1
    force $kdnh_axis1
  class dnh_axis2
    force $kdnh_axis2
  class dch_axis1
    force $kdch_axis1
  class dcaco_axis1
    force $kdcaco_axis1
end

parameters
  nbonds
    repel = 1.2   ! scale factor for vdW radii = 1 ( L-J radii)
    rexp  = 2     ! exponents in (r^irex - R0^irex)^rexp
    irex  = 2
    rcon  = $rcon ! actually set the vdW weight
    nbxmod = 3
    wmin  = 0.01 ! warning off
    cutnb = 100  ! nonbonded cutoff
    tolerance = 45
  end
end

constraints
    interaction (not name ca) (all)
    weights * 1 angl 0.4 impr 0.1 vdw 0 elec 0 end
    interaction (name ca) (name ca)
    weights * 1 angl 0.4 impr 0.1 vdw 1.0 end
end

dynamics verlet
    nstep=1000      !
    timestep=0.001  !
    iasvel=maxwell      firsttemp= $init_t
    tcoupling = true
    tbath = $init_t
    nprint=25
    iprfrq=0
    ntrfr = 99999999
end
```

```
! Restrained Molecular Dynamics

parameters
  nbonds
    rexp = 2        ! exponents in (r^irex - R0^irex)^rexp
    irex = 2
    nbxmod = 3
    wmin  = 0.01 ! warning off
    cutnb = 4.5     ! nonbonded cutoff
    tolerance = 0.5
  end
end

evaluate ($kcdi = 20)

restraints dihed
  scale $kcdi
end

evaluate ($final_t = 100)     { K }
evaluate ($tempstep = 25)     { K }
evaluate ($bath = $init_t)

evaluate ($ncycle = ($init_t-$final_t)/$tempstep)
evaluate ($nstep = int($cool_steps/$ncycle))

evaluate ($ini_rad = 0.9)        evaluate ($fin_rad = 0.80)
evaluate ($ini_con = 0.004)       evaluate ($fin_con = 4.0)
evaluate ($ini_ncs = 0.01)       evaluate ($fin_ncs = 10)
evaluate ($ini_ang = 0.4)        evaluate ($fin_ang = 1.0)
evaluate ($ini_imp = 0.1)        evaluate ($fin_imp = 1.0)

evaluate ($ini_noe = 2.0)        evaluate ($fin_noe = 50.0)
evaluate ($ini_sani = 0.0001)     evaluate ($fin_sani = 1.00)

evaluate ($radius=  $ini_rad)
evaluate ($radfact = ($fin_rad/$ini_rad)^(1/$ncycle))
evaluate ($k_ncs = $ini_ncs)
evaluate ($ncs_fac = ($fin_ncs/$ini_ncs)^(1/$ncycle))
evaluate ($k_vdw = $ini_con)
evaluate ($k_vdwfact = ($fin_con/$ini_con)^(1/$ncycle))
evaluate ($k_ang = $ini_ang)
evaluate ($ang_fac = ($fin_ang/$ini_ang)^(1/$ncycle))
evaluate ($k_imp = $ini_imp)
evaluate ($imp_fac = ($fin_imp/$ini_imp)^(1/$ncycle))
evaluate ($knoe = $ini_noe)
evaluate ($noe_fac = ($fin_noe/$ini_noe)^(1/$ncycle))

evaluate ($ksani = $ini_sani)
evaluate ($sani_fac = ($fin_sani/$ini_sani)^(1/$ncycle))

evaluate ($kdnh_axis1 = 0.32*$ksani)
evaluate ($kdnh_axis2= 0.15*$ksani)
evaluate ($kdch_axis1 = 0.16*$ksani)
evaluate ($kdcaco_axis1 = 9.0*$ksani)

vector do (vx = maxwell($bath)) (all)
vector do (vy = maxwell($bath)) (all)
vector do (vz = maxwell($bath)) (all)

evaluate ($i_cool = 0)
while ($i_cool < $ncycle) loop cool
    evaluate ($i_cool=$i_cool+1)
```

```
   evaluate ($bath  = $bath  - $tempstep)
   evaluate ($k_vdw = min($fin_con,$k_vdw*$k_vdwfact))
   evaluate ($radius= max($fin_rad,$radius*$radfact))
   evaluate ($k_ang = $k_ang*$ang_fac)
   evaluate ($k_imp = $k_imp*$imp_fac)
   evaluate ($knoe  = $knoe*$noe_fac)
   evaluate ($ksani = $ksani*$sani_fac)

evaluate ($kdnh_axis1= 0.32*$ksani)
evaluate ($kdnh_axis2 = 0.15*$ksani)
evaluate ($kdch_axis1 = 0.16*$ksani)
evaluate ($kdcaco_axis1 = 9.0*$ksani)

   constraints interaction (all) (all) weights
    * 1 angles $k_ang improper $k_imp
   end end

   parameter nbonds
    cutnb=4.5 rcon=$k_vdw nbxmod=3 repel=$radius
   end     end


   noe
    scale dist $knoe
   end

   dipo
   class dnh_axis1
    force $kdnh_axis1
   class dnh_mob_axis1
    force $kdnh_axis1
   class dnh_axis2
    force $kdnh_axis2
   class dch_axis1
    force $kdch_axis1
   class dcaco_axis1
    force $kdcaco_axis1
   end

   dynamics  verlet
   nstep=$nstep timestep=0.002 iasvel=current firsttemp= $bath
    tcoupling = true  tbath = $bath   nprint=$nstep iprfrq=0
    ntrfr = 99999999
   end
end loop cool
!-------------------------------------------------------------------------
! Minimization

mini powell nstep= 250 nprint= 50 end
!-------------------------------------------------------------------------
! Minimize: CA-CA

evaluate ($knoe  = 2.0)     ! slope of NOE potential
evaluate ($ksani = 0.0001)    ! dipolar coupling force constant
evaluate ($kcdi = 10.0)     ! torsion angles

noe
  scale     dist $knoe
end

evaluate ($kdnh_axis1 = 0.32*$ksani)
evaluate ($kdnh_axis2 = 0.15*$ksani)
evaluate ($kdch_axis1 = 0.16*$ksani)
evaluate ($kdcaco_axis1 = 9.0*$ksani)
```

```
dipo
  class dnh_axis1
    force $kdnh_axis1
  class dnh_mob_axis1
    force $kdnh_axis1
  class dnh_axis2
    force $kdnh_axis2
  class dch_axsi1
    force $kdch_axis1
  class dcaco_axis1
    force $kdcaco_axis1
end
restraints dihed
  scale $kcdi
end

evaluate ($rcon  = 1.0)

parameters
  nbonds
    repel = 1.2   ! scale factor for vdW radii = 1 ( L-J radii)
    rexp = 2     ! exponents in (r^irex - R0^irex)^rexp
    irex  = 2
    rcon = $rcon ! actually set the vdW weight
    nbxmod = 3
    wmin  = 0.01  ! warning off
    cutnb  = 100   ! nonbonded cutoff
    tolerance = 45
  end
end

constraints
    interaction (not name ca) (all)
    weights * 1 angl 0.4 impr 0.1 vdw 0 elec 0 end

    interaction (name ca) (name ca)
    weights * 1 angl 0.4 impr 0.1 vdw 1.0 end
end

mini powell nstep= 250 nprint= 50 end
!-------------------------------------------------------------------------
! Minimize: All-All Loop
parameters
  nbonds
    rexp  = 2    ! exponents in (r^irex - R0^irex)^rexp
    irex  = 2
    nbxmod = 3
    wmin  = 0.01  ! warning off
    cutnb = 4.5  ! nonbonded cutoff
    tolerance = 0.5
  end
end

evaluate ($kcdi = 20)

restraints dihed
  scale $kcdi
end

evaluate ($i_cool = 0)
while ($i_cool < $ncycle) loop minim
  evaluate ($i_cool=$i_cool+1)

  evaluate ($bath  = $bath  - $tempstep)
  evaluate ($k_vdw=min($fin_con,$k_vdw*$k_vdwfact))
  evaluate ($radius=max($fin_rad,$radius*$radfact))
  evaluate ($k_ncs = $k_ncs*$ncs_fac)
```

```
evaluate ($k_ang = $k_ang*$ang_fac)
evaluate ($k_imp = $k_imp*$imp_fac)
evaluate ($knoc = $knoe*$noe_fac)
evaluate ($ksani = $ksani*$sani_fac)

evaluate ($kdnh_axis1 = 0.32*$ksani)
evaluate ($kdnh_axis2 = 0.15*$ksani)
evaluate ($kdch_axsi1 = 0.16*$ksani)
evaluate ($kdcaco_axis1 = 9.0*$ksani)

    constraints interaction (all) (all) weights
       * 1 angles $k_ang improper $k_imp
    end end

    parameter nbonds
       rcon=$k_vdw
       repel=$radius
    end end

noe
    scale dist $knoe
end

dipo
class dnh_axis1
    force $kdnh_axis1
class dnh_mob_axis1
    force $kdnh_axis1
class dnh_axsi2
    force $kdnh_axis2
class dch_axsi1
    force $kdch_axis1
class dcaco_axis1
    force $kdcaco_axis1
end
mini powell nstep= 100 nprint= 50 end
end loop minim
! Finish
{====>}          {*Name(s) of the family of final structures.*}
evaluate ($file = "protein" + encode ($count) + ".pdb")

write coor output= $file end

end loop main

stop
```

```
----------------------------------------------------
Example of dnh_axis1.tbl

assign     ( resid 500 and name OO      )
           ( resid 500 and name Z       )
           ( resid 500 and name X       )
           ( resid 500 and name Y       )
           ( resid 4   and name N       )
           ( resid 4   and name HN      ) -3.422 0 0
----------------------------------------------------
Example of dnh_mob_axis1.tbl

assign     ( resid 500 and name Z       )
           ( resid 500 and name X       )
           ( resid 500 and name Y       )
           ( resid 2   and name N       )
           ( resid 2   and name HN      ) -7.350 50 0
----------------------------------------------------
Example of dnh_axis2.tbl

assign     ( resid 600 and name OO      )
           ( resid 600 and name Z       )
           ( resid 600 and name X       )
           ( resid 600 and name Y       )
           ( resid 5   and name N       )
           ( resid 5   and name HN      ) -13.116 0 0

----------------------------------------------------
Example of dch_axis1.tbl

assign     ( resid 500 and name OO      )
           ( resid 500 and name Z       )
           ( resid 500 and name X       )
           ( resid 500 and name Y       )
           ( resid 4   and name CA      )
           ( resid 4   and name HA      ) 19.855 0 0
----------------------------------------------------
Example of dcaco_axis1.tbl

assign     ( resid 500 and name OO      )
           ( resid 500 and name Z       )
           ( resid 500 and name X       )
           ( resid 500 and name Y       )
           ( resid 4   and name CA      )
           ( resid 4   and name C       ) -8.442 0 0

----------------------------------------------------
```

Fig. 2F. Example of an input file for protein structure calculation with RDC restraints using XPLOR. Examples of the format used for dipolar couplings restraints in two different alignment media and for NH amide atoms undergoing fast internal motions are shown.

is obtained from the best fit between the experimental data and the existing structure. If the distribution cannot be considered uniform, RDC_{rms} may be calculated as (24,25):

$$RDC_{rms} = \left\{ D_a{}^2 \left[4 + 3\left(D_r / D_a \right)^2 \right] \Big/ 5 \right\}^{1/2} \qquad (9)$$

D_a and D_r are the axial and rhombic components of a traceless second rank diagonal tensor related to the alignment tensor.

In analogy to the free R factor *(26)*, Q has to be calculated for a set of restraints not used in the structure refinement. For example, Q_{NH} would be calculated by using a set of structures that have not been refined with NH RDC restraints.

3. Conclusions

3.1. Measurement of RDC in Proteins

RDCs depend on the gyromagnetic ratios of the nuclei involved in the dipolar interaction and the interatomic distance. Therefore, the intrinsic magnitudes of the RDC of C_α–C′ and C′–N pairs are approximately five and nine times smaller than that of an N–H pair, respectively. In contrast, the intrinsic magnitude of the C_α–H_α RDC is approximately twice that of N–H RDC. These differences should be considered to decide which NMR experiment is best suited to measure a particular dipolar coupling with enough accuracy. For example, C_α–H_α RDC measured with significantly large errors could still be structurally useful, because the RDC is expected to spread in a larger range. The RDC of C′–N pairs is small and should be measured with high accuracy. NMR experiments using magnetization intensity modulated by J + RDC allow RDC measurements with smaller errors than when chemical shift differences are used. Accurate values of the RDC of C′–N pairs can be obtained by using an HNCO-based experiment in which the intensity signal is a sine function of J, or J + RDC when the sample is under isotropic and anisotropic conditions, respectively *(27)*.

When the protein rotational correlation time exceeds 15 ns, the upfield component of the N–H doublet is usually very broad. In such cases, the N–H J-coupling, or J + RDC can be obtained by comparing the chemical shifts of a transverse relaxation optimized spectroscopy (TROSY) spectrum *(28)*, where the narrowest component of the splitting is selected, and that of a ^{1}H-decoupled ^{15}N-HSQC *(2)*.

3.2. RDC in Protein Structure Calculation

Dipolar couplings of methylene and methyl moieties of amino acid sidechains have been applied to protein structure calculation *(16)*. In a side chain with a C–CH_3 moiety, rapid rotation of the methyl group results in an averaged C–H dipolar vector aligned in the direction of the C–C bond. This rapid rotation scales the real value of the C–H dipolar by a factor of approx –1/3. Because for methyl groups the dipolar splitting that can be measured accounts for three protons, its magnitude is the same and its sign is opposite to that predicted for a C–H pair located in the C–CH_3 bond. Thus, dipolar couplings restraints of methyl groups restrict the orientation of the C–CH_3 bond rather than the C–H bond in protein structure determination *(16)*.

The NMR experiments most commonly used to measure dipolar couplings of methylene sites in amino acid sidechains provide the sum of the RDC of both C–H$_1$ and C–H$_2$ vectors *(16)*. To use these data in the structure calculation the energy function that is minimized has the form *(16)*:

$$E_{dip} = k_{dip}\left[\left(RDC_{cal\,CH1} + RDC_{cal\,CH2} - RDC_{obs\,CH1} - RDC_{obs\,CH2}\right)\right]^2 \qquad (10)$$

RDC_{cal} and RDC_{obs} are the calculated and observed dipolar couplings, respectively.

In the absence of quantitative information on side chain dynamics, the measured dipolar couplings must represent lower limits of the magnitude of the real couplings. This can be done, as in the case of backbone dipolar couplings of bond–vectors undergoing large-amplitude fast motions, by applying a penalty function with the shape of a half-open square well.

^{1}H–^{1}H dipolar couplings can be used as structural restraints as well. Nevertheless, except for methylene groups, the distance between protons is an unknown parameter. The internuclear distance, which is fixed for one-bond heteronuclear dipolar couplings, can be included as a variable in the energy term. If the sign of the ^{1}H–^{1}H dipolar coupling cannot be determined, the calculation is carried out using only the magnitude. The energy term then takes the form *(29)*:

$$E_{dip} = k_{dip}\left(\left|RDC_{cal}\right| - \left|RDC_{obs}\right|\right)^2 \qquad (11)$$

RDC can also be applied to nucleic acid structure determination, provided that certain modifications are done in the NMR experiments and structure calculation protocols *(30)*.

References

1. Lipari, G. and Szabo, A. (1982) Model-free approach to the interpretation of nuclear magnetic resonance relaxation in macromolecules, I: theory and range of validity. *J. Am. Chem. Soc.* **104**, 4546–4559.
2. Bax, A., Kontaxis, G., and Tjandra, N. (2001) Dipolar couplings in macromolecular structure determination. *Methods Enzymol.* **339**, 127–174.
3. Tjandra, N. and Bax, A. (1997) Direct measurement of distances and angles in biomolecules by NMR in a dilute liquid crystalline medium. *Science* **278**, 1111–1114.
4. Clore, G. M., Starich, M. R., and Gronenborn, A. M. (1998) Measurement of residual dipolar couplings of macromolecules aligned in the nematic phase of colloidal suspension of rod-shaped viruses. *J. Am. Chem. Soc.* **120**, 10,571–10,572.
5. Hansen, M. R., Mueller, L., and Pardi, A. (1998) Tunable alignment of macromolecules by filamentous phage yields dipolar coupling interactions. *Nat. Struct. Biol.* **5**, 1065–1074.

6. Bodenhausen, G. and Ruben, D. J. (1980) Natural abundance ^{15}N NMR by enhanced heteronuclear spectroscopy. *Chem. Phys. Lett.* **69,** 185–189.

7. Bax, A., Ikura, M., Kay, L. E., Torchia, D. A., and Tschudin, R. (1990) Comparison of different modes of two-dimensional reverse-correlation NMR for the study of proteins. *J. Magn. Reson.* **86,** 304–318.

8. Ottiger, M. and Bax, A. (1998) Characterization of magnetically oriented phospholipid micelles for measurement of dipolar couplings in macromolecules. *J. Biomol. NMR* **12,** 361–372.

9. Tycko, R., Blanco, F. J., and Ishii, Y. (2000) Alignment of biopolymers in strained gels: a new way to create detectable dipole–dipole couplings in high-resolution biomolecular NMR. *J. Am. Chem. Soc.* **122,** 9340, 9341.

10. Sass, H. J., Musco, G., Stahl, S. J., Wingfield, P. T., and Grzesiek, S. (2000) Solution NMR of proteins within polyacrylamide gels: diffusional properties and residual alignment by mechanical stress or embedding of oriented purple membranes. *J. Biomol. NMR* **18,** 303–309.

11. Ishii, Y., Markus, M. A., and Tycko, R. (2001) Controlling residual dipolar couplings in high-resolution NMR of proteins by strain induced alignment in a gel. *J. Biomol. NMR* **21,** 141–151.

12. Chou, J. J., Gaemers, S., Howder, B., Louis, J. M., and Bax, A. (2001) A simple apparatus for generating stretched polyacrylamide gels, yielding uniform alignment of proteins and detergent micelles. *J. Biomol. NMR* **21,** 377–382.

13. Ottiger, M., Delaglio, F., and Bax, A. (1998) Measurement of J and dipolar couplings from simplified two-dimensional NMR spectra. *J. Magn. Reson.* **131,** 373–378.

14. Ikura, M., Kay, L. E., and Bax, A. (1990) A novel approach for sequential assignment of ^{1}H, ^{13}C, and ^{15}N spectra of larger proteins: heteronuclear triple-resonance three-dimensional NMR spectroscopy: application to calmodulin. *Biochemistry* **29,** 4659–4667.

15. Santoro, J. and King, G. C. (1992) A constant-time 2D over Bodenhausen experiment for inverse correlation of isotopically enriched species. *J. Magn. Reson.* **97,** 202–207.

16. Ottiger, M., Delaglio, F., Marquardt, J. L., Tjandra, N., and Bax, A. (1998) Measurement of dipolar couplings for methylene and methyl sites in weakly oriented macromolecules and their use in structure calculation. *J. Magn. Reson.* **134,** 365–369.

17. Tjandra, N. and Bax, A. (1997) Large variations in ^{13}C$_\alpha$ chemical shift anisotropy in proteins correlate with secondary structure. *J. Am. Chem. Soc.* **119,** 9576, 9577.

18. de Alba, E., Suzuki, M., and Tjandra, N. (2001) Simple multidimensional NMR experiments to obtain different types of one-bond dipolar couplings simultaneously. *J. Biomol. NMR* **19,** 63–67.

19. Clore, G. M., Gronenborn, A. M., and Bax, A. (1998) A robust method for determining the magnitude of the fully asymmetric alignment tensor of oriented macromolecules in the absence of structural information. *J. Magn. Reson.* **131,** 159–162.

20. Saas, H. J., Musco, G., Stahl, S. J., Wingfield, P. T., and Grzesiek, S. (2001) An easy way to include weak alignment constraints into NMR structure calculations. *J. Biomol. NMR* **21,** 275–280.

21. Tjandra, N., Omichinski, J. G., Gronenborn, A. M., Clore, G. M., and Bax, A. (1997) Use of dipolar ^{1}H-^{15}N and ^{1}H-^{13}C couplings in the structure determination of magnetically oriented macromolecules in solution. *Nat. Struct. Biol.* **4,** 732–738.
22. Schwieters, C. D., Kuszewski, J. J., Tjandra, N., and Clore, G. M. (2003) The Xplor-NIH NMR molecular structure determination package. *J. Magn. Res.* **160,** 65–73.
23. Ottiger, M. and Bax, A. (1999) Bicelle-based liquid crystals for NMR-measurement of dipolar couplings at acidic and basic pH values. *J. Biomol. NMR* **13,** 187–191.
24. Clore, G. M., Starich, M. R., Bewley, C. A., Cai, M. L., and Kuszewski, J. (1999) Impact of residual dipolar couplings on the accuracy of NMR structures determined from a minimal number of NOE restraints. *J. Am. Chem. Soc.* **121,** 6513, 6514.
25. Clore, G. M. and Garret, D. S. (1999) R-factor, Free R, and complete cross-validation for dipolar coupling refinement of structures. *J. Am. Chem. Soc.* **121,** 9008–9012.
26. Brunger, A. T., Clore, G. M., Gronenborn, A. M., Saffrich, R., and Nilges, M. (1993) Assessing the quality of solution nuclear magnetic resonance structures by complete cross-validation. *Science* **261,** 328–331.
27. Chou, J. J., Delaglio, F., and Bax, A. (2000) Measurement of one-bond ^{15}N-^{13}C′ dipolar couplings in medium sized proteins. *J. Biomol. NMR* **18,** 101–105.
28. Pervushin, K., Riek, R., Wider, G., and Wuthrich, K. (1997) Attenuated T$_2$ relaxation by mutual cancellation of dipole–dipole coupling and chemical shift anisotropy indicates an avenue to NMR structures of very large biological macromolecules in solution. *Proc. Natl. Acad. Sci. USA* **94,** 12,366–12,371.
29. Tjandra, N., Marquardt, J. L., and Clore, G. M. (2000) Direct refinement against proton-proton dipolar couplings in NMR structure determination of macromolecules. *J. Magn. Reson.* **142,** 393–396.
30. de Alba, E. and Tjandra, N. (2002) NMR dipolar couplings for the structure determination of biopolymers in solution. *Prog. Nucl. Magn. Reson. Spectrosc.* **40,** 175–197.
31. Baber, J. L., Libutti, D., Levens, D., and Tjandra, N. (1999) High precision solution structure of the C-terminal KH domain of heterogeneous nuclear ribonucleoprotein K, a c-myc transcription factor. *J. Mol. Biol.* **289,** 949–962.

8

Projection Angle Restraints for Studying Structure and Dynamics of Biomolecules

Christian Griesinger, Wolfgang Peti, Jens Meiler, and Rafael Brüschweiler

Summary

This chapter presents a methodology that allows for the structural and dynamic characterization of biomolecules by means of projection restraints obtained from residual magnetic dipolar couplings. Dipolar couplings reflect the projection of individual internuclear vectors onto the alignment tensor. This technique allows determination of the dynamics of the protein backbone on time-scales, namely, between the rotational tumbling correlation time and approx 50 μs. This range of time-scales has been previously inaccessible by other nuclear magnetic resonance (NMR) techniques. In addition, information about the anisotropy of the motion is obtained.

Key Words: Dipolar coupling; protein dynamics; time scale; anisotropic motion; ubiquitin; projection angle restraints; alignment.

1. Introduction

The traditional nuclear magnetic resonance (NMR) parameters have been augmented recently by the measurement and interpretation of anisotropic parameters, such as dipolar couplings *(1,2)* and cross-correlated relaxation *(3)*. Cross-correlated relaxation has been recently reviewed *(4–6)*. Here, we will recapitulate the theory of dipolar couplings briefly, review methods of alignment as well as NMR experiments to determine dipolar couplings, and discuss the application of dipolar couplings for the determination of protein dynamics. An approach related to the one presented here has recently been described by Tolman et al. *(7–9)*.

The following Hamiltonian describes the dipolar coupling between two nuclei (k and l) expressed in energy units *(10)*:

From: *Methods in Molecular Biology, vol. 278: Protein NMR Techniques*
Edited by: A. K. Downing © Humana Press Inc., Totowa, NJ

Table 1
Tensor Operators in the Rotating Frame and Modified Spherical Harmonics for the Dipole-Dipole Coupling

Tensor operators for the dipolar interaction

$$b_{kl} = -\mu_0 \frac{\gamma_k \gamma_l \hbar}{4\pi r_{kl}^3}$$

Modified spherical harmonics

M	$A_{2,M}(I_k, I_l)$	$Y_{2,M}(\theta, \phi)$
2	$\sqrt{\dfrac{3}{8}} I_k^- I_l^-$	$\sqrt{\dfrac{3}{2}} \sin^2 \theta \exp(+2i\phi)$
1	$\sqrt{\dfrac{3}{8}} I_{k,z} I_l^-$	$\sqrt{6} \sin\theta \cos\theta \exp(+i\phi)$
1	$\sqrt{\dfrac{3}{8}} I_k^- I_{l,z}$	$\sqrt{6} \sin\theta \cos\theta \exp(+i\phi)$
0	$I_{k,z} I_{l,z}$	$3\cos^2 \theta - 1$
0	$\dfrac{1}{4}(I_k^+ I_l^- + I_k^- I_l^+)$	$3\cos^2 \theta - 1$
-1	$\sqrt{\dfrac{3}{8}} I_{k,z} I_l^+$	$\sqrt{6} \sin\theta \cos\theta \exp(-i\phi)$
-1	$\sqrt{\dfrac{3}{8}} I_{k,z} I_l^+$	$\sqrt{6} \sin\theta \cos\theta \exp(-i\phi)$
-2	$\sqrt{\dfrac{3}{8}} I_k^+ I_l^+$	$\sqrt{\dfrac{3}{2}} \sin^2 \theta \exp(-2i\phi)$

Tensor Operators in the Rotating Frame and Modified Spherical Harmonics for the Dipole-Dipole Coupling. The calibration has been chosen such that $\dfrac{1}{4\pi} \int_0^{2\pi} \int_0^{\pi} Y_{2,M}(\theta, \phi) Y_{2,-M}(\theta, \phi) d\cos(\theta) d\phi = \dfrac{4}{5}$ independent of M.

$$H_{kl} = b_{kl} \sum_{M=-2}^{2} Y_{2M}(\theta_{kl}, \phi_{kl}) A(I_k . I_l) \tag{1}$$

using the terms given in **Table 1**.

Making the appropriate averaging over all the orientations in a weakly anisotropic phase, the dipolar coupling becomes:

$$D_{kl} = \frac{b_{kl}}{10}\left(A_p \frac{3\cos^2\theta_{kl} - 1}{2} + \frac{3}{4} A_r \sin^2\theta_{kl} \cos 2\phi_{kl}\right) \tag{2}$$

or after setting: $D_{zz} = \dfrac{b_{kl}}{20} A_p$ and $R = A_r/A_p$:

$$\frac{D_{kl}}{D_{zz}} = Y_{20}(\theta, \phi) + \sqrt{\frac{3}{8}} R\left(Y_{22}(\theta, \phi) + Y_{2-2}(\theta, \phi)\right) \tag{3}$$

By expressing this equation in the coordinate system of the molecular frame for a specific alignment medium i and leaving away the kl index for simplicity, a general expression for the dipolar coupling is found:

$$\frac{D_i^{\exp}}{D_{i,zz}} = \sum_{M=-2}^{2} F_{i,M}\left\langle Y_{2,M}\left(\theta^{mol}, \phi^{mol}\right)\right\rangle \tag{4}$$

in which the angular brackets indicated internal motional averaging. The F-matrix is defined in the following way by the Wigner rotation matrices originating from the rotation from the molecular frame into the alignment frame *(11)*:

$$F_{i,M} = D_{M0}^{(2)}(\alpha_i, \beta_i, \gamma_i) + \sqrt{\frac{3}{8}} R_i\left(D_{M2}^{(2)}(\alpha_i, \beta_i, \gamma_i) + D_{M-2}^{(2)}(\alpha_i, \beta_i, \gamma_i)\right) \tag{5}$$

To invert **Eq. 4**, we need to know the F-matrix that contains the order matrices of all the alignment media. We could show *(11)* that using the rigid protein structure we do not obtain directly the matrix $F_{i,M}$ but rather a matrix $\tilde{F}_{i,M}$ that incorporates the average scaling of the dipolar couplings $S^{overall}$. Therefore the F-matrix is reduced compared to the "true" F-matrix $\tilde{F}_{i,M} = S^{overall} F_{i,M}$ by the $S^{overall}$ scaling factor.

$$\frac{D_i^{\exp}}{D_{i,zz}} = \sum_{M=-2}^{2} \tilde{F}_{i,M}\left\langle \tilde{Y}_{2,M}\left(\theta^{mol}, \phi^{mol}\right)\right\rangle \tag{6}$$

In analogy, we obtain by inversion of **Eq. 6**:

$$\left\langle \tilde{Y}_{2,M}\left(\theta^{mol}, \phi^{mol}\right)\right\rangle = \sum_{i} \tilde{F}_{i,M}^{-1} \frac{D_i^{\exp}}{D_{i,zz}} \tag{7}$$

the scaled spherical harmonics that relate to the "true" spherical harmonics by:

$$\left\langle Y_{2,M}\left(\theta^{mol}, \phi^{mol}\right)\right\rangle = \left\langle \tilde{Y}_{2,M}\left(\theta^{mol}, \phi^{mol}\right)\right\rangle \Big/ S^{overall} \tag{8}$$

provided the $\tilde{F}_{i,M}$-matrix contains sufficiently linearly independent eigenvectors.

Equation 4 can be considered as projecting out from the five-dimensional vector formed from the five averaged spherical harmonics (there are only five values despite the complex values of $Y_{2,M}$ because $Y_{2,-M} = Y_{2,M}^*$) a single component along one axis in this five-dimensional space. Under the assumption that in a protein the lengths of equivalent bonds, such as NH bonds, are the same and in the absence of motion the sum of the absolute values over the five spherical harmonics, which is the order parameter $\tilde{S}_{rdc}^2$, should be 1:

$$\tilde{S}_{rdc}^2 = \frac{1}{4} \sum_{M=-2}^{2} \langle \tilde{Y}_{2,M} \rangle \langle \tilde{Y}_{2,-M} \rangle = 1 \tag{9}$$

In such a rigid situation we would also find for the nonscaled rdc derived order parameter S_{rdc}^2:

$$S_{rdc}^2 = \frac{1}{4} \sum_{M=-2}^{2} \langle Y_{2,M} \rangle \langle Y_{2,-M} \rangle \tag{10}$$

and thereby the overall order parameter $S^{overall}$ would be 1. Similarly, the Lipari Szabo order parameters S_{LS}^2 *(12,13)* of all NH vectors would be 1 in this case.

Let us now assume that the NH vectors undergo internal motion, which is expected to be the case in real systems. This will reduce on average the measured dipolar couplings by the factor $S^{overall}$. The inversion of **Eq. 6** then provides order parameters $\tilde{S}_{rdc}^2$ that, when averaged over all amides in the protein, will again be 1. This is true because the average of the motion is absorbed in the reduced $\tilde{F}_{i,M}$. Scaling with $S^{overall}$ will reduce them to S_{rdc}^2 according to:

$$\tilde{S}_{rdc}^2 S^{overall^2} = S_{rdc}^2 \tag{11}$$

Although the value of $S^{overall}$ cannot be independently determined, an upper bound can be obtained because of the fact that $S_{rdc}^2 \leq S_{LS}^2$.

Thus, the orientational distribution of each NH vector translates into reduced projections onto the five axes reflecting a reduced order parameter S_{rdc}^2. Because of the averaging of the dipolar couplings over the whole range of time-scales up to the chemical shift time-scale of approx 10 ms, the orientational distributions on time-scales previously inaccessible by other NMR methods between the rotational tumbling correlation time and approx 50 μs can now be measured. This is summarized in **Fig. 1**.

Once **Eq. 6** is inverted, it is not only S_{rdc} that can be recovered but also four other parameters characterizing other aspects of the orientational distribution. This can be best accomplished by transforming from the molecular frame into the coordinate frame x', y', z' in which the average orientation of the vector is along the newly defined z'-axis, characterized by the two polar angles θ and ϕ. Mathematically, the z'-axis is found by maximizing $\tilde{Y}_{2,0}(\theta', \phi')$, where θ' and ϕ' refer to the polar coordinates in the new frame. Anisotropy of the orientational distribution can be characterized by an amplitude η:

$$\eta = \sqrt{\frac{\sum\limits_{M=-2,2} \langle \tilde{Y}_{2M}(\theta', \phi') \rangle \langle \tilde{Y}_{2-M}(\theta', \phi') \rangle}{4\tilde{S}_{rdc}^2}} \tag{12}$$

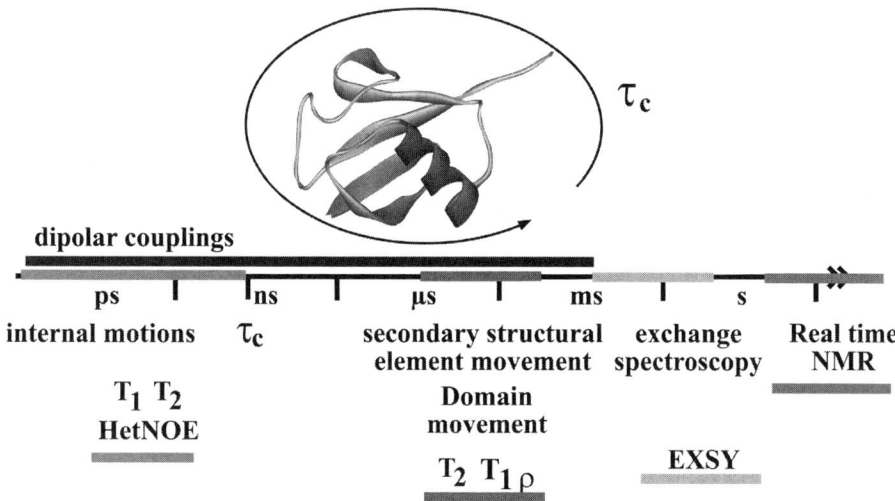

Fig. 1. Time scales that can be detected by various methods in NMR: The fast motions ps to ns are measurable by relaxation of, e.g., ^{15}N, the motion between approximately 50 μs and 10 ms by $T_{1\rho}$ measurement, the slower time scales by exchange spectroscopy and even slower ones by real time NMR. Dipolar couplings cover all time scales from ps to ms and therefore can report also about the "dark" region of time scales.

as well as a further angle $\overline{\phi}'$ that describes the orientation of the anisotropic orientation distribution on the sphere as illustrated in **Fig. 2**:

$$\overline{\phi}' = \frac{1}{2}\arctan\frac{\left\langle \tilde{Y}_{22}(\theta',\phi') \right\rangle - \left\langle \tilde{Y}_{2-2}(\theta',\phi') \right\rangle}{i\left(\left\langle \tilde{Y}_{22}(\theta',\phi') \right\rangle + \left\langle \tilde{Y}_{2-2}(\theta',\phi') \right\rangle\right)}$$

Summarizing these findings, it is possible to extract five parameters: S_{rdc}^2, θ, ϕ η and $\overline{\phi}'$. This allows a model free description of the isotropic and anisotropic motion present in proteins.

2. Materials

In this paragraph, a summary of several methods used to obtain weak alignment for aqueous protein solutions is presented. The weak alignment is induced by steric, electrostatic, or binding interactions between the highly oriented alignment medium and the biomacromolecule. In general, weak binding of the macromolecule to the alignment medium is unfavorable because of the increased apparent correlation time that leads to shortening of the transverse spin relaxation times that affects most severely the upfield shifted components of the ^{15}N doublet.

Five values: <Y$_{2m}$ >

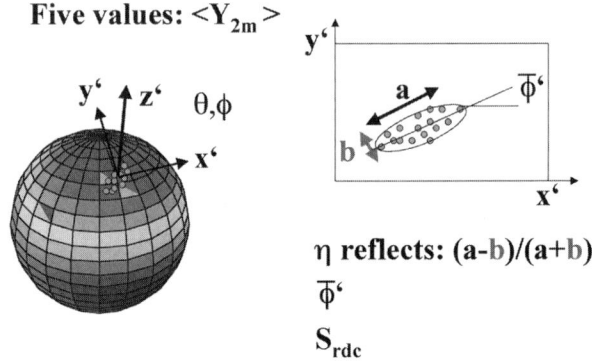

θ,ϕ

η reflects: (a-b)/(a+b)

$\overline{\phi}$'

S$_{rdc}$

Fig. 2. Transformation of the averaged spherical harmonics $\left\langle Y_{2,M}\left(\theta^{mol},\phi^{mol}\right)\right\rangle$ into the coordinate system constructed of z', x', and y'. z' is the axis that points along the average orientation of the vector indicated by the angles θ and ϕ. S_{rdc} has the meaning as explained in the text. η is the amount of anisotropy of the orientational distribution in the x', y' plane and $\overline{\phi}'$ is the orientation of the anisotropy of this orientational distribution in the x', y' plane.

For the study of orientational distributions derived from dipolar couplings, we used rdc data obtained in almost all alignment media that were available at the time. Bicelles can be created from "long-arm" glycolipids (dimyristoylphosphatidylcholine [DMPC], dilauroylphosphatidylcholine [DLPC]) and "short-arm" glycolipids (dihexanoylphosphatidylcholine [DHPC]) *(14–16)* or 3-[(3-cholamidopropyl)dimethylammonio]-2-hydroxy-1-propanesulfonate (CHAPSO) *(1,2,14,17)*. Many different bicelle mixtures for different sample conditions have been proposed—for example, different temperatures *(18)* or different pH values *(19)*. Additives for stabilization and different charging (cetyltrimethylammonium bromide [CTAB], sodium dodecyl sulfate [SDS]) of bicelles have also been introduced *(20,21)*. Ruckert et al. presented new uncharged bicelles *(22)* made of *n*-Alkyl-poly(ethylene glycol)/*n*-alkyl alcohol and glucopone/*n*-hexanol mixtures. The uncharged poly(ethylene glycol)-based systems are insensitive to pH. Most of the interactions are purely from spatial contacts between the bicelles and the biomacromolecule. This alignment medium is also quickly prepared and stable for months.

Other media for alignment are phages *(23–26)* or purple membranes *(27,28)*, which are useful for oligonucleotides or acidic proteins. In these media, the alignment is not only caused by steric interactions, but also by either weak transient binding or large electrostatic interactions of the protein with the aligning medium. This can lead to a severe reduction of the effective transverse relaxation times as compared to the isotropic solution.

Other proposed alignment media are cellulose crystallites *(29)*, surfactant liquid crystals (Helfrich phases) *(30,31)*, and anisotropic aqueous polymeric

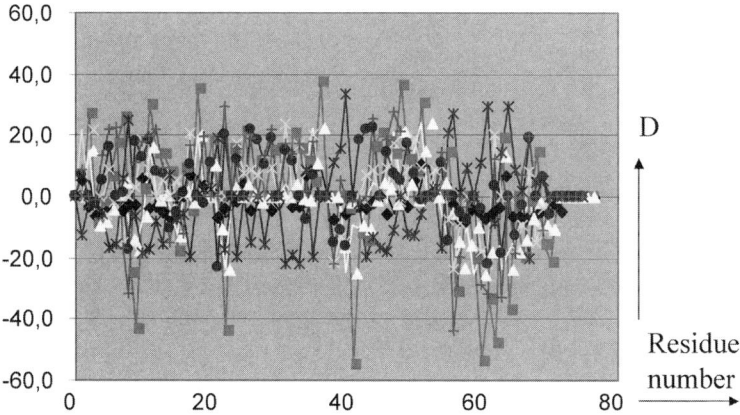

Fig. 3. Distribution of ^{15}N, ^{1}H dipolar couplings of ubiquitin derived from the different alignment media.

gels *(32–34)* that can be used in high-resolution NMR spectroscopy. Inorganic materials *(35)* have also been presented for the alignment of molecules.

3. Methods

Spectra for the measurement of the NH dipolar couplings have been recorded for ubiquitin in isotropic phase and then in anisotropic phase by using S^{3}E-HSQC experiments *(36)* in a total of 11 different alignment media *(37)*. The distribution of the dipolar couplings is depicted in **Fig. 3**, which shows changes of the dipolar couplings between different alignment media. A quantitative measure for the independence of the different media are the singular values of the $F_{i,M}$-matrix that have been back-calculated from the known structure of ubiquitin as well as the dipolar couplings of the individual alignments. The first five are 4.909, 2.260, 0.844, 0.675, 0.594 for all 11 media (DMPC/DHPC*, DMPC/DHPC/CTAB, purple membrane fragments*, CHAPSO/DLPC/SDS*, CHAPSO/DLPC, CHAPSO/DLPC/CTAB 4%, CHAPSO/DLPC/CTAB 5%, polyacrylamide gel, Helfrich phase*, Pf-1 phages*, *n*-dodecyl-penta(ethylene glycol)/*n*-hexanol*), whereas we obtain the singular values: 3.373, 1.898, 0.723, 0.600, 0.546 if we pick only the 6 dipolar coupling sets marked (*) out of the 11 media. The condition number that corresponds to the ratio between the largest and the smallest value is 8.3 for 11 media and 6.2 for only 6 media. The increase of the condition number for the larger number of media is because of the fact that some strongly aligning media have been left out, which decreases the largest eigenvalue more than the smallest eigenvalue when going to 6 media only. For the error propagation, the absolute value of the smallest singular value is the most important number.

The inversion of the $\tilde{F}$ matrix provides, as explained in **Subheading 1**, the scaled spherical harmonics $\tilde{Y}$ for each NH vector. We have shown that the

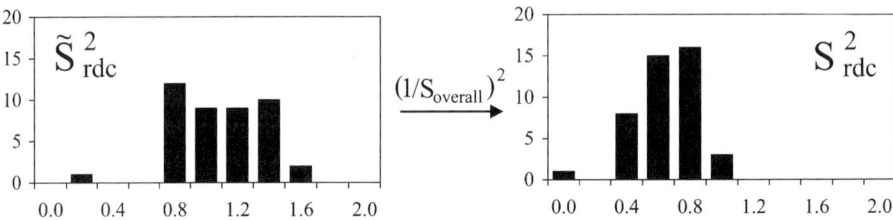

Fig. 4. Determination of the overall scaling factor: $S^{overall}$ from the distribution of $\tilde{S}_{rdc}$. The histogram indicates that the average of $\tilde{S}_{rdc}$ is 1. Rescaling such that the largest order parameters are smaller than 1 yields the maximal values for the unscaled S_{rdc} values. On average S_{rdc} is found to be 0.78, whereas the average S_{LS} amounts to 0.9. (From **ref. 37** with permission.)

average $\tilde{S}_{rdc}^2 = \dfrac{1}{4} \displaystyle\sum_{M=-2}^{2} \left\langle \tilde{Y}_{2,M} \right\rangle \left\langle \tilde{Y}_{2,-M} \right\rangle$ taken over all NH vectors is 1. The value of

$S^{overall}$ can then be determined by the requirement that the largest $\tilde{S}_{rdc}^2$ should not be larger than the Lipari Szabo-derived order parameters. This rescaling leads to an $S^{overall}$ value of 0.78 in the case of ubiquitin. Independent analysis performed by J. R. Tolman has come to the same result recently *(8)*. The histogram of the scaling is shown in **Fig. 4**.

The rescaled order parameters for ubiquitin are shown in **Fig. 5**. It is obvious that the S_{rdc}^2 is considerably smaller in the β-sheet regions as well as in the loops as compared to S_{LS}^2 whereas the helix shows the highest rdc-derived order parameters. These extra motions must occur on time-scales that are slower than the overall rotational tumbling correlation time of 4 ns. Independent support of this extra mobility has recently come from cross-correlated relaxation of adjacent NH vectors. Given a rigid conformation, the order parameter derived from autocorrelated relaxation and cross-correlated relaxation should be the same. However, Bodenhausen et al. *(38)* find that the order parameters for dipole–dipole cross-correlated relaxation of neighboring NH vectors is considerably different from the autocorrelated relaxation. This suggests the presence of conformational transitions on a time-scale slower than the rotational tumbling involving different projection angles *(38)*.

As explained in **Subheading 1**, it is possible to transform the individual spherical harmonics into a new frame in which the anisotropy can be studied (**Fig. 2**). On application of this transformation, we can compare the result for the angle $\overline{\phi}'$ derived from the rdcs to the result from molecular dynamics calculations. The correlation diagram between the experimentally derived angles $\overline{\phi}'$ and the angle $\overline{\phi}'$ derived from the molecular dynamics calculation is shown in **Fig. 6**. There is a fair correlation between these two angles. It should be stressed that

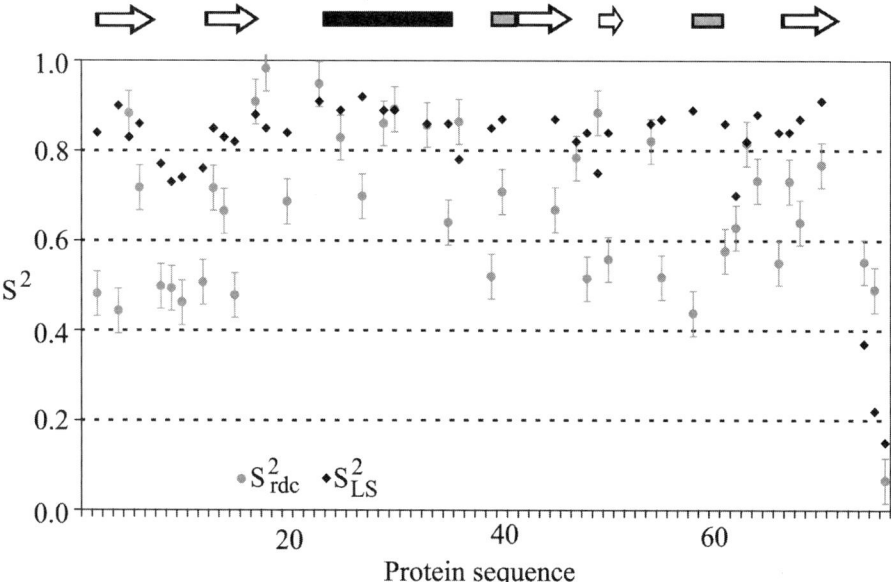

Fig. 5. Comparison of the S_{rdc} values (•) with the relaxation derived values S_{LS} (♦). The rdc derived values are mostly smaller than the Lipari Szabo derived ones since they reflect motion not only up to the rotational tumbling correlation time but also slower motion. (Adapted from **ref. *37*.**)

this good correlation can only be obtained for those NH vectors that are involved in internal hydrogen bonds, whereas those exposed to the water do not reproduce very well *(37)*.

Furthermore, the analysis can be extended to the amino acid residues in the central helix of ubiquitin *(39)*. Here, we find by analyzing the six residues of ubiquitin that have a sufficiently low condition number that their orientational distribution is anisotropic with the angle $\bar{\phi}'$ pointing in the same direction for all analyzed residues as depicted in **Fig. 7**. On average, the data suggest that the amplitude of motion along the more mobile axis is ±21°, whereas the amplitude drops to ±12° along the orthogonal direction. The significance of the anisotropy is checked in the following way: When exchanging the amplitudes between the two orthogonal directions x' and y', the dipolar couplings are less well reproduced as the histogram in **Fig. 8** puts into evidence. There, the 66 experimental couplings are compared with the back-calculated couplings assuming the fitted anisotropy, no anisotropy, and the inverted anisotropy. The agreement with the fitted anisotropy is significantly better than using the inverted anisotropy.

The uniform anisotropy of the orientational distributions of the NH vectors along the helix might reflect rigid body motion of the helix as a whole. It should

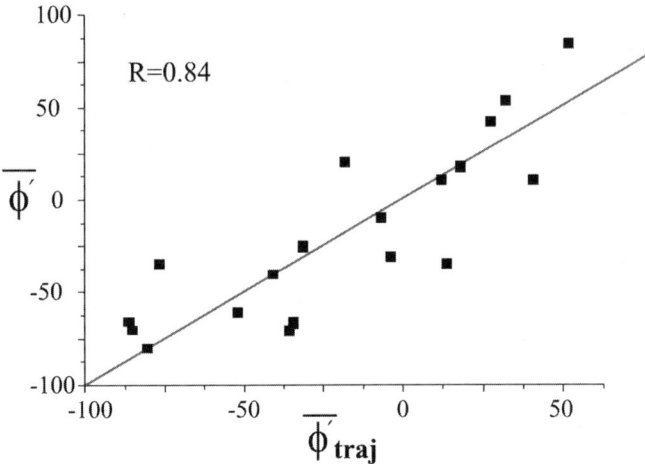

Fig. 6. Correlation of the directions of the anisotropy $\overline{\phi}'$ derived experimentally with the ones derived from molecular dynamics. There is a good correlation of these orientations for the secondary structure elements while this is not the case if NH vectors are included that are not involved in internal hydrogen bonds. (From **ref. 37** with permission.)

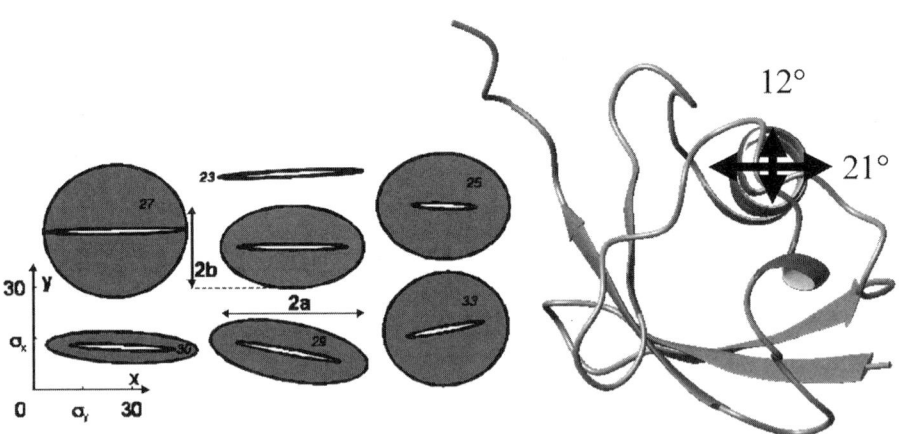

Fig. 7. Anisotropic motions of the NH vectors of the ubiquitin helix. The amplitude of the motion of 6 vectors of the helix that have condition numbers smaller than 10 are uniform. Averaging over the anisotropies the large amplitude motion is found to be ± 21°, while the small amplitude orthogonal is ±12°. The ellipses for the individual NH vectors are shown including the propagated statistical error of the dipolar couplings. The inner ellipse marks one extreme and the outer ellipse the other extreme. (From **ref. 39** with permission.)

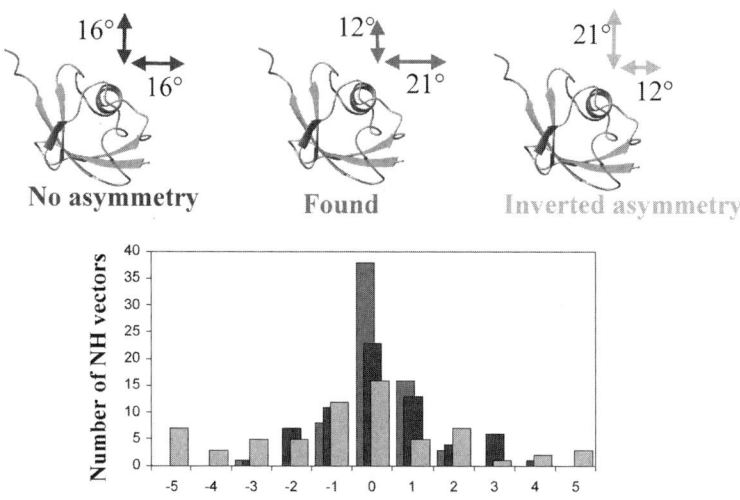

Fig. 8. Comparison of the back calculated ^{15}N, ^1H dipolar couplings from the 6 helix residues of ubiquitin in 11 media assuming the fitted anisotropy (dark gray), no anisotropy (black) and opposite anisotropy (light gray). From the observed deviations of the fitted vs back-calculated values, the fitted anisotropy yields significantly fewer violations of the dipolar couplings.

be stressed that the dipolar couplings give no experimental proof of such a proposition because the data can be equally well explained if one assumes that the vectors move independently from each other. Nonetheless, molecular modeling indicates that a correlated motion of the helix residues is energetically feasible and that the anisotropy observed experimentally follows the energetics of ubiquitin on distortion of the helix (**Fig. 9**).

4. Conclusions

The use of dipolar couplings for the analysis of motion relies on measurements in five different media that yield linearly independent alignment tensors. There are two major assumptions that enter the analysis:

The structure of the molecule must not be changed by the different alignment media. In this case, one would not see orientational distributions of the molecule but rather the softness of certain regions in the protein with respect to others. Hus et al. have investigated this behavior and have found at least for the secondary structure elements that there is little change of the structure in the different alignment media *(40)*.

The second assumption is that the internal motion of the molecule is not correlated with the motional modulation of the alignment tensor. This means that

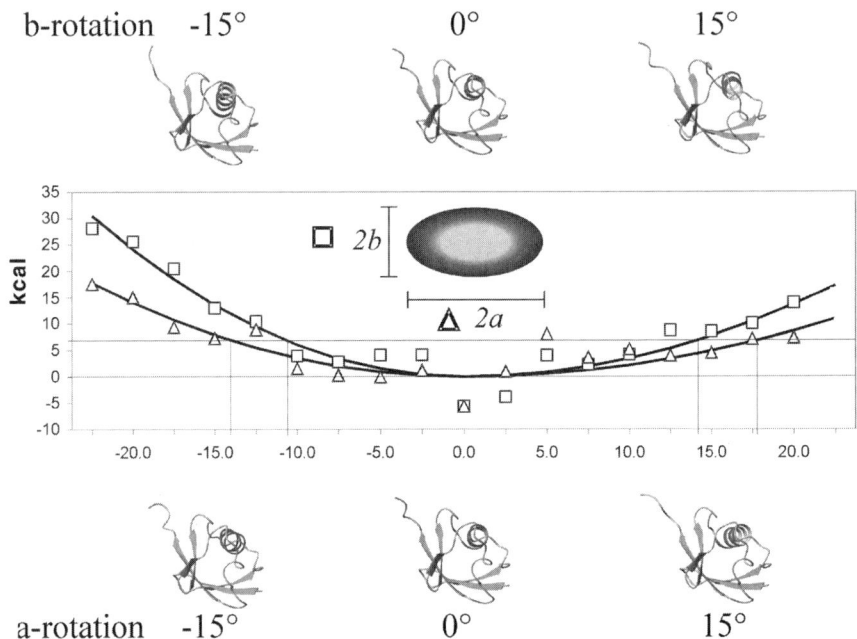

Fig. 9. Molecular modeling of rotations of the whole helix of ubiquitin. Rotation about the axis by the number of degrees indicated at the structures that relate to the large amplitude (bottom structures, triangles) yields a smaller energy increase upon rotation about a given angle, while rotation about the axis with the small amplitude (top structures, squares) yields a steeper increase of the energy indicating that correlated motion of the helix is conceivable. It should be mentioned that the structures with the rotated helix are energy relaxed by a short energy minimization after rotation of the helix and do not show NOE violations by more than 0.2 Å. (Adapted from **ref. *39*** with permis-

the same alignment tensor can be assumed irrespective of the orientation of an individual NH vector. Although this is highly probable for uncorrelated motion, it may not hold for correlated motions of whole secondary structure elements such as the correlated helical motion that could occur based on the uniform anisotropies of the NH vectors in the helix of ubiquitin.

Similar assumptions are being made by the Direct Interpretation of Dipolar Coupling (DIDC) approach *(7)* pursued by J. R. Tolman who uses at least five alignment media to independently determine the orientation of the NH vectors in a protein and assesses the mobility of each NH vector at the same time. Similar to the results obtained by us, the RDC-derived order parameter is lower on average than the one derived from ^{15}N relaxation data. However, by contrast to the findings described in this review, Tolman finds a larger decrease in the helix than we do and less in the β-sheet regions for ubiquitin *(8,9)*. Thus, some details of the dynamics are still unresolved and need further investigation.

Therefore, it is desirable to obtain additional alignments, including alignments induced by paramagnetic tags. Such research is going on in our laboratory *(41)* as well as in others *(42)* and will further augment the analysis of dynamics detected from dipolar couplings in the future.

References

1. Tolman, J. R., Flanagan, J. M., Kennedy, M. A., and Prestegard, J. H. (1995) Nuclear magnetic dipole interactions in field-oriented proteins—information for structure determination in solution. *Proc. Natl. Acad. Sci. USA* **92**, 9279–9283.
2. Tjandra, N. and Bax, A. (1997) Direct measurement of distances and angles in biomolecules by NMR in a dilute liquid crystalline medium. *Science* **278**, 1111–1114.
3. Reif, B., Hennig, M., and Griesinger, C. (1997) Direct measurement of angles between bond vectors in high-resolution NMR. *Science* **276**, 1230–1233.
4. Griesinger, C., Hennig, M., Marino, J. P., Reif, B., Richter, C., and Schwalbe, H. (1999) Methods for the determination of torsion angle restraints in biomacromolecules. In *Biological Magnetic Resonance, vol. 16* (Krishna, N. R. and Berliner, L. J., eds.), Kluwer Academic/Plenum Press, New York, pp. 259–367.
5. Reif, B., Diener, A., Hennig, M., Maurer, M., and Griesinger, C. (2000) Cross-correlated relaxation for the measurement of angles between tensorial interactions. *J. Magn. Reson.* **143**, 45–68.
6. Schwalbe, H., Carlomagno, T., Hennig, M., et al. (2001) Cross-correlated relaxation for measurement of angles between tensorial interactions. In *Nuclear Magnetic Resonance of Biological Macromolecules, Pt A, vol. 338*, pp. 35–81.
7. Tolman, J. R. (2002) A novel approach to the retrieval of structural and dynamic information from residual dipolar couplings using several oriented media in biomolecular NMR spectroscopy. *J. Am. Chem. Soc.* **124**, 12,020–12,030.
8. Briggman, K. B. and Tolman, J. R. (2003) De novo determination of bond orientations and order parameters from residual dipolar couplings with high accuracy. *J. Am. Chem. Soc.* **125**, 10,164, 10,165.
9. Tolman, J. R. and Al-Hashimi, J. M. (2003) NMR studies of biomolecular dynamics and structural plasticity using residual dipolar couplings. In *Annual Reports in NMR Spectroscopy, vol. 49* (Webb, G. A., ed.), Elsevier-Science, London, UK.
10. Griesinger, C., Meiler, J., and Peti, W. (2003) Angular restraints from residual dipolar couplings for structure refinement. In *Protein NMR for the Millennium* (Krishna, N. R. and Berliner, L. J., eds.), Kluwer Academic/Plenum, New York.
11. Meiler, J., Prompers, J. J., Peti, W., Griesinger, C., and Brüschweiler, R. (2001) Model-free approach to the dynamic interpretation of residual dipolar couplings in globular proteins. *J. Am. Chem. Soc.* **123**, 6098–6107.
12. Lipari, G. and Szabo, A. (1989a) Model-free approach to the interpretation of nuclear magnetic resonance relaxation in macromolecules. 1. Theory and range of validity. *J. Am. Chem. Soc.* **104**, 4546–4559.
13. Lipari, G. and Szabo, A. (1989b) Model-free approach to the interpretation of nuclear magnetic resonance relaxation in macromolecules: 2, analysis of experimental results. *J. Am. Chem. Soc.* **104**, 4559–4570.

14. Sanders, C. R. and Schwonek, J. P. (1992) Characterization of magnetically orientable bilayers in mixtures of dihexanoylphosphatidylcholine and dimyristoylphosphatidylcholine by solid-state NMR. *Biochemistry* **31,** 8898–8905.

15. Ottiger, M. and Bax, A. (1998) Characterization of magnetically oriented phospholipid micelles for measurement of dipolar couplings in macromolecules. *J. Biomol. NMR* **12,** 361–372.

16. Sanders, C. R., Hare, B. J., Howard, K. P., and Prestegard, J. H. (1994) Magnetically-oriented phospholipid micelles as a tool for the study of membrane-associated molecules. *Prog. NMR Spec.* **26,** 421–444.

17. Wang, H., Eberstadt, M., Olejniczak, E. T., Meadows, R. P., and Fesik, S. W. (1998) A liquid crystalline medium for measuring residual dipolar couplings over a wide range of temperatures. *J. Biomol. NMR* **12,** 443–446.

18. Cavagnero, S., Dyson, H. J., and Wright, P. E. (1999) Improved low pH bicelle system for orienting macromolecules over a wide temperature range. *J. Biomol. NMR* **13,** 387–391.

19. Ottiger, M. and Bax, A. (1999) Bicelle-based liquid crystals for NMR-measurement of dipolar couplings at acidic and basic pH values. *J. Biomol. NMR* **13,** 187–191.

20. Losonczi, J. A., and Prestegard, J. H. (1998) Improved dilute bicelle solutions for high-resolution NMR of biological macromolecules. *J. Biomol. NMR* **12,** 447–451.

21. Zweckstetter, M. and Bax, A. (2000) Prediction of sterically induced alignment in a dilute liquid crystalline phase: aid to protein structure determination by NMR. *J. Am. Chem. Soc.* **122,** 3791, 3792.

22. Ruckert, M. and Otting, G. (2000) Alignment of biological macromolecules in novel nonionic liquid crystalline media for NMR experiments. *J. Am. Chem. Soc.* **122,** 7793–7797.

23. Clore, G. M., Starich, M. R., and Gronenborn, A. M. (1998) Measurement of residual dipolar couplings of macromolecules aligned in the nematic phase of a colloidal suspension of rod-shaped viruses. *J. Am. Chem. Soc.* **120,** 10,571–10,572.

24. Hansen, M. R., Mueller, L., and Pardi, A. (1998) Tunable alignment of macromolecules by filamentous phage yields dipolar coupling interactions. *Nat. Struct. Biol.* **5,** 1065–1074.

25. Hansen, M. R., Rance, M., and Pardi, A. (1998) Observation of long-range H-1–H-1 distances in solution by dipolar coupling interactions. *J. Am. Chem. Soc.* **120,** 11,210–11,211.

26. Ojennus, D. D., Mitton-Fry, R. M., and Wuttke, D. S. (1999) Induced alignment and measurement of dipolar couplings of an SH2 domain through direct binding with filamentous phage. *J. Biomol. NMR* **14,** 175–179.

27. Koenig, B. W., Hu, J. S., Ottiger, M., Bose, S., Hendler, R. W., and Bax, A. (1999) NMR measurement of dipolar couplings in proteins aligned by transient binding to purple membrane fragments. *J. Am. Chem. Soc.* **121,** 1385, 1386.

28. Sass, J., Cordier, F., Hoffmann, A., Cousin, A., Omichinski, J. G., Lowen, H., et al. (1999) Purple membrane induced alignment of biological macromolecules in the magnetic field. *J. Am. Chem. Soc.* **121,** 2047–2055.

29. Fleming, K., Gray, D., Prasannan, S., and Matthews, S. (2000) Cellulose crystallites: a new and robust liquid crystalline medium for the measurement of residual dipolar couplings. *J. Am. Chem. Soc.* **122,** 5224, 5225.

30. Prosser, R. S., Losonczi, J. A., and Shiyanovskaya, I. V. (1998) Use of a novel aqueous liquid crystalline medium for high-resolution NMR of macromolecules in solution. *J. Am. Chem. Soc.* **120,** 11,010–11,011.

31. Barrientos, L. G., Dolan, C., and Gronenborn, A. M. (2000) Characterization of surfactant liquid crystal phases suitable for molecular alignment and measurement of dipolar couplings. *J. Biomol. NMR* **16,** 329–337.

32. Tycko, R., Blanco, F. J., and Ishii, Y. (2000) Alignment of biopolymers in strained gels: a new way to create detectable dipole–dipole couplings in high-resolution biomolecular NMR. *J. Am. Chem. Soc.* **122,** 9340, 9341.

33. Ishii, Y., Markus, M. A., and Tycko, R. (2001) Controlling residual dipolar couplings in high-resolution NMR of proteins by strain induced alignment in a gel. *J. Biomol. NMR* **21,** 141–151.

34. Sass, H. J., Musco, G., Stahl, S. J., Wingfield, P. T., and Grzesiek, S. (2000) Solution NMR of proteins within polyacrylamide gels: diffusional properties and residual alignment by mechanical stress or embedding of oriented purple membranes. *J. Biomol. NMR* **18,** 303–309.

35. Desvaux, H., Gabriel, J. C. P., Berthault, P., and Camerel, F. (2001) First use of a mineral liquid crystal for measurement of residual dipolar couplings of a nonlabeled biomolecule. *Angew. Chem. Int. Ed. Engl.* **40,** 373–376.

36. Meissner, A., Duus, J. O., and Sorensen, O. W. (1997) Integration of spin-state-selective excitation into 2D NMR correlation experiments with heteronuclear ZQ/2Q pi rotations for (1)J(XH)-resolved E.COSY-type measurement of heteronuclear coupling constants in proteins. *J. Biomol. NMR* **10,** 89–94.

37. Peti, W., Meiler, J., Bruschweiler, R., and Griesinger, C. (2002) Model-free analysis of protein backbone motion from residual dipolar couplings. *J. Am. Chem. Soc.* **124,** 5822–5833.

38. Pelupessy, P., Ravindranathan, S., and Bodenhausen, G. (2003) Correlated motions of successive amide N-H bonds in proteins. *J. Biomol. NMR* **25,** 265–280.

39. Meiler, J., Peti, W., and Griesinger, C. (2003) Dipolar couplings in multiple alignments suggest alpha helical motion in ubiquitin. *J. Am. Chem. Soc.* **125,** 8072, 8073.

40. Hus, J. C., Peti, W., Griesinger, C., and Bruschweiler, R. (2003) Self-consistency analysis of dipolar couplings in multiple alignments of ubiquitin. *J. Am. Chem. Soc.* **125,** 5596, 5597.

41. Ikegami, T., Verdier, L., Sakhaii, P., Grimme, S., Pescatore, B., Saxena, K., et al. (2004) Novel techniques for weak alignment of proteins in solution using a chemical tags coordinating lanthanide ions. *J. Biomol. NMR,* **29,** 339–349.

42. Gaponenko, V., Altieri, A. S., Li, J., and Byrd, R. A. (2002) Breaking symmetry in the structure determination of (large) symmetric protein dimers. *J. Biomol. NMR* **24,** 143–148.

9

Characterizing Domain Interfaces by NMR

Luke M. Rooney, Sachchidanand, and Jörn M. Werner

Summary

The combination of chemical shift, residual dipolar coupling, and backbone relaxation data can be used to characterize the nature of a domain interface in a multidomain protein. Comparison of the parameters obtained from isolated domains and domain pairs provides insight into the composition of the interface as well as into interdomain dynamics. The interface between the 13th and 14th F_3 module from fibronectin is used as an example.

Key Words: Fibronectin; domain; module; interface; dynamics; chemical shift; residual dipolar coupling; ^{15}N relaxation.

1. Introduction

A large proportion of proteins consist of multiple independently folding domains *(1)*. Dissection of these proteins into their constitutive domains has been a successful strategy for understanding their structures and functions *(2)*. From these studies, domains emerge as versatile scaffolds whose surfaces can be adapted to multiple functions. At the same time, it has been revealed that interaction surfaces often extend over more than a single domain and that the nature of the interdomain interface plays an important role in regulation of protein function. Thus, the analysis of the nature of the interfaces in multiple domain proteins or protein fragments leads to a better understanding of protein structure, dynamics and function. High-resolution nuclear magnetic resonance (NMR) is uniquely suited to study the nature of multidomain fragments in the solution state by combining the use of chemical shift, ^1H–^1H nuclear Overhauser effects (NOEs), heteronuclear relaxation, and residual dipolar couplings (RDCs). The strategy we propose relies on the comparison of data collected from individual domains in isolation and as part of a multidomain construct.

From: *Methods in Molecular Biology, vol. 278: Protein NMR Techniques*
Edited by: A. K. Downing © Humana Press Inc., Totowa, NJ

Chemical shifts are highly sensitive to their local environment, a fact that is generally used in identifying binding sites in protein ligand interactions (*see* Chapter 14). The technique can also be applied to characterize domain–domain interfaces. The chemical shifts of a set of nuclei are measured in the isolated domains and as part of a multidomain construct under otherwise identical conditions. The differences in chemical shifts are then mapped onto the structure. Regions of the domains that interact are expected to show significant chemical shift changes, whereas the other regions will be less affected. This approach assumes that the structures of the individual domains in the domain pair are similar to those in the isolated domains. This assumption can be validated by using a range of NMR parameters, such as the NOE patterns and intensities from the assignment of the different constructs, ^{13}C secondary chemical shift data, and RDCs. The identification of an interdomain interface by chemical shift perturbation, however, is usually not sufficient to obtain an average interdomain orientation. For well-ordered domain–domain interfaces, the observation of interdomain ^1H–^1H NOEs may determine an interdomain orientation. However, this may be biased toward the conformation of closest contact resulting from the strongly nonlinear distance dependence of the ^1H–^1H NOE.

The development of techniques for measuring dipolar couplings of weakly aligned molecules provides a set of structural parameters that is highly complimentary to the local information contained in NOEs and J-couplings (*see* Chapter 7). The angular dependence of a residual dipolar coupling between two directly coupled nuclei, such as in an NH bond, is given by:

$$D_{NH}(\theta, \phi) = D_{max}^{NH}\left\{(3\cos^2\theta - 1)A_a + 3/4A_r(\sin^2\theta\cos 2\phi)\right\} \qquad (1)$$

where θ is the angle between the NH bond vector and A_{zz}, ϕ is the angle between the projection of the NH bond vector on the x–y plane, and A_{xx} in which A_{xx}, A_{yy}, A_{zz} are the principal components of the alignment tensor. $A_a = 3/2A_{zz}$ is the axial and the $A_r = (A_{xx}-A_{yy})$ rhombic component of the alignment tensor. $D_{max}^{NH} = -(\mu_0 h/16\pi^3)\gamma_N\gamma_H[r_{NH}^{-3}]$ with μ_0 the permeability of free space, γ_N and γ_H the gyromagnetic ratios of ^{15}N and ^1H, and $[r_{NH}^{-3}]$ the vibrationally averaged inverse cube of the distance between ^{15}N and ^1H.

The interdomain orientation of a domain pair is derived from the analysis of the alignment tensors of subsets of the molecule corresponding to the whole molecule and individual domains (*3*). In the absence of coordinates for the individual domains, differences in the magnitudes of the principal values of the alignment tensors (e.g., A_a and A_r) of individual domains in a pair can be used to identify interdomain flexibility (*4*). In the presence of atomic coordinates, average interdomain orientations can be deduced from the orientations of the alignment tensors of the individual domains. It has been proposed that quantitative information about the extent of interdomain averaging can be obtained from the determination of multiple alignment tensors in different liquid crystalline media *(5)*.

Heteronuclear relaxation rates not only monitor internal dynamics but also depend on the overall diffusion tensor of the molecule *(6,7)*. The spectral density of an asymmetric diffusion tensor with the principal moments D_{xx}, D_{yy}, and D_{zz} and internal dynamics using the model-free approach is given by:

$$J(\omega) = S^2 \sum_{i=1}^{5} \frac{A_i \tau_i}{1 + (\omega \tau_i)^2} + (1 - S^2) \frac{\tau_c}{1 + (\omega \tau_c)^2} \qquad (2)$$

where $A_1 = 3y^2z^2$, $A_2 = 3x^2z^2$, $A_3 = 3x^2y^2$, $A_4 = d + e$, $A_5 = d - e$,

with $d = \frac{1}{4}\{3(x^4 + y^4 + z^4) - 1\}$

and $e = \frac{1}{12}\{\delta_x(3x^4 + 6y^2z^2 - 1) + \delta_y(3y^4 + 6x^2z^2 - 1) + \delta_z(3z^4 + 6y^2x^2 - 1)\}$,

$\delta_i = (D_i - D) / \sqrt{D^2 - L^2}, D = \frac{1}{3}(D_{xx} + D_{yy} + D_{zz})$,

$L^2 = \frac{1}{3}(D_{xx}D_{yy} + D_{xx}D_{zz} + D_{yy}D_{zz})$;

and the correlation times are $\tau_1 = (4D_{xx} + D_{yy} + D_{zz})^{-1}, \tau_2 = (4D_{yy} + D_{xx} + D_{zz})^{-1}$,

$\tau_3 = (4D_{zz} + D_{xx} + D_{yy})^{-1}, \tau_4 = [6(D + \sqrt{D^2 - L^2})]^{-1}, \tau_5 = [6(D + \sqrt{D^2 - L^2})]^{-1}$,

$\tau_c^{-1} = 6D + \tau_e^{-1}$. The direction cosines *x, y, z* are defined with respect to the principal axis of the diffusion tensor. For an axially symmetric diffusion tensor with the parallel component $D_{||} = D_{zz}$ and perpendicular component $D_\perp = \frac{1}{2}(D_{xx} + D_{yy})$ the equations simplify to $A_1 = (1.5 \cos^2 \alpha - 0.5)^2$, $A_2 = 3 \sin^2 \alpha \cos^2 \alpha$, $A_3 = 0.75 \sin^4 \alpha$, with α being the angle of the N-H bond vector with $D_{||}$ and the correlation times $\tau_1 = (6D_\perp)^{-1}$, $\tau_2 = (D_{||} + 5D_\perp)^{-1}$, $\tau_3 = (4D_{||} + 2D_\perp)^{-1}$ and $\tau_c^{-1} = 6D + \tau_e^{-1}$; and S^2 is the generalized order parameter. For a spherical diffusion tensor $(D_{xx} = D_{yy} = D_{zz})$ the first term in the equation *(2)* reduces to a single (isotropic) correlation time, $\tau_m = (6D)^{-1}$ with $\sum_i A_i = 1$. Fast motions are characterized by an order parameter S^2 and a correlation time τ_e.

Several methods have been proposed to obtain interdomain orientations and flexibility from relaxation data *(8–10)*. In this case, the strategy is to measure heteronuclear relaxation time constants of the isolated domains and the individual domains in the domain pair *(11,12)*. A set of residues is identified that is representative of the overall motions of the molecules. These residues are then used to determine the diffusion tensors of the various constructs under identical conditions. Comparison of the diffusion tensors of the isolated domains and the diffusion tensors of the individual domains in the domain pair allows qualitative assessment of interdomain motion. More quantitative information can be obtained by analyzing data acquired at multiple magnetic field strengths *(13,14)*.

This chapter, uses ^{13}F3, ^{14}F3, and ^{13}F3^{14}F3 module pairs from fibronectin to illustrate the method. This fragment contains surface-exposed heparin binding sites *(15)* as well as a proposed binding site for $\alpha_4\beta_1$ integrin *(16)*. The X-ray structure of a four-module fragment spanning ^{12}F3–^{15}F3 showed that the ^{12}F3–^{14}F3 modules form an elongated structure and that the proposed integrin binding site is buried in the interface between the ^{13}F3 and ^{14}F3 modules *(17)*. According to the X-ray data, a large conformational change would be required to make the integrin binding site available. The NMR results suggest that in the solution state the composition of the interdomain interface is similar to that in the crystal state, but the average interdomain orientation is twisted by about 25° with respect to the crystal structure, and there is significant flexibility between the domains. According to these data, integrin binding could be achieved by stabilizing those possible interdomain conformations in which the binding site is accessible.

2. Materials

1. Purified samples of ^{15}N isotopically enriched protein dissolved in 95%/5% $H_2O/^2D_2O$. In our example, the chemical shift comparison and relaxation analysis was based on spectra recorded on samples of ^{13}F3, ^{14}F3, and ^{13}F3^{14}F3 at pH 4.5 in 150 mM NaCl/50 mM acetate buffer (*see* **Notes 1** and **2**).

2. Suitable alignment media (*see* Chapter 6). A stock solution of 15% (w/v) 1,2-*O*-ditridecanyl-*sn*-glycero-3-phosphocholine (DTDPC) and 1,2-*O*-dihexyl-*sn*-glycero-3-phoshocholine (DHOPC) with a molar ratio of 3:1 in sodium phosphate buffer was prepared as follows: 0.1221 g of DTDPC and 0.0279 g DHOPC were weighed out and 800 µL of H_2O, 100 µL of D_2O, and 100 µL of 1.5 M NaCl/0.5 M phosphate buffer was added. The resulting bicelle solution was then mixed by agitation and left at room temperature for a minimum of 24 h until the solution became clear at which point it was cooled below 5°C and the pH was adjusted using small additions of 1 M NaOH and 1 M HCl. Cetyltrimethylammonium bromide (CTAB) was added to the samples to final molar ratio of 30:1 of DTDPC to CTAB. A 300 µL, 3% bicelle sample in 150 mM NaCl/50 mM phosphate buffer, pH 7.0, was produced for measurement of the aligned state of the protein (*see* **Note 2**).

3. Modern high-field NMR spectrometers (typically ≥500 MHz) equipped with tri-axial triple-resonance probe heads were used for the acquisition of two- and three-dimensional (2D and 3D) [^{1}H]–^{15}N correlated spectra.

4. A high-speed computer, with suitable operating system, together with C and FOR-TRAN compilers is required to run various software packages. These include

NMR spectral analysis software, such as Felix 2.3 (MSI, San Diego, CA) and NMRview *(18)*; relaxation data analysis software, such as TENSOR *(19)*, HYDRONMR for hydrodynamic bead modeling *(20)*, and RDC analysis software such as SSIA *(21)* and MODULE *(22)*; and molecular visualization software such as Molmol *(23)*.

3. Domain Assembly by NMR

3.1. Chemical Shift Perturbation and Interface Identification

1. Backbone assignment of ^{13}F3, ^{14}F3, and ^{13}F3^{14}F3 was carried out using standard procedures based on 3D-^{15}N edited ^{1}H–^{1}H nuclear Overhauser effect and 3D-^{15}N edited ^{1}H–^{1}H total correlation spectra *(24)*. The available assignments of the isolated domains greatly simplified assigning the module pair.

2. High-resolution ^{1}H–^{15}N heteronuclear single-quantum coherence spectra were recorded under identical conditions (*see* **Note 3**). The chemical shift difference, $\Delta\delta$, between the ^{13}F3 and ^{14}F3 in isolation and as part of the ^{13}F3^{14}F3 construct was calculated as

$$\Delta\delta = \left|\Delta\delta(^1H)\right| + \frac{1}{a}\left|\Delta\delta(^{15}N)\right|$$

with $a = 8$ for glycine and $a = 6$ otherwise (*see* **Fig. 1**). A value above 0.05 ppm was considered significant, clearly identifying residues E15, T16, T17, N40, L61, Q62, G64, D66, and Y67 in the ^{13}F3, and I91, D92, R115, A116, T119, Q168, and K169 in the ^{14}F3. (I91 should be discounted as the alteration in the preceding amino acid between the isolated ^{14}F3 and, as part of the ^{13}F3^{14}F3 module pair, is likely to contribute to the observed shift perturbation.) These residues are at the interface between ^{13}F3 and ^{14}F3 in the X-ray structure indicating that the composition of the interface in solution is similar to that in the crystal.

3.2. RDCs and Interdomain Orientation

1. ^{1}H–^{15}N dipolar couplings were measured by 2D experiments based on spin state selective ^{1}H–^{15}N correlation spectra resulting in two spectra encoding the downfield and upfield components separated by the apparent ^{15}N–^{1}H J-coupling ($^1J^{app}$) *(25)*. The downfield and upfield components were measured in an isotropic solution and in the bicelle medium with acquisition times of 51.2 ms (^{1}H) and 105 ms (^{15}N).

2. The data were processed with mild resolution enhancement in both dimensions and zero-filled to yield a final resolution of 9.8 and 0.24 Hz per point in the proton and nitrogen dimensions, respectively (*see* **Note 4**). The RDC D_{NH} was obtained as follows: $D_{NH} = {}^1J^{app}(aligned\ state) - {}^1J^{app}(isotropic\ state)$. The error on each measurement is obtained from the ratios of the line widths (*LW*) and the signal-to-noise ratios (*SN*) in the respective spectra: $error = LW/SN$ *(26)*. RDCs of 36 and 32 residues in ^{13}F3 and ^{14}F3, respectively, were located in well-structured regions of the molecule and did not show relaxation rates indicative of significant fast or slow motion.

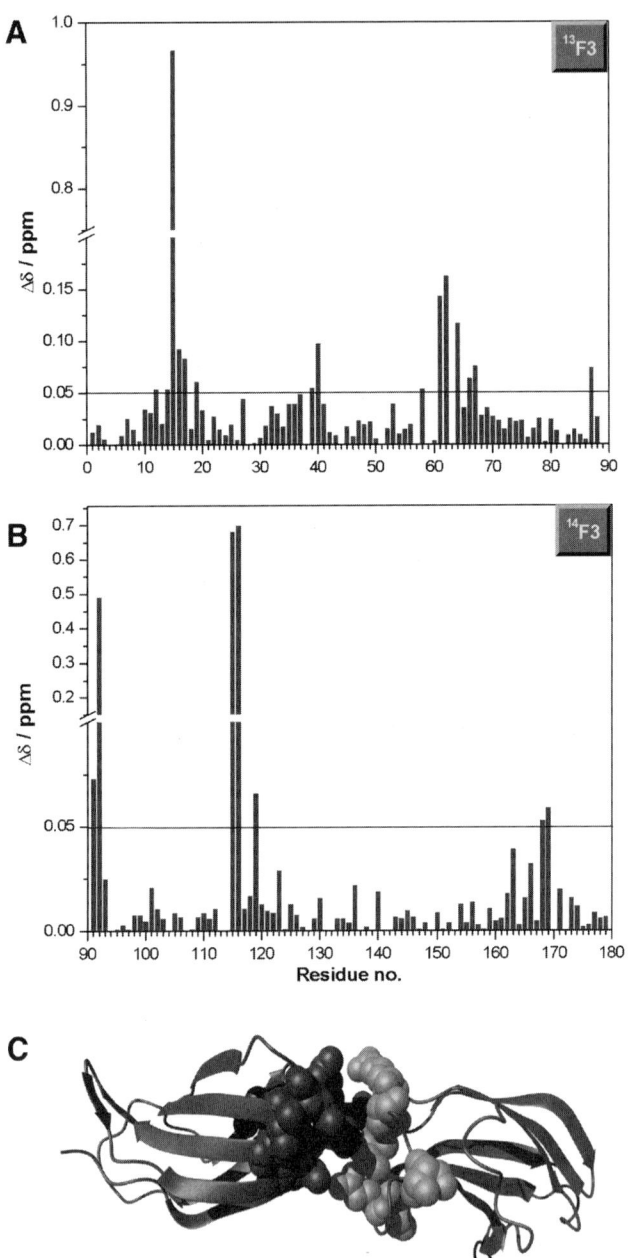

Fig. 1. Identification of the interdomain interface by chemical shift comparison. (**A**) Difference of NH and ^{15}N chemical shifts between the isolated ^{13}F3 and ^{13}F3 as part of the ^{13}F3^{14}F3 module pair. (**B**) same as (**A**) for ^{14}F3. (**C**) Ribbon diagram of the crystal structure of ^{13}F3^{14}F3. Residues identified in (**A**) and (**B**) are shown as CPK. The figure was prepared using Molmol *(23)*.

Table 1
Alignment Tensors for ^{13}F3, ^{14}F3 and ^{13}F3^{14}F3 at pH 7.0

	A_a (10^{-4})	A_r (10^{-4})	α (deg)	β (deg)	γ (deg)	$\chi^2_{per\ residue}$	R
^{13}F3	15.1 ± 0.3	1.7 ± 0.3	15.4 ± 0.4	50.7 ± 0.5	-55.7 ± 0.6	2.1	0.20
^{14}F3	15.7 ± 0.3	0.9 ± 0.3	-11.9 ± 1.0	46.0 ± 0.5	-64.4 ± 0.7	2.1	0.19
^{13}F3^{14}F3	15.3 ± 0.2	0.9 ± 0.2	10.0 ± 0.7	48.7 ± 0.5	-58.9 ± 0.6	5.2	0.23

3. The alignment tensors were determined by fitting the RDCs of the residues in well-structured regions that were not affected by significant slow or fast motion to the X-ray structure (*see* **Note 5**). Residues of ^{13}F3 and ^{14}F3 were used to determine alignment tensors for each module individually and the same sets of residues from both modules were used to obtain a single alignment tensor of the module pair. This was done with the program MODULE *(22)*. The results are shown in **Fig. 2A** and **Table 1**. Although the axial components of the alignment tensors are the same for both modules, the rhombic components show a small difference. Similarly the angle, β, between the z-axes of the alignment and molecular frames is similar in both modules, but the rotations around the other axes are somewhat larger. As a result, the fits of the individual tensors are better than the fit to the whole structure, suggesting that the average interdomain orientation in solution deviates from the crystal structure.

4. The quality of the data can be assessed using an *R*-factor *(27)* defined as follows:

$$R = \sqrt{\sum_i \left(D_i^{obs} - D_i^{calc} \right)^2 \Big/ 2 \sum_i (D_i^{obs})^2} \tag{3}$$

where for each amino acid i D_i^{obs} is the observed RDC and D_i^{calc} is the calculated value using the X-ray structure and the respective alignment tensor. It can be seen in **Table 1** that the R factors are lower when the domains are fitted individually, indicating that the average domain orientation in solution is not quite in agreement with the crystal structure.

5. The individual domains were rotated in their common alignment frame using the program MODULE *(22)*. Residue I91 in the middle of the linker was chosen as the center of rotation, and the solution that was compatible with the covalent bonding structure was retained. The difference between the X-ray and solution structure is shown in **Fig. 2B**. The interdomain reorientation in solution deviates by about $4.7 \pm 0.7°$ with respect to the long axes of the molecules (tilt angle) and by about $25 \pm 1.4°$ in the plane perpendicular to the z-axis (approximately the twist angle).

3.3. Relaxation and Interdomain Dynamics

1. Ideally, all of the data are collected using one sample in one experimental session. The observation of a significant number of T_1 and T_2 values outside the boundary

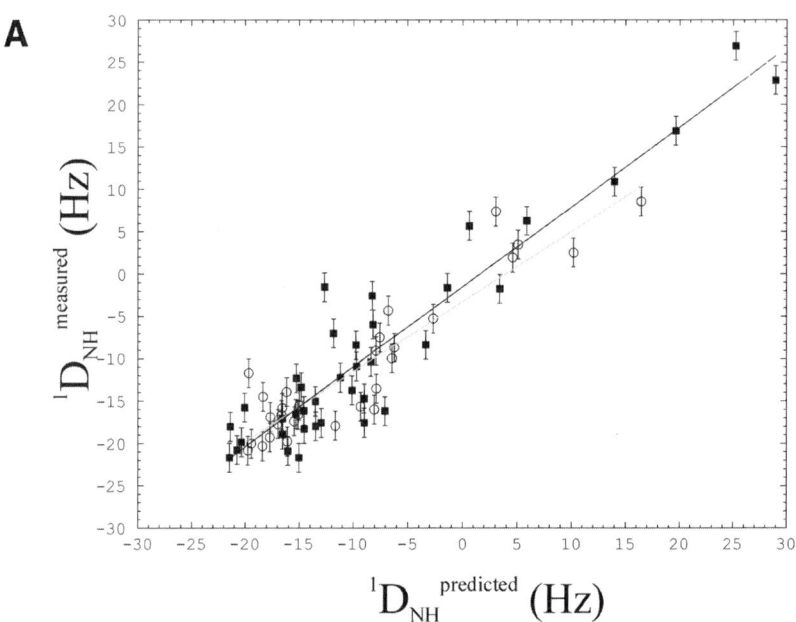

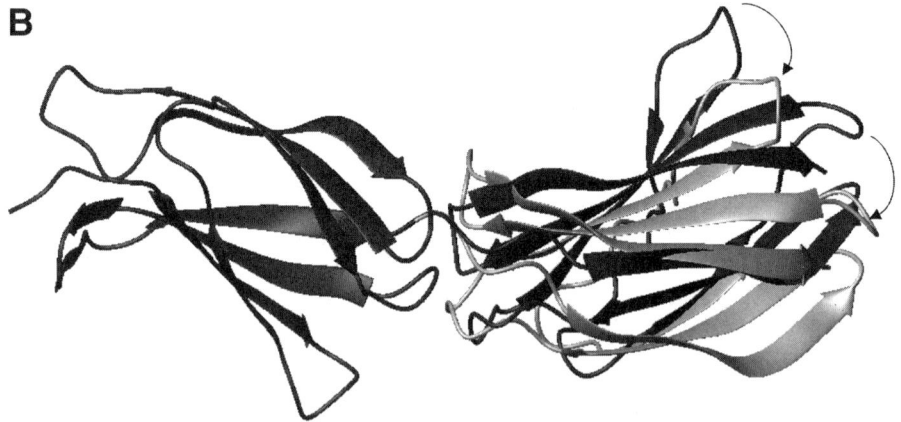

Fig. 2. Interdomain orientation in the solution state inferred from RDCs. (A) Correlation between calculated and measured NH RDCs for ^{13}F3 *(open circles)* and ^{14}F3 *(filled squares)* by fitting a separate alignment tensor to the couplings of each module. (B) Comparison of the interdomain orientation of the X-ray structure of the ^{13}F3^{14}F3 module pair *(black)* and the average solution structure *(gray)*. The two structures were overlaid on the coordinates of ^{13}F3. *Arrows* indicate the overall repositioning of prominent loop regions. The figure was prepared using Molmol *(23)*.

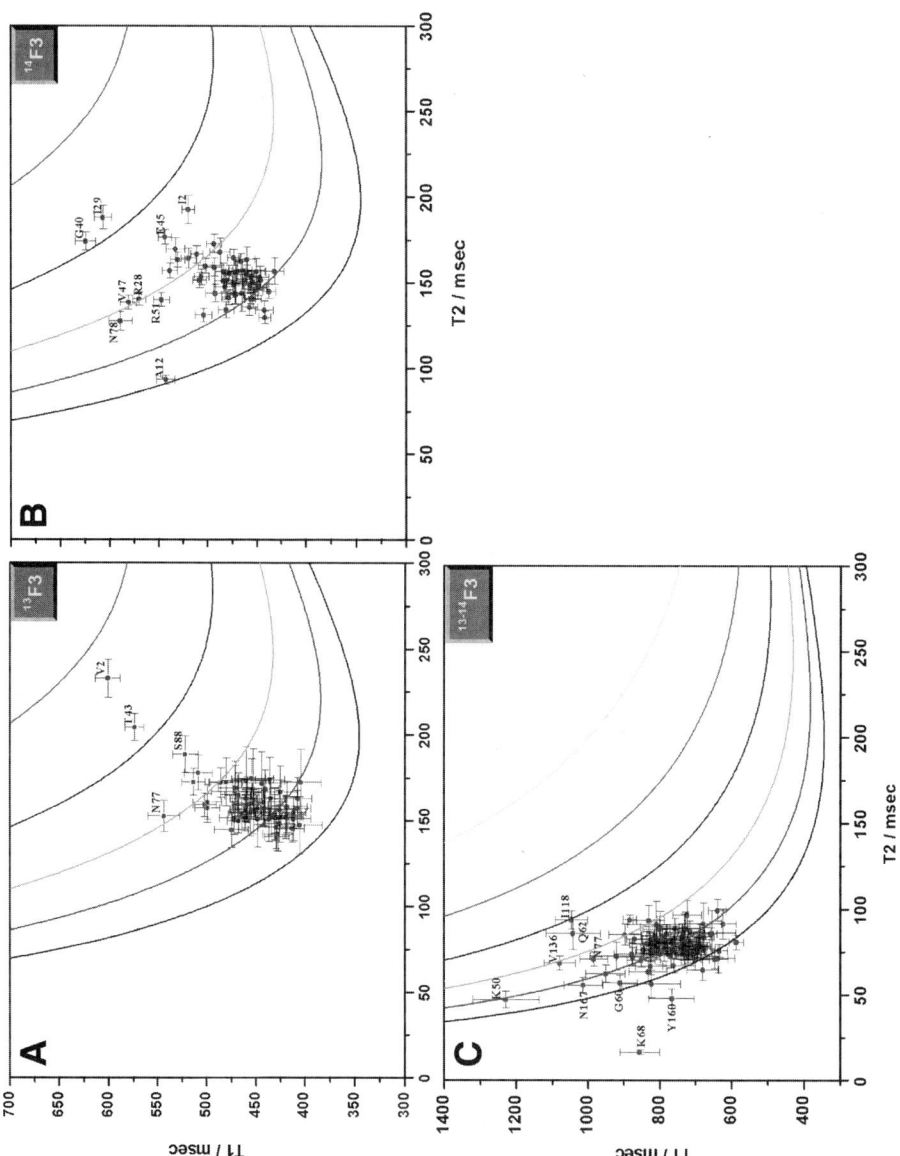

Fig. 3. Dynamics of the module pair at pH 4.5 and 35°C. Backbone ^{15}N relaxation time constants for (**A**) the [13]F3 module, (**B**) the [14]F3 module, and (**C**) the [13]F3[14]F3 module pair measured at 14.1 T are shown as a T_2 vs. T_1 plot. The *continuous lines* are calculated T_1 and T_2 values as a function of correlation time and the order parameter, S^2, using the isotropic Lipari and Szabo model (**6**). Curves with values of S^2 equaling 0.5, 0.6, 0.7, 0.8, 0.9, and 1.0 are shown.

of the parametric curves of the Lipari Szabo model (**Fig. 3**) may indicate substantial chemical exchange either because of the presence of unspecific aggregation, or pervasive μs–ms motion *(28)*. In the first case, repeating the experiment at a lower sample concentration may help. In the second case, recording data with different CPMG delays may provide insight into the nature of the motion. (*see also* **Subheading 3.3., step 3**).

2. The T_1 and T_2 experiments are collected as a series of [^{1}H]–^{15}N autocorrelation spectra incorporating preparation periods for the spin states of interest followed by delays during which relaxation occurs *(29,30)*. Transverse relaxation time constants (T_2) are typically measured using a spin-echo sequence with a refocusing interval of less than approx 1 ms between ^{15}N pulses to prevent the evolution of antiphase magnetization *(31)*. Dipolar and chemical shift anisotropy cross-correlation are removed by application of proton 180° pulses every few ms (T_1) and in the middle of the basic CPMG block (T_2) *(32,33)*. The delays are chosen to sample the intensity decay for up to about 1.5 times the maximal relaxation time constant. This decay is monitored by following the intensity of the NH peak intensities in the first block of the [^{1}H]–^{15}N autocorrelation spectrum at increasing delay times. Typically a series consists of 8–12 spectra. Errors in peak intensities are estimated from the root-mean-deviation of the baseline noise or by collecting spectra in duplicate. Relaxation time constants and associated errors are extracted from the intensity decays by fitting monoexponential decay functions (*see* **Note 6**).

3. Contributions to the line width from motions on the μs–ms time-scale should be investigated by additional experiments (*see* **Note 7**).

4. Estimation of the diffusion tensor. The quantitative description of the overall diffusion tensor is derived from the T_1/T_2 ratios and the orientations of the N–H bond vectors that are usually supplied by an atomic coordinate file (*see* **Note 8**).

5. A combination of structure- and dynamics-based criteria are used to establish a subset of residues that are representative of overall diffusion (*see* **Note 9**). The initial set contains residues in secondary structural elements. This set is further refined by exclusion of residues with low [^{1}H]–^{15}N NOEs or T_2 values that indicate sub-nanosecond or μs to ms time-scale motion, respectively.

6. Models of increasing complexity are fitted to the relaxation data of the selected residues using χ^2 minimization (*see* **Note 10**). Three models are tested: isotropic ($D = D_{xx} = D_{yy} = D_{zz}$), anisotropic ($D_\perp = D_{xx} = D_{yy}$ and $D_\parallel = D_{zz}$), and fully asymmetric ($D_{xx} \neq D_{yy} \neq D_{zz}$) (**Tables 2** and **3**). The χ^2 values of the isotropic fit, χ^2_{iso}, and the axially symmetric fit, χ^2_{axial}, are used to calculate Q_1, denoting the probability that the improvement in χ^2_{axial} is obtained by chance. Similarly, Q_2 is used to validate the asymmetric against the axial model (**Table 2**). It was found that the isolated ^{13}F3 and ^{14}F3 domains are asymmetric and prolate, respectively, with similar values for the axial ratios $2(D_{xx} + D_{yy})/D_{zz}$. When analyzed individually as part of the ^{1314}F3 module pair, both domains have prolate diffusion tensors with axial ratios $2(D_{xx} + D_{yy})/D_{zz}$ that are similar to that of the isolated domains. This indicates that the addition of another module does not change the shape of the molecule significantly. The difference in tilt between the long axes (D_{zz}) of each

Table 2
Experimentally Derived Diffusion Tensors[a] for Isolated ^{13}F3 and ^{14}F3 Modules and ^{13}F3 and ^{14}F3 in ^{13}F3^{14}F3 at pH 4.5

	$(6D)^{-1}$(ns)	$2D_{zz}/(D_{xx}+D_{yy})$	D_{xx}/D_y	θ (deg)	φ (deg)	ψ (deg)	χ^2	Q_1	Q_2
^{13}F3									
isotropic	4.2 ± 0.2						38res / 0.31		
prolate	4.2 ± 0.4	1.27 ± 0.19		127 ± 21	61 ± 21		0.20	2.6×10^{-7}	
oblate	4.3 ± 0.3	0.84 ± 0.09		75 ± 24	143 ± 21		0.23		
asymmetric	4.3 ± 0.2	1.30 ± 0.08	1.12 ± 0.05	140 ± 9	63 ± 6	18 ± 22	0.18		0.002
^{14}F3									
isotropic	4.6 ± 0.2						38res / 1.20		
prolate	4.7 ± 0.3	1.49 ± 0.19		161 ± 8	24 ± 18		0.86	2.5×10^{-4}	
oblate	4.6 ± 0.2	0.94 ± 0.07		128 ± 29	174 ± 44		1.17		
asymmetric	4.7 ± 0.3	1.49 ± 0.13	0.99 ± 0.08	161 ± 3	25 ± 8	14 ± 10	0.86		1.00
^{13}F3 in ^{13}^{14}F3									
isotropic	9.2 ± 0.1						24res / 0.45		
prolate	9.3 ± 0.1	1.23 ± 0.15		95±26	76 ± 21		0.30	2.9×10^{-4}	
oblate	9.3 ± 0.1	0.87 ± 0.10		87 ± 24	171 ± 31		0.36		
asymmetric	9.3 ± 0.1	1.23 ± 0.14	1.02 ± 0.12	95 ± 26	77 ± 20	48 ± 7	0.30		0.96
^{14}F3 in ^{13}^{14}F3									
isotropic	9.6 ± 0.1						28res / 1.00		
prolate	9.6 ± 0.1	1.35 ± 0.14		140 ± 14	139 ± 23		0.67	3.2×10^{-7}	
oblate	9.6 ± 0.1	0.81 ± 0.09		107 ± 20	49 ± 20		0.84		
asymmetric	9.6 ± 0.3	1.36 ± 0.04	1.10 ± 0.05	145 ± 11	141 ± 9	−7 ± 15	0.66		0.97

[a]Parameters are obtained by fitting experimental T_2/T_1 values to the crystal structures of ^{13}F3 and ^{14}F3. Symbols as explained in the text.

133

Table 3
Comparison of Experimental and Modeled Diffusion Tensors

	$(6D)^{-1}(ns)$	$2D_{zz}/(D_{xx}+D_{yy})$	% difference of $(6D)^{-1}$	% difference of $2D_{zz}/(D_{xx}+D_{yy})$
^{13}F3exp/^{13}F3 model	4.2 ± 0.2/4.5	1.3 ± 0.2/1.3	6	0
^{14}F3exp/^{14}F3model	4.7 ± 0.2/4.6	1.5 ± 0.2/1.3	−0.2	15
^{13}F3(in ^{13}F3^{14}F3)exp / ^{13}F3^{14}F3model	9.3 ± 0.2/11.9	1.2 ± 0.2/2.1	22	42
^{14}F3(in ^{13}F3^{14}F3)exp / ^{13}F3^{14}F3model	9.6 ± 0.2/11.9	1.4 ± 0.2/2.1	19	19

module is 45 ± 29° and the difference in the rotation around the z-axis is 63 ± 31°. The second angle is more difficult to determine from relaxation data because of the rotational symmetry of the molecule. Both tilt and twist angles are larger than that obtained from the RDCs. Because the tilt and twist angles determined from relaxation data at pH 7 (data not shown) are significantly closer to the angles obtained from the RDC data, the difference may be the result of the difference in pH of both measurements. In addition, differences in the time-scales of the averages considered may be important.

7. The degree of interdomain motion is estimated by comparing experimentally derived diffusion tensors and diffusion tensors derived from hydrodynamic bead models of rigid molecules (**Table 3**). Diffusion tensors are obtained for the isolated domains and the domain pair using (a) the method specified in **Subheading 3.3., step 6** and (b) hydrodynamic bead models *(20)*. A hydrodynamic model is constructed by placing beads at the Cα positions of all except the very flexible N-terminal and C-terminal residues using the coordinates of ^{13}F3, ^{14}F3, or ^{13}F3^{14}F3 from the X-ray structure. The best overall agreement with the isotropic correlation times of the isolated domains was achieved by using a bead radius of 3.2 Å. It is clear from **Table 3** that the experimentally derived anisotropies and those modeled using a rigid bead model agree reasonably well for the isolated domains but show large deviations for the domain pair. Both the experimentally determined anisotropies and the correlation times are between 20 and 40% smaller than those obtained for a rigid domain pair. This indicates that there is significant interdomain flexibility between ^{13}F3 and ^{14}F3 in the ^{13}F3^{14}F3 domain pair.

4. Notes

1. The numbering of residues is as follows: N1-T89 and A90-T179 refer to ^{13}F3 and ^{14}F3, respectively, and correspond to N1813–T1901 and A1902–T1991 of human fibronectin (SWISS-PROT entry Finc_HUMAN; primary accession number: P02751). The coordinates were obtained from the X-ray structure of the fibronectin fragment ^{12}F3–^{14}F3, Protein Data Bank (PDB) entry 1FNH *(17)*.

2. Chemical shift comparison spectra were recorded on samples containing 1 m*M* protein, whereas the relaxation data were recorded on samples containing 0.25 m*M* protein to ensure that they were not affected by sample aggregation. For the same reason the RDC experiments were performed on samples containing 0.25 m*M* protein. The assignment and relaxation experiments were performed at 35°C and the RDC experiments at 40°C.

3. It is recommended to use the same stock of buffer solution and to perform the experiments on the same spectrometer, using the same temperatures and protein concentrations. In each case a 10–20 μ*M* 2,2-dimethyl-2-silapentane-5-sulfonic acid should be added to the sample to serve as a chemical shift reference.

4. Extensive zero filling in the dimension of the measurement (i.e., ^{15}N) should be used to ensure that the digital resolution is much higher than the resolution given by the acquisition time. Accurate phasing of the spectra is important to minimize systematic errors on estimating $^1J^{app}$.

5. Missing protons in the coordinates set in the PDB file were added using Molmol *(23)*.

6. The analysis can conveniently be performed using the relaxation analysis tool in NMRview *(18)*.

7. Methods to obtain quantitative information on motions in the μs to ms time-scale are reviewed by Loria et al. (*see* Chapter 12).

8. In the absence of structural information, the magnitude of the diffusion tensor may be estimated from the distribution of the T_2/T_1 ratios *(9)*.

9. Some authors have suggested the use of entirely dynamics-based selection criteria *(34)* or approaches based on Bayesian statistics *(35)*.

10. The analysis can be performed using the program TENSOR *(19)*.

Acknowledgments

We would like to thank David Staunton for support with expression and purification of the samples and Jonathan Boyd and Nick Soffe for maintaining the NMR facility. Sachchidanand thanks the Felix Foundation, and J.M. Werner. the Wellcome Trust, for financial support.

References

1. Vitkup, D., Melamud, E., Moult, J., and Sander, C. (2001) Completeness in structural genomics. *Nat. Struct. Biol.* **8**, 559–566.

2. Bork, P., Downing, A. K., Kieffer, B., and Campbell, I. D. (1996) Structure and distribution of modules in extracellular proteins. *Q. Rev. Biophys.* **29**, 119–167.

3. Fischer, M. W., Losonczi, J. A., Weaver, J. L., and Prestegard, J. H. (1999) Domain orientation and dynamics in multidomain proteins from residual dipolar couplings. *Biochemistry* **38**, 9013–9022.

4. Braddock, D. T., Cai, M., Baber, J. L., Huang, Y., and Clore, G. M. (2001) Rapid identification of medium- to large-scale interdomain motion in modular proteins using dipolar couplings. *J. Am. Chem. Soc.* **123**, 8634, 8635.

5. Tolman, J. R. (2002) A novel approach to the retrieval of structural and dynamic information from residual dipolar couplings using several oriented media in biomolecular NMR spectroscopy. *J. Am. Chem. Soc.* **124,** 12,020–12,030.

6. Lipari, G. and Szabo, A. (1982) Model-free approach to the interpretation of nuclear magnetic resonance relaxation in macromolecules, 1: theory and range of validity. *J. Am. Chem. Soc.* **104,** 4546–4559.

7. Woessner, D. E. (1962) Nuclear spin relaxation in ellipsoids undergoing rotational Brownian motion. *J. Chem. Phys.* **37,** 647–654.

8. Bruschweiler, R., Liao, X., and Wright, P. E. (1995) Long-range motional restrictions in a multidomain zinc-finger protein from anisotropic tumbling. *Science* **268,** 886–889.

9. Clore, G. M., Gronenborn, A. M., Szabo, A., and Tjandra, N. (1998) Determining the magnitude of the fully asymmetric diffusion tensor from heteronuclear relaxation data in the absence of structural information. *J. Am. Chem. Soc.* **120,** 4889, 4890.

10. Ghose, R., Fushman, D., and Cowburn, D. (2001) Determination of the rotational diffusion tensor of macromolecules in solution from NMR relaxation data with a combination of exact and approximate methods: application to the determination of interdomain orientation in multidomain proteins. *J. Magn. Reson.* **149,** 204–217.

11. Fushman, D., Xu, R., and Cowburn, D. (1999) Direct determination of changes of interdomain orientation on ligation: use of the orientational dependence of ^{15}N NMR relaxation in Abl SH(32). *Biochemistry* **38,** 10,225–10,230.

12. Smith, S. P., Hashimoto, Y., Pickford, A. R., Campbell, I. D., and Werner, J. M. (2000) Interface characterization of the type II module pair from fibronectin. *Biochemistry* **39,** 8374–8381.

13. Tugarinov, V., Liang, Z., Shapiro, Y. E., Freed, J.H., and Meirovitch, E. (2001) A structural mode-coupling approach to ^{15}N NMR relaxation in proteins. *J. Am. Chem. Soc.* **123,** 3055–3063.

14. Baber, J. L., Szabo, A., and Tjandra, N. (2001) Analysis of slow interdomain motion of macromolecules using NMR relaxation data. *J. Am. Chem. Soc.* **123,** 3953–3959.

15. Ingham, K. C., Brew, S. A., and Atha, D. H. (1990) Interaction of heparin with fibronectin and isolated fibronectin domains. *Biochem. J.* **272,** 605–611.

16. Mould, A. P. and Humphries, M. J. (1991) Identification of a novel recognition sequence for the integrin alpha 4 beta 1 in the COOH-terminal heparin-binding domain of fibronectin. *EMBO J.* **10,** 4089–4095.

17. Sharma, A., Askari, J. A., Humphries, M. J., Jones, E. Y., and Stuart, D. I. (1999) Crystal structure of a heparin- and integrin-binding segment of human fibronectin. *EMBO J.* **18,** 1468–1479.

18. Johnson, B. A. and Blevins, R. A. (1994) NMRView—a computer-program for the visualization and analysis of NMR data. *J. Biomol. NMR* **4,** 603–614.

19. Dosset, P., Hus, J. C., Blackledge, M., and Marion, D. (2000) Efficient analysis of macromolecular rotational diffusion from heteronuclear relaxation data. *J. Biomol. NMR* **16,** 23–28.

20. Garcia de la Torre, J., Huertas, M. L., and Carrasco, B. (2000) HYDRONMR: prediction of NMR relaxation of globular proteins from atomic-level structures and hydrodynamic calculations. *J. Magn. Reson.* **147,** 138–146.

21. Zweckstetter, M. and Bax, A. (2001) Characterization of molecular alignment in aqueous suspensions of Pf1 bacteriophage. *J. Biomol. NMR* **20,** 365–377.

22. Dosset, P., Hus, J. C., Marion, D., and Blackledge, M. (2001) A novel interactive tool for rigid-body modeling of multi-domain macromolecules using residual dipolar couplings. *J. Biomol. NMR* **20,** 223–231.

23. Koradi, R., Billeter, M., and Wuthrich, K. (1996) MOLMOL: A program for display and analysis of macromolecular structures. *J. Mol. Graph.* **14,** 51–55.

24. Driscoll, P. C., Clore, G. M., Marion, D., Wingfield, P. T., and Gronenborn, A. M. (1990) Complete resonance assignment for the polypeptide backbone of interleukin 1 beta using three-dimensional heteronuclear NMR spectroscopy. *Biochemistry* **29,** 3542–3556.

25. Lerche, M. H., Meissner, A., Poulsen, F. M., and Sorensen, O. W. (1999) Pulse sequences for measurement of one-bond $^{(15)}$N–$^{(1)}$H coupling constants in the protein backbone. *J. Magn. Reson.* **140,** 259–263.

26. Bax, A., Kontaxis, G., and Tjandra, N. (2001) Dipolar couplings in macromolecular structure determination. *Meth. Enzymol.* **339,** 127–174.

27. Clore, G. M. and Garret, D. S. (1999) R-factor, free R, and complete cross-validation for dipolar coupling refinement of NMR structures. *J. Am. Chem. Soc.* **121,** 9008–9012.

28. Werner, J. M., Campbell, I. D., and Downing, A. K. (2002) Shape and dynamics of a calcium-binding protein investigated by nitrogen-15 NMR relaxation. In *Methods in Molecular Biology, vol. 173: Calcium-Binding Protein Protocols, vol. 2* (Vogel, H. J. ed.). Humana, Totowa, NJ, pp. 285–300.

29. Kay, L. E., Torchia, D. A., and Bax, A. (1989) Backbone dynamics of proteins as studied by ^{15}N inverse detected heteronuclear NMR-spectroscopy—application to staphylococcal nuclease. *Biochemistry* **28,** 8972–8979.

30. Farrow, N. A., Zhang, O. W., Formankay, J. D., and Kay, L. E. (1994) A heteronuclear correlation experiment for simultaneous determination of ^{15}N longitudinal decay and chemical-exchange rates of systems in slow equilibrium. *J. Biomol. NMR* **4,** 727–734.

31. Vold, R. R. and Vold, R. L. (1976) Transverse relaxation in heteronuclear coupled spin systems: AX, AX2, AX3, and AXY. *J. Chem. Phys.* **64,** 320–332.

32. Boyd, J., Hommel, U., and Campbell, I. D. (1990) Influence of cross-correlation between dipolar and anisotropic chemical-shift relaxation mechanisms upon longitudinal relaxation rates of ^{15}N in macromolecules. *Chem. Phys. Lett.* **175,** 477–482.

33. Kay, L. E., Nicholson, L. K., Delaglio, F., Bax, A., and Torchia, D. A. (1992) Pulse sequences for removal of the effects of cross-correlation between dipolar and chemical-shift anisotropy relaxation mechanism on the measurement of heteronuclear T_1 and T_2 values in proteins. *J. Magn. Reson.* **97,** 359–375.

34. Barbato, G., Ikura, M., Kay, L. E., Pastor, R. W., and Bax, A. (1992) Backbone dynamics of calmodulin studied by ^{15}N relaxation using inverse detected 2-dimensional NMR-spectroscopy—the central helix is flexible. *Biochemistry* **31,** 5269–5278.
35. Andrec, M., Inman, K. G., Weber, D. J., Levy, R. M., and Montelione, G. T. (2000) A Bayesian statistical method for the detection and quantification of rotational diffusion anisotropy from NMR relaxation data. *J. Magn. Reson.* **146,** 66–80.

10

Characterization of the Overall Rotational Diffusion of a Protein From ^{15}N Relaxation Measurements and Hydrodynamic Calculations

Jennifer Blake-Hall, Olivier Walker, and David Fushman

Summary

In this chapter, we discuss experimental and theoretical methods for characterizing the over-·all rotational diffusion of molecules in solution. The methods are illustrated for the B3 domain of protein G, a small protein with rotational anisotropy of $D_{par}/D_{perp} = 1.4$. The rotational diffusion tensor of the protein is determined directly from ^{15}N relaxation measurements. The experimental data are treated assuming various possible models for the overall tumbling: isotropic, axially symmetric, and fully anisotropic, and the results of these analyses are compared to determine an adequate diffusion model for the protein. These experimentally derived characteristics of the protein are compared with the results of theoretical calculations of the diffusion tensor using various hydrodynamic models, to find optimal models and parameter sets for theoretical predictions. We also derive model-free characteristics of internal backbone motions in the protein, to show that different models for the overall motion can result in significantly different pictures of motion. This emphasizes the necessity of accurately characterizing the overall tumbling of a molecule to determine its local dynamics.

Key Words: Anisotropic rotational diffusion; hydrodynamic calculations; nuclear relaxation; protein dynamics; protein G.

1. Introduction

Knowledge of the overall rotational properties of a molecule is an essential prerequisite for accurate analysis of its internal dynamics by solution nuclear magnetic resolution (NMR), for example *(1)*. Because of the restricted amplitudes and fast correlation times of local motions, the overall rotation makes the dominant contribution to the measured values of ^{15}N relaxation rates (R_1 and R_2) in proteins. To obtain an accurate picture of protein dynamics, the much

From: *Methods in Molecular Biology, vol. 278: Protein NMR Techniques*
Edited by: A. K. Downing © Humana Press Inc., Totowa, NJ

weaker contributions from local motions must be correctly deconvolved from those originating from the overall tumbling. This deconvolution is severely complicated if the rotational diffusion of a molecule is anisotropic, which is quite general for proteins and reflects deviation in the shape of the molecule from a perfect sphere. The microdynamic parameters derived from these data could be in error if the rotational anisotropy is not taken into account—as pointed out in *(1–3)*. As shown in *(1,3–5)*, a relaxation data analysis that does not include rotational anisotropy could result in spurious conformational exchange motions. Accurate identification of these motions in a protein is of particular interest in view of their potential relevance to biological function *(6)*.

Measurements of overall rotational characteristics can provide valuable structural information on domain orientation in multidomain systems *(7,8)* and their structural rearrangements on ligand binding *(7)*, which could hold the key to our understanding of regulation of various cellular events, including ligand recognition, enzyme catalysis, signal transduction, and so forth. Other related applications include structure refinement of proteins based on orientation constraints derived from relaxation anisotropy *(9)* and characterization of the size and shape of protein–protein complexes formed during aggregation *(10)*.

When molecular tumbling is anisotropic, ^{15}N spin relaxation rates depend on the orientation of the backbone amide N–H vectors with respect to the principal axes frame of the rotational diffusion tensor. If the protein structure is known, this allows the determination of the rotational diffusion tensor of a protein from the orientational dependence of the measured spin–relaxation rates. Several approaches to this problem have been suggested in the past *(7,11–16)*. We have recently developed an efficient method for a full, direct characterization of the rotational diffusion tensor from ^{15}N relaxation measurements *(16a)*. Here, we apply this method to a small protein and compare the derived diffusion tensor with theoretical predictions using various hydrodynamic models.

2. Materials

To illustrate the method, we use a small, 56-amino acid protein, the B3 domain of streptococcal protein G *(17–21)* further called GB3. The structure of the protein has been solved by both X-ray crystallography *(22)* and NMR *(23)* and shows a well-packed hydrophobic core formed by a four-stranded β-sheet and a four-turn α-helix (**Fig. 1**). Details of the protein sequence, sample/experimental conditions, and particular relaxation experiments can be found in *(3)*.

Relaxation data for these studies comprise the rates of ^{15}N longitudinal (R_1) and transverse (R_2) relaxation and the rate of $^{15}N–^{1}H$ cross-relaxation measured via steady-state $^{15}N\{^{1}H\}$ nuclear Overhauser effect (NOE). The experiments were performed using standard approaches described elsewhere *(3,24)*.

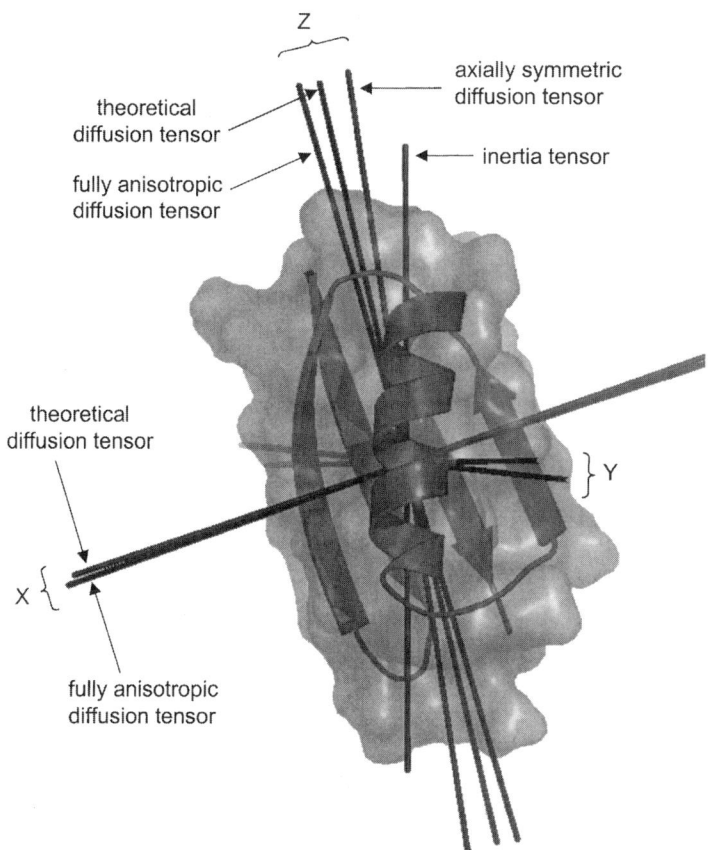

Fig. 1. Cartoon representation of the tertiary structure and the van der Waals surface of the GB3 domain, generated using PyMOL *(39)*. Various rods represent the orientation of the diffusion tensor axes (as indicated) obtained directly from ^{15}N relaxation data for the axially symmetric and fully anisotropic models and predicted theoretically using HYDRONMR, along with the unique axis of the inertia tensor (a shorter rod). Atom coordinates are from the crystal structure (1IDG) *(22)*. The orientations of all three diffusion tensors are similar within the experimental errors. The z-axis of the axially symmetric tensor makes an 8° angle with the z-axes of the fully anisotropic tensors, both measured and predicted. The angle between the z-axes of the fully anisotropic and the theoretical tensor is 3°. All these z-axes are oriented approximately along the α-helix axis: the tilt angle is 23°, 30°, and 28°, for the axially symmetric, fully anisotropic, and the HYDRONMR-predicted tensors. Similar angles with respect to the unique axis of the inertia tensor are 10°, 18°, and 17°; this axis is tilted by 15° from the helix axis.

3. Methods

3.1. Determination of the Rotational Diffusion Tensor Using ^{15}N Relaxation Measurements in Solution

3.1.1. Preliminary Analysis of the Raw ^{15}N Relaxation Data: What We Can See With a "Naked Eye"

Fifty-five resolved backbone amide cross-peaks were observed in the two-dimensional (2D) spectra. Though they could be sufficiently resolved for assignment, residues Glu[15] and Asn[35] and Thr[25] and Glu[27] are not included in relaxation analysis as their signal intensities could be affected by spectral overlap. Gln[2] was excluded from anisotropic analyses because it is not present in the available protein coordinates. The relaxation parameters for the GB3 domain are shown in **Fig. 2**.

The R_1, R_2, and NOE data (**Fig. 2A–C**) show simultaneous decrease in the $\beta1/\beta2$ and $\alpha/\beta3$ loops, thus indicating flexible regions in the protein. There is no such decrease in any of the relaxation parameters in the $\beta2/\alpha$ loop, however, and in the loop connecting strands $\beta3$ and $\beta4$ there is a decrease in R_1 and R_2 but not in the NOE.

Noticeably elevated R_2 values are observed for the entire α-helix (**Fig. 2B**). The NOEs here are also somewhat higher than in the rest of the backbone, whereas the R_1 values are at about the same level as, or slightly higher than, in the other elements of the secondary structure. An elevation in R_2 observed in the α-helix could be indicative of conformational exchange on the microsecond–millisecond time-scale. However, given the structure of GB3, it is difficult to imagine a physical model that would account for every residue in the helix (including those not facing the β-sheet) being involved in conformational exchange. The orientation dependence of the transverse relaxation rate could account for the elevation in R_2 for residues in the α-helix if the helix axis of the GB3 domain is aligned parallel to the longitudinal axis of the prolate rotational diffusion tensor. This orientation will align the NH vectors in the α-helix along the axis of fast overall rotation—as a result, they will experience slower rates of overall tumbling (hence higher R_2s) compared to the rest of the protein.

3.1.2. Characterization of the Diffusion Tensor From ^{15}N Relaxation Data: Description of the Method

We derive the overall rotational diffusion tensor of a molecule directly from the ratio of the measured parameters:

$$\rho = \left(\frac{2R_2'}{R_1'} - 1\right)^{-1} \tag{1}$$

where the primes indicate that the relaxation rates R_2 and R_1 were modified to

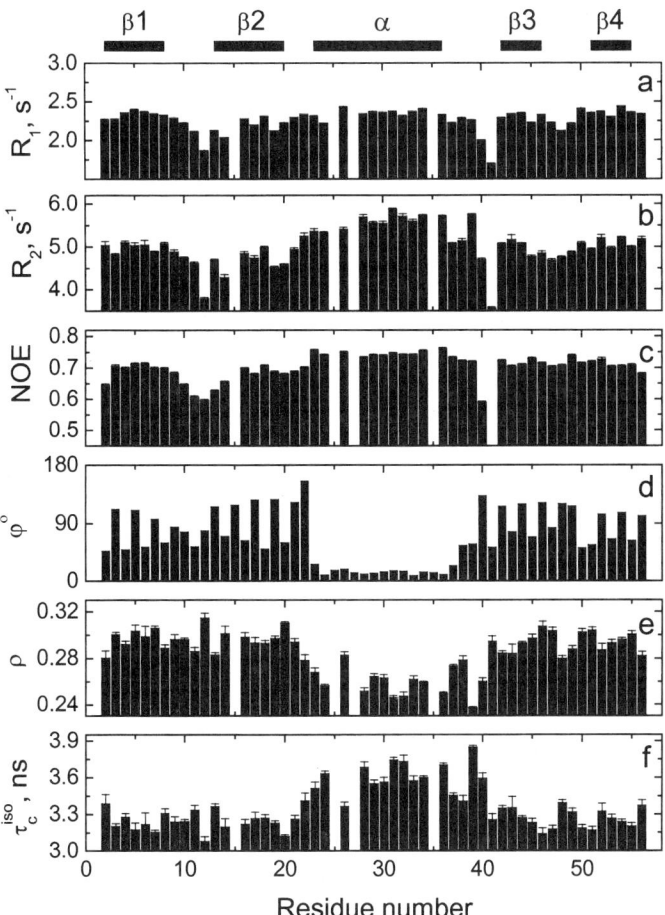

Fig. 2. Backbone ^{15}N relaxation rates at 14.1 Tesla, (**A**) R_1, (**B**) R_2, and (**C**) $^{15}N\{^1H\}$ NOE vs residue number for GB3. Also shown are (**D**) the angles ϕ between each NH vector and the α-helix axis; (**E**) the ratio ρ of the relaxation parameters (**Eq. 1**); and (**F**) the "isotropic" overall rotational correlation time τ_c^{iso} experienced by the individual amides in the backbone, calculated using **Eq. 5.** The error bars represent standard errors in the parameters because of experimental noise. The horizontal bars on the top indicate the positions of the secondary structure elements in the protein sequence.

subtract the high-frequency components of the spectral density (*see*, for example **ref. 25**):

$$R_1' = R_1\left[1 - 1.249|\gamma_N/\gamma_H|(1 - NOE)\right] \qquad (2)$$

$$R_2' = R_2 - 1.079|\gamma_N/\gamma_H|R_1(1 - NOE) \qquad (3)$$

Here γ_H and γ_N are the gyromagnetic ratios for ^1H and ^{15}N. As shown in *(7)*, the ratio ρ contains information on the principal values of the tensor and on the orientation of the NH vector with respect to the diffusion tensor frame.

In the most general case of a completely anisotropic diffusion tensor, six parameters have to be determined: the principal values of the tensor (D_x, D_y, D_z) and the Euler angles (α,β,γ) that define the orientation of the principal axes frame of the tensor with respect to the protein coordinate (Protein Data Bank) frame. In the axially symmetric case, the number of parameters is reduced to four: $D_{||}(=D_z)$, $D_{\perp}(=D_x=D_y)$ and the angles (α,β). The determination of the diffusion tensor parameters involves an optimization search in a six-dimensional (or four-dimensional) space performed by minimizing the target function

$$\chi^2 = \sum_{i=1}^{N_r}\left(\frac{\rho_i^{exp} - \rho_i^{calc}}{\sigma_i}\right)^2 \qquad (4)$$

where N_r is the total number of residues included in the analysis and σ_i denotes the experimental error in ρ_i for residue i. ρ^{exp} is directly derived from the measured relaxation parameters, according to **Eqs. 1–3**, and ρ^{calc} is calculated based on the current parameters of the diffusion tensor. This method is implemented in our program ROTDIF, based on an efficient Levenberg–Marquardt minimization algorithm. Further details of the method are outlined elsewhere *(16b)*.

Note that the ratio ρ is independent, to a first-order approximation, of the values of the order parameter (when using NH groups belonging to the core of a protein) and of the site-specific variations in the ^{15}N chemical shift anisotropy and in the NH bond length *(26,27)*. This also applies to the diffusion tensor derived using this approach. Those amides involved in conformational exchange and therefore having significant R_{ex} contributions to R_2 have to be excluded from the analysis. A detailed discussion of how these groups can be identified can be found in *(3)*.

After the characteristics of the diffusion tensor are obtained, they can be used together with the relaxation rates and hetero-NOE, as input parameters for our program DYNAMICS *(3,28)* to determine the model-free parameters for each amide group in a protein.

3.1.3. Characterization of the Diffusion Tensor From ^{15}N Relaxation Data: Data Analysis

3.1.3.1. ISOTROPIC OVERALL TUMBLING MODEL

First we consider the isotropic tumbling model. In this case, the overall correlation time experienced by each NH group can be directly calculated from the ratio of relaxation parameters *(29)*:

$$\tau_c^{iso} = \frac{1}{2\omega_N^2} \sqrt{\frac{6R_2'}{R_1'} - 7} = \frac{1}{2\omega_N^2} \sqrt{\frac{3}{\rho} - 4} \qquad (5)$$

The observed values of τ_c^{iso} in GB3 (**Fig. 2E**) show significant variation (mean 3.36 ns, SD 0.18 ns) from site to site that correlates with the orientation of the NH vector (cf. **Fig. 2D**). This behavior is inconsistent with the isotropic model, where all groups should experience the same overall correlation time, and it calls for more realistic, anisotropic models for the protein.

3.1.3.2. ANISOTROPIC OVERALL TUMBLING MODELS

The rotational diffusion tensor of the protein was derived from the orientational dependence (**Fig. 2F,D**) of the ratio, ρ of ^{15}N relaxation rates as described in **Subheading 3.1.2.** Only those residues belonging to well-defined elements of the secondary structure were used in this analysis. Residues in the flexible loops and in the termini, where the N–H bond orientation might not be well defined, were not included. Also excluded from this analysis was Val[39], showing R_{ex} contributions. The characteristics of the rotational diffusion tensor of GB3 derived using various models are presented in **Table 1**.

3.1.3.2.1. Axially Symmetric Tumbling Model For the axially symmetric model, the rotational diffusion tensor of GB3 is characterized by $D_{\parallel}/D_{\perp} = 1.37 \pm 0.06$ and $\tau_c = 3.34 \pm 0.11$ ns. The orientation of its unique principal axis with respect to the crystal structure (**Fig. 1**) is characterized by $\alpha = 94°\pm7°$ and $\beta = 69°\pm12°$. A small angle between the axis of the diffusion tensor and the α-helix axis (**Fig. 1**) would then explain the observed elevated R_2 values in this part of the protein.

The agreement between the experimental and calculated (fitting curve) values of ρ is shown in **Fig. 3**. The vertical spread of the data around the fitting curve in **Fig. 3B**, most pronounced near the maximum (at θ approx 90°), indicates that there is a certain degree of rhombicity in the diffusion tensor, not accounted for by the axially symmetric model (*see* next section).

3.1.3.2.2. Fully Anisotropic Tumbling Model The characteristics of the fully anisotropic diffusion tensor are very similar to those derived in the previous

A

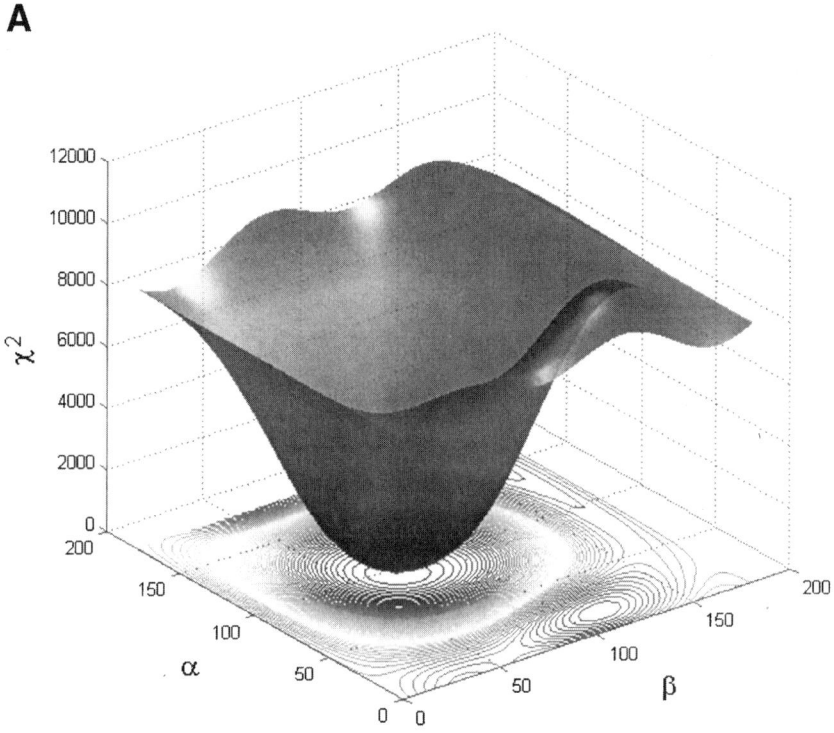

B

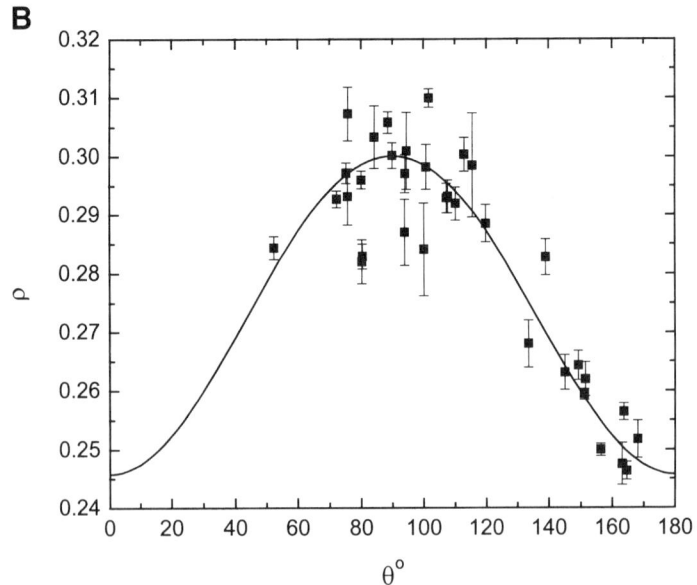

Fig. 3A,B.

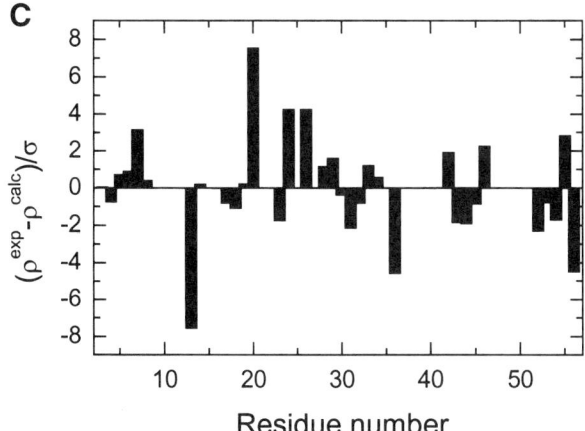

Fig. 3. Analysis of the experimental data assuming the axially symmetric model. (**A**) Surface representation of the angular dependence of the target function χ^2 indicating a global minimum in the coordinates α and β; (**B**) the experimental (symbols) and calculated (line) values of ρ as a function of the NH vector orientation (angle θ) with respect to the unique axis of the diffusion tensor; (**C**) the residuals of fit.

section (**Table 1**, **Fig. 1**). The fitting curve is shown in **Fig. 4B** that represents the experimental and calculated values of ρ as a function of the (polar) angle θ between the NH vector and the longitudinal (associated with D_z) axis of the diffusion tensor. In the case of full anisotropy this representation is not sufficient, as the orientation (angle ϕ) of the NH vector with respect to the other two axes of the diffusion tensor also has an effect on the relaxation parameters. A more detailed assessment of the quality of the fit can be obtained from the three-dimensional (3D) hilly surface shown in **Fig. 4A** and representing the theoretical values of ρ as a function of θ and ϕ. The actual heights of the "hills" and depths of the "valleys" of this surface depend on the principal values $\{D_x, D_y, D_z\}$ of the diffusion tensor. The difference in height between the hilltops and the saddle points (in the valleys) is proportional to $(D_y - D_x)$ and disappears in the axially symmetric case.

For the axially symmetric model, this whole surface is projected onto an area on the $\rho(\theta)$ plot (cf. **Figs. 3B** and **4B**). The upper and lower boundaries of $\rho(\theta)$ are given by $\phi = 0$ and $\phi = \pi/2$ and correspond to the cases where the NH vector B parallel to the D_x–D_z or D_y–D_z plane. The gap between the two boundaries varies with the angle θ: negligible for θ close to 0, it increases with the deviation of the NH vector from the z-axis and reaches maximum when $\theta = \pi/2$ (*16b*). The top points of these curves correspond to the two limiting cases of the NH vector along the D_x axis ($\theta = \pi/2$, $\phi = 0$, upper boundary) or along the D_y axis ($\theta = \pi/2$, $\phi = \pi/2$, lower boundary). This then explains the vertical spread in the data points around the fitting curve in **Fig. 3B**.

Table 1
Hydrodynamic Characteristics of the GB3 Domain Derived From ^{15}N Relaxation Data Using Various Models of the Overall Tumbling and From Hydrodynamics Calculations

Model of overall motion	D_x^a	D_y^a	D_z^a	α^b	β^b	γ^b	τ_c^c	Anisotropy^d	Rhombicity^e	χ^2/df^f	p^g
Isotropic	4.86 (0.04)	4.86 (0.04)	4.86 (0.04)	–	–	–	3.43 (0.03)	1	0	102.5	–
Axial symmetry	4.45 (0.11)	4.45 (0.11)	6.07 (0.33)	94 (7)	69 (12)	– (0.11)	3.34 (0.06)	1.37	0	8.4	4 10^{-17}
Full anisotropy	4.13 (0.24)	4.60 (0.18)	6.25 (0.34)	85 (10)	68 (7)	179 (14)	3.34 (0.10)	1.43 (0.09)	0.37 (0.24)	6.2	5 10^{-3}
Bead model, "dry" protein^h	4.36	4.99	6.01	70	84	152	3.22	1.29	0.71		
Bead model, "wet" protein^i	4.35	4.49	5.98	75	61	172	3.23	1.35	0.13		
Shell model^j	4.43	4.64	6.36	87	64	175	3.31	1.40	0.17		

Numbers in parentheses represent standard deviations.

^{a}Principal values (in 10^7 s^{-1}) of the rotational diffusion tensor, ordered so that $D_x = D_y = D_z$.

^{b}Euler angles {α, β, γ} (in degrees) describe the orientation of the principal axes frame of the rotational diffusion tensor with respect to protein coordinate frame.

^{c}Overall rotational correlation time (in ns) of the molecule, $\tau_c = 1/[2\,Tr(D_{||})]$.

^{d}The degree of anisotropy of the diffusion tensor, $2D_z/(D_x + D_y)$.

^{e}The rhombicity of the diffusion tensor, $1.5(D_y - D_x)/[D_z - 1/2(D_x + D_y)]$.

^{f}Residuals of the fit (χ^2) divided by the number of degrees of freedom.

^{g}Probability that the reduction in χ^2 (compared to the model in the row directly above it) could occur by chance. Both axially symmetric and fully anisotropic models are statistically a much better fit than the isotropic model.

^{h}The results of hydrodynamic calculations using the bead model for protein alone; the bead radius was set to 1.4 Å.

^{i}The results of hydrodynamic calculations using the bead model and hydration shell; the bead radius was set to 1.0 Å with a hydration shell of width 1.3 Å.

^{j}The results of hydrodynamic calculations using HYDRONMR program (37); parameter a was set to 2.6 Å.

A

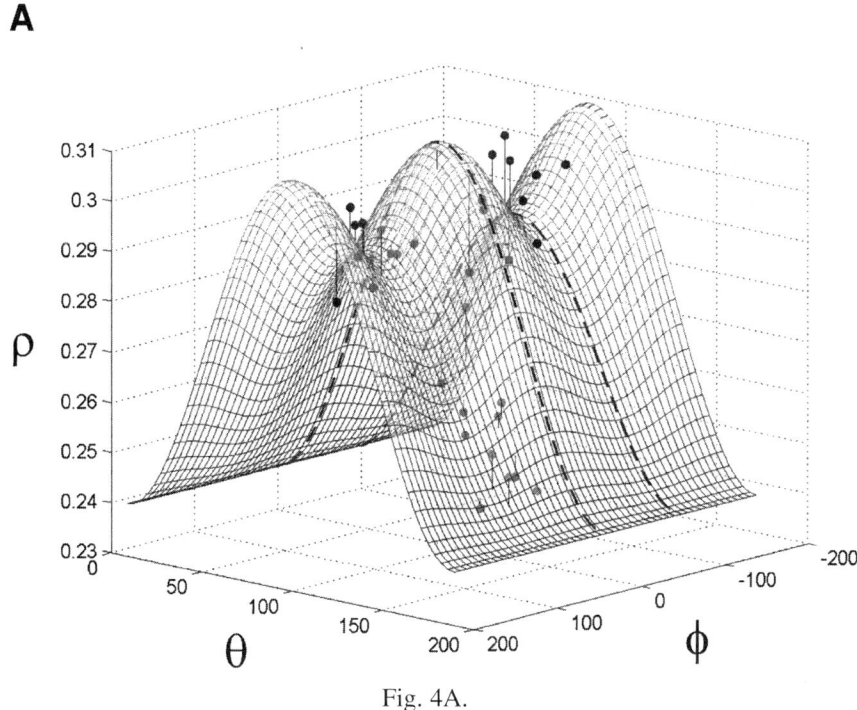

Fig. 4A.

B

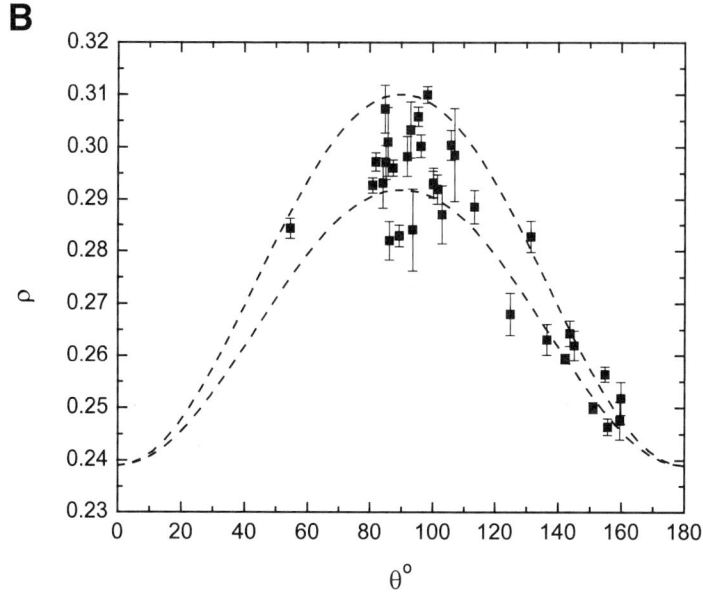

Fig. 4B.

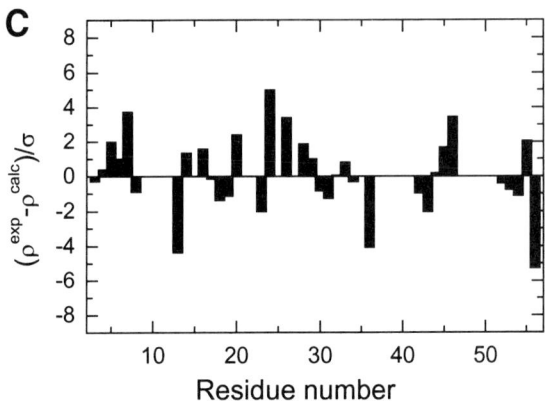

Fig. 4. Analysis of the experimental data for the fully anisotropic model. (**A**) a three-dimensional representation of the agreement between the experimental (symbols) and calculated (hilly surface) values of ρ as a function of the NH vector orientation (angles θ, ϕ) in the principal axes frame of the diffusion tensor; (**B**) a two-dimensional projection (onto the $\phi = 0$ plane) of the surface shown in panel **A**, representing a comparison similar to that depicted in **Fig. 2B**; (**C**) the residuals of fit. The dashed curves in **B** (also shown in **A**) represent $\rho(\theta)$ at $\phi = 0$ and $-90°$, respectively, and correspond to the upper and lower boundaries on the spread of data because of the presence of the rhombic component in the tensor. Vertical lines connecting symbols with the surface in (**A**) represent deviations of the data points from their theoretical values.

3.1.3.3. WHICH OVERALL TUMBLING MODEL IS A BETTER FIT?

To assess which model fits the experimental data best, we used the statistical F-test *(30)* that determines whether an observed improvement in the fit is statistically justified and not merely a result of the increase in the number of fitting parameters when going from the isotropic to axially symmetric and then to fully anisotropic model (one, four, and six parameters, respectively). Based on this analysis, both anisotropic tumbling models agree with the experimental data significantly better than the isotropic model. For the axially symmetric model the F-statistics analysis of the fit results in the probability $p = 5 \cdot 10^{-17}$ that this could occur by chance. The full anisotropy model provides an improvement in the fit over the axially symmetric model, with the confidence level of 99.5% ($p = 0.005$). This improvement, however, is not as dramatic as when going from the isotropic to axially symmetric model. These data suggest that GB3 tumbling in solution is not perfectly axially symmetric, which is also supported by the results of hydrodynamic calculations (**Table 1**).

3.2. Comparison With Theoretical Predictions of the Diffusion Tensor

Theoretical prediction of the rotational properties of proteins in solution is complex, largely because one has to account for the actual shape of the protein and for the unknown size and shape of the hydration shell formed by nearby water molecules moving together with the tumbling protein molecule. In addition, large-amplitude dynamics (e.g., of the loops and/or termini) can alter the shape of the molecule, which will then render inaccurate any predictions based on a rigid-body approximation. The range of these latter variations could, in principle, be estimated using an ensemble of NMR structures, assuming that these provide good statistical sampling of the various shapes of the molecule in solution. Here, we consider several approaches to theoretical prediction of the overall rotational diffusion tensor of the molecule, based on different representations of its shape, starting with the simplest and then increasing the level of complexity in the representation of the details of protein structure.

The rotational properties of a rigid rotor are characterized to a large extent by its inertia tensor *(31)*. The inertia tensor of GB3 calculated using the coordinates of the heavy atoms from the crystal structure (1IDG) has the ratio of the principal values of 1.80:1.79:1.00, indicating that the protein can be modeled as an axially symmetric rotor. The unique axis of the tensor is approximately parallel to the α-helix axis as shown in **Fig. 1**. The relationship between the inertia tensor of a protein molecule and its rotational diffusion tensor, relatively straightforward for rigid objects of simple shape *(32,33)*, becomes more complex for more realistic representations of a protein *(13)*.

3.2.1. Simple Models

A rough theoretical estimate of the principal components, D_i (i = x, y, or z), of the diffusion tensor and of the overall correlation time, τ_c, of GB3 can be made assuming the Stokes–Einstein–Debye hydrodynamics model in which the protein is approximated as a rigid rotor in the shape of a sphere, cylinder, or prolate ellipsoid of revolution.

3.2.1.1. SPHERE MODEL

The simplest model is to represent a protein by a sphere. Although this approximation contradicts with the experimental data (**Figs. 2**, **3B**, and **4A,B**) it is instructional to consider it here. Using the Debye equation: $\tau_c = \eta V/kT$, we obtain 1.64 ns. Here, η is the solvent viscosity and V is the volume of the molecule, which we estimated from the molecular weight of the protein assuming that the specific volume is 0.73 cm^3/g.

3.2.1.2. CYLINDER MODEL

The cylinder approximation using empirical relationships from the literature *(34)* resulted in $D_{||}/D_{\perp} = 1.45$ and $\tau_c = 2.35$ ns, assuming solvent viscosity of 0.91 cpoise at 24°C. The sizes of the molecule in the relevant dimensions (27Å in the z and 16Å in both x and y, the axial ratio 1.69) were obtained from the crystal structure. The axial ratio for the cylinder back-calculated from the inertia tensor values is 1.64.

3.2.1.3. ELLIPSOID MODEL

For a prolate ellipsoid model, the ratio of the principal values of the diffusion tensor is approximately given by $D_{||}/D_{\perp} \approx (I_{\perp}/I_{||})^{1/\sqrt{2}}$ *(13)*, where $I_{||}$ and $I_{\perp}$ are the principal components of the inertia tensor. Using this simple hydrodynamics model, we obtained $D_{||}/D_{\perp} = 1.51$ and $\tau_c = 2.64$ ns under identical solvent conditions.

3.2.2. Atomic-Resolution Models

We will now consider more sophisticated models that take into account atomic-resolution details of the shape of the protein. Simple hydrodynamic calculations **(Subheadings 3.2.1.2. and 3.2.1.3.)** based on the size of a "dry" protein molecule provide a reasonable estimate (within 6%) of the anisotropy of the tensor but significantly (by 30–40%) overestimate the rate of molecular rotation. Here, we will take into account the actual structure of the protein and will include the effect of the solvent. Because a detailed picture of the protein's interactions with the surrounding water molecules is not available, these are modeled by including a hydration layer that tumbles together with the protein. A more detailed theoretical analysis should consider specific interactions between water molecules and protein atoms and the friction effects because of the roughness of the protein surface *(35)*.

Having experimental data for the rotational diffusion tensor of the protein enables us to test if the current theoretical models are capable of reproducing these results. The actual parameters for the hydration layer model are not known *a priori* and could be adjusted based on the comparison with the experimental data. Here, we focus on two characteristics of the diffusion tensor: its anisotropy and the overall correlation time. We select these parameters because of the opposite character of their dependence on the size of the hydration shell: Adding a layer of water molecules will increase τ_c (as the rotating body is now larger) and decrease the $D_{||}/D_{\perp}$ (as the hydration shell-enclosed protein is more rounded than the protein alone). Note that, in contrast to τ_c, the anisotropy $(D_{||}/D_{\perp})$ is a dimensionless quantity, which is size-independent and should reflect only the shape of the molecule including the hydration shell. Therefore, a simultaneous comparison of the predictions for both characteristics of the tensor could provide insights into the optimal settings for theoretical hydrodynamic models.

3.2.2.1. BEAD MODEL: "DRY" PROTEIN

The so-called bead model approximates protein molecule by a series of beads *(36)* placed at the coordinates of heavy atoms and with the bead size representing the average atomic radius. First, we consider a "dry protein." For the B3 domain of protein G we could reproduce the experimentally obtained value of τ_c for an atomic radius of 1.45Å, and the experimentally obtained value of $D_\parallel/D_\perp$ for a radius of 1.1Å. Theoretically, one average atomic radius should reproduce both experimental parameters of overall rotational diffusion. Therefore, we conclude that the "dry protein" model is not adequate.

3.2.2.2. BEAD MODEL: "WET" PROTEIN

We then add hydration shells of increasing thickness (0–5Å) to the protein bead model to test if this could reproduce the values of both experimental parameters $(D_\parallel/D_\perp$ and $\tau_c)$ for one bead size and one shell thickness. For these calculations, the protein molecule was "soaked" in water using a 5Å shell following standard procedure in Insight, and then only those waters within a given distance (shell thickness) from the protein atoms were considered. It turns out that several combinations of bead size and hydration shell thickness are consistent with the experimental values (*see* **Fig. 5A**), given the experimental uncertainties. The optimal bead sizes ranged from 0.8 to 1.2Å and the corresponding values of the shell thickness from 1.2 to 1.5Å. The latter values are somewhat smaller than those typically used in hydrodynamic calculations.

3.2.2.3. SHELL MODEL OF THE HYDRATION LAYER

Another, more recent method for calculating surface effects of molecules in solution, uses a strategy known as "shell modeling" *(37)*, where the hydration effects are represented by a shell covering the surface of the protein. This model is characterized by a single parameter *a* that represents the sum of the thickness of the hydration shell and the average atomic van der Waals radius in the molecule. For the B3 domain of protein G we were able to reproduce the experimentally obtained values for both τ_c and $D_\parallel/D_\perp$ for an *a* between 2.5 Å and 2.8Å (**Fig. 5B**). As shown in **Table 1**, both the principal values and the orientation of the calculated diffusion tensor are in remarkable agreement with the experimental data. Assuming an average atomic van der Waals radius of a heavy atom in the protein is about 1.5Å, the hydration shell should have a thickness of between 1.0Å and 1.3Å. This is generally consistent with the results of the bead model, where the hydration shell thickness was calculated to be between 1.2Å and 1.5Å.

A

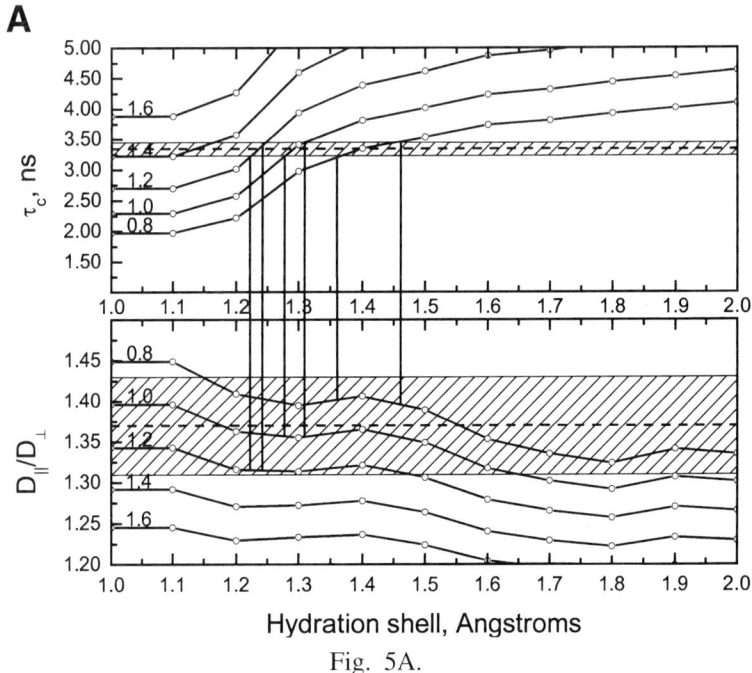

Fig. 5A.

Values of parameter a between 2.5Å and 2.8Å are consistent with the results obtained by de la Torre et al. *(37)*, who calculated rotational diffusion tensors for a variety of a values for 15 proteins covering a range of molecular weights from 2.93 to 26.7 kDa. They found that in most cases experimental values of τ_c were reproduced with values of a between 2Å and 4Å.

4. Discussion

4.1. Effect of the Overall Tumbling Model on the Derived Picture of Local Dynamics

To illustrate how the choice of the overall rotational model could affect the derived picture of local motions, we determined backbone model-free parameters (S^2, τ_{loc}) *(38)* and R_{ex} contributions (if any) for GB3 using all three models of the overall rotational diffusion tensor: isotropic, axially symmetric, and fully anisotropic. The characteristics of the rotational diffusion tensor for both anisotropic models were derived as described in the previous sections and then used to obtain residue-specific values of the microdynamic parameters (*see* **Subheading 3.1.2.**). The values of S^2 and R_{ex} for each residue are summarized in **Fig. 6**.

All three models show simultaneous decrease in the order parameters in the $\beta1/\beta2$ and $\alpha/\beta3$ loops indicating particularly flexible regions. In the $\beta2/\alpha$ and

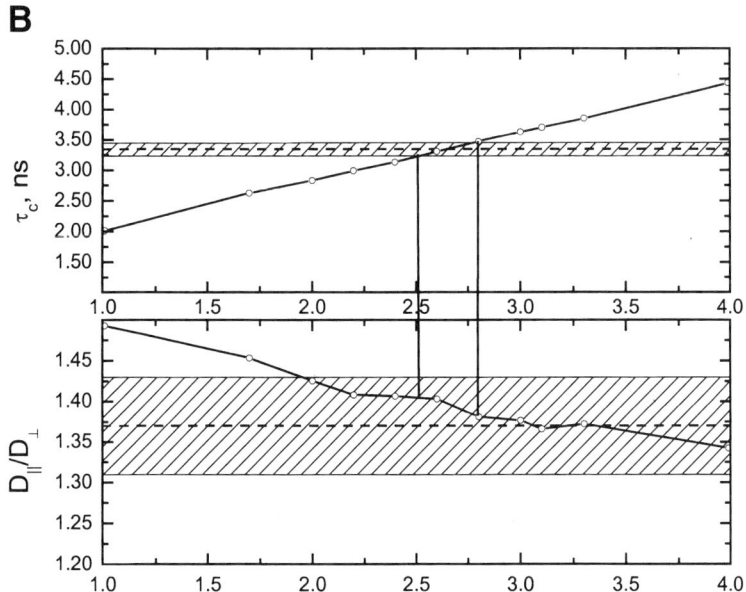

Fig. 5B. Comparison of the characteristics of the diffusion tensor, τ_c and $D_\parallel/D_\perp$, derived from ^{15}N relaxation data with the results of hydrodynamic calculations. (**A**) "Bead model": dependence of these parameters on the thickness of hydration shell for various "bead" sizes, indicated by the corresponding numbers for each line. (**B**) "Shell model": dependence on the parameter *a* (*see* **Subheading 3.2.2.3.**). The dashed lines represent the experimental values of the diffusion tensor characteristics derived for the axially symmetric model, whereas the hatched areas represent their 68.3%-confidence region. The vertical bars in panel **A** mark the regions that are inside the experimental errors for both measured parameters ($D_\parallel/D_\perp$ and t_c) for a particular bead size. Similar bars in **B** indicate the range of shell thickness that is consistent with both experimental parameters. The "bead" model calculations were performed using an in-house Matlab program based on the method outlined in **ref. 36**. The "shell" model calculations were done using program HYDRONMR *(37)*.

β3/β4 loops, there is a small decrease in order parameters, indicating that these regions are more flexible than the elements of secondary structure but less flexible than the other, more extended loops. All three models yield somewhat elevated order parameters in the region of the α-helix. The fully anisotropic and axially symmetric models predict slightly higher values for the order parameters in this region than does the isotropic model. Overall, the difference in order parameters derived using the isotropic and anisotropic models is observable but does not represent dramatically different pictures of motions. For most of the NH groups, the order parameters derived using the two anisotropic models are practically indistinguishable from each other.

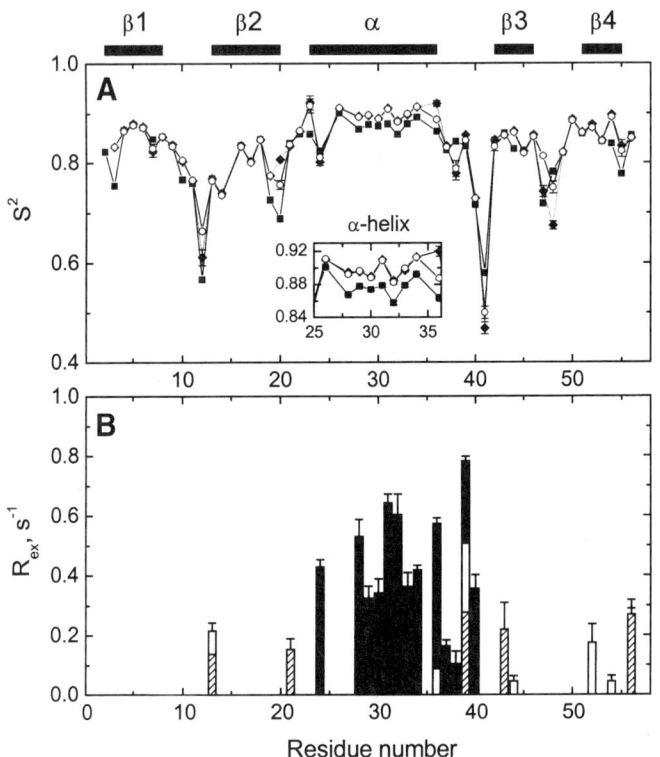

Fig. 6. Comparison of the **(A)** order parameters (S^2) and **(B)** conformational exchange terms (R_{ex}) for backbone dynamics in GB3, derived using the three models of the overall tumbling. In panel **A**, different symbols represent squared order parameters for isotropic (*solid squares*), axially symmetric (*open circles*), and anisotropic (*solid triangles*) models. The *bars* representing R_{ex} values in **B** correspond to the isotropic (*solid*), axially symmetric (*open*), and fully anisotropic (*hatched*) models of overall rotational diffusion. Insert in panel **A** is a blowup of the order parameters in the region of the α-helix. The horizontal bars on the top of panel **A** indicate the elements of secondary structure.

The most striking difference between the isotropic and anisotropic models is in the resulting pictures of conformational exchange motions (**Fig. 6B**). The isotropic model predicts conformational exchange in a stretch of 13 residues, Glu[24], Lys[28]–Ala[34], and Asp[36]–Asp[40], covering the entire α-helix (note that Thr[25], Glu[27], and Asn[35] were excluded because of signal overlap), whereas a significantly fewer number of sites show this type of motion when rotational anisotropy is taken into account. Only for Val[39] do all three models agree, making it a likely candidate for conformational exchange (see a more detailed discussion in **ref. 3**). A detailed analysis including additional, model-independent approaches *(3)* confirmed the absence of conformational exchange contributions in the α-helix. These results illustrate the necessity of using the right overall rotational model for accurate analysis of protein dynamics.

Acknowledgments

Supported by NIH grant GM65334 to D. Fushman.

References

1. Fushman, D. and Cowburn, D. (1998) Studying protein dynamics with NMR relaxation. In *Structure, Motion, Interaction and Expression of Biological Macromolecules* (Sarma, R. and Sarma, M., eds.). Adenine Press, Albany, NY, pp. 63–77.
2. Schurr, J. M., Babcock, H. P., and Fujimoto, B. S. (1994) A test of the model-free formulas: effects of anisotropic rotational diffusion and dimerization. *J. Magn. Reson.* **B105,** 211–224.
3. Hall, J. B. and Fushman, D. (2003) Characterization of the overall and local dynamics of a protein with intermediate rotational anisotropy: differentiating between conformational exchange and anisotropic diffusion in the B3 domain of protein G. *J. Biomol. NMR* **27,** 261–275.
4. Tjandra, N., Wingfield, P., Stahl, S., and Bax, A. (1996) Anisotropic rotational diffusion of predeuterated HIV protease from ^{15}N NMR relaxation measurements at two magnetic fields. *J. Biomol. NMR* **8,** 273–284.
5. Luginbuhl, P., Pervushin, K. V., Iwai, H., and Wuthrich, K. (1997) Anisotropic molecular rotational diffusion in ^{15}N spin relaxation studies of protein mobility. *Biochemistry* **36,** 7305–7312.
6. Eisenmesser, E. Z., Bosco, D. A., Akke, M., and Kern, D. (2002) Enzyme dynamics during catalysis. *Science* **295,** 1520–1523.
7. Fushman, D., Xu, R., and Cowburn, D. (1999) Direct determination of changes of interdomain orientation on ligation: use of the orientational dependence of ^{15}N NMR relaxation in Abl SH(32) *Biochemistry* **38,** 10,225–10,230.
8. Varadan, R., Walker, O., Pickart, C., and Fushman, D. (2002) Structural properties of polyubiquitin chains in solution. *J. Mol. Biol.* **324,** 637–647.
9. Tjandra, N., Garrett, D. S., Gronenborn, A. M., Bax, A., and Clore, G. M. (1997) Defining long range order in NMR structure determination from the dependence of heteronuclear relaxation times on rotational diffusion anisotropy. *Nat. Struct. Biol.* **4,** 443–449.
10. Akerud, T., Thulin, E., Van Etten, R. L., and Akke, M. (2002) Intramolecular dynamics of low molecular weight protein tyrosine phosphatase in monomer-dimer equilibrium studied by NMR: a model for changes in dynamics upon target binding. *J. Mol. Biol.* **322,** 137–152.
11. Tjandra, N., Feller, S. E., Pastor, R. W., and Bax, A. (1995) Rotational diffusion anisotropy of human ubiquitin from ^{15}N NMR relaxation. *J. Am. Chem. Soc.* **117,** 12,562–12,566.

12. Lee, L. K., Rance, M., Chazin, W. J., and Palmer, A. G., III. (1997) Rotational diffusion anisotropy of proteins from simultaneous analysis of ^{15}N and ^{13}C alpha nuclear spin relaxation. *J. Biomol. NMR* **9,** 287–298.

13. Copie, V., Tomita, Y., Akiyama, S. K., et al. (1998) Solution structure and dynamics of linked cell attachment modules of mouse fibronectin containing the RGD and synergy regions: comparison with the human fibronectin crystal structure. *J. Mol. Biol.* **277,** 663–682.

14. Blackledge, M., Cordier, F., Dosset, P., and Marion, D. (1998) Precision and uncertainty in the characterization of anisotropic rotational diffusion by ^{15}N relaxation. *J. Am. Chem. Soc.* **120,** 4538, 4539.

15. Dosset, P., Hus, J. C., Blackledge, M., and Marion, D. (2000) Efficient analysis of macromolecular rotational diffusion from heteronuclear relaxation data. *J. Biomol. NMR* **16,** 23–28.

16. Ghose, R., Fushman, D., and Cowburn, D. (2001) Determination of the rotational diffusion tensor of macromolecules in solution from NMR relaxation data with a combination of exact and approximate methods—application to the determination of interdomain orientation in multidomain proteins. *J. Magn. Reson.* **149,** 214–217.

16a. Fushman, D., Varadan, R., Assfalg, M., et al. (2004). Determining domain orientation in macromolecules by using spin-relaxation and residual dipolar coupling measurements. *Prog. NMR Spectr.*, in press.

16b. Fushman, D., Varadan, R., Assfalg, M., and Walker, O. (2004) Determining domain orientation in macromolecules by using spin-relaxation and residual dipolar coupling measurements. *Prog. NMR Spectr.*, in press.

17. Gronenborn, A. M., Filpula, D. R., Essig, N. Z., et al. (1991) A novel highly stable fold of the immunoglobulin binding domain of streptococcal protein G. *Science* **253,** 657–661.

18. Achari, A., Hale, S. P., Howard, A. J., et al. (1992) 1.67-A X-ray structure of the B2 immunoglobulin-binding domain of streptococcal protein G and comparison to the NMR structure of the B1 domain. *Biochemistry* **31,** 10,449–10,457.

19. Derrick, J. P. and Wigley, D. B. (1992) Crystal structure of a streptococcal protein G domain bound to a Fab fragment. *Nature* **359,** 752–754.

20. Derrick, J. P., Wigley, D. B., Lian, L. Y., et al. (1993) Structure and mechanism of streptococcal protein G. *Biochem. Soc. Trans.* **21,** 333S.

21. Gallagher, T., Alexander, P., Bryan, P., and Gilliland, G. L. (1994) Two crystal structures of the B1 immunoglobulin-binding domain of streptococcal protein G and comparison with NMR. *Biochemistry* **33,** 4721–4729.

22. Derrick, J. P. and Wigley, D. B. (1994) The third IgG-binding domain from streptococcal protein G: an analysis by X-ray crystallography of the structure alone and in a complex with Fab. *J. Mol. Biol.* **243,** 906–918.

23. Lian, L. Y., Derrick, J. P., Sutcliffe, M. J., Yang, J. C., and Roberts, G. C. (1992) Determination of the solution structures of domains II and III of protein G from Streptococcus by ^{1}H nuclear magnetic resonance. *J. Mol. Biol.* **228,** 1219–1234.

24. Fushman, D., Cahill, S., and Cowburn, D. (1997) The main chain dynamics of the dynamin pleckstrin homology (PH) domain in solution: analysis of ^{15}N relaxation with monomer/dimer equilibration. *J. Mol. Biol.* **266,** 173–194.

25. Fushman, D., Tjandra, N., and Cowburn, D. (1999) An approach to direct determination of protein dynamics from ^{15}N NMR relaxation at multiple fields, independent of variable ^{15}N chemical shift anisotropy and chemical exchange contributions. *J. Am. Chem. Soc.* **121**, 8577–8582.

26. Fushman, D. (2002) Determination of protein dynamics using ^{15}N relaxation measurements. In *BioNMR in Drug Research* (Zerbe, O., ed.). Wiley-VCH, New York, pp. 283–308.

27. Fushman, D. and Cowburn, D. (2002) Characterization Of inter-domain orientations in solution using the NMR relaxation approach. In *Protein NMR for the Millenium: Biological Magnetic Resonance, vol. 20* (Krishna, N. R., ed.). Kluwer, Dordrecht, The Netherlands, pp. 53–78.

28. Fushman, D., Cahill, S., and Cowburn, D. (1997) The main-chain dynamics of the dynamin pleckstrin homology (PH) domain in solution: analysis of ^{15}N relaxation with monomer/dimer equilibration. *J. Mol. Biol.* **266**, 173–194.

29. Fushman, D., Weisemann, R., Thüring, H., and Rüterjans, H. (1994) Backbone dynamics of ribonuclease T1 and its complex with 2'GMP studied by two-dimensional heteronuclear NMR spectroscopy. *J. Biomol. NMR* **4**, 61–78.

30. Draper, N. R. and Smith, H. (1981) *Applied Regression Analysis.* Wiley, New York.

31. Goldstein, H. (1980) *Classical Mechanics.* 2nd ed. Addison-Wesley, Boston.

32. Koenig, S. (1975) Brownian motion of an ellipsoid. Correction to Perrin's results. *Biopolymers* **14**, 2421–2423.

33. Cantor, C. R. and Schimmel, P. R. (1980) *Biophysical Chemistry.* 3 vols. Freeman, New York.

34. Tirado, M. M. and de la Torre, J. G. (1980) Rotational dynamics of rigid, symmetric top macromolecules: application to circular cylinders. *J. Chem. Phys.* **73**, 1986–1993.

35. Fushman, D. (1990) Surface fractality of proteins from theory and NMR data. *J. Biomol. Struct. Dyn.* **7**, 1333–1344.

36. de la Torre, J. G., Navarro, S., Martinez, M. C. L., Diaz, F. G., and Cascales, J. J. L. (1994) HYDRO: A computer program for the prediction of hydrodynamic properties of macromolecules. *Biophys. J.* **67**, 530, 531.

37. de la Torre, J. G., Huertas, M. L., and Carrasco, B. (2000) HYDRONMR: prediction of NMR relaxation of globular proteins from atomic-level structures and hydrodynamic calculations. *J. Magn. Reson.* **B147**, 138–146.

38. Lipari, G. and Szabo, A. (1982) Model-free approach to the interpretation of nuclear magnetic resonance relaxation in macromolecules. 1. Theory and range of validity. *J. Am. Chem. Soc.* **104**, 4546–4559.

39. DeLano, W. L. (2002) *The PyMOL Molecular Graphics System.* DeLano Scientific, San Carlos, CA.

11

TROSY-Based NMR Experiments for the Study of Macromolecular Dynamics and Hydrogen Bonding

Guang Zhu, Youlin Xia, Donghai Lin, and Xiaolian Gao

Summary

Transverse relaxation-optimized spectroscopy (TROSY)-based nuclear magnetic resonance (NMR) experiments can be exploited to obtain chemical shift assignment and values of J-coupling constants, residual dipolar couplings, and nuclear Overhauser effects (NOEs) for structural studies of proteins, as discussed in Chapter 5. Furthermore, the application of TROSY-based NMR experiments can be extended to the measurements of molecule dynamics, amide proton exchange rates, and hydrogen bonds. This chapter describes these experiments.

Key Words: TROSY; T1-TROSY; T2-TROSY; NOE-TROSY; SEA-TROSY; HNN-TROSY.

1. Introduction

In addition to the transverse relaxation-optimized spectroscopy (TROSY)-based nuclear magnetic resonance (NMR) experiments described in Chapter 5, the application of TROSY has been extended to NMR studies of protein dynamics *(1–3)* and protein–protein interactions. The saturation transfer method, which is very similar to the fluorescence resonance energy transfer method in optical spectroscopy, can be used to map the interface of protein–protein interaction *(4)*. To reduce spectral overlap, solvent exposed amides (SEA)-TROSY may be employed to observe the ^{1}H–^{15}N moieties that are on the protein surfaces and in active exchange with water *(5)*. A modified version can be readily used to measure exchange rates *(6)* and can be applied to study protein folding and protein–protein interactions.

TROSY NMR methods have also been applied to the direct detection of hydrogen bonds in nucleic acids and proteins *(7–9)*. The reduced line width associated with TROSY allows direct observation of scalar coupling across hydrogen bonds in nucleic acid basepairs by the measurement of resolved

From: *Methods in Molecular Biology, vol. 278: Protein NMR Techniques*
Edited by: A. K. Downing © Humana Press Inc., Totowa, NJ

multiplets *(10)*. In more complex coherence transfer experiments *(11,12)* couplings are identified based on small chemical shift differences, which may be measured using TROSY-type Exclusive COrrelation SpectroscopY (E.COSY) *(10,13)* and by correlation peak intensity, which is measured using TROSY-type J quantitative correlation experiments *(14–16)*. Similarly, residual dipolar couplings can be readily measured by TROSY-based $^1H/^{15}N/^{13}CO$ (HNCO)-type three-dimensional (3D) experiments and serve as important structural constraints for the highly deuterated proteins *(17,18)*. TROSY has also been applied to measure other J-couplings to obtain constraints for structural determination of proteins and nucleic acids *(19)*.

2. TROSY-Based Experiments for NMR Studies of Protein Dynamics

Complete understanding of protein function requires detailed studies of protein dynamics *(20–22)*. NMR spectroscopy is a powerful tool for investigating nanosecond to picosecond and millisecond to microsecond dynamics of backbone atoms of proteins via ^{15}N relaxation studies *(23–26)*. These studies include measurements of the longitudinal relaxation time (T_1) and transverse relaxation time (T_2) of backbone ^{15}N nuclei, and the heteronuclear nuclear Overhauser effect (NOE) between backbone ^{15}N and the attached ^1H by the well-established NMR techniques based on the heteronuclear single-quantum coherence (HSQC) experiment. In this section, TROSY-based experiments for the measurement of dynamics parameters measurement are discussed *(1–3)*.

2.1. Protein Dynamics Measurement By Two-Dimensional TROSY-Based Experiments

TROSY has been introduced into HSQC-based two-dimensional (2D) experiments for measuring T_1, T_2, and the heteronuclear NOE *(1)*. Sensitivity and spectral resolution in these TROSY-based NMR experiments for ^{15}N dynamics studies of proteins are significantly improved. Increased sensitivity gain and higher resolution are expected when a deuterated protein sample is used. The methods described should allow investigation of protein dynamics beyond the limitations of the currently used NMR techniques.

The pulse sequences based on TROSY for measuring T_1, T_2, and NOE of $^{15}N–^1H^N$ moieties are depicted in **Fig. 1**. In HSQC-based pulse sequences for ^{15}N relaxation studies *(27)*, a reverse insensitive nucleus enhancement by polarization transfer (INEPT) sequence is used to produce the required antiphase magnetization of ^{15}N before the preservation of equivalent path (PEP) sequence *(28)*. In the TROSY-based sequences, this INEPT sequence is eliminated and the PEP sequence is replaced by the single transition to single transition polarization transfer sequence *(29)*. The introduction of TROSY components in this way not only effectively selects the slowest relaxing component of the $^{15}N–^1H^N$ moiety but also

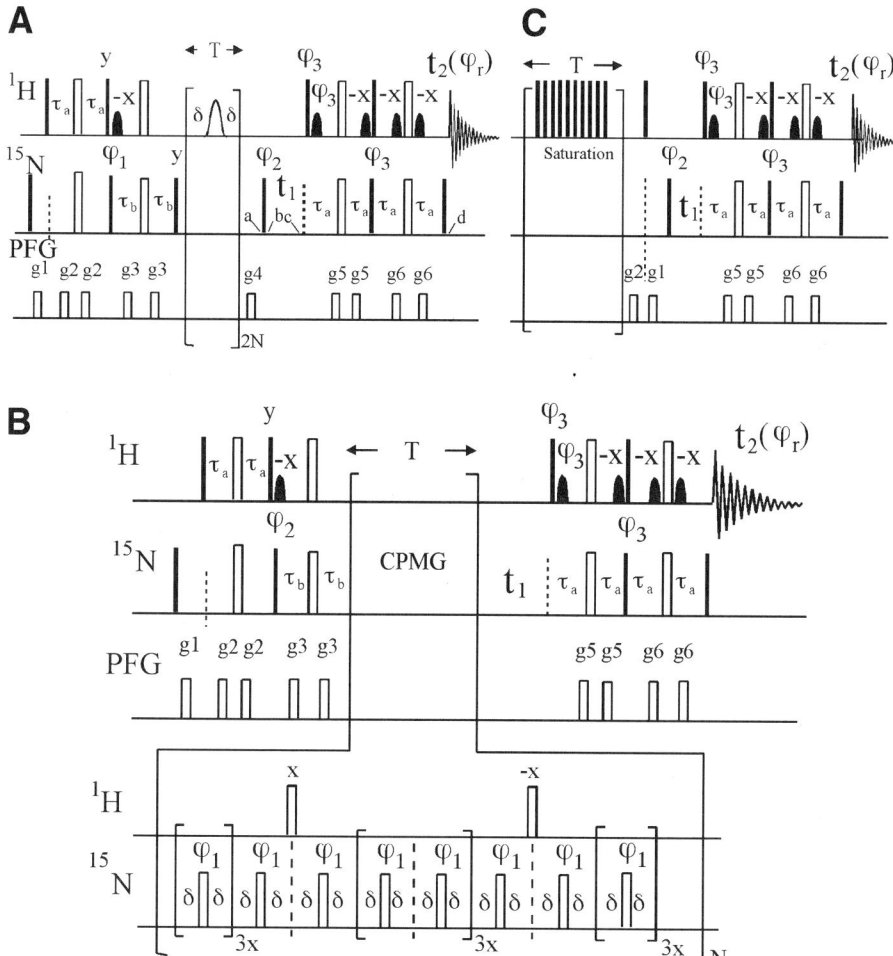

Fig. 1. Pulse sequences for the measurement of (**A**) ^{15}N T_1 (T1–TROSY), (**B**) ^{15}N T_2 (T2–TROSY), and (**C**) ^{1}H–^{15}N NOE (NOE–TROSY). In all sequences, *filled bars* and *open bars* represent 90° and 180° pulses, respectively. Filled shaped pulses are 1.1 ms sinc-modulated rectangular 90° pulses to selectively excite the water resonance. Default phases are x. In the T1–TROSY sequence (**A**), phase cycling is as follows: $\varphi_1 = 4(x), 4(-x)$; $\varphi_2 = (y, x, -y, -x)$; $\varphi_3 = (y)$; $\varphi_r = (x, -y, -x, y, -x, y, x, -y)$. In T2–TROSY (**B**), phase cycling is: $\varphi_1 = (y, x)$; $\varphi_2 = (y, -x, -y, x)$; $\varphi_3 = (y)$; $\varphi_r = (x, -y, -x, y)$. In NOE–TROSY (**C**), phase cycling is: $\varphi_2 = (y, x, -y, -x)$; $\varphi_3 = (y)$; $\varphi_r = (x, -y, -x, y)$.

has the benefit of shortening the pulse sequences by 5.4 ms (= $1/2^1J_{HN}$) and results in a great increase in sensitivity. The data-fitting procedures and the analysis of the protein dynamics remain unchanged with respect to the conventional HSQC-based relaxation experiments. An extensive discussion of a related topic about the characterization of chemical exchange phenomena *(3)* can be found in Chapter 12.

2.1.1. Application to Proteins

Protein sample used: 1 mM ^{15}N-labeled calmodulin (147 a.a.) in 90% H_2O/10% D_2O at pH 6.8.

Equipment: 750-MHz NMR spectrometer with three channels.

Experimental parameters and results:

1. Temperature: 5°C.
2. Experimental parameters: In **Fig. 1**, τ_a = 2.25 ms and τ_b = $1/(4^1J_{NH})$ ≈ 2.75 ms. For the three experiments, echo/antiecho selections during t_1 are done by reversing φ_3 and the even-number phases of the receiver. To remove axial peaks in the ω_1 dimension, phase φ_2 and the receiver phase φ_r are inverted for every second t_1 increment. The durations and strengths of the gradients are of conventional values. In the T1–TROSY sequence (**Fig. 1A**), an even number of shaped pulses, 333 μs cosine-modulated rectangular 180° pulses with excitation maximum positioned at 4 ppm downfield from the carrier (on water), are applied every 5 ms (δ = 2.5 ms) during the recovery time T. The experimental recovery delay is 1.1 s. In T2–TROSY (**Fig. 1B**), δ = 0.55 ms, and the experimental recovery delay is 2 s. To reduce heating during the CPMG portion, the ^{15}N pulse power is decreased by 6 dB compared with that in **Fig. 1B** and **C**. In NOE–TROSY (**Fig. 1C**), ^{1}H saturation is achieved by the application of 120° pulses spaced at 5 ms intervals for 3 s prior to the first ^{15}N pulse. An overall delay between scans of 5 s was employed in both the NOE and NONOE experiments. The duration and power of GARP decouplings on the ^{15}N channel are 100 ms and 3.5 kHz, respectively.
3. To compare the dynamic parameters obtained from the conventional HSQC and the proposed TROSY methods, the corresponding two sets of spectra were recorded. In all experiments, 256×1024 complex points were acquired in the time domain with spectral widths of 3000 Hz and 10,500 Hz, respectively. The T_1 values were measured from spectra recorded with eight different relaxation delays with T = 10.7, 85.3, 192.0, 320.0, 480.0, 693.3, 981.3, and 1440.0 ms. The T_2 values were determined from spectra recorded with relaxation delays T = 19.4, 38.8, 58.3, 77.7, 97.1, 136.0, 174.8, and 213.7 ms. The NOE values were determined from spectra recorded in the presence (NOE) and absence (NONOE) of a proton presaturation period of 3 s. The numbers of scans were 16, 16, and 64 for T_1, T_2, and NOE experiments, respectively. All data processing was performed using the NMRPipe software package *(30)*. All spectral matrices were 4096×4096 and were processed using identical parameters
4. Results: The sensitivity and line widths of the two spectra obtained by T1–TROSY and the corresponding HSQC experiment with the relaxation delay T = 10.7 ms are

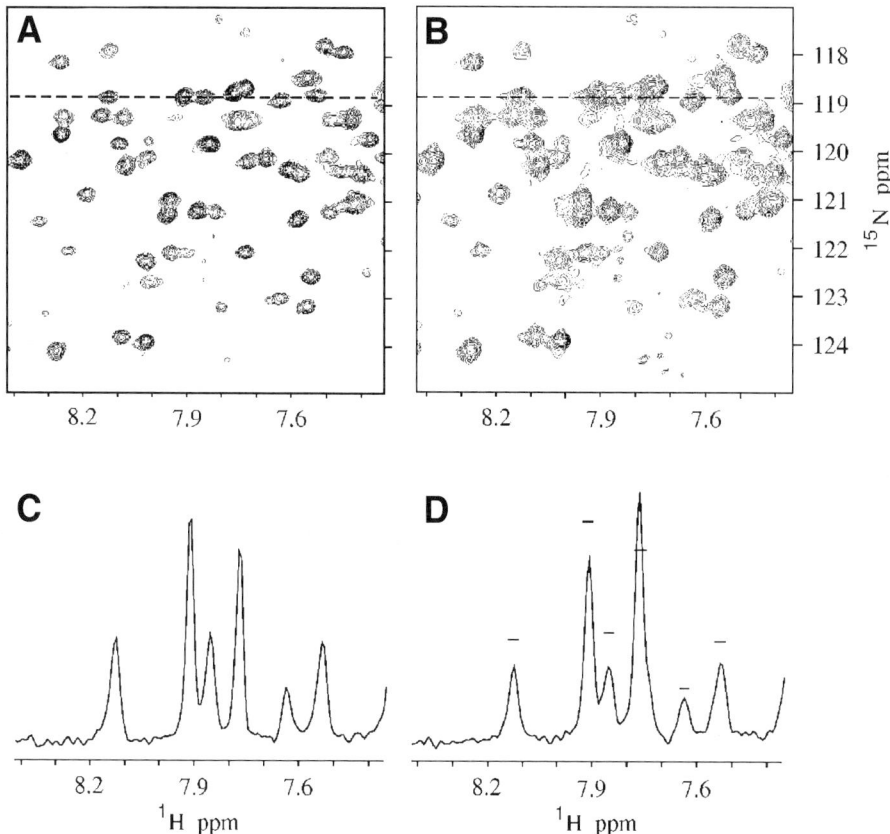

Fig. 2. Sections of T1–TROSY spectrum (**A**) and the corresponding T1–HSQC spectrum (**B**) of ^{15}N-labeled calmodulin. Both spectral sections were recorded and processed with the same parameters. They are plotted at the same levels with a contour spacing of 1.2. For the convenience of comparison, the spectrum (**A**) is shifted 45 Hz in both the ^{1}H and ^{15}N dimensions. (**C**) and (**D**) are the 1D slices taken from spectra (**A**) and (**B**), respectively, at the *dotted line* positions. The *short bars* in spectrum (**D**) mark the heights of the corresponding peaks in (**C**).

compared. Two corresponding sections of the 2D spectra are displayed in **Fig. 2A** and **B**. The peaks in the T1–TROSY spectrum are, on average, 13% more intense and 2–14 Hz narrower than the corresponding peaks in the T1–HSQC spectrum, when 80 isolated peaks are selected for the comparison. Statistical analysis shows excellent agreement between the protein dynamics parameters obtained using the TROSY-based and the corresponding HSQC-based experiments (*1*).

These studies also show that the TROSY experiments are only about 5% less sensitive (rather than the 50% expected based on observation of a single component of the multiplets) at 35°C than the corresponding HSQC experiments under the experimental conditions described in the two preceding paragraphs. This sensitivity gain is the result of the following factors. First, because of the TROSY effect, the cross-peaks in the TROSY-based spectra are 3–5 Hz narrower in line width in both dimensions when compared with the corresponding cross-peaks in the HSQC-based spectra. Second, in the TROSY-based pulse sequences, the time during which magnetization resides in the transverse plane is 5.4 ms shorter than that in the corresponding HSQC pulse sequences, resulting in a great sensitivity gain. These effects are clearly demonstrated in **Fig. 2A–D**, where the T1–TROSY and T1–HSQC experiments have been recorded on calmodulin at 5°C, a temperature at which the molecule tumbles at an estimated rate of about 18 ns assuming isotropic tumbling. This is equivalent to the tumbling rate of a protein about 40–50 kDa at room temperature. The sensitivity gain will be more pronounced when larger and deuterated biomolecules (>25 kDa) are employed in similar NMR relaxation studies.

2.2. Protein Dynamics Measurement By 3D TROSY–HNCO Experiment

TROSY-based 2D NMR experiments for the measurement of dynamics parameters can significantly improve signal sensitivity and spectral resolution *(1)*. However, as the molecular weight of proteins under NMR investigation increases, spectral overlap becomes more severe in 2D TROSY spectra and hampers accurate analysis of the protein dynamics. An apparent solution is to expand these 2D TROSY-based experiments to 3D TROSY-based experiments. In parallel, the 3D HSQC–HNCO-based experiments should be effective for nondeuterated proteins with smaller molecular weights and seriously overlapped 2D HSQC spectra. Several novel 3D TROSY–HNCO-based pulse sequences *(2)* are described for measuring ^{15}N T_1 and T_2, and ^1H–^{15}N NOE to overcome spectral overlap problems in the corresponding 2D spectra.

Fig. 3 depicts the pulse sequences used for the 3D TROSY–HNCO-based NMR experiments. HNCO is the most sensitive triple-resonance experiment; therefore TROSY–HNCO has been selected and modified to measure the dynamics parameters of larger deuterated proteins. Because about 12 to 18 3D spectra must be recorded for the measurement of T_1, T_2, and NOE values, a minimal number of phase-cycling steps is desired to finish all these 3D NMR experiments in a reasonable time. Therefore a two-step phase-cycling scheme has been adopted for these experiments.

2.2.1. Application to Proteins

Protein sample used: uniformly 100% ^{15}N, 100% ^{13}C, and 70% ^2H-labeled 1.0 m*M* trichosanthin (approx 27 kDa) in 20 m*M* Na_2HPO_4, pH 6.8, 95% H_2O/5% D_2O.

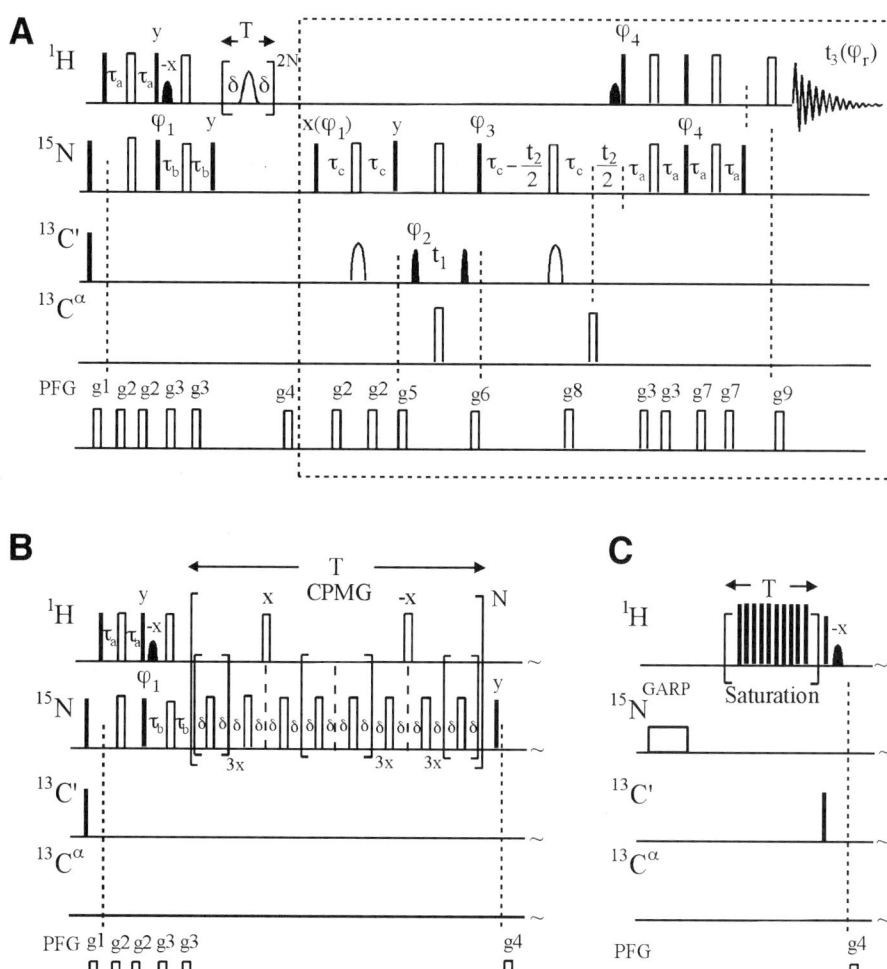

Fig. 3. Pulse sequences for the measurement of ^{15}N T_1 (**A**) (T1–TROSY–HNCO), ^{15}N T_2 (**B**) (T2–TROSY–HNCO), and ^{1}H–^{15}N NOE (**C**) (NOE–TROSY–HNCO). The sections not plotted in (**B**) and (**C**) are the same as in the dashed box of (**A**). In all sequences, filled bars and open bars represent 90° and 180° pulses, respectively. Filled shaped pulses are 1.1 ms sinc-modulated rectangular 90° pulses to selectively excite the water resonance. Default phases are x. For the three experiments based on 3D TROSY–HNCO, phase cycling is as follows: $\varphi_1 = (x, -x)$; $\varphi_2 = (x)$; $\varphi_3 = (y)$; $\varphi_4 = (y)$; $\varphi_r = (x, -x)$. For the 90° pulse labeled x (φ_1) in (**A**), its phase is x for measuring T_1 and T_2, and φ_1 for measuring the NOE. (Reproduced with permission from **ref. 2**.)

Equipment: 750-MHz NMR spectrometer with three channels.
Experimental parameters and results:

1. Temperature: 30°C.
2. Experimental and processing parameters: ^{1}H, ^{15}N, and ^{13}C′ carrier frequencies were centered at 4.7, 120, and 174 ppm, respectively. ^{13}C′ 90° (filled shaped) and 180° (open shaped) pulses are sinc-modulated rectangular pulses with pulse widths of 71.2 and 64.4 μs, respectively. ^{13}C$^\alpha$ 180° pulses are phase-modulated rectangular pulses, whose pulse widths, 39 μs, satisfy the relation of $\sqrt{3}/\Delta$, where Δ is the separation between the centers of the ^{13}C$^\alpha$ and ^{13}C′ chemical shift regions. $\tau_a =$ 2.50 ms, $\tau_b = 1/(4^1J_{NH}) \approx 2.70$ ms, $\tau_c = 15$ ms, and $\tau_d = 5.4$ ms. Quadrature components in t_1 are acquired through altering φ_2 in a States-time proportional phase increment (TPPI) manner; echo/antiecho selections during t_2 are done by inverting the sign of φ_4 and gradient g8. Axial peaks in the F_1 and F_2 dimensions are removed by inverting (φ_2,φ_r) and (φ_3,φ_r) for every second t_1 and t_2, respectively. The durations and strengths of the gradients are of conventional values. Other parameters are the same as that in 2D T1–TROSY, T2–TROSY, and NOE–TROSY experiments (**Fig. 1**).
3. Data matrices in the time domain were composed of 42×60×1024 complex points, with spectral widths 2400×2400×10,500 Hz. The number of scans for each transient was two. The recovery delay between scans was 0.8 s, $T = 10.6$ ms, and the total experimental time for acquiring a 3D spectrum was 5 h and 44 min. Cosine-square bell window functions were used to obtain the 3D spectra before Fourier transformation, which were composed of 128×256×1024 points. All data were processed using the NMRPipe software package *(30)*.
4. Results: **Fig. 4A** and **B** show small regions of spectra recorded using 2D T1–HSQC and 2D T1–TROSY *(1)*, respectively. The 2D TROSY spectrum has better sensitivity and resolution than the 2D HSQC spectrum; however, some peaks, such as peaks 117 and 132, are not resolved even in the 2D TROSY spectrum. Two small 2D slices taken from the 3D T1–TROSY–HNCO spectrum, with CO chemical shifts of 174.91 ppm and 175.40 ppm, are shown in **Fig. 4C** and **D**, respectively. Because these two peaks have different CO chemical shifts, they are completely resolved in the 3D TROSY–HNCO spectrum.

The 3D TROSY–HNCO-based pulse sequence has been used successfully to measure ^{15}N T_1, T_2, and the ^{1}H–^{15}N NOE. Because the number of phase-cycling steps in these pulse sequences is limited to two, all experiments can be finished within an appropriate period of time (less than 7 d). TROSY-based measurements can significantly enhance signal sensitivity in dynamics studies of large deuterated protein molecules. For the tested sample at 30°C, the sensitivity of the 3D TROSY–HNCO-based technique is, on the average, enhanced by 72% compared to that of the 3D HSQC–HNCO-based measurements. For fully deuterated or larger proteins, the TROSY effect should be even stronger. Statistical analysis

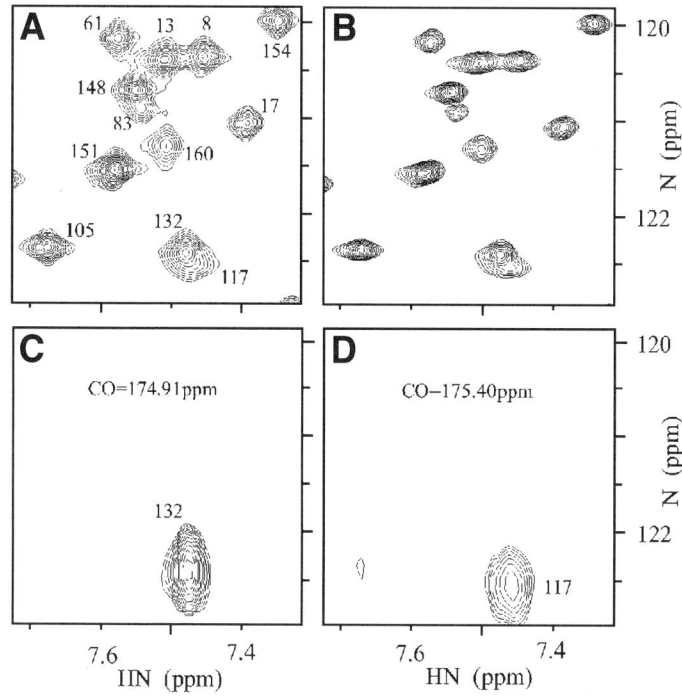

Fig. 4. Small regions of spectra recorded by 2D T1–HSQC (**A**) and 2D T1–TROSY (**B**). Small 2D slices taken from a 3D TROSY–HNCO spectrum at CO chemical shifts of 174.91 ppm (**C**) and 175.40 ppm (**D**). The lowest contours for spectra (**A**) and (**B**) are drawn at the same level, and they are identical for (**C**) and (**D**). Contours are spaced by a factor of 1.3. The numbers in (**A**) are peak numbers. (Reproduced with permission from **ref. 2**.)

confirmed that the results from the 3D TROSY–HNCO-based experiments are in good agreement with those measured with the traditional 2D HSQC techniques.

3. TROSY Experiments Based on Amide Proton Exchange

3.1. SEA–TROSY

Spectral overlap is usually one of the major obstacles for structure determination of very large proteins. A SEA–TROSY experiment (*5*) has been proposed to partly resolve this problem. Because the solvent exposed amide protons can exchange rapidly with water, they are expected to appear in the SEA–TROSY spectra as stronger peaks, although those buried in the interior of the protein will disappear or appear as weaker peaks. Changes in peak intensity allow some overlapped peaks to be resolved.

Figure 5A shows the pulse sequence for SEA–TROSY experiments. The pulse sequence starts with a ^{15}N double filter *(31,32)* that serves to eliminate all the magnetization generated from amide protons. Water magnetization is not affected by the ^{15}N filter and subsequently is returned to the *z*-axis by the last 90° ^{1}H pulse prior to a variable mixing time τ_m. At this time-point, water *z* magnetization is allowed to exchange with amide protons during τ_m. Backbone amides that are exposed to the solvent will acquire magnetization from the solvent that can be subsequently detected with a TROSY-type experiment *(29,33)*.

The peak intensity in the SEA–TROSY spectrum is related to the exchange rate of the amide proton. The exchange rate depends on a number of factors, including the local environment of the residue in the protein, the amino acid type, and its solvent accessibility *(34,35)*. In general, residues located in loop regions on the protein surface are in very fast exchange with water and appear as stronger peaks in the spectrum, whereas residues located in hydrogen-bonded secondary structures or buried in the interior of proteins are in relatively slow exchange with water and appear as weaker peaks or do not appear in the SEA–TROSY spectrum.

The SEA element can be combined with triple-resonance TROSY-type experiments, performed on triply ^{13}C/^{2}H/^{15}N-labeled samples, as in SEA–^{1}H/^{15}N/^{13}CA (HNCA)–TROSY *(5)*, to obtain backbone resonance assignments for solvent exposed loop regions in very large proteins.

3.2. Clean SEA–TROSY

Perdeuteration of the protein sample is a prerequisite for SEA–TROSY experiments to work properly. For partially deuterated or protonated proteins and protein complexes, SEA–TROSY spectra may contain NOE contributions from aliphatic protons. Moreover, SEA–TROSY spectra may be contaminated with exchange-relayed NOE contributions from fast exchanging hydroxyl or amine protons *(36)* and artifacts owing to longitudinal relaxation processes during τ_m *(37)*.

The Clean SEA–TROSY experiment *(6)*, a modified version of SEA–TROSY, not only effectively eliminates these NOE contributions but also suppresses the longitudinal relaxation contributions by using an appropriate phase-cycling scheme. Therefore, perdeuteration of the sample is not required for Clean SEA–TROSY experiments. The peak intensities in a Clean SEA–TROSY spectrum, *I*, are related to their exchange rates, k_{ex}, according to the following equation *(38,39)*:

$$I = \frac{fX_A k_{ex}}{\left(R_{1A,app} + k_{ex} - R_{1B,app}\right)} \left(\exp^{\left(-R_{1B,app}\tau_m\right)} - \exp^{\left[-\left(R_{1A,app} + k_{ex}\right)\tau_m\right]} \right) \qquad (1)$$

A

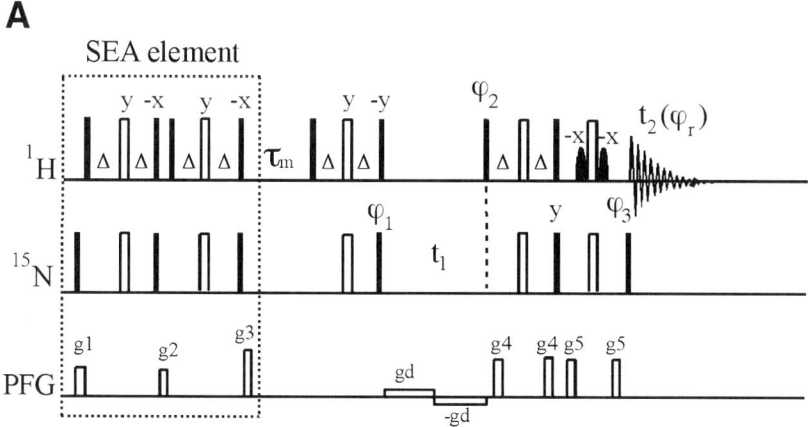

B

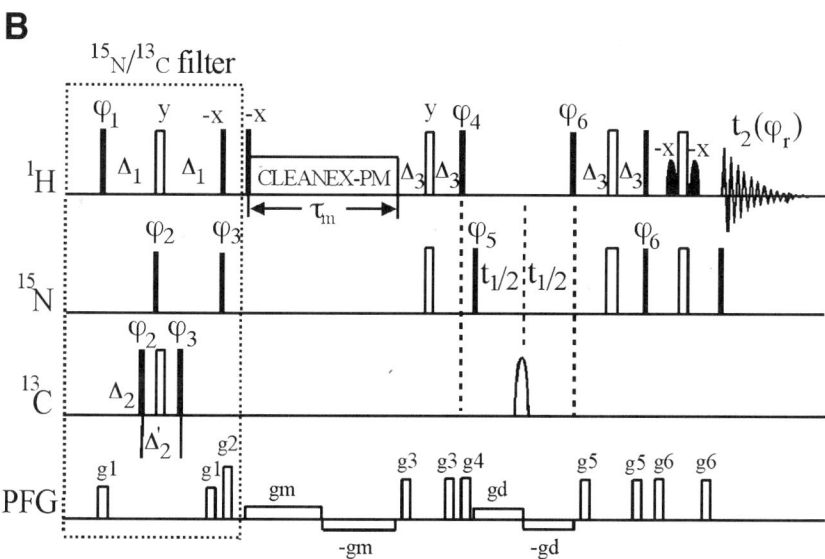

Fig. 5. (**A**) SEA–TROSY pulse sequence for ^{15}N/^{2}H-labeled proteins. Narrow and wide squares denote 90° and 180° hard pulses, respectively. Unless indicated, pulse phases are applied along x. The durations and strengths of pulsed field gradients are of conventional values. The bipolar gradient **gd** is 0.5 G/cm and it is used to avoid radiation damping effects during t_1. The delay Δ is set to 2.7 ms. The phase cycle is as follows: $\varphi 1 = y, -y, -x, x$; $\varphi 2 = y$; $\varphi 3 = x$; $\varphi_r = x, -x, -y, y$. A phase sensitive spectrum in the ^{15}N dimensional is obtained by recording a second FID for each t_2 value, with $\varphi 1 = -y, y, -x, x$, $\varphi 2 = -y$, and $\varphi 3 = -x$. The SEA element is outlined by the dashed rectangle. (**B**) Clean SEA–TROSY pulse sequence for ^{15}N/^{13}C-labeled proteins. The phase cycle is as follows: $\varphi 1 = 16(x), 16(-x)$; $\varphi 2 = 4(x), 4(-x)$; $\varphi 3 = 8(x), 8(-x)$; $\varphi 4 = 16(y), 16(-y)$; $\varphi 5 = y, x, -y, -x$; $\varphi 6 = y$; $\varphi r = x, -y, -x, y$.

where $R_{1A,app}$ is the combination of longitudinal relaxation and transverse relaxation rates of amide protons, depending on the trajectory of magnetization, and $R_{1B,app}$ is that of water molecules; k_{ex} is the normalized exchange rate between amide protons and water ($k_{ex} = k_A X_B = k_B X_A$, in which k_A and k_B are the forward and backward exchange rates from amide protons to water, X_A and X_B are the molar fractions of the protein and water, $X_B \approx 1$), and f is the proportionality factor between the molar fraction of the protein and the peak intensity.

Figure 5B shows the Clean SEA–TROSY pulse sequence for $^{15}N/^{13}C$-labeled proteins. This pulse scheme starts with a double $^{15}N/^{13}C$ filter *(37)*, which serves to eliminate all of the magnetization generated from protons attached to nitrogen and carbon atoms. A pair of gradient pulses (g1) bracket the double ^{15}N filter to prevent water magnetization loss caused by radiation damping. After the filter, a gradient pulse (g2) is used to remove non-zero-order coherences. Water magnetization is allowed to pass the filter and exchange with amide protons during the subsequent mixing period. The CLEANEX-PM mixing scheme *(36,38)* is used to eliminate both exchange-relayed NOE contributions from the fast exchanging hydroxyl or amine protons and intramolecular NOE contributions from aliphatic protons. Radiation damping during τ_m is removed by a weak bipolar gradient (gm).

For ^{15}N-labeled proteins, a spin-echo filter *(39,40)* integrated with a double ^{15}N filter *(31,32)* is used instead *(6)*. Accordingly, the duration of the double ^{15}N filter is extended to 33 ms ($6\Delta_1 = 3/^1J_{HN}$). A bipolar gradient (g1) is applied to suppress radiation damping in the spin-echo filter. During the filter, most of the protein magnetization decays away completely because of J-coupling evolution and a much shorter longitudinal relaxation time compared with that of water protons.

3.2.1. Application to Proteins

Protein sample used: 1.5 mM uniformly ^{15}N-labeled sample of dihydrofolate reductase (162 a,a) complexed with unlabeled methotrexate in an NMR buffer (200 mM KCl, 50 mM KH$_2$PO$_4$, pH 6.5, in 90% H$_2$O/10% D$_2$O).

Equipment: 500-MHz NMR spectrometer with three channels.

Experimental parameter and results:

1. Temperature: 35°C.
2. Experimental parameters: In **Fig. 5B**, g1 (0.5 ms × 1.1 G/cm), g2 (0.5 ms × 25.8 G/cm), gm (0.05 G/cm), and gd (0.05 G/cm) are applied. A water-selective flip-back 90° pulse (one-lobe sinc function, 1–2 ms) is used in the WATERGATE sequence. The delays and strengths of the other gradient pulses are of conventional values. The delays used in the pulse sequence are $\Delta_1 = 5.5$ ms, $\Delta_2 = 3.7$ ms, $\Delta'_2 = 3.6$ ms, $\Delta_3 = 2.3$ ms. The length of the double ^{15}N filter is set to $2\Delta_1 = 1/^1J_{NH} = 11.0$ ms. The proton carrier is switched during the CLEANEX-PM

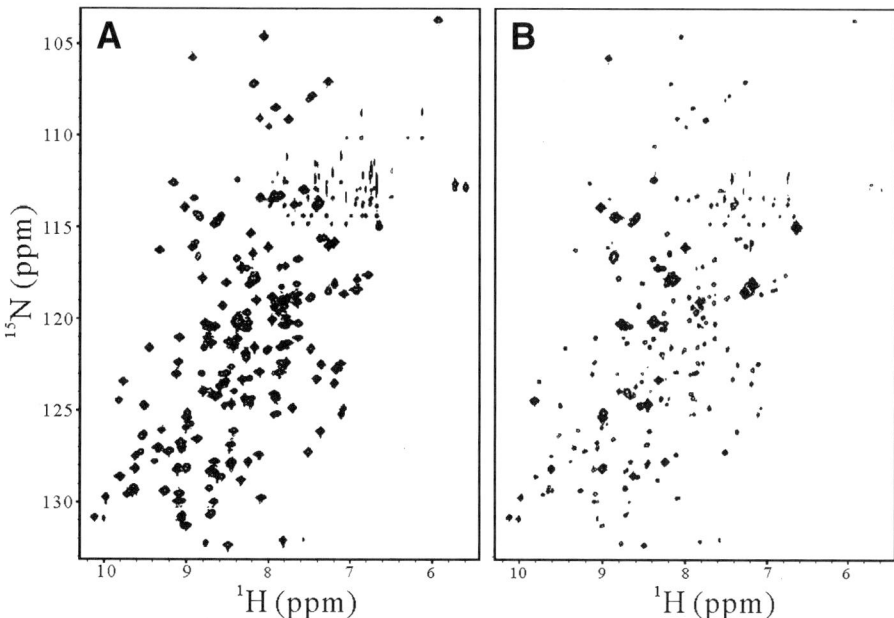

Fig. 6. **(A)** TROSY spectrum recorded on ^{15}N-labeled sample of DHFR complexed with unlabeled MTX. **(B)** Clean SEA-TROSY spectrum recorded on the same sample. Spectrum B was drawn with a contour level 50% of that in spectrum A.

pulse train from the water resonance to the middle of the NH range. The carbon carrier is placed on the aliphatic carbons (35 ppm) at the beginning of the pulse sequence, then shifted to the center between $^{13}C_\alpha$ and ^{13}CO (116 ppm) after the double $^{15}N/^{13}C$ filter. A selective ^{13}C 180° pulse on $^{13}C_\alpha$ and ^{13}CO is applied to refocus ^{15}N–^{13}C couplings during t_1. For ^{15}N-labeled proteins, the double ^{13}C filter is substituted with a spin-echo filter. Accordingly, the length of the double ^{15}N filter is extended to $6\Delta_1 = 33$ ms. A weak bipolar gradient, g1 (0.05 G/cm), is applied to remove radiation damping in the filter. The mixing time and predelay were 100 ms and 2.3 s, respectively.

3. Results: **Fig. 6A** is a normal TROSY spectrum, and **Fig. 6B** is a Clean SEA–TROSY spectrum with a mixing time of 100 ms. A comparison between the two spectra shows that the degree of spectral overlap in Clean SEA–TROSY spectra is greatly reduced.

Clean SEA–TROSY experiments can be used to accurately measure the exchange rates of amide protons with water by fitting the relative growth of peak intensity to **Eq. 1**. A series of mixing times (10–100 ms) should be used to obtain acceptable experimental sensitivities and to avoid spin diffusion effects. An HSQC version of Clean SEA-type experiments, Clean SEA-HSQC (*6*) has been used to measure the amide proton exchange rates in a 1.0 mM ^{15}N-labeled sample of the human acidic fibroblast growth factor (74 residues) in 100 mM phosphate buffer and 200 mM ammonium sulfate in 90% H$_2$O/10% D$_2$O at pH 6.0 *(41)*.

For binding studies only amides that are exposed to the solvent are of interest, whereas those buried in the interior of the protein are not likely to be involved in intermolecular interactions. Therefore, SEA–TROSY experiments are particularly useful to study these intermolecular interactions.

4. Hydrogen-Bond Measurement By TROSY-Based Experiments
4.1. TROSY-Based Experiments for Hydrogen Bond Measurement of Proteins

Hydrogen bonds are essential for the stability of biomolecular structures, substrate binding, and enzymatic reactions. The discovery of J couplings across the hydrogen bond in proteins and nucleic acids *(7,9)* has enabled direct detection of hydrogen bonds and the opportunity for new constraints in structural determination of biological macromolecules. There are several J couplings across hydrogen bonds that can be measured in proteins, such as $^{3h}J_{C\alpha HN}$ (<1.4 Hz) *(42)*, $^{2h}J_{COHN}$ (<1.3) *(7,8,43)*, and $^{3h}J_{NCO}$ (<1 Hz) *(14,44)*. Correlation through the H-bond allows unambiguous identification of the pairs of atoms involved in a given H-bond, unlike methods dependent on the observation of slowly exchanging amide protons. Because these J-coupling constants are small, it requires longer de- and rephase durations for coherence transfer in pulse sequences. Conventional HSQC-based triple-resonance experiments will suffer tremendous sensitivity loss because of the shorter transverse relaxation time of proteins. TROSY-based experiments instead can be readily applied to overcome the shortcomings in the conventional NMR experiments, even for relatively smaller proteins, either by using TROSY-based J quantitative correlation or the TROSY-based E.COSY experiments *(14,15,43)*. Direct correlation between $^{3h}J_{NC'}$ and hydrogen bond length in protein crystal structures has been observed *(45)*. In a TROSY–HNCO experiment *(14)* $^{3h}J_{NC'}$ across a hydrogen bond can be measured with high sensitivity despite its small magnitude in a 30-kDa protein.

4.1.1. Application to Proteins

Protein sample used: 0.7 mM ^{15}N/^{13}C/85% ^{2}H-labeled ribosome inactivating protein MAP30 (30 kDa).
Equipment: 750-MHz NMR spectrometer with three channels.
Experimental parameters and results:

1. Temperature: 40°C.
2. Experimental parameters: In **Fig. 7**,^{13}C pulses have a sinc(x) function shape of the center lobe, and a duration of 150 μs. Shaped ^{13}C pulses at the midpoint of the 2T

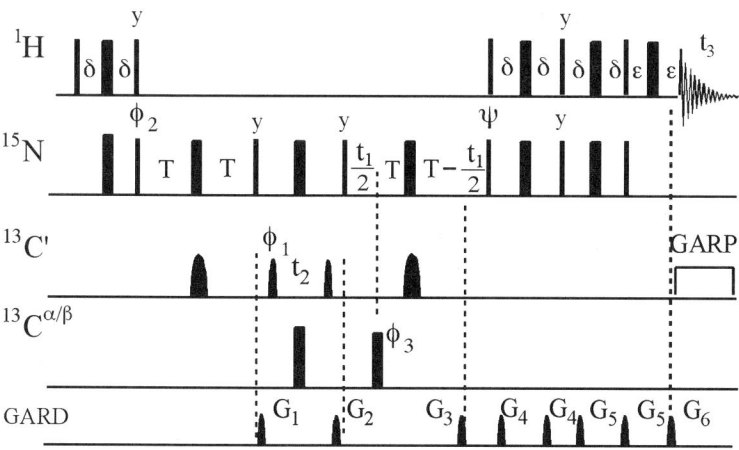

Fig. 7. Pulse sequence of the water-flip-back 3D 3hJNC' HNCO experiment, used to detect through-hydrogen bond J connectivities in medium sized proteins. Narrow and wide pulses correspond to flip angles of 90° and 180°, respectively. All pulse phases are x, unless specified. Phase cycling: $\psi = -x$; $\phi 1 = x, -x$; $\phi 2 = 2(x), 2(-x)$; $\phi 3 = 4(x), 4(-x)$; receiver = x, 2(-x),x.

period are 180°; pulses bracketing the $^{13}C'$ t_2 evolution are of lower power and correspond to 90°. Delay durations: $\delta = 2.65$ ms; $T = 66.6$ ms = $1/^{1}J_{NC'}$. Echo/antiecho selections during t_1 are done by inverting the sign of ψ and gradient G3.

3. The spectrum (**Fig. 8**) was obtained from a 99×32×768 complex point matrix with acquisition times of 65 ms (t_1), 14 ms (t_2), and 79 ms (t_3). Linear prediction methods *(46,47)* were used to extend t_1 and t_2 domains before applying Fourier transformation.

4.2. J_{NN}–TROSY for Hydrogen-Bond Measurement of Nucleic Acids

Several remarkable works have been published reporting the observation of scalar couplings between the hydrogen bond donating and accepting ^{15}N nuclei in Watson–Crick basepairs of ^{15}N-labeled double-stranded RNA *(7)* and DNA ($^{2h}J_{NN} = 2$–8 Hz), and the smaller J-coupling between the amino proton and the hydrogen bond accepting ^{15}N nucleus ($^{h}J_{HN} = 1$–4 Hz) *(10,11)*. Similar couplings were also found in Hoogsteen basepairs of RNA *(48)* and mismatched basepairs of DNA involving amino groups *(49,50)*. These observations provide unambiguous evidence of the existence of hydrogen bonds within these basepairs. This will be of great importance for the structure determination of nucleic acids, for which the proton density, and hence the NOE density, is lower than in proteins.

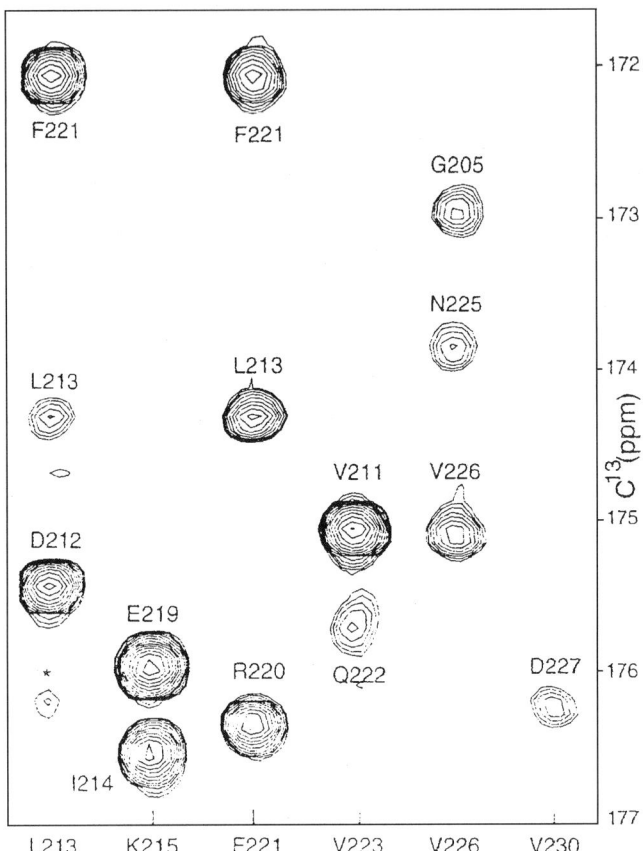

Fig. 8. Strip plot of the 3D 3hJNC' HNCO spectrum of MAP30, showing through-hydrogen-bond J-correlations involved amides of the 9th and 10th β-strands. The strips are labeled by the residue of the amide resonance detected during t3, and cross peaks to carbonyls are marked by their corresponding residues. Correlations resulting from small 2JNC' couplings are observed for L213, F221, and V226.

The J_{NN}-correlated-[^{15}N, ^{1}H]-TROSY *(10)* experiment for the correlation of hydrogen-bonded nuclei and measurement of $^{2h}J_{NN}$ uses the TROSY effect and thus is suitable for large molecules. However, the measurement of $^hJ_{HN}$ using E.COSY-type peaks in this experiment *(10)* is not practical for large molecules, such as transfer RNA (tRNA), in which the fast-relaxing component is too weak to be detected. A novel method, $^hJ_{HN}$-quantitative [^{15}N,^{1}H]-ZQ TROSY has been reported recently to measure the $^hJ_{HN}$ values in a 17-kDa protein–DNA complex *(51)*. This method involves $^hJ_{HN}$ modulating ^{1}H transverse magnetization before detection. Because the smaller $^hJ_{HN}$, compared to $^{2h}J_{NN}$, introduces a longer delay for magnetization modulation and the transverse relaxation of protons is faster than that of ^{15}N, this experiment may not be as sensitive as the J_{NN}-correlated-[^{15}N, ^{1}H]-TROSY experiment. A modified J_{NN}-correlated-[^{15}N, ^{1}H]-TROSY *(10)* has been described for the measurement of the $^{h}J_{HN}$ coupling constant of large biomolecules, and it has been applied to tRNA^{Trp}. In this experiment, only the slow-relaxing component is used during the long delay of 4T for both peaks used for $^{h}J_{HN}$ measurement, by using a method similar to that reported by Meissner and Sorenson *(43)*.

The J_{NN}-correlated-[^{15}N, ^{1}H]-TROSY *(10)* pulse sequence in **Fig. 9A** is modified for measuring $^{h}J_{NN}$, as depicted in **Fig. 9B**. Two data sets are recorded in an interleaved manner, with two pairs of 90° pulses flanking the t_1 evolution period. The phases of the two 90° pulses within each pair are opposite in the first experiment and the same in the second experiment. Only one component is selected in each experiment. With this arrangement, the first experiment is equivalent to the original J_{NN} HNN–COSY experiment *(7)*. In the second experiment, the proton spin state is exchanged just before and after the t_1 evolution period.

In the proton dimension, the slow-relaxing component is recorded in both experiments. The difference between the peak intensities in these two experiments is introduced only by the different relaxation rates of the magnetization during the t_1 period. The steady-state enhancements are maintained and contribute to the peak intensity equally in both experiments *(29)*.

In the second experiment, the water magnetization is in the –z direction during the t_1 interval. To avoid the recovery of the water magnetization from –z because of radiation damping, a pair of bipolar gradient pulses are used. It has been shown that a weak gradient can effectively suppress the radiation damping effect *(52)*. Also, the phases of the second pair of proton 90° pulses at the end of the t_1 interval are set to be opposite in phase to the first pair of 90° pulses to compensate for an incomplete conversion of water magnetization to +z as a result of imperfect pulse widths. A shorter 3–9–19 composite pulse instead of long soft selective pulses is used in the WATERGATE sequence to obtain better sensitivity for nucleic acids.

4.2.1. Application to Nucleic Acids

Nucleic acid sample used: ^{15}N-labeled *Bacillus subtilis* tRNA^{Trp} A73G mutant (76 nucleotides) in 5 m*M* MgCl$_2$, 90% H$_2$O/10D$_2$O.

Equipment: 500-MHz NMR spectrometer with three channels.

Experimental parameters and results:

1. Temperature: 25°C.
2. Experimental parameters: In **Fig. 9**, $\delta_1 = 2.25$ ms, $\delta_2 = 2.7$ ms, T = 15 ms. A 3–9–19 composite pulse is used for water suppression. Shaped pulses for the selective excitation of water resonance are 1.7 ms sinc 90° pulses. The relaxation recovery delay

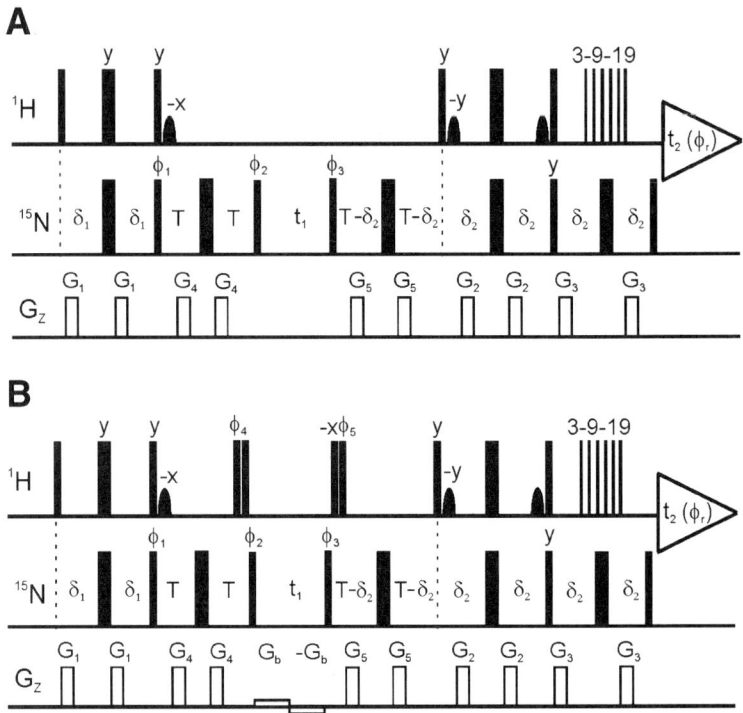

Fig. 9. Pulse sequences used for the measurement of $^hJ_{HN}$ of ^{15}N-labeled tRNA^{Trp}. In all sequences, *narrow and wide bars* represent 90° and 180° pulses, respectively. Default phases are x. **(A)** The E.COSY type $^hJ_{NN}$-correlation-$[^{15}N, ^1H]$-TROSY is similar to that reported by Pervushin et al. *(10)*, except that only one experiment is carried out to observe both components with the same phase. The phase cycling is $\phi1 = y, -y$; $\phi2 = -x, -x, x, x$; $\phi3 = 4(-x), 4(x)$; $\phi r = y, -y$. **(B)** Modified $^hJ_{NN}$-correlation-$[^{15}N, ^1H]$-TROSY. Two data sets are measured in an interleaved manner with the following phase cycling schemes: (I) $\phi1 = x, y, -x, -y$; $\phi2 = y, -x, -y, x, -y, x, y, -x$; $\phi3 = -y, x, y, -x, -y, x, y, -x, y, -x, -y, x, y, -x, -y, x$; $\phi4 = -x$; $\phi5 = x$; $\phi r = x, y, -x, -y$. (II) $\phi4 = x$; $\phi5 = -x$, whereas the other phases are the same as in (I). The resulting two spectra after Fourier transformation display cross-peaks displaced along ω_1 by the proton–nitrogen coupling constants, $^1J_{HN}$ and $^hJ_{HN}$, for direct and relayed correlation cross-peaks, respectively. (Reprinted from **ref. *13*** with permission from Elsevier.)

is 1.0 s. The durations and strengths of the gradients along the z-axis are of conventional values; Gb are 0.2 G/cm bipolar gradients. Quadrature detection in t_1 for all experiments is achieved through altering $\phi1$ and $\phi2$ in a States-TPPI manner.

3. The spectra were acquired with spectral widths and complex data points of 2027 Hz, 128 for ω_1 and 12,000 Hz, 2048 for ω_2, respectively. The total acquisition time for each experiment was about 23 h.

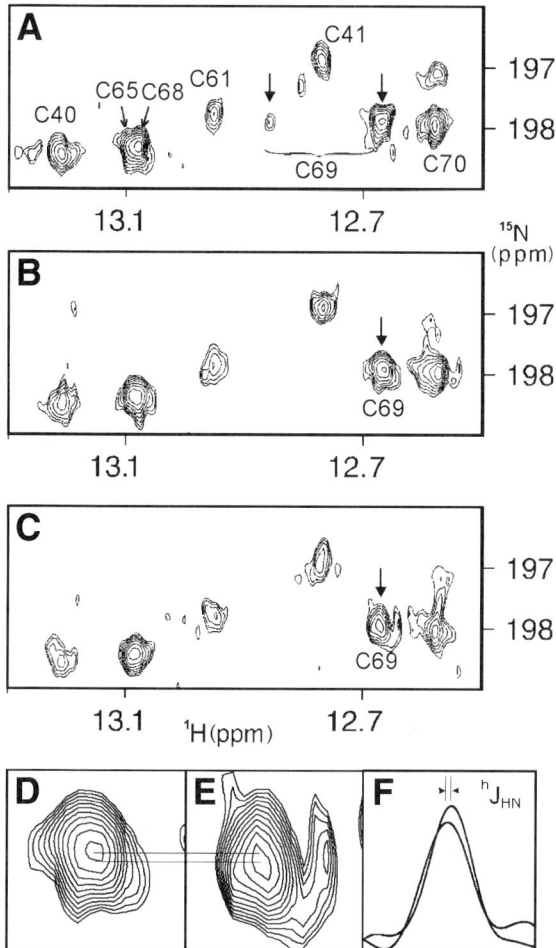

Fig. 10. Comparison of the measurement of $^{h}J_{HN}$-couplings using (**A**) E.COSY-type $^{h}J_{NN}$-correlation-[^{15}N, ^{1}H]-TROSY and (**B, C**) modified $^{h}J_{NN}$-correlation-[^{15}N, ^{1}H]-TROSY. Shown here are the GC basepair region spectra of the *B. subtilis* tRNA^{Trp} A73 mutant. (**D, E**) Enlargement of the cross-peak of the G4C69 pair from (**B**) and (**C**), respectively. (**F**) Traces along ω_{1} taken from (**D**) and (**E**), with the difference of the peak position between two slices giving the $^{h}J_{HN}$ value. (Reprinted from **ref. 13** with permission from Elsevier.)

4. Results: The spectrum acquired using the pulse sequence depicted in **Fig. 9A** is shown in **Fig. 10A**. The experiment in **Fig. 9A** is similar to the original $^{h}J_{NN}$-correlation-[^{15}N, ^{1}H]-TROSY *(10)*, but with two component peaks in one spectrum for simplicity. In **Fig. 10A**, both the fast- and slow-relaxing components used for $^{h}J_{HN}$ measurement of ^{15}N-labeled tRNA^{Trp} exhibit very different peak intensities and line widths because of the different transverse relaxation rates. As a result,

only one pair of peaks is clearly observable in the GC pair region as indicated in **Fig. 10A**. **Figure 10B,C** shows the results obtained with the modified $^hJ_{NN}$-correlation-[^{15}N, ^{1}H]-TROSY experiment by using the pulse sequence depicted in **Fig. 9B**. The measurements of $^hJ_{HN}$ from peaks taken from **Fig. 10B,C** are shown in **Fig. 10D–F**. The $^hJ_{HN}$-coupling constants were calculated by inverse Fourier transformation of the slices corresponding to each peak along ω_1, zero filling, then Fourier transformation and data fitting. The results show that the modified method is a better way to measure the $^hJ_{HN}$ coupling constants of base-pairs in tRNA and other large nucleic acid molecules.

5. Conclusions

Introduction of TROSY greatly increases the sensitivity and resolution of NMR experiments on large proteins (>25 kDa) and nucleic acids. TROSY-based multidimensional NMR has been applied not only to protein backbone assignment but also to the protein dynamics studies, protein–protein interactions, studies of exchange phenomena, and measurement of hydrogen bonds in proteins and nucleic acids. With the use of high-field instruments and cryoprobes, NMR studies of biomolecules are no longer limited to the smaller proteins and nucleic acids at high sample concentrations (>0.7 m*M*). These advanced NMR methods and instrumentation provide scientists with tremendous opportunities in structural biology and functional genomics, and the impact of these new methods will soon be seen. As the number of high-field magnets increases, many larger proteins and protein complexes will be studied by TROSY-based NMR experiments.

References

1. Zhu, G., Xia, Y. L., Nichoson, L. K., and Sze, K. (2000) Protein dynamic measurements by TROSY-based NMR experiments. *J. Magn. Reson.* **143,** 423–426.
2. Xia, Y., Sze, K. H., Li, N., Shaw, P. C., and Zhu, G. (2002) Protein dynamics measurements by 3D HNCO based NMR experiments. *Spectroscopy* **16,** 1–13.
3. Loria, J. P., Rance, M., and Palmer, A. G. (1999) A TROSY CPMG sequence for characterizing chemical exchange in large proteins. *J. Biomol. NMR* **15,** 151–155.
4. Takahashi, H., Nakanishi, T., Kami, K., Arata, Y., and Shimada, I. (2000) A novel NMR method for determining the interfaces of large protein–protein complexes. *Nat. Struct. Biol.* **7,** 220–223.
5. Pellecchia, M., Meininger, D., Shen, A. L., Jack, R., Kasper, C. B., and Sem, D. S. (2001) SEA–TROSY (Solvent exposed amides with TROSY): A method to resolve the problem of spectral overlap in very large proteins. *J. Am. Chem. Soc.* **123,** 4633, 4634.
6. Lin, D. H., Sze, K. H., Cui ,Y. F., and Zhu, G. (2002) Clean SEA-HSQC: a method to map solvent exposed amides in large non-deuterated proteins with gradient-enhanced HSQC. *J. Biomol. NMR* **23,** 317–322.

7. Cordier, F., Rogowski, M., Grzesiek, S., and Bax, A. (1999) Observation of through hydrogen-bond $^{2h}J_{HC'}$ in a perdeuterated protein. *J. Magn. Reson.* **140**, 510–512.

8. Cordier, F. and Grzesiek, S. (1999) Direct observation of hydrogen bond in proteins by interresidue $3^{h}J_{NC'}$ scalar couplings. *J. Am. Chem. Soc.* **121**, 1601–1602.

9. Dingley, A. J. and Grzesiek, S. (1998) Direct obsveration of hydrogen bonds in nucleic acid base pairs by internucleotide $^{2}J_{NN}$ coupling. *J. Am. Chem. Soc.* **120**, 8293–8297

10. Pervushin, K., Ono, A., Fernandez, C., Szyperski, T., Kainosho, M., and Wüthrich, K. (1998) NMR scalar couplings across Watson–Crick base pair hydrogen bonds in DNA observed by transverse relaxation optimized spectroscopy. *Proc. Natl. Acad. Sci. USA* **95**, 14,147–14,151.

11. Dingley, A. J., Masse, J. E., Peterson, R. D., Barfield, M., Feigon, J., and Grzesiek, S. (1999) Internucleotide scalar couplings across hydrogen bonds in Watson–Crick and Hoogsteen base pairs of a DNA triplex. *J. Am. Chem. Soc.* **121**, 6019–6027.

12. Majumdar, A., Kettani, A., Skripkin, E., and Patel, D. J. (1999) Observation of internucleotide NH . . . N hydrogen bonds in the absence of directly detectable protons. *J. Biomol. NMR* **15**, 207–211.

13. Yan, X., Kong, X., Xia, Y. L., Sze, K., and Zhu, G. (2000) Determination of internucleotide $^{h}J_{HN}$ couplings by the modified 2D JNN-correlated [^{15}N,^{1}H] TROSY. *J. Magn. Reson.* **147**, 357–360.

14. Wang, Y. X., Jacob, F., Cordier, F., Wingfield, P., Stahl, S. J., Kaufman, J. D., et al. (1999) Measurement of $3^{h}J_{NC'}$ connectivities across hydrogen bonds in a 30kDa protein. *J. Biomol. NMR* **14**, 181–184.

15. Bax, A., Vuister, G. W., Grzesiek, S., Delaglio, F., Wang, A. C., Tschudin, R., et al. (1994). Measurement of homo- and heteronuclear J couplings from quantitative J correlation. *Methods Enzymol.* **239**, 79–105.

16. Pervushin, K. (2000) Impact of transverse relaxation optimized spectroscopy (TROSY) on NMR as a technique in structural biology. *Q. Rev. Biophys.* **33**, 161–197.

17. Yang, D. W., Venters, R. A., Mueller, G. A., Choy, W. Y., and Kay, L. E. (1999) TROSY-based HNCO pulse sequences for the measurement of (HN)-H-1-N-15, N-15-(CO)-C-13, (HN)-H-1-(CO)-C-13, (CO)-C-13-C-13(alpha) and (HN)-H-1-C-13(alpha) dipolar couplings in N-15, C-13, H-2-labeled proteins. *J. Biomol. NMR* **14**, 333–343.

18. Permi, P., Rosevear, P. R., and Annila, A. (2000) A set of HNCO-based experiments for measurement of residual dipolar couplings in N-15, C-13, (H-2)-labeled proteins. *J. Biomol. NMR* **17**, 43–54.

19. Kover, K. K. and Batta, G. (2001) J modulated TROSY experiment extends the limits of homonuclear coupling measurements for larger proteins. *J. Magn. Resn.* **151**, 60–64.

20. Amadei, A., Linssen, A. B. M., and Berendsen, H. J. C. (1993) Essential dynamics of proteins. *Proteins* **17**, 412–425.

21. Feher, V. A. and Cavanagh, J. (1999) Millisecond-timescale motions contribute to the function of the bacterial response regulator protein SpO0F. *Nature* **400,** 289–293.

22. Karplus, M. and Petsko, G. A. (1990) Molecular-dynamics simulations in biology. *Nature* **347,** 631–639.

23. Barbato, G., Ikura, M., Kay, L. E., Pastor, R. W., and Bax, A. (1992) Backbone dynamics of calmodulin studied by N-15 relaxation using inverse detected 2-dimensional NMR-spectroscopy—the central helix is flexible. *Biochemistry* **31,** 5269–5278.

24. Kay, L. E., Torchia, D. A., and Bax, A. (1989) Backbone dynamics of proteins as studied by N-15 inverse detected heteronuclear NMR-spectroscopy—application to staphylococcal nuclease. *Biochemistry* **28,** 8972–8979.

25. Kay, L. E., Nicholson, L. K., Delaglio, F., Bax, A., and Torchia, D. A. (1992) Pulse sequences for removal of the effects of cross-correlation between dipolar and chemical-shift anisotropy relaxation mechanism on the measurement of heteronuclear T_1 and T_2 values in proteins. *J. Magn. Reson.* **97,** 359–375.

26. Milburn, M. V., Tong, L., de Vos, A. M., Brunger, A., Yamaizumi, Z., Nishimura, S., et al. (1990) Molecular switch for signal transduction—structural differences between active and inactive forms of protooncogenic Ras proteins. *Science* **247,** 939–945.

27. Farrow, N. A., Muhandiram, R., Singer, A. U., Pascal, S. M., Kay, C. M., Gish, G., et al. (1994) Backbone dynamics of a free and a phosphopeptide-complexed Src homology-2 domain studied by N-15 NMR relaxation. *Biochemistry* **33,** 5984–6003.

28. Cavanagh, J. and Rance, M. (1993) Sensitivity-enhanced NMR techniques for the study of biomolecules. *Annu. Rev. NMR Spectrosc.* **27,** 1–58.

29. Pervushin, K., Wider, G., and Wüthrich, K. (1998) Single transition-to-single transition polarization transfer (ST2-PT) in [N-15,H-1]-TROSY. *J. Biomol. NMR* **12,** 345–348.

30. Delaglio, F., Grzesiek, S., Vuister, G. W., Zhu, G, Pfeifer, J., and Bax, A. NMRPIPE—a multidimensional spectral processing system based on Unix pipes. *J. Biomol. NMR* **6,** 277–293.

31. Otting, G. and Wüthrich, K. (1990) Heteronuclear filters in two-dimensional [^{1}H,^{1}H]-NMR spectroscopy: combined use with isotope labeling for studies of macromolecular conformation and intermolecular interactions. *Q. Rev. Biophys.* **23,** 39–96.

32. Breeze, A. L. (2000) Isotope-filtered NMR methods for the study of biomolecular structure and interactions. *Prog. NMR Spectrosc.* **36,** 323–372.

33. Zhu, G., Kong, X. M., and Sze, K. H. (1999) Gradient and sensitivity enhancement of 2D TROSY with water flip-back, 3D NOESY-TROSY and TOCSY-TROSY experiments. *J. Biomol. NMR* **13,** 77–81.

34. Bai, Y., Milne, J. S., Mayne, L., Englander, S. W. (1993) Primary structure effects on peptide group hydrogen exchange. *Proteins* **17,** 75–86.

35. Dempsey, C. E. (2001) Hydrogen exchange in peptides and proteins using NMR spectroscopy. *Prog. NMR Spectrosc.* **39,** 135–170.

36. Hwang, T.-L., Mori, S., Shaka, A. J., and van Zijl, P. C. M. (1997) Application of phase-modulated CLEAN chemical Exchange Spectroscopy (CLEANEX-PM) to detect intermolecular NOE. *J. Am. Chem. Soc.* **119,** 6203, 6204.
37. Gemmecker, G., Jahnke, W., and Kessler, H. (1993) Measurement of fast proton exchange rates in isotopically labeled compounds. *J. Am. Chem. Soc.* **115,** 11, 620–11,621.
38. Hwang, T.-L., van Zijl, P. C. M., and Mori, S. (1998) Accurate quantitation of water-amide proton exchange rates using the phase-modulated CLEAN chemical EXchange (CLEANEX-PM) approach with a Fast-HSQC (FHSQC) detection scheme. *J. Biomol. NMR* **11,** 221–226.
39. Mori, S., Berg, J. M., and van Zijl, P. C. M. (1996) Application of phase-modulated CLEAN chemical Exchange Spectroscopy (CLEANEX-PM) to detect intermolecular NOE. *J. Biomol. NMR* **7,** 77–82.
40. Mori, S., Abeygunawardana, C., Berg, J. M., and van Zijl, P. C. M. (1997) NMR study of rapidly exchanging backbone amide protons in staphylococcal nuclease and the correlation with structural and dynamic properties. *J. Am. Chem. Soc.* **119,** 6844–6852.
41. Chi, Y. -H., Kumar, T. K. S., Kathir, K. M., Lin, D. H., Zhu, G., Chiu, I. M., et al. (2002) Investigation of the structural stability of the human acidic fibrooblast growth factor by hydrogen-deuterium exchange. *Biochemistry* **41,** 15,350–15,359.
42. Meissner, A. and Sorensen O. W. (2000) ^{3h}J coupling between C^a and H_N across hydrogen bonds in proteins. *J. Magn. Reson.* **143,** 431–434.
43. Meissner, A. and Sorensen O. W. (2000) New techniques for the measurement of C'N and C'HN J coupling constants across hydrogen bonds in proteins, *J. Magn. Reson.* **143,** 387–390.
44. Cordier, F. and Grzesiek, S. (1999) Direct observation of hydrogen bond in proteins by interresidue $^{3h}J_{NC'}$ scalar couplings. *J. Am. Chem. Soc.* **121,** 1601–1602.
45. Cornilescu, G., Ramires, B. E., Frank, M. K., Clore, M. G., Gronenborn, A. M., and Bax, A. (1999) Correlation between $^{3h}J_{NC'}$ and hydrogen bond length in proteins, *J. Am. Chem. Soc.* **121,** 6275–6279.
46. Zhu, G. and Bax, A. (1990) Improved linear prediction for truncated signals of known phase. *J. Magn. Reson.* **90,** 405–410.
47. Zhu, G. and Bax, A. (1992) Improved linear prediction of truncated damped sinusoids using modified backward-forward linear prediction. *J. Magn. Reson.* **100,** 202–207.
48. Wöhnert, J., Dingley, A. J., Stoldt, M., Göhlach, M., Grzesiek, S., and Brown, L. R. (1999) Direct identification of NH . . . N hydrogen bonds in non-canonical base pairs of RNA by NMR spectroscopy. *Nucleic Acids Res.* **27,** 3104–3110.
49. Majumdar, A., Kettani, A., and Skripkin, E. (1999) Observation and measurement of internucleotide (2)J(NN) coupling constants between N-15 nuclei with widely separated chemical shifts. *J. Biomol. NMR* **14,** 67–70.

50. Majumdar, A., Kettani, A., Skripkin, E., and Patel, D. J. (1999) Observation of internucleotide NH . . . N hydrogen bonds in the absence of directly detectable protons. *J. Biomol. NMR* **15,** 207–211.

51. Pervushin, K., Fernández, C., Riek, R., Ono, A., Kainosho, M., and Wüthrich. K. (2000) Determination of (h2)J(NN) and (h1)J(HN) coupling constants across Watson–Crick base pairs in the Antennapedia homeodomain-DNA complex using TROSY. *J. Biomol. NMR.* **16,** 39–46.

52. Sklenar, V. (1995) Suppression of radiation damping in multidimensional NMR experiments using magnetic-field gradients. *J. Magn. Reson. Ser. A.* **114,** 132–135.

12

Measurement of Intermediate Exchange Phenomena

James G. Kempf and J. Patrick Loria

Summary

Understanding the crucial role of protein motions in the function of biological macro-molecules requires methods for their characterization. These motions lead to noticeable alterations in the decay of nuclear spin magnetization. Recent advances in solution nuclear magnetic resonance (NMR) make quantitative connections between μs–ms motions and nuclear spin relaxation in proteins. The techniques serve as useful probes of motional kinetics and thermo-dynamics and their relation to function. Here, we review the two most common experimental methods for characterizing conformational motions in proteins: the relaxation-compensated Carr–Purcell–Meiboom–Gill (rcCPMG) and off-resonance $R_{1\rho}$ experiments.

Key Words: Chemical exchange; conformational exchange; protein dynamics; spin relaxation; rcCPMG, $R_{1\rho}$; off-resonance spin relaxation; solution NMR; relaxation dispersion.

1. Introduction

Intramolecular conformational motions, ubiquitous in biological processes, occur over a wide range of time-scales. Thus, it is particularly advantageous that solution nuclear magnetic resonance (NMR) experiments are sensitive to motional processes over a large, relevant window. Recent advances allow characterization of protein dynamics in the μs–ms time regime, where processes referred to as chemical or conformational exchange occur, and which are often similar to catalytic rates and thus intriguing for direct involvement in biological processes. Here, we detail the setup, usage, applicability, and limitations of NMR experiments designed to detect and quantify chemical-exchange events. An outline of our treatment follows.

We begin with a brief review of the theory of chemical-exchange phenomena (**Subheading 2.1.**) and provide particular analysis of exchange characterization

From: *Methods in Molecular Biology, vol. 278: Protein NMR Techniques*
Edited by: A. K. Downing © Humana Press Inc., Totowa, NJ

based on transverse relaxation (**Subheading 2.2.**) and spin-locked relaxation (**Subheading 2.3.**). Following this, we describe common aspects of NMR methods (i.e., pulse sequences) for characterization of chemical exchange and discuss practical considerations for their implementation (**Subheading 3.1.1.**). The subsequent two sections detail the techniques for characterization of µs–ms motions via measurements of (a) the transverse spin-relaxation rate, R_2, (**Subheading 3.2.**) and (b) the off-resonance rotating-frame rate, $R_{1\rho}$ (**Subheading 3.3.**). We include discussions on several methods for data analysis and on many practical concerns for accurate determination by either method. Throughout the text, we provide several examples to illustrate the experimental approaches. Generally, discussion is limited to investigation of ^1H–^{15}N moieties in proteins. However, adaptations to other isolated *I–S* spin systems are often a straightforward extension of the topics considered in detail here, and particular examples are cited.

2. Chemical Exchange and NMR

2.1. Theory

Molecular motion that transfers a nuclear spin between two magnetically inequivalent sites on a time-scale similar to the chemical shift difference between these sites results in measurable changes of NMR observables, such as the resonance line shape and the spin-relaxation rates for transverse (R_2) and spin-locked ($R_{1\rho}$) magnetization. The primary focus of this chapter concerns the effects of conformational exchange on R_2 and $R_{1\rho}$ in proteins. Variations in R_2 are also manifest as changes in the observed NMR line shape (*see* **Fig. 1**), and methods for line-shape analysis have demonstrated utility in monitoring protein-folding rates and may be amenable to other dynamics studies. However, in this chapter, we do not treat the use of such methods as a means of conformational-exchange characterization. Readers with interest in this topic are encouraged to further reading *(1)* and should be aware of an excellent computer program recently developed for line shape analysis in terms of motional dynamics *(2)*.

The theory of chemical-exchange effects in NMR has been treated in detail elsewhere *(3–5)*. Here, we present some of the most salient points. To begin, exchange is most simply defined by the equilibrium

$$A \underset{k_{-1}}{\overset{k_1}{\rightleftharpoons}} B \tag{1}$$

between two conformers or chemical states, A and B. NMR is sensitive to any such process (e.g., intra- or intermolecular, or ligand-binding) in which the magnetic environment of the nuclei under investigation changes (*see* **Note 1**). The rate law for this system can be written as

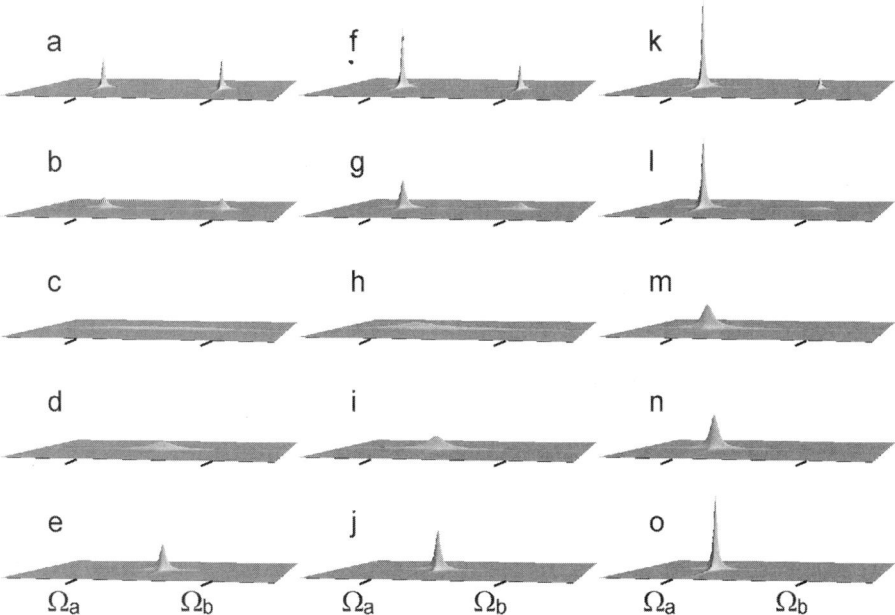

Fig. 1. Two-dimensional NMR spectra depicting two-site ^{15}N chemical exchange for (**A–E**) $p_A = p_B$; (**F–J**) p_A, p_B = 0.7, 0.3; (**K–O**) p_A, p_B = 0.9, 0.1. Simulated results incorporated $\Delta\omega(^{15}N)$ = 120 Hz and (top to bottom in each column) k_{ex} = 40, 200, 500, 2000, and 10,000 s^{-1}. Spectra are Fourier transforms of the time-domain signal calculated using expressions derived from **Eq. 6** and the assumption of exponential signal decay. For simplicity, we took $R_{2A} = R_{2B}$, and exchange was assumed to affect only ^{15}N evolution, as indicated by the constant 7 Hz ^{1}H line width. ω_A and ω_B are the ^{15}N chemical shifts at sites **A** and **B**. Calculations and graphics were executed using a software package written in *Mathematica* (Wolfram Research, Inc., Champaign, IL).

$$\frac{d}{dt}\begin{pmatrix}[A](t)\\ [B](t)\end{pmatrix} = \begin{pmatrix} -k_1 & k_{-1} \\ k_1 & -k_{-1}\end{pmatrix}\begin{pmatrix}[A](t)\\ [B](t)\end{pmatrix} \qquad (2)$$

or, in compact notation,

$$\frac{d\mathbf{A}(t)}{dt} = \mathbf{K}\mathbf{A}(t). \qquad (3)$$

The utility of **Eq. 3** is that it is readily generalized to n-site exchange involving sites A_i (i = 1 to n) connected by elementary reaction steps $A_i \rightarrow A_j$ at rate k_{ij}. In this case, diagonal and off-diagonal elements of the rate matrix **K** are expressed as $K_{ii} = -\sum_{j=1}^{n} k_{ij}$ and $K_{ij} = k_{ji}$, where $i \neq j$. The effects of multisite exchange in NMR are complex. Thus, for simplicity, this chapter considers only two-site mechanisms, whereas here we note relevant selections from the literature for treatment of n-site or more general exchange models (*6–11*).

NMR evolution for a system in equilibrium chemical exchange may be described according to the modified Bloch equations (i.e., the McConnell equations) (*4,12*), provided that scalar-coupling constants for a given spin are not affected by the exchange process (*13*). The chemical parameters required in this formalism are the equilibrium site populations (p_A and p_B), the chemical exchange rate constants, normally condensed to a single exchange rate, $k_{ex} = (k_1 + k_{-1})$, or equivalently, $k_{ex} = k_1/p_B = k_{-1}/p_A$, and finally, the chemical shift difference, $\Delta\omega = (\omega_A - \omega_B)$, between the magnetically distinct sites. The relationship between $\Delta\omega$ and k_{ex} defines the commonly encountered regimes of the chemical-exchange time-scale:

$$k_{ex} > \Delta\omega \rightarrow \text{fast exchange,}$$
$$k_{ex} \sim \Delta\omega \rightarrow \text{intermediate exchange,} \qquad (4)$$
$$k_{ex} < \Delta\omega \rightarrow \text{slow exchange.}$$

The spectral consequences of exchange events across these time-scales are depicted in **Fig. 1**. The results indicate the changes in line widths, amplitudes, and position that accompany exchange between inequivalent sites at rates over the slow-to-fast regimes and for various values of p_A and p_B. Interpretation of such changes is the basis for the noted method of line-shape analysis.

The primary route to characterizing the chemical parameters of exchange involves measurements of R_2 and/or $R_{1\rho}$. Provided that $\Delta\omega > 0$, these rates increase as a signature of exchange processes because of the additional contribution of the exchange-induced spin-relaxation rate, R_{ex}. Both R_2 and $R_{1\rho}$ are sensitive to events with a motional rate k_{ex} that is comparable in magnitude to an effective field applied to the spin system during relaxation. For $R_{1\rho}$, the maximum practical value of such a field, using current instrumentation and care to avoid sample damage, is $\omega_1(^{15}\text{N}) = -\gamma B_1$ approx 5 kHz (*14*) in on-resonance experiments, but it can be much larger in off-resonance measure of $R_{1\rho}$ (**Subheading 2.2.**). For R_2, the effective field, $\left(\sqrt{12}\nu_{CP}\right) \leq 3.5$ kHz, is relevant (*14*), where ν_{CP} is the repetition rate of π pulses in the Carr–Purcell–Meiboom–Gill (CPMG) sequence used to measure R_2 (*15,16*). Thus, $R_{1\rho}$ methods can access significantly faster motional processes than an R_2-based approach. Nonetheless, R_2 experiments offer relative simplicity in setup and are sensitive to slower motional processes than $R_{1\rho}$ experiments. We analyze both approaches in this chapter.

2.2. Transverse Relaxation

Perhaps the more commonly employed approach to conformational-exchange characterization involves measuring R_2 as a function of the spacing τ_{CP} between successive π pulses in a CPMG echo sequence (*17–21*). As with

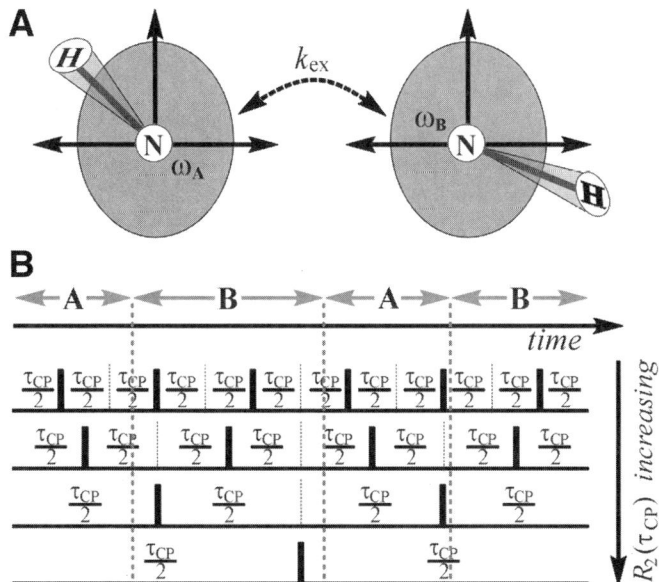

Fig. 2. **(A)** Two-site chemical exchange at rate k_{ex} of an NH bond vector between A and B conformers with distinct ^{15}N NMR frequencies, ω_A and ω_B. **(B)** Exchange time line concurrent with CPMG sequences of π pulses (*solid bars*). *Dashed lines* locate the occurrence of individual echoes in the CPMG train. The average persistence of the A and B states relative to the pulse spacing, τ_{CP}, influences the observed relaxation rate as indicated.

any echo sequence, relaxation is characterized by monitoring decay in the envelope of successive echoes along the π-pulse train. The echoes occur because the pulses refocus coherent spin evolution, whereas decoherent losses from spin–spin relaxation yield the envelope decay. In the context of conformational exchange, we refer to decoherent losses as background relaxation, as they occur regardless of the π pulses. In contrast, exchange contributes to echo decay in a manner dependent on the pulse repetition rate, $\nu_{CP} = \tau_{CP}^{-1}$.

The latter concept is depicted in **Fig. 2**. Part **(A)** of **Fig. 2** shows an example of two-site exchange at rate k_{ex} between orientations A and B of an NH bond vector that exhibit unique ^{15}N chemical shift frequencies, ω_A and ω_B ($\Delta\omega = \omega_A - \omega_B$). The top of **Fig. 2B** is a time line indicating typical spans over which a spin resides in either conformation, and it is shown in contrast to CPMG trains with pulse spacing from short to long relative to this cartoon representation of the average A and B lifetimes. When τ_{CP} is comparable to the typical durations of the A and B states, asymmetric evolution is likely to occur about a given π pulse, thus interfering with echo formation. As a result, the observed relaxation rate, $R_2(\tau_{CP}^{-1})$, combines the background and exchange contributions, R_2^0 and R_{ex}. However, for τ_{CP} short relative to the exchange process, individual echoes are formed without perturbation by exchange events and the observed rate approaches R_2^0. Here, writing R_2 as a function of the inverse pulse spacing, τ_{CP}^{-1}, emphasizes its functional dependence, presented in **Eqs. 6–9**, and keeps with common practice.

From this description, we find that the exchange contribution, R_{ex}, to the transverse relaxation is revealed by the difference

$$\Delta R_2 = [R_2(\tau_{CP}^{-1} \to 0) - R_2(\tau_{CP}^{-1} \to \infty)] \sim R_{ex}. \tag{5}$$

between rates measured at the extremes of τ_{CP} values. Often, it is useful to quickly assess for the occurrence or absence of μs–ms dynamics at sites across the protein sequence by measuring just these two rates for each resolved peak in an heteronuclear single-quantum coherence (HSQC) spectrum. With this simple approach, $\Delta R_2 > 0$ indicates the influence of such dynamics on the corresponding residue. For example, $\Delta R_2 (\tau_{CP} = 1.0, 64.8 \text{ ms})$ uncovers motion on time-scales from about 100 s^{-1} to 3000 s^{-1}, provided there is a sufficient ^{15}N chemical shift difference $\Delta\omega$ between conformers *(5)*, and as seen by noting that R_2 is sensitive to events with k_{ex} on the order of the effective field $\left(\sqrt{12}\nu_{CP}\right)$. Nonetheless, obtaining detailed information on the motion and chemical states of the A and B conformers requires measuring the shape of $R_2(\tau_{CP}^{-1})$. This rate variation with τ_{CP} is referred to as relaxation dispersion.

The general expression for the transverse relaxation rate, encompassing all chemical exchange time-scales, is *(6,22,23)*

$$R_2\left(\tau_{CP}^{-1}\right) = \frac{1}{2}\left(R_{2A}^0 + R_{2B}^0 + k_{ex} - \tau_{CP}^{-1} \cosh^{-1}\left[D_+ \cosh\left(\eta_+\right) - D_- \cos\left(\eta_-\right)\right]\right), \tag{6}$$

where

$$
\begin{aligned}
D_{\pm} &= \frac{1}{2}\left[\pm 1 + \frac{\psi + 2\Delta\omega^2}{\left(\psi^2 + \xi^2\right)^{1/2}}\right], \\
\eta_{\pm} &= \frac{\tau_{CP}}{\sqrt{2}}\left[\pm\psi + \left(\psi^2 + \xi^2\right)^{1/2}\right]^{1/2}, \\
\psi &= \left(R_{2A}^0 - R_{2B}^0 - p_A k_{ex} + p_B k_{ex}\right)^2 - \Delta\omega^2 + 4 p_A p_B k_{ex}^2, \\
\xi &= 2\Delta\omega\left(R_{2A}^0 - R_{2B}^0 - p_A k_{ex} + p_B k_{ex}\right),
\end{aligned}
\tag{7}
$$

and $R_{2A(B)}^0$ refers to the exchange-free transverse relaxation rate of site A (B). However, the complexity of **Eqs. 6** and **7** both complicates interpretation of observed rate variation and obscures any intuitive feel for this dependence, as presented in the discussion of **Fig. 2**. This is more readily seen in the approximate form

$$R_2\left(\tau_{CP}^{-1}\right) = R_2\left(\tau_{CP}^{-1} \to \infty\right) + \frac{p_A p_B \Delta\omega^2 k_{ex}}{k_{ex}^2 + \left(p_A^2 \Delta\omega^4 + 144\tau_{CP}^{-4}\right)^{1/2}}, \tag{8}$$

which is valid at all time-scales provided that $p_A \gg p_B$. Another simplification, valid for the limit of fast exchange without restriction on (p_A/p_B), yields *(17,24)*

$$R_2\left(\tau_{CP}^{-1}\right) = R_2\left(\tau_{CP}^{-1} \to \infty\right) + \left(p_A p_B \Delta\omega^2 / k_{ex}\right)\left[1 - \frac{2\tanh\left(k_{ex}\tau_{CP}/2\right)}{k_{ex}\tau_{CP}}\right]. \tag{9}$$

Fits of observed $R_2(\tau_{CP}^{-1})$ dispersion profiles to either of **Eqs. 8** and **9** yield the motional (k_{ex}) and chemical ($p_A p_B$ and $\Delta\omega$) parameters to which we previously alluded. More detailed discussion of this process is provided in **Subheading 3.2.7.3.**

2.3. Off-Resonance Spin-Locked Relaxation

Exchange characterization via $R_{1\rho}$ is perhaps both more powerful (in terms of the accessible dynamics time-scale) and more challenging to implement than CPMG-based (i.e., pure-R_2) approaches. $R_{1\rho}$ is measured as the decay rate of coherence aligned (i.e., spin-locked), with an effective field, ω_e, in the rotating frame of the applied radio frequency (RF) field. The decay rate depends on the parameters of exchange ($p_A p_B$, $\Delta\omega$ and k_{ex}) and on the magnitude and orientation of $\omega_e = \sqrt{\Delta\Omega^2 + \omega_1^2}$, obtained from the RF field amplitude, ω_1, and the offset, $\Delta\Omega = (\omega - \omega_0)$, of its frequency ω from ω_0, the population-weighted average of the chemical shifts. This form for ω_e is derived easily from the rotating-frame view of spin magnetization. By definition, ω_1 is static in this frame, whereas transverse spin magnetization precesses according to the frequency offset, that is, as if it experiences a field $\Delta\Omega$ along the z-axis of the static field. The effective field is then the vector sum of these quantities, with a tilt angle $\theta = \tan^{-1}(\omega_1/\Delta\Omega)$ relative to the z-axis.

The off-resonance spin-relaxation rate is

$$R_{1\rho} = R_1 \cos^2\theta + \left(R_2^0 + R_{ex}\right)\sin^2\theta, \tag{10}$$

with weights $\cos^2\theta$ and $\sin^2\theta$ for the longitudinal and transverse components of the spin magnetization. Similar to our discussion of CPMG-based approaches, R_{ex} is the exchange contribution given by the general form, valid over all time-scales of exchange *(25)*,

$$R_{ex} = \frac{p_A p_B \Delta\omega^2 k_{ex}}{\left(\omega_{ae}^2 \omega_{be}^2 / \omega_e^2\right) + k_{ex}^2}, \tag{11}$$

where the ω_{xe} are effective fields at sites A and B. In fast-exchange *(26,27)*, $\omega_{ae}^2 \omega_{be}^2 / \omega_e^2 \sim \omega_e^2$, and **Eq. 11** reduces to

$$R_{ex} = \frac{p_A p_B \Delta\omega^2 k_{ex}}{\omega_e^2 + k_{ex}^2}. \tag{12}$$

Measuring $R_{1\rho}$ as a function of ω_e provides access to the parameters of exchange, via $R_{1\rho}$ dispersion, much like that available from τ_{CP}-variation in CPMG experiments. To qualitatively grasp the content of **Eqs. 10–12**, first consider the independence of longitudinal relaxation from exchange, as indicated by the grouping of terms in **Eq. 10**. Intuitively, this independence can be understood by considering that R_1 results from energy transfer at frequencies similar to ω_0, which is negligibly altered by exchange because $\Delta\omega \ll \omega_0$ for all cases relevant to protein NMR. Meanwhile, the direct influence of exchange on transverse relaxation places R_2 and R_{ex} together in **Eq. 10**, as one might easily anticipate from **Subheading 2.2.** and, in particular, the discussion of **Fig. 2**.

Thus, we have established that only R_{ex} varies with ω_e in **Eq. 10**, but for the trivial angular dependence, which is known from the resonance positions of spectral peaks (providing $\Delta\Omega$) and a well-calibrated ω_1 (**Subheading 3.3.5.**). It is satisfying then, to finally intuit the ω_e-dependence of R_{ex} in the $R_{1\rho}$ case, as we attempted with CPMG. Imagine that, as ω_e increases (via either $\Delta\Omega$ or ω_1), the chemical shift changes that define exchange yield diminishing variation in either the magnitude or direction of ω_e. Correspondingly, exchange-induced dephasing is mitigated, and so $R_{1\rho}(\omega_e)$ provides a dispersion profile from which one obtains $p_A p_B$, $\Delta\omega$, and k_{ex}. For further detail on fitting dispersion profiles obtained from $R_{1\rho}(\omega_e)$, we refer the reader to **Subheading 3.2.7.3.**

2.4. CPMG or $R_{1\rho}$?

At the end of **Subheading 2.1.**, we touched on the pros and cons of CPMG (pure-R_2) and off-resonance spin-lock ($R_{1\rho}$) methods of exchange characterization. The most thorough investigation of exchange dynamics will combine these methods to probe a larger range of motional time-scales, with CPMG's coverage over $100 \text{ s}^{-1} \leq k_{ex} \leq 3000 \text{ s}^{-1}$, and that of $R_{1\rho}$ up to k_{ex} approx 50,000 s^{-1}, but with difficulty for $k_{ex} < 1000 \text{ s}^{-1}$. Poorer access to slow time-scale dynamics with off-resonance $R_{1\rho}$ is because of the challenge of calibration when $B_1 \leq 500 - 1000$ Hz and the lower limit of $\Delta\Omega$ is approx 1000–2000 Hz if adiabaticity is to be maintained in alignment of spins with ω_e (*see* **ref. 28** and **Subheading 3.3.3.**). Nonetheless, the much broader high-end-exchange time-scale accessible with off-resonance $R_{1\rho}$ does encompass rates relevant to biological processes *(29–31)*, and so the utility of $R_{1\rho}$ is both unique and significant. Thus, some foreknowledge of the biochemical system under study may aid in choosing between or prioritizing CPMG and $R_{1\rho}$, whereas a general prescription for their implementation would be naïve.

Finally, the difficulty of accessing $k_{ex} < 1000$ s^{-1} with off-resonance $R_{1\rho}$ may be circumvented, in addition to the CPMG route, by using nominally resonant versions in which the RF carrier for the relaxing spins is placed in the center of their chemical shift range. Such near-resonance measure of $R_{1\rho}$ has shown utility in studies of protein dynamics *(23,32)*. However, the method is not necessarily amenable to the methods presented in this chapter for aligning spin coherence with ω_e via adiabatic RF field sweeps as adiabaticity may not be satisfied if $\Delta\Omega$ is small (**Subheading 3.3.3.**). A separate alignment method is readily applicable to near-resonance $R_{1\rho}$ *(4,33–35)*, and although more complex in concept, it is straightforward to implement and can be as efficient provided the spread of chemical shifts (hence $\Delta\Omega$ values) is not too great *(4,28,34)*. However, there is no particular advantage of near-resonance $R_{1\rho}$ over CPMG-based methods for exchange characterization, and so we omit further discussion of the former from this chapter, referring the reader to existing thorough treatments *(4,33–35)*.

2.5. Exchange-Like Phenomena in NMR

Rotational diffusion is often anisotropic in macromolecules with non-spherical shape. This quality can lead to elevated values of R_2 relative to the average isotropic value. The effective-field-dependent characterization of R_2 or $R_{1\rho}$ in **Subheadings 2.2.** and **2.3.** will uncover only true exchange contributions to elevated rates. However, another approach to dynamics characterization at fast (ps–ns) time-scales is often interpreted using terms intended to simultaneously account for slow dynamics. This approach interprets dynamics from so-called model-free analysis of observed R_1 (longitudinal) and R_2 relaxation rates, and nuclear Overhauser enhancements (NOEs) *(36,37)*. Because the latter approach does not necessarily rely on measurement of rate variation with the effective or static magnetic fields to uncover R_{ex} contributions, it is susceptible to misassociation of diffusional-anisotropy effects with chemical exchange (*see* **ref. 38** and Chapter 10 of this volume).

A robust method based on relaxation interference measurements can distinguish the two effects and unambiguously identify exchanging sites using data primarily geared to ps–ns dynamics. The method is applicable to *I–S* spin systems, such as ^{1}H–^{15}N, with significant contribution to the relaxation from *S*-spin chemical shift anisotropy (CSA). When motions inducing *I–S* dipole-mediated relaxation and *S*-spin CSA-mediated relaxation are identical or correlated, the two mechanisms interfere with each other. Cross-relaxation between longitudinal (I_z) and two-spin order ($2I_zS_z$) coherences (or between S_y and $2I_zS_y$) resulting from this interference is sensitive to diffusional anisotropy and independent of chemical exchange *(39,40)*. Thus, comparison of longitudinal (η_z) and transverse (η_{xy}) cross-relaxation rates with longitudinal and transverse autorelaxation rates (R_1 and R_2) provides a means to properly attribute the source of elevated rates *(39)*.

The longitudinal and transverse CSA/dipolar cross-relaxation rates for ^1H–^{15}N two-spin systems are

$$\eta_z = -\sqrt{3}\; cd\, P_2(\cos\beta)\, J(\omega_N),$$ (13)

and

$$\eta_{xy} = -\frac{\sqrt{3}}{6} c\, d\, P_2(\cos\beta)\big[4J(0) + 3J(\omega_N)\big],$$ (14)

in which $d = (\mu_0 h \gamma_H \gamma_N)/(8\pi^2 r_{NH}^3)$ and $c = \left(\gamma_N B_0 \Delta\sigma/\sqrt{3}\right)$ are the dipolar and CSA coupling constants, μ_0 is the permeability of free space, h is Planck's constant, $\gamma_{H(N)}$ are the ^1H and ^{15}N gyromagnetic ratios, r_{NH} is the distance between the two nuclei, B_0 is the static magnetic field strength, $\Delta\sigma$ is the value of the axially symmetric ^{15}N CSA tensor, β is the angle between the symmetry axes of the dipolar and chemical shift tensors, and $P_2(x) = (3x^2 - 1)/2$. The quantities $J(\omega = 0, \omega_N)$ are the spectral densities (i.e., fractional components in frequency space) of fluctuations in the relaxation-inducing interactions, and ω_N is the ^{15}N Larmor frequency. The ratio of cross-relaxation rates relates to that of the autorelaxation rates according to *(39)*

$$\frac{R_2 - 1.079\sigma_{NH}}{R_1 - 1.249\sigma_{NH}} = \frac{\eta_{xy}}{\eta_z} + \frac{R_{ex}}{R_1 - 1.249\sigma_{NH}},$$ (15)

where σ_{NH} is the heteronuclear cross-relaxation rate. **Equation 15** indicates that protein sites with R_2/R_1 significantly greater than η_{xy}/η_z are subject to conformational-exchange processes. Its rearrangement provides the exchange-free ($R_{ex} = 0$) transverse relaxation rate

$$R_2^0 = \left(R_1 - 1.249\sigma_{NH}\right)\frac{\eta_{xy}}{\eta_z} + 1.079\sigma_{NH}.$$ (16)

Thus, residues with $R_2 > R_2^0$ identified in this manner can be confidently predicted to be involved in a conformational exchange process.

3. Methods

3.1. Experimental Characterization of Conformational Exchange

3.1.1. Experimental Setup (Instrument Preparation)

As with all NMR experiments proper instrument preparation is crucial to obtaining high-quality data. Related issues are not the main focus of this work, but brief discussion and literature references are provided given their importance. In addition, we discuss some aspects of the experimental setup that are of unique or particular concern in relaxation studies.

To begin, proper shimming of the magnet is essential to ensure adequate solvent suppression, in lieu of which saturation of the spectrometer preamps and/or striping, roll, or excessive noise in the frequency spectrum can prevent accurate determination of the peak heights or volumes used to assess signal decay. Typically, automated gradient shimming, combined with the methods outlined in the text by Cavanagh et al. *(41)*, work quite well.

Proper calibration of pulse widths (i.e., of $\pi/2$ and π pulse times) is also of particular concern in using sequences aimed at NMR relaxation. In particular, imperfections in the pulses of CPMG trains can be a source of decay in the echo envelope that has nothing to do with relaxation. Furthermore, the degree of this contribution will depend on the number of pulses in the train, thus providing a τ_{CP}-dependent loss that skews interpretation of $R_2(\tau_{CP}^{-1})$. Pulse errors result from both imperfect calibration and the unavoidable variation of pulse performance with frequency offset. Fortunately, established error-compensation protocols *(16,42)* yield canceling errors in neighboring pulses of CPMG sequences. This compensation requires an even-numbered CPMG train consisting of pulses phase-shifted 90° relative to the $\pi/2$ preparation pulse preceding the relaxation period or 180° phase alternation throughout the train. In practice, ^{15}N pulse times can be calibrated to within ±2% at the carrier frequency, whereas error compensation sufficiently accounts for additional imperfection away from the carrier, although off-resonance effects have been implicated in systematic R_2 variations *(43)*.

3.1.2. Pulse Sequences for Relaxation Measurements (General Features)

Finally, it is worthwhile to describe the common and most basic aspects of NMR pulse sequences for measurement of either $R_2(\tau_{CP}^{-1})$ or $R_{1\rho}$ relaxation. Consistent with the rest of this chapter, our description is in the context of ^{1}H–^{15}N heteronuclear sequences, of which an example is shown in **Fig. 3**. The sequence provides a standard HSQC spectrum with ^{1}H and ^{15}N chemical shifts in orthogonal dimensions. It begins with a delay for recovery of spin magnetization to equilibrium, followed by the element combining a $\pi/2$ pulse and z-gradient G_1 to purge initial ^{15}N magnetization (*see* **Note 2**). The subsequent insensitive nuclei-enhanced polarization transfer (INEPT) element *(44)* partially transfers magnetization from the amide ^{1}H to the attached ^{15}N, whereas the next ^{1}H $\pi/2$ pulse stores the resulting coherence as $2I_zS_z$. The gradient G_3 purges any residual magnetization not in this state (including ^{1}H$_2$O). In addition to the usual sensitivity gain from INEPT transfer, this preparation, with the subsequent ^{15}N $\pi/2$ pulse, provides a definite and consistent magnitude of $2I_zS_x$ as the starting point for the relaxation period. Such procedure is standard in NMR experiments, but particularly important when monitoring coherence decay.

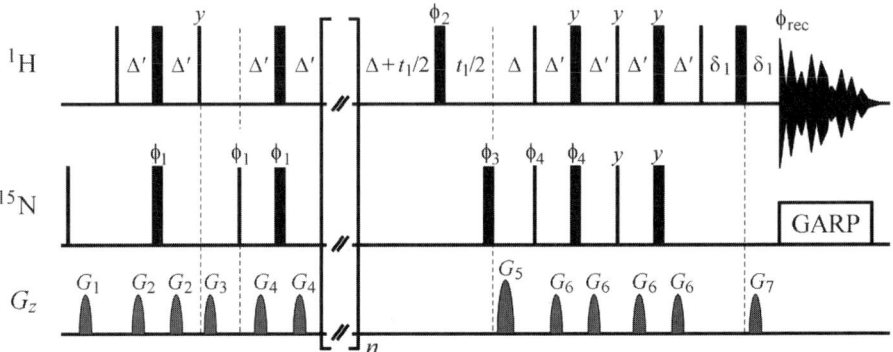

Fig. 3. General ^{1}H–^{15}N HSQC pulse sequence of the type used in relaxation experiments presented throughout Chapter 12. As for all subsequent sequences shown, *thin (broad) bars* represent $\pi/2$ (π) pulses with RF phase x unless otherwise indicated, and *dashed lines* are a guide to the eye in aligning individual elements. Typical full-power 90° pulse times are $t_{\pi/2}(^1H) = 8.0$ μs, $t_{\pi/2}(^{15}N) = 25.0$ μs, whereas lower amplitude pulses may be used during relaxation, as discussed in the text. GARP decoupling *(47)* is applied during acquisition with 1.2 kHz RF field strength. Delays are $\Delta = 2.7$ ms approx $(4J_{NH})^{-1}$, where J_{NH} approx 93 Hz is the amide one-bond scalar-coupling constant, $\Delta' = 2.5$ ms, $\delta_1 = 400$ μs, and t_1 is the incremented ^{15}N evolution time. The reduced value of Δ' relative to Δ is intended to maximize the magnitude of the final coherence state by reducing losses from relaxation at the expense of an incomplete transfer. Although specific implementations will vary through the chapter, the sequence shown is to be used with signal averaging over the phase cycle: $\phi_1 = \{-x, x\}$, $\phi_2 = \{x, x, y, y, -x, -x, -y, -y\}$, $\phi_3 = \{x, x, -x, -x\}$, $\phi_4 = \{x\}$, $\phi_{rec} = \{x, -x, -x, x\}$, whereas inversion of ϕ_1 and the receiver phase on successive t_1 increments *(87)* plus repetition of the experiment with ϕ_4 and G_5 inverted *(45)* provides frequency discrimination and coherence selection *(41)*. Gradients are along the z-axis, and in this case occur at magnitudes 8.7, 13.0, 14.1, 11.9, 30.0, 10.9, 15.0 G/cm and durations 1.0, 0.6, 1.0, 0.6, 1.25, 0.7, and 0.251 ms corresponding to G_1 through G_7. Finally, the recycle delay is approx 3.0 s here, allowing adequate recovery of the ^{1}H$_2$O magnetization to avoid saturation transfer (signal loss) from exchange of the water and amide protons. Other sequences using flip-back techniques that leave ^{1}H$_2$O along the z-axis during most of the experiment require much shorter recycle delays of 0.5 s.

Following this preparatory stage, the sequence of **Fig. 3** incorporates a second INEPT transfer and a relaxation period, which is highlighted by the square-bracketed enclosure. This combination with an INEPT period is not universal to sequences of this chapter, but it is often used along with the relaxation period for modification of the specific coherence type monitored. We treat variations of specific implementation of the relaxation period on a case-by-case basis. Following the relaxation periods, a ^{1}H-decoupled evolution period encodes the ^{15}N chemical shift. Various schemes are used for coherence selection and frequency discrimination during t_1, whereas the example of **Fig. 3** includes delays for placement of the gradient G_5 for Rance–Kay coherence selection *(45)*. As a side benefit, this selection scheme suppresses any ^1H_2O contribution to the ultimate signal, although the additional delays yield corresponding decay that decreases the signal-to-noise ratio (S/N) relative to an earlier approach that does not rely on gradients for coherence selection *(46)*. Back-to-back, or refocused, INEPT periods transfer postevolution coherence in a linear combination of the $2I_zS_x$ and $2I_zS_y$ states from the amide ^{15}N to the attached ^1H. Finally, the sequence is terminated with the simple spin-echo element incorporating G_7 to decode the desired coherence by refocusing the prior action of G_5. ^1H magnetization is detected while decoupling ^{15}N using the GARP phase-cycling scheme *(47)*. Note that, to avoid sample heating during acquisition times totaling approx 100 ms, decoupling fields (approx 1 kHz) are substantially reduced relative to the RF amplitude (approx 10 kHz) used for $\pi/2$ and π pulses. Other sequence details are noted in the caption to **Fig. 3**. We also note that, where significant variations from this general HSQC scheme occur, we have made an effort to explain their workings and purpose in the text.

More extensive discussion of the basic issues of experimental preparation is provided by Brereton in vol. 60 of this series *(48)* and by Cavanagh et al. *(41)*. The topic of pulse calibration in the context of spin locking requires more detailed presentation of the relevant experiments and is thus included in the section on $R_{1\rho}$ measurement. Further details specific to experiments for $R_{1\rho}$ and R_2 measurement are included in subsequent sections of this chapter.

3.2. CPMG Experiments for Transverse Relaxation and Conformational Exchange

3.2.1. Relaxation-Compensated and Related CPMG Sequences

The most general family of CPMG experiments for characterization of $R_2(\tau_{CP}^{-1})$ are shown in **Fig. 4**. Each sequence in **Fig. 4** is intended for use over a distinct portion of the τ_{CP} time line, and used together, the set of experiments can provide the most complete description of $R_2(\tau_{CP}^{-1})$ dispersion profiles. The relaxation-compensated CPMG (rcCPMG) sequence *(19)* of **Fig. 4A** is useful over the range $\tau_{CP} = 0.80$ to 10.0 ms (*see* **Note 3**) whereas the even- and Hahn-echo experiments *(20)* in parts **(B)** and **(C)** of the figure are applicable at $\tau_{CP} = m/(J_{NH})$ approx ($m \times 10.8$ ms) and $\tau_{CP} = 2m/J_{NH}$ approx ($m \times 21.6$ ms), respectively, where m is an integer and J_{NH} approx 93 Hz is the amide one-bond scalar-coupling constant. Understanding the origins of this prescription requires working knowledge of two important requirements for the relaxation periods: that CSA/dipolar cross-correlated relaxation be suppressed *(49)* and that the monitored coherence spend consistent fractions of its time as in- and antiphase

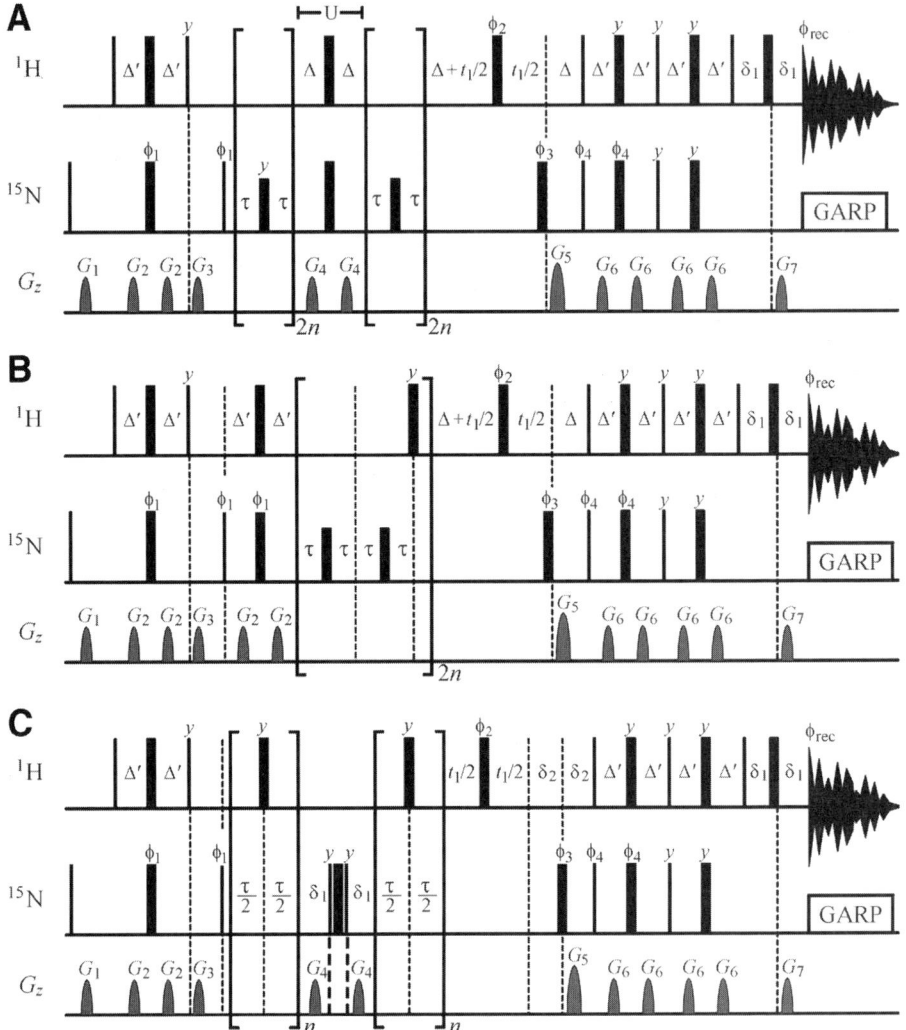

Fig. 4. NMR pulse sequences for measurement of the amide ^{15}N $R_2(\tau_{CP}^{-1})$ via (**A**) the relaxation-compensated CPMG (rcCPMG), and the J-compensated (**B**) even-echo CPMG and (**C**) Hahn-echo experiments. In each case, the relaxation period consists of the segment(s) between square brackets with delays $\tau = \tau_{CP}/2$. During relaxation in (**A**) and (**B**), low-power ^{15}N pulses corresponding to $t_\pi(^{15}N) = 100.0$ μs are used. The Hahn-echo experiment (**C**) is for two-point rate determination with $n = 0$, 1 only (**Subheading 3.2.7.**). Delays are $\Delta = 2.7$ ms, $\Delta' = 2.5$ ms, $\delta_1 = 400$ μs, and $\delta_2 = 2.05$ ms, whereas the recycle delay is 3.0 s. Each sequence uses the phase cycle: $\phi_1 = \{-x, x\}$, $\phi_2 = \{x, x, y, y, -x, -x, -y, -y\}$, $\phi_3 = \{x, x, -x, -x\}$, $\phi_4 = \{x\}$, $\phi_{rec} = \{x, -x, -x, x\}$. Inversion of ϕ_1 and the receiver phase on successive t_1 increments (**87**) plus repetition of the experiment with ϕ_4 and G_5 inverted (**45**) provides frequency discrimination and coherence selection (**41**). Gradients G_1 through G_7 correspond to magnitudes 8.7, 13.0, 14.1, 8.7, 30.0, 10.9, 15.0 G/cm and durations 1.0, 0.6, 1.0, 0.5, 1.25, 0.7, and 0.251 ms.

(e.g., in the S_x and $2I_zS_x$ states, respectively) *(19,49)*. The three implementations achieve these ends by distinct means, with rcCPMG averaging in-phase/antiphase residency times to leave τ_{CP} unrestricted, but with suppression of CSA/dipolar cross-relaxation that requires increments to the relaxation time at $t = n \times 4\tau_{CP}$ ($n = 0$, 1, 2, etc.). In contrast, the τ_{CP}-restricted Hahn-echo sequence (**Fig. 4C**) allows increments $t = 0$ or τ_{CP}, and thus is favored for τ_{CP} > 20 ms. The τ_{CP}-restricted even-echo sequence (**Fig. 4B**) is limited, like rcCPMG, to $t = n \times 4\tau_{CP}$, but improves on the latter in the quality of cross-relaxation suppression (**Subheading 3.2.3.**) and thus may be preferable to rcCPMG as τ_{CP} approaches $1/J_{NH}$ approx 10.8 ms.

3.2.2. Methods for Averaging In-Phase/Antiphase Magnetization

Averaging in- and antiphase residencies is the more significant of the noted constraints. The necessity stems from the inequivalence of the corresponding relaxation rates, $R_2^{in} < R_2^{anti}$, and the dominant role of scalar coupling in evolution during the relaxation period. Because the latter exchanges coherence between the two types, both R_2^{in} and R_2^{anti} contribute to the observed average background relaxation rate *(19,49)*. For each sequence in **Fig. 4**, tracking evolution of the coherence input into the relaxation period and computing the fractional residency times, in- and antiphase types, yields the time-averaged background relaxation rate,

$$R_{2,av} = (1 - \varepsilon)R_2^{anti} + \varepsilon R_2^{in}, \tag{17}$$

where

$$\varepsilon = 1 - \mathrm{sinc}(\pi J_{NH}\tau_{CP}). \tag{18}$$

The simplest route to R_{ex} and the chemical parameters that determine $R_2(\tau_{CP}^{-1})$, demands that ε, and hence $R_{2,av}$, not vary with the relaxation time. In the rcCPMG sequence of **Fig. 4A**, this is achieved by separating halves of the relaxation period by an intervening U element. The latter is simply an INEPT period situated to accept in- (anti-) phase coherence from the preceding relaxation period and convert it to an anti- (in-) phase state at the start of the subsequent one. In effect, this exchanges the coefficients of R_2^{anti} and R_2^{in} in **Eq. 17** by exactly counterbalancing the corresponding residence times from the two periods that make up the total relaxation time. The resulting $R_{2,av} = (R_2^{anti} + R_2^{in})/2$ is constant regardless of the total relaxation time, $t = 4n\tau_{CP}$, and so τ_{CP} is unrestricted.

A different tactic is used in the sequences of **Fig. 4B** and **C**. There, the durations of the relaxation period are fixed at values for which scalar-coupled evolution provides equal time as in- and antiphase coherence. This corresponds to replacing ε with 0.5 in **Eq. 17** and yields the acceptable value

$R_{2,av} = (R_2^{anti} + R_2^{in})/2$, exactly as obtained in the unrestricted rcCPMG case. That larger increments of τ_{CP} are required for the sequence of **Fig. 4C** than in **Fig. 4B** is because of the mode each uses to suppress CSA/dipolar cross-relaxation, as presented in the following.

3.2.3. Methods for Suppressing CSA/Dipolar Cross-Relaxation

CSA/dipolar cross-correlated relaxation is significant when the two interactions are of similar magnitude, as with ^{15}N protein sites at field strengths $\geq$ 500 MHz ^{1}H frequency, but less so for ^{13}C$^\alpha$ and nonaromatic ^{13}C side-chain sites *(49)*. CSA/dipolar cross-correlation is troublesome in CPMG experiments because it adds time-dependent terms to the overall relaxation rate, complicating the observed decay and preventing accurate assessment of dispersion profiles with **Eqs. 6, 8,** or **9**. The heteronuclear nature of the phenomenon provides a means for its suppression: incorporation of ^{1}H π pulses to reverse its effects midway through the relaxation period or at regular intervals therein. Such pulses invert the heteronuclear dipole interaction but leave the CSA of ^{15}N (or other heteronucleus) untouched, and thus reverse the sense of the interference.

The rcCPMG sequence subtly uses the ^{1}H π pulse in the U period, already required for in-phase/antiphase averaging, to suppress the cross-relaxation. The sequence of **Fig. 4B** achieves this goal with a ^{1}H π pulse that follows even repetitions of the ^{15}N spin echoes. This particular placement prevents interference of the ^{1}H pulse with the pulse-error-correction capacity of the CPMG train, which we recall must include an even number of π pulses. Finally, the Hahn-echo sequence of **Fig. 4C** places ^{1}H π pulses for cross-relaxation suppression midway through each half-relaxation period. That placement suppresses the CSA/dipolar mechanism individually in each half-relaxation period and avoids interfering with in-phase/antiphase averaging by arriving with the monitored coherence as purely in-phase S magnetization otherwise immune to a ^{1}H pulse.

Other schemes have been used in which averaging the residence time for in- and antiphase coherences is unnecessary. In a recent example *(21)*, their interconversion via scalar-coupled evolution was prevented by WALTZ-16 *(50,51)* decoupling of ^{1}H spin during relaxation, rendering $\varepsilon = 1$ in **Eq. 17**. Such decoupling is also known to suppress CSA/dipolar cross-correlated relaxation *(52)*. However, the resulting suppression has been shown to be inferior relative to schemes relying on ^{1}H π pulses *(49)*, as in **Fig. 4**. Furthermore, decoupling by continuous irradiation also has been shown to yield an undesirable increase in the background relaxation rate *(49,53)*. Thus, the sequences of **Fig. 4** seem better suited for quantitation of the chemical parameters of exchange-induced relaxation. Orekhov et al. *(18)* used another approach in which CPMG relaxation was followed without explicit account for in-phase/antiphase residencies. Instead, pulse spacing was kept to $\tau_{CP} \pm 2.0$ ms, which limits the antiphase residency to <10% of the total relaxation time such that ε approx 1 in **Eq. 17**. This correspondingly limits the detectable range of exchange rates to >1000 s^{-1}, whereas variation of ε with τ_{CP} limits interpretation of $R_2(\tau_{CP}^{-1})$ solely in terms of chemical exchange.

3.2.4. rcCPMG Methods With Enhanced Sensitivity to Chemical Exchange

Wang et al. *(54)* have developed a set of pulse sequences (**Fig. 5**) based on the rcCPMG family of the previous section, but which incorporate a novel feature that subtracts the effects of ^{1}H–^{1}H dipolar relaxation from the relevant background. To emphasize their advance from and relation to the previous experiments, we refer to the new sequences as "background-subtracted" rcCPMG experiments. The sequences in **Fig. 5A–C** each cover a distinct portion of the τ_{CP} time line, with the same benefits and restrictions on τ_{CP} described for the corresponding sequences in **Fig. 4A–C**.

The main advantage of the background-subtracted versions is in the fact that the new background relaxation rate is separately measurable, thus reducing the number of parameters required to fit the measured R_2 (τ_{CP}^{-1}) dispersion profile. This is particularly important for sites exchanging with $k_{ex} > 1250$ s^{-1}. In such cases, the practical minimum pulse spacing, τ_{cp}^{min} approx 0.8 ms, is insufficient to reach the limit $R_2(\tau_{CP}^{-1} \rightarrow \infty)$ in **Eqs. 8** and **9**, and fits to the dispersion profile may yield incorrect estimates of $\Delta\omega$ and the chemical parameters, p_A, p_A, k_{ex}. That $R_2(\tau_{CP}^{-1} \rightarrow \infty)$ is unattainable can be shown from **Eqs. 8** and **9** where inserting the noted limits results in an exchange contribution still at approx 10% of its maximum (i.e., of R_{ex}).

The background subtraction in these sequences is achieved by incorporating the relaxation periods labeled $2I_zS_z$ in **Fig. 5**, which provide for relaxation of the corresponding coherence. The usual relaxation periods for decay of transverse (in- and antiphase) coherence are implemented in the standard manner, where increments of n increase the total transverse relaxation time t. The duration δ of the $2I_zS_z$ is linked to t by

$$\delta = (T - t) / 2 \qquad (19)$$

where T is the maximum of t. In this way, the sequences achieve a constant degree of relaxation at rates related to the $2I_zS_z$ coherence, yielding the overall signal decay

$$I(t) = \left(I_0 e^{-R_1^{IS}T/2}\right) \times \exp\left[\left(R_2\left(\tau_{cp}^{-1}\right) - R_1^{IS}\right)t\right], \qquad (20)$$

where I_0 is the intensity in the absence of relaxation (i.e., both $T = 0$ and $t = 0$). The constancy of $2I_zS_z$ relaxation enables its subtraction from the decay monitored over t and its appearance in the initial intensity, $I(t = 0)$. The subtraction of R_1^{IS} beneficially increases the relative contribution of R_{ex} to $R_2(\tau_{CP})$ and

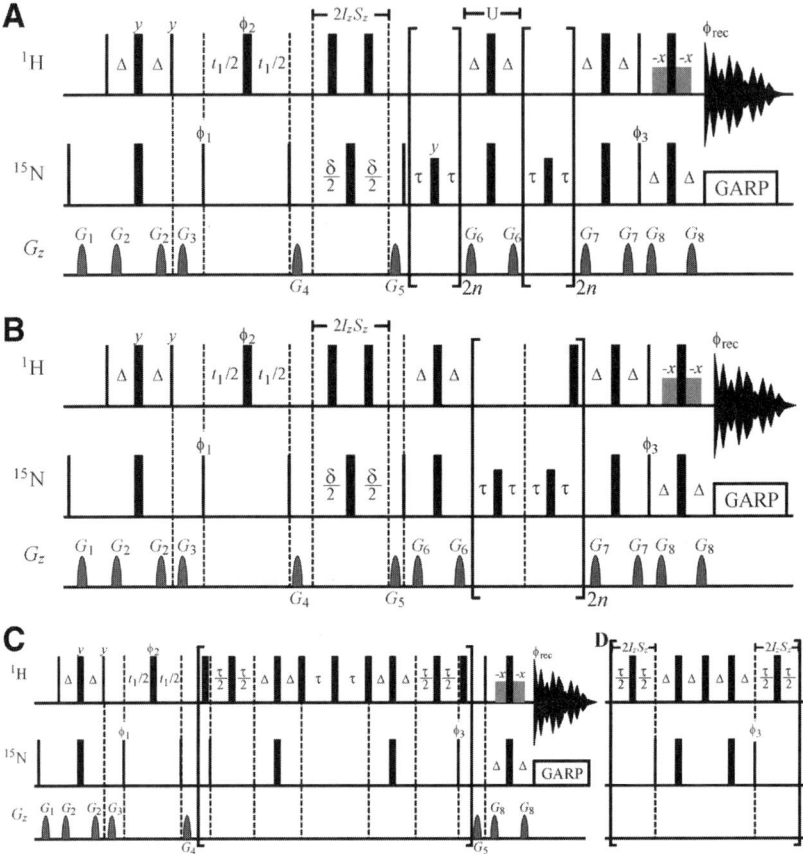

Fig. 5. Pulse sequences with enhanced sensitivity to chemical exchange obtained by subtraction of background relaxation. Each measures the amide ^{15}N $R_2(\tau_{CP}^{-1})$ via sequences named by analogy to those in **Fig. 4**: (**A**) the background-subtracted rcCPMG, and J-compensated experiments, (**B**) the background-subtracted even-echo CPMG, and (**C**) the background-subtracted Hahn-echo, which uses the element in (**D**). In (**A**) and (**B**), the primary relaxation periods consist of the segments between square brackets with delays $\tau = \tau_{CP}/2$ and low-power ^{15}N pulses $[t_\pi(^{15}N) = 100.0 \ \mu s]$. In addition, the period in each labeled $2I_zS_z$ incorporates relaxation of that coherence for time $\delta = (T - t)/2$, where t is the total relaxation time [(**A**) $t = 4n \ \tau_{CP}$ and (**B**) $t = 2n \ \tau_{CP}$] for in- and antiphase coherences and T is the maximum value of t used to monitor relaxation. For the experiment of (**C**) and (**D**), only two relaxation time-points are collected (**Subheading 3.2.7.**), one executed as shown in (**C**), for which $t = 2\tau_{CP}$, and the other with the element in (**D**) replacing the square-bracketed portion in (**C**), for which $t = 0$. As for (**A**) and (**B**), $\tau = \tau_{CP}/2$. (**C**) and (**D**) differ slightly from the presentation of Wang et al. (*54*) owing to a typographical error in their figure. In all sequences $\Delta = 2.7$ ms, the recycle delay is 3.0 s, and the phase cycle: $\phi_1 = \{x, -x\}$, $\phi_2 = \{x, x, -x, -x\}$, $\phi_3 = \{4x, 4(-x)\}$, $\phi_{rec} = \{x, -x, x, -x, -x, x, -x, x\}$. The States-TPPI scheme (*87*) is used for frequency discrimination and coherence selection, whereby ϕ_1 and the receiver phase are inverted on successive t_1 increments, executed both for the noted phase cycle and a second with $\phi_1 = \{y, -y\}$ (*41*). Gradients G_1 through G_8 correspond to magnitudes 4.4, 3.0, 12.0, 9.0, 4.5, 3.0, 3.0, and 9.0 G/cm and durations 2.0, 0.6, 2.0, 2.0, 1.0, 0.6, 0.6, and 1.0 ms.

detrimentally reduces the initial intensity by $\exp(-R_1^{IS}T/2)$. In spite of the S/N loss, Wang et al. *(54)* showed that the new sequences matched the accuracy of the original rcCPMG family for quantifying exchange in basic pancreatic trypsin inhibitor protein.

Finally, the background-subtracted rcCPMG sequences provide an improved estimate of R_{ex} from $\Delta R_2 = [R_2(1/\tau_{cp}^{max}) - R_2(1/\tau_{cp}^{min})]$ because of the increased contrast between R_{ex} and the background. Recall from the previous section that ΔR_2 is useful for performing a fast survey for μs–ms dynamics across the protein sequence. Furthermore, improved estimates of R_{ex} from ΔR_2 enable more accurate assessment of the chemical-exchange time-scale using the field-dependent method of Millet et al. *(20)*. Therein (dR_{ex}/dB_0) approx $(d\Delta R_2/dB_0)$ can reveal dynamics information without resorting to complete determination of the $R_2(\tau_{CP}^{-1})$ dispersion.

3.2.5. Dynamics in Large Proteins: Transverse Relaxation-Optimized (TROSY) CMPG

TROSY methodology, introduced by Pervushin et al. *(55)* extends the size limit of proteins practically accessible by NMR from 30 kDa to ≥100 kDa and can be useful for proteins ≥15 kDa *(56)*. The limitation of non-TROSY experiments stems from the slow tumbling exhibited by larger proteins, which correspondingly increases transverse relaxation rates and thus broadens NMR transitions to reduce sensitivity and resolution. TROSY relies on the interference of CSA and dipolar relaxation mechanisms, and so, as with our previous suppression of CSA/dipolar cross-correlated relaxation, is most useful for study of the amide NH sites at fields ≥600 MHz ^{1}H frequency (TROSY effects and their application is covered in much more detail in Chapters 5 and 11 of this volume). ^{1}H–^{15}N HSQC spectra in the absence of heteronuclear decoupling in either dimension reveal a scalar-coupled quartet of peaks for each residue. According to long-recognized physics *(57,58)*, each of the four peaks exhibits a distinct line width because the relative states of the ^{1}H and ^{15}N spins determines both the quartet splitting pattern and the sign of the dipolar interaction. The sign variation yields one of four resonances for which the correlated CSA and dipolar interactions cancel in both spectral dimensions to yield slower relaxation and a narrow, high-amplitude resonance. The other three peaks exhibit constructive CSA/dipolar interference in one or both dimensions to yield much broader resonances. Pervushin et al. designed NMR sequences in which the favored multiplet component survives a novel pulse and phase-cycling scheme that cancels the undesirable peaks, whereas others have updated the original sequence for enhanced sensitivity *(59–62)*.

TROSY–CPMG sequences (**Fig. 6**) for characterization of ^{15}N exchange have been developed to capitalize on and extend the benefits of TROSY *(63)*. Along with the usual two-dimensional (2D) resolution and sensitivity advantages, this approach mitigates background relaxation by employing TROSY coherence selection during the relaxation period. In application to the 54 kDa protein, triosephosphate isomerase, TROSY–rcCPMG yielded a 2.5-fold reduction in the average $R_{2,av}$ relative to the standard rcCPMG. Thus, contrast between the relaxation rates of exchanging and static residues increased by a factor of 2.5, improving both qualitative and quantitative aspects of the dynamics characterization. The TROSY–rcCPMG sequence of **Fig. 6** uses similar protocols for in-phase/antiphase averaging and suppression of CSA/dipolar cross-correlated relaxation as the rcCPMG sequence of **Fig. 4A**. In addition, the sequence uses the fast FHSQC scheme for water suppression *(64)*, whereby ^1H_2O magnetization is dephased and then refocused by gradients G_6 during t_1, and subsequently maintained along the z-axis to reduce saturation transfer and allow a shorter recycle delay.

3.2.6. CPMG–INEPT Signal-to-Noise (S/N) Enhancement

Unavoidably, characterization of μs–ms dynamics in proteins encounters residues that offer very poor S/N as an exact result of exchange processes. This is seen most clearly in **Fig. 1**, where all spins, except those in the absolute slow- or fast-exchange limits, have decreased amplitude and increased line width. Avoidably, signal loss related to exchange occurs in the ubiquitous INEPT transfers of coherence between ^1H and ^{15}N (or other) spins. This loss can be alleviated using the same CPMG methods discussed earlier in the context of measuring exchange. There, it was shown that the exchange contribution to $R_2(\tau_{CP}^{-1})$ vanishes as $\tau_{CP}^{-1} \rightarrow \infty$. Standard INEPT transfers are essentially one-pulse CPMG trains with $\tau_{CP}^{-1} = 2J_{NH}$ approx 185 Hz, thus losses are expected when k_{ex} is of this order or larger, which includes almost all events characterized by transverse relaxation.

Improvements on the standard approach have been implemented by Mueller et al. *(65)* in NOESY experiments for RNA structure determination, and by Mulder et al. *(66)* for use in amide or amino ^1H–^{15}N HSQC experiments, whereas Palmer et al. *(4)* suggested their use for characterizing exchange. These advances rely on the novel $xy16$ Carr–Purcell sequence *(67)*. Like common CPMG sequences, $xy16$ consists of a series of π pulses with intervening delays that, for consistency, we refer to as τ_{CP}. The pulses in this 16-member train have sequential phases: $\{x, y, x, y, y, x, y, x, -x, -y, -x, -y, -y, -x, -y, -x\}$, preventing accumulation of errors in subsequent pulses. Gullion et al. *(67)* demonstrated that the error-correction capacity of $xy16$ is nearly on par with the usual CPMG modification of the Carr–Purcell approach. More important, this avoids compounding our desired suppression of exchange with losses because of accumulated pulse errors.

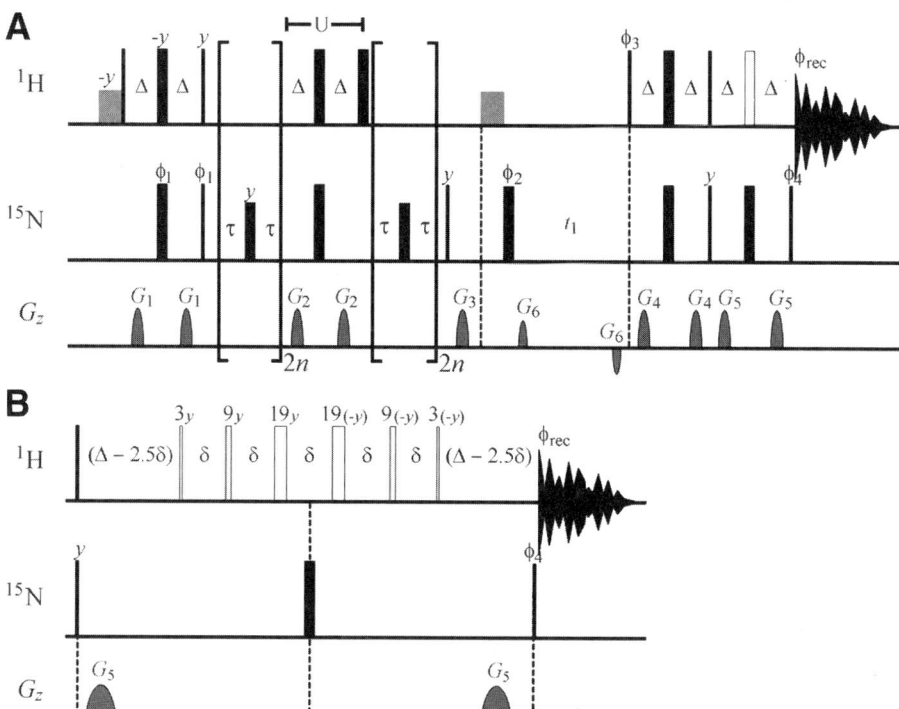

Fig. 6. **(A)** TROSY–rcCPMG pulse sequence for measurement of the amide ^{15}N $R_2(\tau_{CP}^{-1})$ in large proteins. The relaxation period consists of the segments between square brackets with delays $\tau = \tau_{CP}/2$ and incorporates a U element similar to that of **Fig. 4A.** *Short gray bars* are $^{1}H_2O$ selective $\pi/2$ pulses of approx 1 ms duration and low-power ^{15}N pulses at $t_\pi(^{15}N) = 100.0$ µs are used during relaxation. Delays are $\Delta = 2.7$ ms and the recycle delay is 3.0 s. Signals were collected using the PEP scheme in which echo and antiecho signals are acquired at each t_1 point *(62,88,89)*. In the echo case the phase cycle is $\phi_1 = \{4(-x), 4x\}$, $\phi_2 = \{y, -y, x, -x\}$, $\phi_3 = \{-y\}$, $\phi_4 = \{x\}$, $\phi_{rec} = \{x, -x, y, -y, -x, x, -y, y\}$, whereas antiecho is acquired with $\phi_1 = \{4x, 4(-x)\}$, $\phi_2 = \{y, -y, x, -x\}$, $\phi_3 = \{y\}$, $\phi_4 = \{-x\}$, $\phi_{rec} = \{-x, x, y, -y, x, -x, -y, y\}$. Phases are specific to Varian spectrometers, and y, $-y$ must be interchanged for use with Bruker instruments *(62)*. Gradient magnitudes 9.0, 6.2, 9.7, 10.4, and 24.2 G/cm and durations 0.4, 0.3, 0.35, 0.4, and 0.5 ms correspond to G_1 through G_5. The bipolar gradient pair, G_6 approx 4 G/cm, 320 µs, should be set to sufficiently suppress radiation damping by $^{1}H_2O$ magnetization, and is omitted entirely when $t_1 < 640$ µs. The *open bar* on the ^{1}H time line represents the 3-9-19 selective pulse element *(90)*, shown within **(B)**, an expanded view of the final INEPT transfer in **(A)**. With irradiation resonant with $^{1}H_2O$, the 3-9-19 element yields null water excitation and maximum excitation at frequency Δv, set to the approx 8 ppm center of the amide ^{1}H region. Successive pulses in the sequence are of duration $t_p = (3/13), (9/13), (19/13), (19/13), (9/13)$ and $(3/13)$ in units of the ^{1}H $\pi/2$ time and are separated by delays $\delta = (2 \Delta v)^{-1}$.

Here, we have modified and tested the performance of the rcCPMG sequence of **Fig. 4A** by incorporating $xy16$-derived CPMG–INEPT transfers as shown in **Fig. 7A**. Enhanced S/N is obtained by simultaneous application of the $xy16$ sequence to ^{1}H and ^{15}N spins. In effect, this breaks the standard INEPT sequence of duration $(2J_{NH})^{-1}$ into 16 individual transfers with $\tau_{CP}^{-1} = 16$ $\times (2J_{NH})^{-1}$ approx 1.1 kHz. Thus, exchange losses are prevented to much higher k_{ex}. Enhancements, defined as the relative S/N change from the sequence in **Fig. 7A** to that in **Fig. 4A**, are shown in **Fig. 7B** for peaks in the spectrum of ribonuclease A (RNase A). Several assigned residues with well-characterized exchange behavior *(68)* stand out from the general pattern of enhancements. Surprisingly, even exchange-free residues exhibit, on average, significant S/N enhancement, whereas losses might be expected from the residual errors from 16 pulses vs the single (^{1}H and ^{15}N) π of INEPT. We hypothesize that the average gain is because of a combination of uncompensated error in the lone INEPT π pulse and/or improved water suppression with CPMG–INEPT, which could lower the noise floor, as has been observed.

Among enhanced peaks, we present full $R(\tau_{CP}^{-1})$ dispersion data and analysis for K66 (54% enhancement) and one unassigned peak, X4 (107% enhancement), in **Fig. 7C**. The dispersion reveals $k_{ex} = 1270 \pm 350$ s^{-1} and 1470 ± 830 s^{-1} for K66 and X4, respectively, both in the predicted range for enhancement. Uncertainty in the value at K66 is reduced by 17% over previous results *(68)*, and the new k_{ex} value (previously 3380 ± 1520 s^{-1}) groups the motion with that of other residues in RNase A suspected to exhibit concerted dynamics with a functional role. Meanwhile X4 may be among the 15 nonproline residues that (at pH 7.5) were previously inaccessible, perhaps because of exchange losses during INEPT transfers. Finally, that enhancements X4 >> K66, mesh with the dispersion results, whereby X4 ($R_{ex} = 24.2 \pm 1.6$ s^{-1}, $\Delta\omega = 1450 \pm 110$ Hz) should experience greater relaxation losses in the absence of CPMG–INEPT than K66 ($R_{ex} = 6.2 \pm 0.6$ s^{-1} and $\Delta\omega = 297 \pm 40$ Hz).

3.2.7. Methods of Rate Measurements and Dispersion Analysis

3.2.7.1. Two-Point Data Sampling

Rate determination is achieved using one of two methods, applicable with each of the sequences described in this section. The first, termed the two-point method, relies on measuring the intensity (peak heights or volumes) at just two values of the total relaxation time, $t = 0$ and the sampling time, t_s. Straightforward rearrangement of the expression

$$I(t_s) = I(0)\exp[-R_2(\tau_{CP}^{-1})\, t_s], \qquad (21)$$

for the intensity at t_s, yields

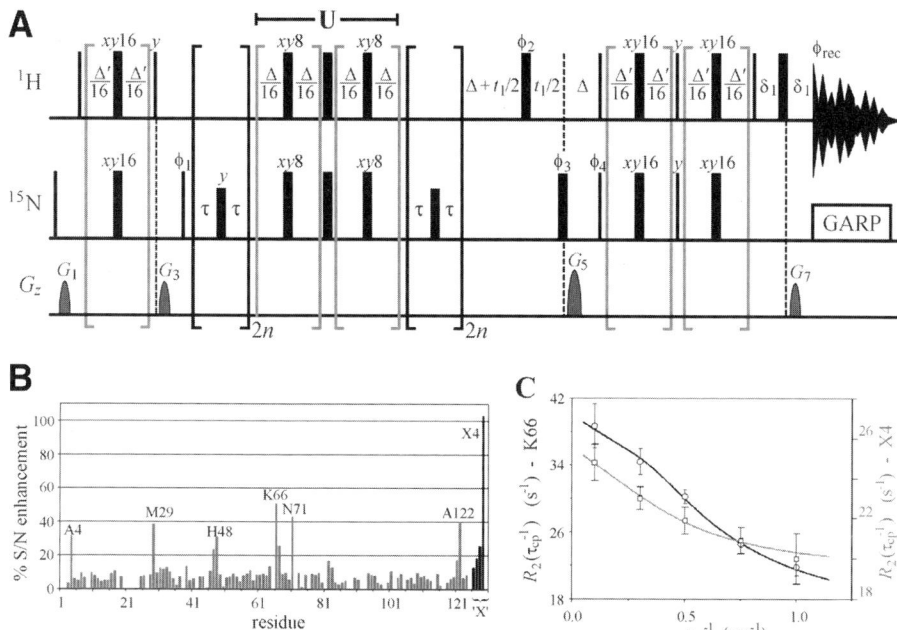

Fig. 7. (**A**) CPMG-INEPT enhanced rcCPMG pulse sequence. Delays, gradients, pulse amplitudes, phases and incrementation of ϕ_1 with t_1 are identical to those in **Fig. 4**, excepting $\phi_4 = \{-x\}$ and $\{x\}$ are combined with positive and negative G_5, respectively, which is opposite the scheme for coherence selection in the sequences of **Fig. 4**. CPMG–INEPT periods enclosed in *gray square brackets* are 16 repetitions of the echo shown with pulse phases stepping through the *xy*16 cycle defined in the text. For the U period two 8-step sequences are used with *xy*8 (the first half of the *xy*16 cycle) to allow placement of the intervening pair of π pulses, of which the ¹H π is required for suppression of CSA/dipolar cross-relaxation. (**B**) Sensitivity enhancements from CPMG–INEPT in RNase A (600 μ*M*, 50 m*M* NaCl, pH 7.5). Peaks with particularly large enhancements are labeled with the residue assignment, whereas the right-most four values correspond to unassigned peaks "X1-4." (**C**) Dispersion of $R(\tau_{CP}^{-1})$ values for K66 and X4. Individual rates were obtained using the sequence in (**A**) at $\tau_{CP} = 1.0, 1.334, 2.0, 5.0,$ and 10.0 ms and the multiple time-point approach described in **Subheading 3.2.7.** of the text. Dispersion profiles were fit to the full-exchange expression of **Eq. 8** using a program generously provided by Prof. Arthur G. Palmer.

$$R_2\left(\tau_{CP}^{-1}\right) = \left(t_s^{-1} \ln\left[\overline{I(0)} \,/\, \overline{I(t_s)}\right] \pm t_s^{-1}\Delta Q\right),\tag{22}$$

where the noted uncertainty derives **(69)** from propagating those of $I(0)$ and $I(t_s)$ via

$$\Delta Q = \sqrt{\Delta I(0)^2 + \Delta I(t_s)^2},\tag{23}$$

the percentage uncertainty in the quotient, $\overline{I(0)}/\overline{I(t_s)}$, of averaged intensities. Averaged intensities are either the mean or trimmed mean of values obtained from repeated experiments at each value of the relaxation time. To avoid systematic errors, it is important to stagger experiments such that points at a common value of t are not collected entirely back to back. The trimmed mean, where highest and lowest values are discarded, is often preferred to avoid contributions from outliers caused by instrument instabilities or other unpredictable factors. The measured uncertainties, $\Delta I(t = 0, t_s)$, are likewise taken as the standard deviation among collected values or the trimmed set.

Interestingly, Jones has shown that the two-point approach is optimal under conditions that are often relevant in protein NMR **(70)**. Selecting points at $t = 0$ and $t = t_s = [\, 0.9\, R_2(\tau_{CP}^{-1})\,]^{-1}$ yields the minimum rate uncertainty. Of course, this optimum may not be a valid choice as some pulse sequences (**Fig. 4B,C** and **Fig. 5B,C**) limit t_s to integer increments of a restricted τ_{CP}. Furthermore, there is no single optimum as $R_2(\tau_{CP}^{-1})$ varies across the protein sequence, for which reason we are interested in dynamical NMR. Jones considered this situation and found the two-point protocol to be generally optimal within guidelines. First, the choice t_s approx $(R_2^{max}R_2^{min})^{-1/2}$ is appropriate for characterizing a range of rates from R_2^{min} to R_2^{max}. Often, using previous protein NMR results, a good guess at the appropriate range is possible, although it is worth noting that elevated, rather than average, rates are of most interest in dynamics studies. Thus, choosing R_2^{min} above the protein average is often warranted. Second, because S/N is less at t_s, approx four times as many points should be collected at t_s than at $t = 0$. This has implications in calculating the rate uncertainty. For example, collecting two and eight experiments at $t = 0$, t_s, fits the protocol, but then $\Delta I(0)$ is uncertain. Assuming consistent absolute uncertainties, i.e., $\Delta I(0) = I(0) \times [\Delta I(t_s)/I(0)]$, circumvents this problem and is valid in the absence of t_1 noise **(71)**.

3.2.7.2. Sampling at Incremental Time Intervals

Finally, rates are often determined from data sets collected at multiple values of t, which are then fit to **Eq. 21** using a nonlinear least-squares routine. This approach avoids the need to guess a single optimum of t_s, but it does require an intelligently selected set of values. One approach is to increment t to yield n evenly spaced peak intensities, $I(t_i)$. This is achieved using values $t = t_i$, where

$$t_i = -\left(1/R_g\right) \ln \left[1.0 - 0.9(i-1)/(n-1)\right], \tag{24}$$

and $i = 1$ to n to yields $I(t_i)$ such that $[I(t_i)/I(0)]$ varies evenly from 1.0 to 0.1 when R_g is a reasonable guess of the actual rate. For example, using $n = 6$ and $R_g = 20 \text{ s}^{-1}$ (typical for transverse amide ^{15}N $R_{2,\text{av}}$ coherence in an approx 15 kDa protein) suggests the set t_i approx 0, 9.9, 22.3, 38.8, 63.6, and 115.1 ms. As with the two-point method, staggering the executed order of time-points avoids introducing systematic error to the rates. Uncertainties in the rates determined from multiple time-points are taken as the standard error or confidence interval returned by the fitting routine.

3.2.7.3. Relaxation Dispersion

For quantitative analysis of the dynamics, the transverse relaxation rate must be measured at various τ_{CP} values, yielding the so-called dispersion profile of $R_2(\tau_{\text{CP}})$. As noted in **Subheadings 3.2.1.** and **3.2.4.**, optimal pulse sequences should be chosen according to the particular τ_{CP} value or range thereof. Thus, NMR dispersion analysis requires a significant time investment. A rapid check for the presence of interesting μs–ms dynamics involves measuring $R_2(\tau_{\text{CP}})$ at the extremes of τ_{CP} ($\tau_{\text{CP}}^{-1} \rightarrow 0, \infty$). A significant difference between the two rates indicates the presence of chemical exchange and that a more detailed study may be warranted. The full dispersion of R_2 as a function of the pulse spacing can be fit to either **Eq. 6, 8,** or **9**. The choice of the proper equation to use requires knowledge of the chemical exchange time-scale. This information can be obtained from field-dependent relaxation data *(20)*. Briefly, exchange broadening is independent of the static magnetic field (B_0), in the slow-exchange limit while varying quadratically with B_0 for fast exchange. The general result for all exchange regimes is

$$\alpha = \frac{d \ln R_{\text{ex}}}{d \ln \Delta \omega} \tag{25}$$

Provided that $p_a \gg p_b$, α varies from 0 to 2 and defines the chemical exchange

$0 \le \alpha < 1$	slow exchange
$\alpha = 1$	intermediate exchange
$1 < \alpha \le 2$	fast exchange (26)

Equation 26 can be reduced to the simplified expression

$$\alpha = \left(\frac{B_{02} + B_{01}}{B_{02} - B_{01}} \right) \left(\frac{R_{\text{ex2}} - R_{\text{ex1}}}{R_{\text{ex2}} + R_{\text{ex1}}} \right)$$

and determined from values (R_{ex1} and R_{ex2}) of R_{ex} determined at two static fields, $B_0 = B_{01}$ and B_{02}. With the exchange time-scale thus illuminated, one can choose the appropriate expression with which to fit the dispersion. Under slow-exchange conditions, either **Eq. 6** or **8** is appropriate, and each yields separate values for $p_A p_B$, k_{ex}, and $\Delta\omega$. For fast-exchange conditions, fits to **Eq. 8** are possible, but that might lead to overinterpretation of the results. Thus, fits to the fast-exchange expression (**Eq. 9**) are preferred in this case, but they only provide values for k_{ex} and the product $p_A p_B \Delta\omega^2$.

3.3. Off-Resonance $R_{1\rho}$ Experiments

3.3.1. $R_{1\rho}$ Pulse Sequence

The basic pulse sequence for measuring the off-resonance $R_{1\rho}$ is shown in **Fig. 8**. There are two approaches to implementing the sequence *(35)*. Each contains the spin-locked relaxation period of duration t between points **a** and **b** on the time line of **Fig. 8**. In the first approach, one omits the subsequent longitudinal relaxation period of duration $(T - t)$ between **b** and **c**. In this case, the observed spin-relaxation rate is given by **Eq. 10** and the angular dependence of the measured rate is removed by resorting to the effective rate

$$R_{\text{eff}} = R_{1\rho}/\sin^2\theta - R_1/\tan^2\theta \qquad (27)$$
$$= R_2^0 + R_{\text{ex}},$$

where the second equality is obtained by subtracting R_1 from **Eq. 10** and dividing by $\sin^2\theta$. Thus, separate determination of R_1 is required for dispersion analysis of R_{eff} using three-parameter (k_{ex}, R_2^0 and the product $p_A p_B \Delta\omega^2$) fits to **Eq. 11** or **12** plus the constant offset, R_2^0. Despite the additional time needed to obtain R_1, this approach may be preferred for study of smaller proteins (<10 kDa), which exhibit large R_1 values and consequent S/N loss when using the following approach.

The alternative $R_{1\rho}$ method employs the sequence of **Fig. 8**, including the longitudinal relaxation period (T – t) between **b** and **c** *(35)*, where T is the maximum value used for rate determination from series vs t. Thus, the total relaxation time is constant at T, and the observed decay rate is

$$\left(R_{1\rho} - R_1 \right) = \left(R_2^0 - R_1 \right) \sin^2\theta + R_{\text{ex}} \sin^2\theta, \qquad (28)$$

which follows from **Eq. 10** and the expression for signal decay:

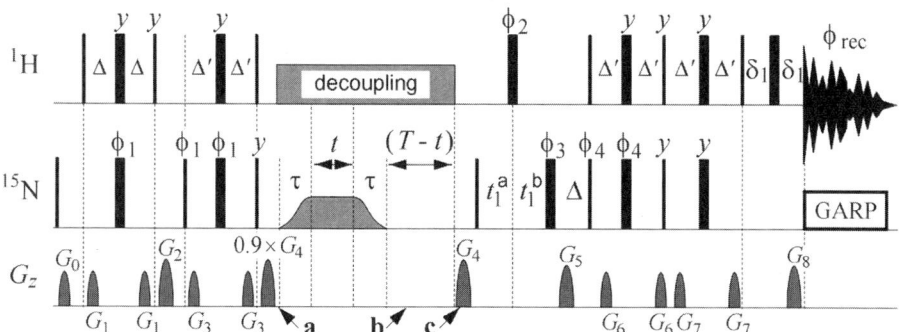

Fig. 8. Pulse sequence used to measure off-resonance $R_{1\rho}$ relaxation rates. The phase cycle $\phi_1 = \{x, -x\}$, $\phi_2 = \{4x, 4(-x)\}$, $\phi_3 = \{x, x, y, y, -x, -x, -y, -y\}$, $\phi_4 = \{y\}$, $\phi_{rec} = \{x, -x, x, -x\}$ provides an echo-modulated signal. Coherence selection is achieved with gradients G_5 and G_8, where ϕ_4 and the amplitude of G_5 and ϕ_4 are inverted to provide the antiecho counterpart, whereas the receiver phase and ϕ_1 are inverted on alternate t_1 increments. ^{1}H decoupling with a 5-6 kHz amplitude RF field is typical. Gradient magnitudes $G_0 - G_8$ are 8.7, 10.8, 32.5, 11.9, 39.0, 21.6, 13.0, 17.4, and 19.63 G/cm with durations 1.0, 0.6, 0.9, 0.6, 1.5, 1.6, 0.6, 0.6, and 0.176 ms. Delays are $\Delta = (4J_{NH})^{-1}$ approx 2.7 ms, Δ' approx 2.5 ms, $t_1{}^a = (\Delta + t_1/2)$, and $t_1{}^b = t_1/2$, whereas $\delta_1 = 250$ μs and the recycle delay is 3.0 ms.

$$I(t) = I_0 e^{-R_{1\rho}T} e^{-R_1\tau} \tag{29}$$
$$= \left(I_0 e^{-R_1 T_{max}}\right) e^{-(R_{1\rho} - R_1)T}.$$

Here, a separate measure of R_1 is unnecessary, as removal of angular dependence requires only division of **Eq. 28** by $\sin^2 \theta$ to yield

$$R_{eff} = \left(R_2{}^0 - R_1\right) + R_{ex}. \tag{30}$$

Three-parameter dispersion analysis, now substituting the offset $(R_2{}^0 - R_1)$ for $R_2{}^0$, is again possible. Also apparent from **Eq. 29** is the drop in S/N by $e^{-R_1 T}$ as a result of constant-time relaxation. For a typical 15 kDa protein (6 ns tumbling at 25°C), $R_1(^{15}N)$ approx 1.8 s^{-1} at 600 MHz ^{1}H frequency, corresponding to S/N decreases of 0.76 and 0.57 for $T = 150$ and 300 ms.

The following sequence details are relevant both to the $R_{1\rho}$-only and the $(R_{1\rho} - R_1)$ implementation. At point **a** in **Fig. 8**, just prior to the start of spin-locked relaxation, the spins reside in the pure ^{15}N state $\pm S_z$. There, the ^{15}N carrier is moved to the desired off-resonance value, whereas that of the ^{1}H is moved from ^{1}H$_2$O to the amide center for more efficient decoupling across the dispersion of relevant chemical shifts. Adiabatic rotation of $\pm S_z$ to the effective field at θ is achieved using amplitude and frequency sweeps over time τ, with tanh and tan modulation, respectively, as detailed in **Subheading 3.3.3.** Spin-locked coherence then relaxes at rate $R_{1\rho}$ for time t after which it is returned to the z-axis at **b** by reversal of the first adiabatic sweep. During t, CSA/dipolar cross-relaxation is suppressed with WALTZ-16 *(50)* or other ^1H decoupling scheme (*see* **Subheading 3.3.4.**). More important, both signs, $\pm S_z$, of the initial condition at **a** occur across the phase cycle so that their combination in the final signal subtracts the nonvanishing ^{15}N steady state provided by the ^1H–^{15}N NOE from the decay. Gradients in multiples of G_4 are used to dephase any residual transverse magnetization. In the $(R_{1\rho} - R_1)$ case, longitudinal relaxation for $(T - t)$ follows **b**, whereas, in both versions, the RF carriers are returned to ^1H_2O and ^{15}N amide center at **c**. Subsequent pulse sequence elements are similar to those discussed in the general scheme (**Subheading 3.1.2.**). Finally, RF influence on sample temperature is rendered t-independent by irradiating ^{15}N during the 2.5 s recycle delay (not shown) for time $(T - t)$ at the spin-lock power level and far (approx 12 kHz) off-resonance. Further sequence details are presented in the caption to **Fig. 8**.

3.3.2. TROSY (R1r– R1) for Studies of Large Proteins

Figure 9 is a sequence for measurement of the ^{15}N $(R_{1\rho} - R_1)$ with TROSY coherence selection in both t_1 and t_2 *(72)*. Benefits of this sequence over the $(R_{1\rho} - R_1)$ experiment of **Fig. 8** match those discussed in **Subheading 3.2.5.** in the context of TROSY rcCPMG, accepting that, for a spin-locked relaxation period, TROSY coherence selection to reduce the background relaxation rate is not performed. Thus, rates measured with the sequences of **Figs. 8** and **9** are identical. However, the TROSY sequence has been observed to yield large S/N enhancements, averaging approx 50% among observed ^1H–^{15}N resonances from a 53 kDa perdeuterated protein in a modest 600-MHz field, with corresponding line-width reductions of up to 60% *(72)*. Because this sequence is intended for application to large proteins, which exhibit small R_1, only slight S/N losses result from the constant-time design according to our previous discussion of **Eq. 29**. For example, a 53 kDa (24 ns tumbling) protein at 25°C exhibits $R_1(^{15}N)$ approx 0.5 s^{-1}, corresponding to e^{-R_1T} 0.93 and 0.86 for $T = 150, 300$ ms.

An important consideration for TROSY–$(R_{1\rho} - R_1)$ is that $\theta = \tan^{-1}(\omega_1/\Delta\Omega)$ must be calculated using $\Delta\Omega$ values derived from the non-TROSY positions (Ω^{non}) of monitored NMR peaks, as the TROSY positions (Ω^{TR}) are relevant only to the t_1 and t_2 periods that *follow* relaxation. This suggests a separate collection of a conventional ^1H–^{15}N HSQC to directly obtain Ω^{non}; however, this may not be practical for a large protein. More simply stated, negligible error will be introduced by instead relying on the assumption that $\Omega^{non}(^{15}N)$ approx $[\Omega^{TR}(^{15}N) - (^1J_{NH}/2)]$.

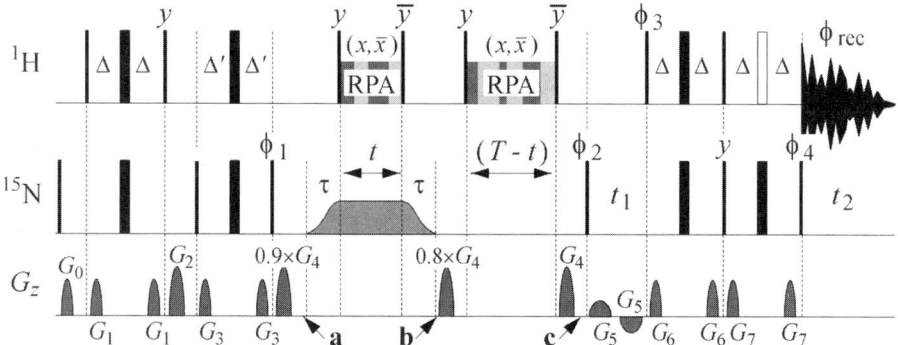

Fig. 9. TROSY-($R_{1\rho}$–R_1) pulse sequence for exchange characterization at ^{15}N sites. TROSY selection in t_1 and t_2 and PEP *(46)* coherence modulation are achieved with a phase cycle yielding an echo signal for $\phi_1 = \{4y, 4(-y)\}$, $\phi_2 = \{y, -y, -x, x, -y, y, x, -x\}$; $\phi_3 = \{y\}$ $\phi_4 = \{x\}$ and $\phi_{rec} = \{x, -x, y, -y\}$, and the antiecho counterpart with inversion of ϕ_3 and ϕ_4 and replacement of ϕ_{rec} with $\{-x, x, y, -y\}$. In each case, ϕ_2 and ϕ_{rec} are inverted on alternate steps in t_1. Phases are specific to Varian spectrometers *(62)*. CSA/dipolar suppression during relaxation periods with durations t and $(T - t)$ is achieved with random phase alternation (RPA) decoupling, as discussed in **Subheadings 3.3.4.** and **4.3.3.** Typically, a 5.8 kHz amplitude decoupling field is used. Gradient magnitudes G_0–G_7 are 8.7, 10.8, 32.5, 11.9, 39.0, 1.5, 13.0, and 17.4 G/cm with durations 1.0, 0.6, 0.9, 0.6, 1.5, 1.0, 0.6, and 0.6 ms. The bipolar pair, G_5, is set sufficient to dephase ^{1}H$_2$O for suppression of radiation damping, though it is omitted entirely when $t_1 < 800$ µs, and applied for 0.4 ms when 0.8 ms $\leq t_1 \leq 2.0$ ms. Delays are Δ, $\Delta' = 2.7$, 2.5 ms and $\tau = 4$ ms. The *open bar* on the ^{1}H time line represents the 3-9-19 selective pulse element elaborated in **Fig. 6B**.

As with the traditional ($R_{1\rho} - R_1$) experiment of **Fig. 8**, the TROSY ($R_{1\rho} - R_1$) sequence incorporates (*i*) ^{15}N and ^{1}H frequency switching at points **a** and **c** in the time line of **Fig. 9**, (*ii*) adiabatic reorientation of the RF field about the spin-locked relaxation period according to the discussion of **Subheading 3.3.3.**, and (*iii*) temperature compensation by far off-resonance ^{15}N irradiation during the recycle delay (not shown). Newly implemented in **Fig. 9** is the random phase alternation (RPA) decoupling scheme for more efficient suppression of CSA/dipolar cross-relaxation. This method is detailed in **Subheading 3.3.4.**, whereas here we note that RPA periods are bracketed by ^{1}H $\pi/2$ pulses in order that ^{1}H$_2$O magnetization be returned to a consistent orientation independent of the random sequence. Additional sequence details are presented in the caption to **Fig. 9**.

3.3.3. Adiabatic-Shaped Pulses

Consistent and accurate determination of the off-resonance $R_{1\rho}$ requires careful alignment of spin magnetization with ω_e across the broad range of chemical shifts in the protein and at the several values of $\Delta\Omega$ employed for dispersion analysis. The sequences of **Figs. 8** and **9** achieve this using the adiabatic (i.e., ideally loss-free) ramps of RF amplitude and frequency that are placed about the spin-lock relaxation period. Just prior to relaxation, with magnetization aligned with the z-axis, the ^{15}N carrier is set to the desired off-resonance value ($\Delta\Omega$) and then reorientation to ω_e at θ is achieved by ramping the RF frequency, $\Delta\Omega(t)$, from ($\Delta\Omega - \Omega_{sw}$) to $\Delta\Omega$ and its amplitude, $\omega(t)$, from 0 to ω_1 over $t = 0, \tau$. The reverse yields the downward ramp that follows relaxation and returns the decayed spin-locked magnetization to the z-axis. Adiabaticity can be obtained using the shapes *(28,73,74)*

$$\omega(t) = \omega_1 \tanh[10t'], \tag{31}$$

and

$$\Delta\Omega(t) = (\Delta\Omega_{sw} / 50) \tan[(1 - t')\tan^{-1}(50)], \tag{32}$$

where $t' = t/\tau$ and $(1 - t/\tau)$ for the ramps up and down, respectively, and $\Delta\Omega_{sw}$ is the width of the frequency sweep. Mulder et al. calculated spin trajectories using these shapes with $\tau = 4$ ms to demonstrate adiabaticity for $| \Delta\Omega | \geq 1000$ Hz and ω_1 up to 2 kHz. Sweep functions other than those of **Eqs. 31** and **32** (e.g., hyperbolic secant amplitude sweeps) may also be used, although the noted tanh/tan combination is designed to provide adiabaticity similar to a numerically optimized case *(28,73,74)*.

3.3.4. CSA/Dipolar Suppression Methods in R1ρ Experiments

Finally, experiments used to measure relaxation rates in an *I–S* spin system in which the *S* spin has a sizable CSA, such as ^{15}N, require suppression of the cross-correlation between the dipole–dipole and CSA relaxation mechanisms, as was described for the rcCPMG sequences in **Subheading 3.2.1.** Traditionally, suppression in $R_{1\rho}$ experiments is achieved by applying evenly spaced ^{1}H π pulses or a ^{1}H decoupling sequence such as WALTZ-16 *(50)* during the relaxation delay of the heteronucleus *(49,52,53)*. Recently, Korzhnev et al. *(75)* demonstrated that these methods lead to erroneous ^{15}N $R_{1\rho}$ values owing to residual cross-relaxation and provided two alternatives that improve on suppression to yield the correct rates. The first of these relies on CW ^{1}H decoupling with RF-phase inversion between x and $-x$ at random time intervals of average duration δ, an approach similar to earlier noise-decoupling techniques *(76)*. The ^{1}H bandwidth of this RPA method exceeds that of constant-interval phase alternation by an amount that is significant in very high-field (≥ 800 MHz) applications because of a corresponding breadth of chemical shift dispersion. Thus, as high-field experiments are especially desirable in the study of large proteins, RPA decoupling is included in the TROSY–$(R_{1\rho} - R_1)$ experiment of **Fig. 9**. RPA is equally amenable to the non-TROSY version of **Fig. 8**, but it may be less often essential if it is used only at lower fields. Nonetheless, even at moderate fields, RPA decoupling may be useful in $R_{1\rho}$ studies of high-salt samples, where RF heating is a particular concern (*see* ref. 77 and **Subheadings 4.1.1.** and **4.2.1.**) In that case, using lower ^1H power can alleviate heating, whereas RPA can compensate for the corresponding loss of ^1H bandwidth. Still, the most effective route to reduced heat deposition is to decouple with ^1H π pulses, as discussed later in this section.

A very important restriction of RPA (or other CW ^1H decoupling sequence) as applied during spin-locked relaxation is avoidance of the Hartmann–Hahn matching condition *(78)*,

$$\sqrt{\left(\Delta\Omega^{1H}\right)^2 + \left(\omega_1^{1H}\right)^2} = \sqrt{\left(\Delta\Omega^{15N}\right)^2 + \left(\omega_1^{15N}\right)^2}, \tag{33}$$

where the superscripted labels indicate isotope-specific RF offsets and amplitudes. When conditions satisfy **Eq. 33**, fast transfer of coherence between ^1H and ^{15}N (or other heteronucleus) is facilitated by normally inactive terms in the heteronuclear dipole interaction, leading to significant and misleading elevation of the apparent spin-relaxation rate. Provided that ω_1^{1H} may be adjusted up without significant increase of the sample temperature or down without excessive loss in decoupling bandwidth, the matching condition can be avoided with careful choice of that value as directed by the current ω_1^{15N} and given ranges of $\Delta\Omega^{1H}$ and $\Delta\Omega^{15N}$. For ^{15}N relaxation, the right and left sides of **Eq. 33** should differ by ≥ 1 kHz to avoid perturbation of the decay by Hartmann–Hahn transfer, whereas a larger difference may be required in ^{13}C studies because the transfer rate increases with the square of the dipole interaction (i.e., the square of the ratio of gyromagnetic ratios, $\left(\gamma^{13C}/\gamma^{15N}\right)^2$ approx 6.15, predicts the relative rates here). Additional details for RPA implementation are presented in **Subheading 4.3.3.**

In the second alternative, the ^1H decoupling scheme provided by Korzhnev et al., varying numbers of ^1H π pulses are applied in an arrangement depending on the duration, *T*, of the relaxation period and estimates of $T_{1\rho} = R_{1\rho}^{-1}$ values. For example, no ^1H π pulses are applied when $0 \leq T \leq (T_{1\rho}/3)$, whereas a single ^1H π pulse is applied at the center of the spin-lock period when $(T_{1\rho}/3) \leq T \leq (2T_{1\rho}/3)$, and two 180° ^1H pulses separated by $T/2$ are applied when $(2T_{1\rho}/3) \leq T \leq T_{1\rho}$. Finally, when $T = T_{1\rho}$, three ^1H π pulses are applied at even intervals during the spin-locking period, and the relaxation time series is not extended beyond this point. Although Korzhnev et al. demonstrated decoupling performance of this variable pulse-spacing scheme on par with RPA and superior to traditional regular-interval ^1H π-pulse decoupling, its greater requirement for user intervention and inability to provide a consistent pulse-spacing recipe to cover the considerable $T_{1\rho}$ variation expected in a protein sample *(72)* renders it less desirable than RPA decoupling.

3.3.5. Calibration of the RF Field

As noted at the end of **Subheading 2.3.**, the angular dependence of $R_{1\rho}$ or $(R_{1\rho} - R_1)$ may be trivially removed provided that ω_1 is accurately calibrated. Failure to implement the correct values for ω_1 will result in skewed rates that cannot be interpreted by the three-parameter dispersion analysis noted in the previous section. Often the RF field strength of a pulse is determined from the length of a calibrated 90° or 360° pulse, where $\omega_1 = (t_{360°})^{-1}$. However, that procedure is likely to yield much greater uncertainty in ω_1 than desired for analysis of $R_{1\rho}$-based rates. Therefore, in this section, we present two more-accurate approaches to ω_1 calibration, whereas a related concern, that of RF field homogeneity, is discussed in **Subheading 4.2.2.**

The first of these depends on variation in the splitting of peaks along the ^{1}H frequency axis because of residual scalar coupling in the presence of an off-resonance CW ^{15}N decoupling field. The method entails collecting two ^{1}H–^{15}N HSQC spectra, one using off-resonance CW constant-phase decoupling at frequency offset Ω_{RF} during t_2, and the other without ^{15}N decoupling during t_2. The latter yields splitting by $^1J_{NH}$, whereas off-resonance decoupling results in splitting scaled to the apparent value, (*see* **ref. *79*** and **Subheading 3.4.3. of ref. *41***)

$$^1J_{NH,app}^i = {}^1J_{NH}^i \frac{\Delta\Omega_i^2}{\sqrt{{\omega_1}^2 + \Delta\Omega_i}} \tag{34}$$

where $\Delta\Omega_i = (\Omega_i - \Omega_{RF})$ is the offset corresponding to i^{th} peak in the spectrum. Rearrangement of **Eq. 34** yields

$$\omega_1/\Delta\Omega_i = \sqrt{\left(\frac{{}^1j_{NH}^i}{{}^1j_{NH,\,app}^i}\right) - 1} \tag{35}$$

Thus, the right-hand side of **Eq. 35** is a measure of $\tan\theta = (\omega_1/\Delta\Omega)$, the slope of a plot of this value against $\Delta\Omega_i$ yields the desired calibration of ω_1, as demonstrated in **Fig. 10A**. There calibration yielded $\omega_1 = (1614 \pm 35)$ Hz, for only approx 2% uncertainty.

The second approach employs adiabatic sweeps of the RF in the context of the $R_{1\rho}$ sequence of either **Figs. 8** or **9** (as appropriate to protein size). In both cases, the longitudinal relaxation period $(T - t)$ is omitted and $t = 0$. Three distinct spectra are collected with these conditions: (a) employing the upward adiabatic ramp only to yield peak intensities I_A^i, (b) employing both upward and downward ramps to yield peak intensities I_{AB}^i, and (c) without either ramp, to

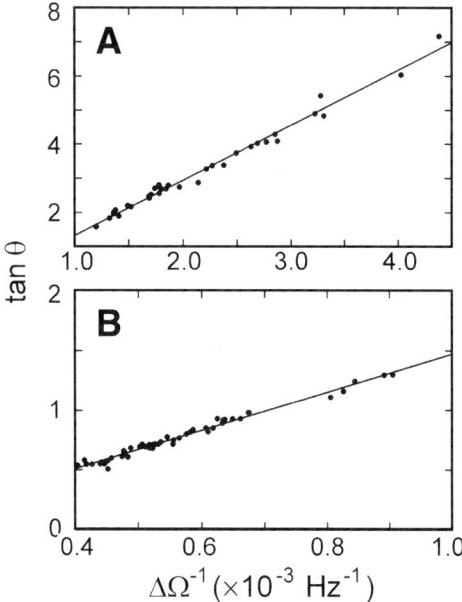

Fig. 10. Measurement of ω_1 field strength. **(A)** Values of tan θ calculated using **Eq. 35** for ^{15}N resonance plotted vs $1/\Omega$. Data was acquired on a Varian Inova 600 MHz instrument. **(B)** Values of tan θ calculated using **Eq. 38** for ^{15}N resonance plotted vs $1/\Delta\Omega$. Data was acquired on a Bruker DRX 600 MHz instrument with a triple-resonance (HCN) probe. Least-squares fitting yielded field strengths of 1614 ± 35 Hz in **(A)** and 1601 ± 32 Hz in **(B)**.

yield intensities $I_0{}^i$. Accounting for spin relaxation at effective rate R_{eff} during ramps of duration τ, these intensities are expressed as

$$I_A^i = I_0^i e^{-2R_{\mathit{eff}}\tau},\tag{36}$$

and

$$I_{AB}^i = I_0^i e^{-R_{\mathit{eff}}\tau}\cos\theta.\tag{37}$$

An approximation to tan θ is made using

$$\tan\theta = \sqrt{\left(\frac{I_0^i + I_{AB}^i}{2I_{AB}^i}\right)^2 - 1},\tag{38}$$

where, from **Eqs. 36** and **37**, we have

$$\left(\frac{I_0^i + I_{AB}^i}{2I_{AB}^i}\right)^2 = \left(\frac{1 + e^{-R_{\mathit{eff}}^i\tau}}{2e^{-R_{\mathit{eff}}^i\tau}\cos\theta}\right)^2$$

$$\sim \left(\frac{1 + \left(1 - R_{\mathit{eff}}^i\tau\right)}{2\left(1 - R_{\mathit{eff}}^i\tau\right)\cos\theta}\right)^2 = \sec^2\theta,\tag{39}$$

which is obtained from Taylor expansion of the exponentials to first order in τ. Inserting the ultimate result of **Eq. 39** into **Eq. 38** results in a trigonometric identity. Thus, in direct analogy to the $^1J_{NH}$-based calibration method, here, the right-hand side of **Eq. 38** is a measure of $\tan \theta = (\omega_1/\Delta\Omega)$ and the slope plotted against $\Delta\Omega_i$ yields ω_1. **Fig. 10B** shows a result obtained using the adiabatic-sweep method with the same sample and conditions as in **Fig. 10A** with the $^1J_{NH}$ approach. Here, the calibration yielded $\omega_1 = (1601 \pm 32)$ Hz, in agreement with the above and again resulting in only approx 2% uncertainty.

The values of J_{NH}^i and $J_{NH, app}^i$ or I_0^i, I_{AB}^i and I_{AB}^i used to approximate $\tan \theta$ in **Eqs. 35** and **38** should be carefully chosen for linear least-squares fitting, as in **Fig. 10**, to yield ω_1. Only well-resolved, high S/N peaks should be selected from the spectrum, and these should span as much of the ^{15}N chemical shift range (hence $\Delta\Omega$) as possible. Relying on average values of the measured quantities taken from repeated spectra will also reduce systematic errors.

Finally, the results of both of the above methods agree with the apparent value (1600 Hz) obtained from pulse-width calibration. However, it is difficult to assess the uncertainty of the latter for propagation to the effective rates (**Eqs. 27** and **30**) used in dispersion analysis, and agreement may be less fortuitous for smaller ω_1. Of these two approaches, the method of adiabatic sweeps is somewhat preferable because it provides ω_1 calibration in the context of the exact pulse sequence used for $R_{1\rho}$ measurement. Furthermore, data analysis is simpler in that approach, as I_0^i, I_{AB}^i and I_{AB}^i may be obtained from peak amplitudes, whereas J_{NH}^i should be determined from fits to the splitting.

3.3.6. R1ρ Methods of Rate Measurement and Dispersion Analysis

As with the majority of NMR spin-relaxation experiments, the $R_{1\rho}$ experiment is designed to measure the decay of magnetization as a function of increasing relaxation time, T. Implementation of the $R_{1\rho}$ experiment requires collecting a series of 2D spectra with varying relaxation delay, t, as shown in **Figs. 8** and **9**. Typically, at least seven different t values should be used with two or more duplicate experiments to assess uncertainties. Peak intensities from each spectrum are extracted and the heights fit to an exponential decay to obtain the rate constant (*see* **Note 4**)

$$I(t) = I(0)e^{-R_{eff}T} \tag{40}$$

in which $I(t)$ is the resonance peak intensity at a particular relaxation time T, $I(0)$ is the peak intensity at $T = 0$, and R_{eff} is the effective relaxation rate. The intensities of the peak heights in each of the spectra are used to determine the effective relaxation rate (R_{eff}). The exact form of R_{eff} depends on whether the conventional ($\tau = 0$) or the constant time version ($\tau = T_{max} - T$) of the pulse sequence in **Figs. 8** and **9** is used. R_{eff} is then determined at various values of the tilt angle, θ. This can be achieved by varying ω_1, the offset of the spin-lock pulse, or both. Fitting of $R_{1\rho}$ values obtained at multiple effective fields, ω_e, requires optimizing three parameters by nonlinear least-squares fitting. In the case of the $R_{1\rho}$ experiment the three parameters are R_2^0, $\Phi_{ex} = p_A p_B \Delta\omega^2$, and k_{ex}, whereas the $R_{1\rho} - R_1$ experiment requires $(R_2^0 - R_1^0)$, Φ_{ex}, and k_{ex}. The exchange-free transverse relaxation rate can be determined by measuring the free-precession line width in an in-phase decoupled HSQC *(80)* or by the $^1H-^{15}N$ dipole/^{15}N CSA interference rates described by **Eq. 16**. The value of R_1^0 can be determined by conventional methods *(81)*. If R_2^0 is measured then in the case of the $R_{1\rho}$ experiment only two parameters, Φ_{ex} and k_{ex}, require optimization; a similar situation arises in the $(R_{1\rho} - R_1)$ experiment if both R_2^0 and R_1^0 are measured.

In the example shown (**Fig. 11**), the offset was varied to achieve the desired nominal tilt angles *(82)*. $R_{1\rho}$ spin-relaxation rates were determined at eight different values of θ for a mutant of the C-terminal domain of calmodulin *(82)*. The different θ values were obtained by varying the frequency of the spin-lock pulse. Measured values of $R_{1\rho}$ vs θ are shown in **Fig. 11A** for selected residues in the calmodulin mutant. **Fig. 11B** depicts the variation of $R_{eff}/\sin^2\theta$ as a function of the effective field squared, ω_e^2. Once the $R_{1\rho}$ dispersion has been obtained, fits to **Eq. 12** yield k_{ex}, p_A, p_B, at detectably exchanging sites. Akke and coworkers used η_{xy} and η_z measurements to determine R_2^0 and nonlinear fitting yielded k_{ex} values ranging from 2.2 to 7.7×10^4 s^{-1} for a calmodulin C-terminal domain mutant *(82)*, consistent with biochemically measured Ca^{2+} dissociation rates.

4. Additional Considerations

4.1. rcCPMG

4.1.1. Sample Temperature

The final factor to consider is the effect of CPMG pulses on the sample temperature. This unfortunate side effect is minimized by reducing the RF power used for the CPMG train and by adjusting the cooling air flow rate *(77,83)*. In addition to these parameters, sample salt content is a major determinant of RF heating, which is primarily because of interaction of ions in solution with the electric component of the RF field. Wang and Bax found an eightfold increase in RF heating at 200 m*M* NaCl relative to salt-free samples *(77)*. However, because salt content may be essential to avoid aggregation or to maintain the native protein fold, we cannot take it as a free parameter in minimizing RF heating, but rather should pay special attention to heating when high-salt content is required. As for freely adjustable parameters, reduction of the RF power should be balanced against excessive pulse duty cycle *(4,19)* and, less significantly, against

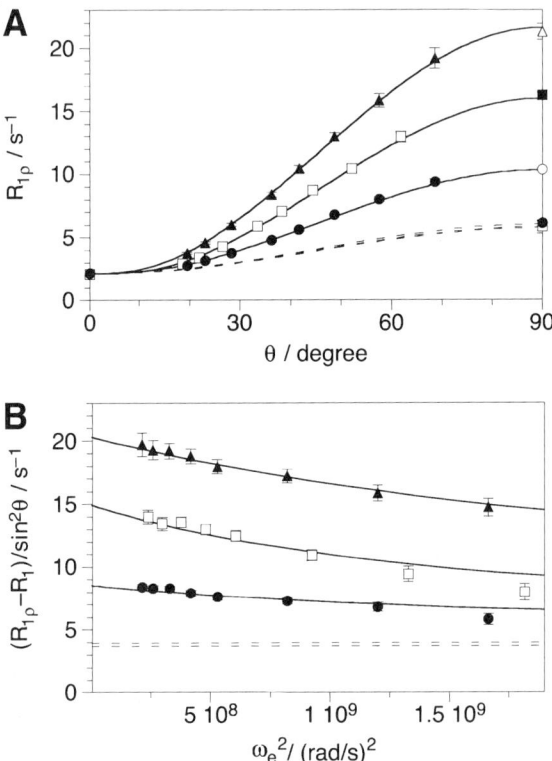

Fig. 11. Off-resonance rotating-frame ^{15}N relaxation data. The rotating-frame ^{15}N relaxation rate constants ($R_{1\rho}$) are shown as a function of **(A)** tilt angle, θ, and **(B)** the effective field squared, ω_e^2. R_1 is included at $\theta = 0°$, and R_2 (inverted symbols) and R_2^0 at $\theta = 90°$. The *solid lines* represent the nonlinear fits as described in the original publication *(82)*. Error bars represent one standard deviation. The *dashed lines* show the curves expected in the absence of conformational exchange, i.e., $R_{ex} = 0$. The optimized values of τ_{ex} and ϕ are 20.3 ± 2.1 μs, (269 ± 27) × 10^3 s^{-1} (R86); 16.9 ± 1.7 μs, (980 ± 104) × 10^3 s^{-1} (S101); 23.3 ± 1.5 μs, (484 ± 36) × 10^3 s^{-1} (D131). (Reproduced from **ref.** *82* with permission from Elsevier.)

reduced spectral coverage by the pulses. Because spins exhibit a distinct relaxation rate during RF pulses, it is advised that t_π be <10% of the smallest τ_{CP} value to be used. For typical $\tau_{cp}^{min} = 1$ ms, this suggests $t_\pi \leq 100$ μs is appropriate. Concerning spectral coverage of the pulses, consider that amide ^{15}N in proteins exhibit a typical chemical shift range of 105 to 135 ppm for a 1.2-kHz spectral width at 14.1 T (600-MHz ^{1}H frequency). To a good approximation, π pulses are effective over the Rabi frequency ± $(4\,t_\pi)^{-1}$, and consequently, it is desirable that π pulses in the CPMG train satisfy $t_\pi \leq 400$ μs, which is not restrictive compared to concerns on RF duty cycle. Experiments instead focused on $^{15}NH_2$ or $^{13}CH_3$ relaxation *(84,85)* require even less spectral coverage than the ^{15}NH case.

To quantify the utility of ^{15}N power reduction for avoiding temperature increases, we varied the repetition rate ν_{CP} of ^{15}N π pulses in a CPMG train and monitored the temperature of a 550-μL sample in a 5-mm diameter NMR tube with 50 m*M* NaCl and 600 μm ^{15}N RNase A at pH 7.5. Experiments were performed on a Varian Inova 600-MHz spectrometer equipped with a *z*-axis gradient triple-resonance probe. The cooling air flow rate was 12 L/m and temperature was set to 25°C. Loening and Keeler recently showed that differences, ΔT, between sample temperature and the monitored coolant temperature are accentuated with flow rates below this level or at temperatures exceeding 25°C *(83)*. Flow rates >12 L/m have been observed to result in distorted NMR spectra, possibly because of mechanical disturbance of the sample holder. In other work, Wang and Bax found significant RF heating with high (14.5%) ^{13}C or, especially, ^{1}H RF duty cycles in experiments intended to mimic the effects of decoupling sequences *(77)*. However, RF applied at ^{15}N frequencies yielded negligible temperature increases in their experiments. Nonetheless, the activated nature of exchange processes, and thus R_{ex} measurement, may require a higher degree of temperature control than other NMR applications, thus warranting the present quantification of RF heating with CPMG sequences.

Results at RF fields corresponding to $t_{\pi}(^{15}N) = 50$ and 100 μs are shown in **Fig. 12** and indicate a small, but significant, dependence of temperature on ν_{CP}. The RF field reduction from ω_1 $(^{15}N) = 5$ to 2.5 kHz similarly decreased sample heating from 0.2°C to less than 0.1°C over the range of ν_{CP} values relevant for rcCPMG characterization of $R_2(\tau_{CP}^{-1})$. The reduced and full-power data yielded slopes vs ν_{CP} of (0.056 ± 0.009) and (0.156 ± 0.009)°C/kHz. Results were obtained by measuring the relative temperature change in the sample via shifts in the $^{1}H_2O$ resonance, observed in a series of rcCPMG-like experiments, vs ν_{CP}.

4.2. $R_{1\rho}$

4.2.1. Sample Temperature

Measurements of $R_{1\rho}$ or $(R_{1\rho} - R_1)$ may be hampered by RF heating from application of strong spin-lock and decoupling fields. Small increases in sample temperature can skew observed rates and lead to erroneous conclusions regarding the nature of the activated motional processes under investigation, whereas large increases in temperature can result in sample loss. Temperature increases manifest in two ways: differential heating resulting from varying lengths, *t*, of the CW spin-lock pulse and overall heating of the sample. As noted in **Subheadings 3.3.1.** and **3.3.2.**, variation in such heating with the duration *t* of

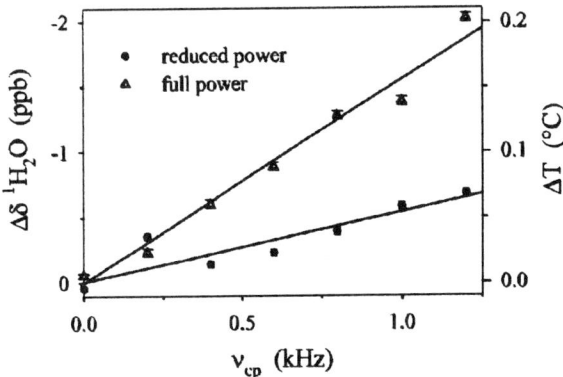

Fig. 12. Temperature increase vs repetition rate, ν_{CP}, of ^{15}N π pulses at full and reduced RF power in the rcCPMG-like experiment discussed in the text. Temperature scaling on the right-hand axis was calculated as described in the text. The pulse sequence consisted of two 20-ms periods with evenly spaced ^{15}N π pulses separated by a 5.5 ms period that mimicked the U element employed with rcCPMG. The sequence was terminated with a low-power, small-angle ^1H pulse to provide the observed ^1H_2O resonance, relative to the nominal 25°C value. As with standard rcCPMG implementation, a 2.5-s recycle delay was used along with a 1.25 kHz ^{15}N decoupling field during detection. Data was collected after executing dummy scans at each ν_{CP} value to allow 5 min of equilibration. The deuterium lock was disengaged throughout the series to prevent its variation of the static field in opposition to temperature-induced shifts. Drift-induced changes in δ ^1H_2O were corrected by bracketing each ν_{CP} series with identical experiments at $\nu_{CP} = 0$ and the assumption of linear drift. The reliability of this approach for measuring relative temperatures was established using similar experiments in which temperature was explicitly varied over the range 22.5–27.5°C. There, the unlocked, drift-corrected ^1H_2O resonance reproduced the known temperature dependence *(41)*, which we used to convert $\Delta\delta$ ^1H_2O to a temperature change.

the spin-locked relaxation period must be compensated by far off-resonance ^{15}N irradiation during the recycle delay for $(T - t)$ *(77,82)*. Overall sample heating is somewhat more problematic. Evenas et al. *(82)* dealt with this issue by measuring the temperature increase (via the difference in DSS and HDO chemical shifts) after accumulating a large number of steady-state scans when the temperature had equilibrated. Observed temperature increases were compensated by changing the spectrometer's temperature setting to match the DSS/HDO-calibrated value with the desired target.

4.2.2. Distribution of RF Fields

Although accurate calibration of the RF field amplitude ω_1 is possible using the methods of **Subheading 3.3.5.**, the value obtained is an average over the sample volume. However, spatial inhomogeneity of ω_1 across the sample is generally non-negligible, and often asymmetric, using present-day probe technology *(86)*. The ω_1 distribution can be determined using the method of Guenneugues et al. *(86)* in which the one-dimensional (1D) ^{15}N nutation spectrum of a single $^1H,^{15}N$ resolved peak is measured on resonance, yielding a maximum at ω_1 and a distribution that reflects the RF inhomogeneity. The noted 1D spectrum is taken as a slice from a 2D spectrum with axes for ^1H chemical shift and ^{15}N nutation frequency. RF field distributions have been reported with full widths at half-maxima from 3 to 7% of the average ω_1 *(4,75)* and with both asymmetric *(4,86)* and nearly symmetric *(75)* shapes. These variations underscore the need to measure the distribution for each particular probe and sample. The significance of ω_1 inhomogeneity for $R_{1\rho}$-based dispersion analysis is discussed by Guenneugues et al. Advantageous narrowing of the distribution and any asymmetry may be possible using reduced sample volume centered on and with dimensions similar to or smaller than the RF coil of the relaxing species. This is achieved using susceptibility-matched sample tubes, such as those available from Shigemi, Inc. (Allison Park, PA).

4.3. Frequency From Phase Modulation for Adiabatic Field Sweeps

In **Subheading 3.3.3.**, adiabatic RF field sweeps were discussed as a method to align spins with the final effective field. In practice, such RF-field modulation is achieved using the shaped-pulse functionality of a spectrometer's sequence programming code. Typically, shaped pulses consist of a table of amplitude and phase values applied in step at given time increments, Δt, over the pulse duration. Thus, frequency modulation must be achieved indirectly by varying the RF phase, which produces an instantaneous offset from the RF carrier (*see* **Note 5**):

$$\Delta\Omega(t) = \frac{d}{dt}\phi(t). \tag{41}$$

In general, phase modulation may be approximated by

$$\phi(t + \Delta t) \sim \phi(t) + \Delta t \left[\frac{d}{dt}\phi(t) \right]_{t=(t+\Delta t)}. \tag{42}$$

Inserting **Eq. 32** for $\Delta\Omega(t)$ into **Eq. 41** and inverting the result yields the form to substitute into **Eq. 42** and obtain the desired adiabatic modulation. C-language programs to generate amplitude and phase tables according to **Eqs. 31, 32,** and **42** are available from the author's website at http://xbeams.chem.yale.edu/~loria. The program's output is intended for Varian spectrometers but is easily adapted for use with Bruker or other instruments provided there is familiarity with the particular format required as input for corresponding shaped-pulse functions.

4.3.1. An Implementation for RPA Decoupling

As noted in **Subheading 3.3.4.**, the RPA decoupling scheme of Kay and coworkers is important for suppression of CSA/dipolar cross-relaxation in $R_{1\rho}$ experiments $\geq$800 MHz ^{1}H frequency. In principle, the method may be used on any spectrometer with access to some form of random number generation either during execution or preparation of the NMR pulse sequence program. In practice, the implementation is a computational challenge with distinct, often subtle, limitations imposed by the format of programming code offered by a given vendor and the simplicity of computing power on which some spectrometers are built. Here, we provide only the barest recipe for RPA generation and encourage the nonexpert to access existing code we have written (http://xbeams.chem.yale.edu/~loria) for use on Varian spectrometers, or to request original code as offered in the original work of Korzhnev et al. *(75)*.

Briefly, Korzhnev et al. recommend using average intervals, $\delta = 2$–10 ms, between each phase-inversion event, although generally they chose to employ $\delta = 10$ ms. A distinct random pattern of alternation should be applied in each transient of each t_1 increment. Although no specific recipe is given in the original reference, individual segment durations were allowed to vary between $\delta \times (1 \pm \Delta\delta)$, with $\Delta\delta = 0.8$ in a recent implementation. As of this writing, we have not investigated the variation of RPA's ^{1}H bandwidth with $\Delta\delta$.

5. Notes

1. Because k_1 is written as a first-order rate in **Eq. 1** and subsequent usage, it is taken to represent a pseudo first-order rate constant for ligand-binding and other bimolecular processes—that is, $k_1 = k_f$ [L] or k_f [P] for a study of spins on the protein (P) or ligand (L), respectively.

2. For experiments with TROSY coherence selection [*(35)*, and see **Subheading 3.2.5.**], initial ^{15}N magnetization contributes constructively to the final result, thus the ^{15}N purge is omitted unless such omission perturbs the relaxation (as in **Subheading 3.3.2.**).

3. The lower τ_{CP} limit for rcCPMG avoids RF duty cycles >10% during relaxation periods with typical ^{15}N π pulse duration approx 80 μs. Higher duty cycles can increase sample temperature and/or skew the observed rate from the free-precession value. The high-τ_{CP} limit depends on the suppression of cross-relaxation discussed in **Subheading 3.2.3.**

4. If the spectral visualization/assignment program, Sparky *(66)* is used, macros for extracting peak heights and calculating relaxation rates are freely available from the author's website (http://xbeams.chem.yale.edu/~loria).

5. The validity of **Eq. 41** is readily understood taking the relative time derivative,

$\left(\dfrac{d}{dt}\sin\left[\omega t + \phi(t)\right]\right)\Big/\cos\left[\omega t + \phi(t)\right]$, of a general RF field, which returns the

angular frequency ω when $\phi(t)$ is constant, and $\omega + \dfrac{d}{dt}\phi(t)$ otherwise.

6. Summary and Conclusions

We have outlined the basic theory and methods for NMR characterization of protein dynamics on the μs– ms time-scale. Approaches based on CPMG (pure-R_2) and spin-locked ($R_{1\rho}$) pulse sequences, plus several varieties within these categories, have been presented in a manner intended for provide clarity for the relative NMR novice, but with sufficient detail and discussion to provide insight to the more expert researcher. These methods are robust and with proper use should lead to novel insight into the functioning of important biological molecules.

Acknowledgments

We thank Dr. Mikael Akke (Lund University) for generously supplying the figure used here as **Fig. 11** (which is reproduced from **ref.** *82* with permission from Elsevier). J.G. Kempf acknowledges NIH F23-GM66599-02. J.P. Loria acknowledges The Camille and Henry Dreyfus Foundation and the NSF CAREER Award.

References

1. Rao, B. D. N. (1989) Determination of equilibrium constants for enzyme-bound reactants and products by nuclear magnetic resonance. *Methods Enzymol.* **177,** 358–375.
2. Burton, R. E., Huang, G. S., Daugherty, M. A., Fullbright, P. W., and Oas, T. G. (1996) Microsecond protein folding through a compact transition state. *J. Mol. Biol.* **263,** 311–322.
3. Wennerström, H. (1972) Nuclear magnetic relaxation induced by chemical exchange. *Mol. Phys.* **24,** 69–80.
4. Palmer, A. G., Kroenke, C. D., and Loria, J. P. (2001) Nuclear magnetic resonance methods for quantifying microsecond-to-millisecond motions in biological macromolecules. *Methods Enzymol.* **339(Part B),** 204–238.
5. Kempf, J. G. and Loria, J. P. (2002) Protein dynamics from solution NMR: theory and applications. *Cell Biochem. Biophys.* **39,** 187–212.
6. Jen, J. (1978) Chemical exchange and NMR T_2 relaxation—the multisite case. *J. Magn. Reson.* **30,** 111–128.
7. McConnell, H. M. (1957) Theory of nuclear magnetic shielding in molecules, I: long-range dipolar shielding of protons. *J. Chem. Phys.* **27,** 226–229.
8. Allan, E. A., Hogben, M. G., Reeves, L. W., and Shaw, K. N. (1972) Multisite chemical exchange. *Pure Appl. Chem.* **32,** 9–25.

9. Allerhand, A. and Thiele, E. (1966) Analysis of Carr-Purcell spin-echo NMR experiments on multiple-spin systems, II: the effect of chemical exchange. *J. Chem. Phys.* **45,** 902–916.

10. Forsen, S. and Hoffman, R. A. (1964) Exchange rates by nuclear magnetic multiple resonance, III: Exchange reactions in systems with several nonequivalent sites. *J. Chem. Phys.* **40,** 1189–1196.

11. Bain, A. D. and Cramer, J. A. (1993) Optimal NMR measurements for slow exchange in two-site and three-site systems. *J. Phys. Chem.* **97,** 2884–2887.

12. McConnell, H. M. (1958) Reaction rates by nuclear magnetic resonance. *J. Chem. Phys.* **28,** 430, 431.

13. Woessner, D. E. (1995) Brownian motion and its effects in NMR chemical exchange and relaxation in liquids. *Concepts Magn. Reson.* **8,** 397–421.

14. Ishima, R. and Torchia, D. A. (1999) Estimating the time scale of chemical exchange of proteins from measurements of transverse relaxation rates in solution. *J. Biomol. NMR* **14,** 369–372.

15. Carr, H. Y. and Purcell, E. M. (1954) Effects of diffusion on free precession in nuclear magnetic resonance experiments. *Phys. Rev.* **94,** 630–638.

16. Meiboom, S. and Gill, D. (1958) Modified spin-echo method for measuring nuclear spin relaxation times. *Rev. Sci. Instrum.* **29,** 688–691.

17. Luz, Z. and Meiboom, S. (1963) Nuclear magnetic resonance study of the protolysis of trimethylammonium ion in aqueous solution—order of the reaction with respect to solvent. *J. Chem. Phys.* **39,** 366–370.

18. Orekhov, V. Y., Pervushin, K. V., and Arseniev, A. S. (1994) Backbone dynamics of (1–71) bacteriorhodopsin studied by two-dimensional ^1H-^{15}N NMR spectroscopy. *Eur. J. Biochem.* **219,** 887–896.

19. Loria, J. P., Rance, M., and Palmer, A. G. (1999) A relaxation-compensated Carr-Purcell-Meiboom-Gill sequence for characterizing chemical exchange by NMR spectroscopy. *J. Am. Chem. Soc.* **121,** 2331, 2332.

20. Millet, O. M., Loria, J. P., Kroenke, C. D., Pons, M., and Palmer, A. G. (2000) The static magnetic field dependence of chemical exchange linebroadening defines the NMR chemical shift time scale. *J. Am. Chem. Soc.* **122,** 2867–2877.

21. Wang, L., Pang, Y., Holder, T., Brender, J. R., Kurochkin, A. V., and Zuiderweg, E. R. (2001) Functional dynamics in the active site of the ribonuclease binase. *Proc. Natl. Acad. Sci. USA* **98,** 7684–7689.

22. Carver, J. P. and Richards, R. E. (1972) A general two-site solution for the chemical exchange produced dependence of T_2 upon the Carr-Purcell pulse separation. *J. Magn. Reson.* **6,** 89–105.

23. Davis, D. G., Perlman, M. E., and London, R. E. (1994) Direct measurements of the dissociation-rate constant for inhibitor-enzyme complexes via the $T_{1\rho}$ and T_2 (CPMG) methods. *J. Magn. Reson., Ser B* **104,** 266–275.

24. Piette, L. H. and Anderson, W. A. (1959) Potential energy barrier determinations for some alkyl nitrites by nuclear magnetic resonance. *J. Chem. Phys.* **30,** 899–908.

25. Trott, O. and Palmer, A. G., III. (2002) R1rho relaxation outside of the fast-exchange limit. *J. Magn. Reson.* **154,** 157–160.

26. Deverell, C., Morgan, R. E., and Strange, J. H. (1970) Studies of chemical exchange by nuclear magnetization relaxation in the rotating frame. *Mol. Phys.* **18,** 553–559.

27. Meiboom, S. (1961) Nuclear magnetic resonance study of the proton transfer in water. *J. Chem. Phys.* **34,** 375–388.

28. Mulder, F. A. A., de Graaf, R. A., Kaptein, R., and Boelens, R. (1998) An off-resonance rotating frame relaxation experiment for the investigation of macromolecular dynamics using adiabatic rotations. *J. Magn. Reson.* **131,** 351–357.

29. Akke, M., Liu, J., Cavanagh, J., Erickson, H. P., and Palmer, A. G. (1998) Pervasive conformational fluctuations on microsecond time scales in a fibronectin type III domain. *Nat. Struct. Biol.* **5,** 55–59.

30. Vugmeyster, L., Kroenke, C. D., Picart, F., Palmer, A. G., and Raleigh, D. P. (2000) ^{15}N R1ρ measurements allow the determination of ultrafast protein folding rates. *J. Am. Chem. Soc.* **122,** 5387, 5388.

31. Eisenmesser, E. Z., Bosco, D. A., Akke, M., and Kern, D. (2002) Enzyme dynamics during catalysis. *Science* **295,** 1520–1523.

32. Szyperski, T., Luginbühl, P., Otting, G., Güntert, P., and Wüthrich, K. (1993) Protein dynamics studied by rotating frame ^{15}N spin relaxation times. *J. Biomol. NMR* **3,** 151–164.

33. Griesinger, C. and Ernst, R. R. (1987) Frequency offset effects and their elimination in NMR rotating-frame cross-relaxation spectroscopy. *J. Magn. Reson.* **75,** 261–271.

34. Yamazaki, T., Muhandiram, R., and Kay, L. E. (1994) NMR experiments for the measurement of carbon relaxation properties in highly enriched, uniformly ^{13}C,^{15}N-labeled proteins: application to ^{13}C$^\alpha$ carbons. *J. Am. Chem. Soc.* **116,** 8266–8278.

35. Akke, M. and Palmer, A. G. (1996) Monitoring macromolecular motions on microsecond–millisecond time scales by $R_{1\rho}$–R_1 constant-relaxation-time NMR spectroscopy. *J. Am. Chem. Soc.* **118,** 911–912.

36. Lipari, G. and Szabo, A. (1982) Model-free approach to the interpretation of nuclear magnetic resonance relaxation in macromolecules, 1: theory and range of validity. *J. Am. Chem. Soc.* **104,** 4546–4559.

37. Lipari, G. and Szabo, A. (1982) Model-free approach to the interpretation of nuclear magnetic resonance relaxation in macromolecules, 2: Analysis of experimental results. *J. Am. Chem. Soc.* **104,** 4559–4570.

38. Mandel, A. M., Akke, M., and Palmer, A. G. (1996) Dynamics of ribonuclease H: temperature dependence of motions on multiple time scales. *Biochemistry* **35,** 16,009–16,023.

39. Kroenke, C. D., Loria, J. P., Lee, L. K., Rance, M., and Palmer, A. G. (1998) Longitudinal and transverse ^{1}H–^{15}N Dipolar/^{15}N chemical shift anisotropy relaxation interference: unambiguous determination of rotational diffusion tensors and chemical exchange effects in biological macromolecules. *J. Am. Chem. Soc.* **120,** 7905–7915.

40. Tjandra, N., Szabo, A., and Bax, A. (1996) Protein backbone dynamics and ^{15}N chemical shift anisotropy from quantitative measurement of relaxation interference effects. *J. Am. Chem. Soc.* **118,** 6986–6991.

41. Cavanagh, J., Fairbrother, W. J., Palmer, A. G., and Skelton, N. J. (1996) *Protein NMR Spectroscopy: Principles and Practice,* Academic Press, San Diego, CA.

42. Slichter, C. P. (1992) Advanced Concepts in Pulsed Magnetic Resonance. *Principles of Magnetic Resonance,* 3rd ed. Springer Series in Solid-State Sciences (Fulde, P., ed.), Springer-Verlag, New York.

43. Korzhnev, D. M., Tischenko, E. V., and Arseniev, A. S. (2000) Off-resonance effects in ^{15}N T_2 CPMG measurements. *J. Biomol. NMR* **17,** 231–237.

44. Morris, G. A. and Freeman, R. (1979) Enhancement of nuclear magnetic-resonance signals by polarization transfer. *J. Am. Chem. Soc.* **101,** 760–762.

45. Kay, L. E., Keifer, P., and Saarinen, T. (1992) Pure absorption gradient enhanced heteronuclear single quantum correlation spectroscopy with improved sensitivity. *J. Am. Chem. Soc.* **114,** 10,663–10,665.

46. Cavanagh, J. and Rance, M. (1990) Sensitivity improvement in isotropic mixing (TOCSY) experiments. *J. Magn. Reson.* **88,** 72–85.

47. Shaka, A. J., Barker, P. B., and Freeman, R. (1985) Computer-optimized decoupling scheme for wideband applications and low-level operation. *J. Magn. Reson.* **64,** 547–552.

48. Brereton, I. M. (1997) Spectrometer calibration and experimental setup. In *Protein NMR Techniques, vol. 60* (Reid, D. G., ed.). Humana, Totowa, NJ, pp. 363–410.

49. Palmer, A. G., Skelton, N. J., Chazin, W. J., Wright, P. E., and Rance, M. (1992) Suppression of the effects of cross-correlation between dipolar and anisotropic chemical shift relaxation mechanisms in the measurement of spin–spin relaxation rates. *Mol. Phys.* **75,** 699–711.

50. Shaka, A. J., Keeler, J., Frenkiel, T., and Freeman, R. (1983) An improved sequence for broadband decoupling: WALTZ-16. *J. Magn. Reson.* **52,** 335–338.

51. Shaka, A. J., Keeler, J., and Freeman, R. (1983) Evaluation of a new broadband decoupling sequence: WALTZ-16. *J. Magn. Reson.* **53,** 313–340.

52. Boyd, J., Hommel, U., and Campbell, I. D. (1990) Influence of cross-correlation between dipolar and anisotropic chemical shift relaxation mechanisms upon longitudinal relaxation rates of nitrogen-15 in macromolecules. *Chem. Phys. Lett.* **175,** 477–482.

53. Kay, L. E., Nicholson, L. K., Delagio, F., Bax, A., and Torchia, D. A. (1992) Pulse sequences for removal of the effects of cross correlation between dipolar and chemical-shift anisotropy relaxation mechanisms on the measurement of heteronuclear T_1 and T_2 values in proteins. *J. Magn. Reson.* **97,** 359–375.

54. Wang, C., Grey, M. J., and Palmer, A. G., III. (2001) CPMG sequences with enhanced sensitivity to chemical exchange. *J. Biomol. NMR* **21,** 361–366.

55. Pervushin, K., Riek, R., Wider, G., and Wuthrich, K. (1997) Attenuated T2 relaxation by mutual cancellation of dipole–dipole coupling and chemical shift anisotropy indicates an avenue to NMR structures of very large biological macromolecules in solution. *Proc. Natl. Acad. Sci. USA* **94,** 12,366–12,371.

56. Salzmann, M., Pervushin, K., Wider, G., Senn, H., and Wuthrich, K. (1998) TROSY in triple-resonance experiments: new perspectives for sequential NMR assignment of large proteins. *Proc. Natl. Acad. Sci. USA* **95,** 13,585–13,590.

57. Griffey, R. H. and Redfield, A. G. (1987) Proton-detected heteronuclear edited and correlated nuclear magnetic resonance and nuclear Overhauser effect in solution. *Q. Rev. Biophys.* **19,** 51–82.

58. Goldman, M. (1984) Interference effects in the relaxation of a pair of unlike spin-1/2 nuclei. *J. Magn. Reson.* **60,** 437–452.

59. Czisch, M. and Boelens, R. (1998) Sensitivity enhancement in the TROSY experiment. *J. Magn. Reson.* **134,** 158–160.

60. Zhu, G., Kong, X., Yan, X., and Sze, K. (1998) Sensitivity enhancement in transverse relaxation optimized NMR spectroscopy. *Angew. Chem. Int. Ed. Engl.* **37,** 2859–2861.

61. Loria, J. P., Rance, M., and Palmer, A. G. (1999) Transverse-relaxation-optimized (TROSY) gradient-enhnaced triple-resonance NMR spectroscopy. *J. Magn. Reson.* **141,** 180–184.

62. Rance, M., Loria, J. P., and Palmer, A. G. (1999) Sensitivity improvement of transverse relaxation-optimized spectroscopy. *J. Magn. Reson.* **136,** 92–101.

63. Loria, J. P., Rance, M., and Palmer, A. G. (1999) A TROSY CPMG sequence for characterizing chemical exchange in large proteins. *J. Biomol. NMR* **15,** 151–155.

64. Mori, S., Abeygunawardana, C., O'Neil Johnson, M., and van Zijl, P. C. M. (1995) Improved sensitivity of HSQC spectra of exchanging protons at short interscan delays using a new fast HSQC (FHSQC) detection scheme that avoids water saturation. *J. Magn. Reson., Ser. B* **108,** 94–98.

65. Mueller, L., Legault, P., and Pardi, A. (1995) Improved RNA structure determination by detection of NOE contacts to exchange-broadened amino protons. *J. Am. Chem. Soc.* **117,** 11,043–11,048.

66. Mulder, F. A. A., Spronk, C. A. E. M., Slijper, M., Kaptein, R., and Boelens, R. (1996) Improved HSQC experiments for the observation of exchange broadened signals. *J. Biomol. NMR* **8,** 223–228.

67. Gullion, T., Baker, D. B., and Conradi, M. S. (1990) New, compensated Carr-Purcell sequences. *J. Magn. Reson.* **89,** 479–484.

68. Cole, R. and Loria, J. P. (2002) Evidence for flexibility in the function of ribonuclease A. *Biochemistry* **41,** 6072–6081.

69. Shoemaker, D. P., Garland, C. W., and Nibler, J. W. (1989) *Experiments in Physical Chemistry.* 5th ed. McGraw-Hill, New York.

70. Jones, J. A. (1997) Optimal sampling strategies for the measurement of relaxation times in proteins. *J. Magn. Reson.* **126,** 283–286.

71. Weitekamp, D. P. (1983) Time-domain multiple-quantum NMR. *Adv. Magn. Reson.* **11,** 111–273.

72. Kempf, J. G., Jung, J., Sampson, N. S., and Loria, J. P. (2003). Off-resonance TROSY $(R_{1\rho}-R_1)$ for quantitation of fast exchange processes in large proteins. *J. Am. Chem. Soc.* **125,** 12,064–12,065.

73. Ugurbil, K., Garwood, M., and Rath, J. R. (1988) Optimization of modulation functions to improve insensitivity of adiabatic pulses to variations in B_1 magnitude. *J. Magn. Reson.* **80,** 448–469.

74. Garwood, M. and Yong, K. (1991) Symmetric pulses to induce arbitrary flip angles with compensation of RF inhomogeneity and resonance offsets. *J. Magn. Reson.* **94,** 511–525.

75. Korzhnev, D. M., Skyrnnikov, N. R., Millet, O., Torchia, D., and Kay, L. E. (2002) An NMR experiment for the accurate measurement of heteronuclear spin-lock relaxation rates. *J. Am. Chem. Soc.* **124,** 10,743–10,753.

76. Ernst, R. R. (1966) Nuclear magnetic double resonance with an incoherent radio-frequency field. *J. Chem. Phys.* **45,** 3845–3861.

77. Wang, A. C. and Bax, A. (1993) Minimizing the effects of radio-frequency heating in multidimensional NMR experiments. *J. Biomol. NMR* **3,** 715–720.

78. Hartmann, S. R. and Hahn, E. L. (1962) Nuclear double resonance in the rotating frame. *Phys. Rev.* **128,** 2042–2053.

79. Shaka, A. J. and Keeler, J. (1987) Broadband spin decoupling in isotropic liquids. *Prog. NMR Spectrosc.* **19,** 47–129.

80. Bax, A., Ikura, M., Kay, L. E., Torchia, D. A., and Tschudin, R. (1990) Comparison of different modes of two-dimensional reverse-correlation NMR for the study of proteins. *J. Magn. Reson.* **86,** 304–318.

81. Skelton, N. J., Palmer, A. G., Akke, M., Kördel, J., Rance, M., and Chazin, W. J. (1993) Practical aspects of two-dimensional proton-detected ^{15}N spin relaxation measurements. *J. Magn. Reson., Ser. B* **102,** 253–264.

82. Evenas, J., Malmendal, A., and Akke, M. (2001) Dynamics of the transition between open and closed conformations in a calmodulin C-terminal domain mutant. *Structure* **9,** 185–195.

83. Loening, N. M. and Keeler, J. (2002) Temperature accuracy and temperature gradients in solution-state NMR spectrometers. *J. Magn. Reson.* **159,** 55–61.

84. Skrynnikov, N. R., Mulder, F. A., Hon, B., Dahlquist, F. W., and Kay, L. E. (2001) Probing slow time scale dynamics at methyl-containing sidechains in proteins by relaxation dispersion NMR measurements: application to methionine residues in a cavity mutant of T4 lysozyme. *J. Am. Chem. Soc.* **123,** 4556–4566.

85. Mulder, F. A., Skrynnikov, N. R., Hon, B., Dahlquist, F. W., and Kay, L. E. (2001) Measurement of slow (micros-ms) time scale dynamics in protein sidechains by (15)N relaxation dispersion NMR spectroscopy: application to Asn and Gln residues in a cavity mutant of T4 lysozyme. *J. Am. Chem. Soc.* **123,** 967–975.

86. Guenneugues, M., Berthault, P., and Desvaux, H. (1999) A method for determining B1 field inhomogeneity: are the biases assumed in heteronuclear relaxation experiments usually underestimated? *J. Magn. Reson.* **136,** 118–126.

87. Marion, D., Ikura, M., Tschudin, R., and Bax, A. (1989) Rapid recording of 2D NMR spectra without phase cycling: application to the study of hydrogen exchange in proteins. *J. Magn. Reson.* **85,** 393–399.

88. Cavanagh, J., Palmer, A. G., Wright, P. E., and Rance, M. (1991) Sensitivity improvement in proton-detected two-dimensional heteronuclear relay spectroscopy. *J. Magn. Reson.* **91,** 429–436.

89. Palmer, A. G., Cavanagh, J., Wright, P. E., and Rance, M. (1991) Sensitivity improvement in proton-detected two-dimensional heteronuclear correlation NMR spectroscopy. *J. Magn. Reson.* **93,** 151–170.

90. Sklenár, V., Piotto, M., Leppik, R., and Saudek, V. (1993) Gradient-tailored water suppression for H-1-N-15 HSQC experiments optimized to retain full sensitivity. *J. Magn. Reson., Ser. A* **102,** 241–245.

13

NMR Studies of Partially Folded Molten-Globule States

Christina Redfield

Summary

Nuclear magnetic resonance (NMR) spectroscopy is a powerful technique for the study of the structure, dynamics, and folding of proteins in solution. It is particularly powerful when applied to dynamic or flexible systems, such as partially folded molten-globule states of proteins, which are not usually amenable to X-ray crystallography. This chapter describes NMR methods suitable for the characterization of molten-globule states. These include pulsed-field-gradient NMR techniques for the measurement of the hydrodynamic radius, bulk and site-specific hydrogen–deuterium exchange experiments for the identification of regions of secondary structure, and ^{15}N-edited NMR experiments carried out in increasing concentrations of denaturants, which allow the stability of different regions of the molten globule to be probed. Examples of the application of these methods to the study of the low-pH molten globule of human α-lactalbumin are presented.

Key Words: Molten globule; partially folded protein; protein folding; protein denaturation; hydrogen–deuterium exchange; circular dichroism; ANS fluorescence; hydrodynamic radius; HSQC; α-lactalbumin; PG–SLED; resonance assignment.

1. Introduction

Most biomolecular nuclear magnetic resonance (NMR) spectroscopists would be disappointed to obtain an NMR spectrum such as that shown in **Fig. 1A** for a protein that they wished to characterize in detail. The poor chemical-shift dispersion shown in this spectrum is not what would be expected for a native globular protein (**Fig. 1B**). On the other hand, the broad lines are not characteristic of a completely unfolded protein (**Fig. 1C**). The spectrum shown in **Fig. 1A** belongs to human α-lactalbumin (α-LA) at pH 2. Under these conditions α-LA adopts a partially folded conformation known as a molten globule. This chapter describes NMR methods suitable for the study of molten-globule states of proteins.

From: *Methods in Molecular Biology, vol. 278: Protein NMR Techniques*
Edited by: A. K. Downing © Humana Press Inc., Totowa, NJ

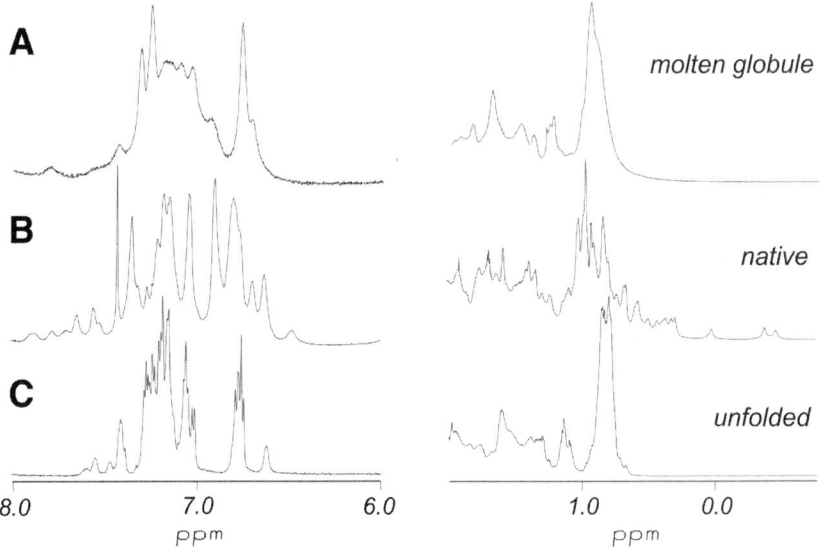

Fig. 1. The aromatic (*left*) and methyl (*right*) regions of 600 MHz ^{1}H NMR spectra of human α-LA in its (**A**) molten globule (pH 2, 20°C), (**B**) native (pH 6.3, Ca^{2+}, 20°C), and (**C**) unfolded (10 *M* urea, pH 2, 20°C) states.

1.1. What Is a Molten Globule?

The folding of a protein to its functional native state from the information encoded in its amino acid sequence is a key feature of the conversion of genetic information into biological activity. Although in some cases protein folding has been found to be highly cooperative, in others partially structured species have been observed to form in the early stages of refolding prior to the formation of the native state. These partially folded species are compact, have extensive native-like secondary structure, but lack the specific sidechain packing that is characteristic of native structures (*1,2*). These species are known as molten globules (*3,4*); the term *molten* refers to the lack of rigid tertiary packing and *globule* refers to the compact shape. For numerous proteins, similar species have been found to be stable at equilibrium under mildly denaturing conditions, such as acidic pH, low concentrations of urea or guanidine hydrochloride, or elevated temperatures (*1,2*). Numerous studies have suggested a close similarity between molten-globule states observed at equilibrium and those formed during the early stages of refolding of several proteins, including α-LA (*5–7*) and apomyoglobin (*8,9*). Thus, insights into the transient kinetic molten globule and the determinants of the protein fold can be gained through detailed studies of the stable equilibrium molten globules of proteins.

1.2. Which Proteins Form Molten Globules?

Molten globules have been identified for numerous proteins under mildly denaturing conditions or on removal of a cofactor. Low pH is a particularly effective method for generating molten globules *(1,2)*. At low pH, α-LA, equine lysozyme, ribonuclease HI from *Escherichia coli*, and β-lactoglobulin all form molten globules. For some proteins, including cytochrome *c* and staphylococcal nuclease, high salt concentrations are required at low pH to stabilize the molten globule. Myoglobin forms a molten globule at pH 4 on removal of the heme group to form apomyoglobin. The NMR methods described in this chapter are illustrated with examples from human α-LA.

1.3. The Human α-Lactalbumin Molten Globule

The molten globule formed by α-LA at low pH is one of the most extensively studied partially folded proteins. α-LA is a 14-kDa, Ca^{2+}-binding protein whose native structure is divided into two lobes; one is largely helical, the α-domain, and the other has a significant β-sheet content, the β-domain *(10)*. The protein contains four disulfide bonds, two in the α-domain, one in the β-domain, and one linking the two domains (**Fig. 2**). α-LA undergoes partial unfolding to form a molten globule at low pH or by the removal of Ca^{2+} at neutral pH and slightly denaturing conditions *(1,2)*. The α-LA molten globule is compact, with a hydrodynamic radius close to that of the native state *(7,11,12)*. It has significant native-like helical secondary structure but is characterized by a high degree of conformational disorder, which is likely to result from a lack of fixed tertiary interactions *(13,14)*. The α-domain of the α-LA molten globule has a native-like fold and a hydrophobic core formed by the helices, whereas the β-domain appears to be less ordered and lacks the propensity for a native-like fold *(15,16)*.

1.4. How Can Molten Globules Be Studied?

The lack of fixed tertiary interactions in most molten globules results in a fluctuating ensemble of structures, which interconvert on a millisecond-to-microsecond time-scale *(13,17)*. This heterogeneous structural character makes detailed structural studies using conventional NMR or X-ray crystallographic methods difficult. Although extreme line broadening often precludes direct NMR structural studies of the molten globule, detailed residue-specific information may be gained using indirect NMR methods. In the following sections, methods suitable for the initial identification of molten-globule species, for the measurement of the hydrodynamic radius, for the identification of regions of secondary structure, and for the study of the stability of the molten globule are described.

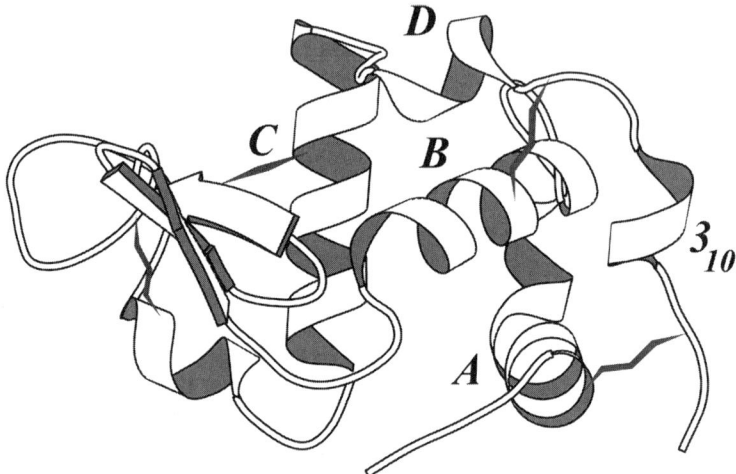

Fig. 2. Schematic representation of the native structure of human α-LA *(10)*. The α-domain helices are labeled and the disulfide bonds are represented as *zigzag lines*.

2. Materials

Pure protein is required for all the methods described in the following sections. Unlabeled protein can be used for circular dichroism (CD) and 8-anilonaphthalene-1-sulfonic acid (ANS) fluorescence spectroscopy, for pulsed-field-gradient NMR measurements of the hydrodynamic radius, and for bulk hydrogen-exchange experiments. Residue-specific hydrogen-exchange experiments can be carried out with unlabeled protein if correlated spectroscopy (COSY) or total correlation spectroscopy (TOCSY) spectra are used for the detection of protected amides. However, ^{15}N-labeled protein is required if protected amides are detected using heteronuclear single-quantum coherence (HSQC) methods. ^{15}N-labeled protein is required for the HSQC urea titration experiments that are used to probe the stability of the molten-globule state.

CD spectra are collected with a protein concentration in the range of 20–50 μM. Cuvets with 0.1- and 1.0-cm path lengths and solution volumes of approx 300 µL and approx 3 mL are used for the far- and near-ultraviolet (UV) CD spectra, respectively. ANS fluorescence experiments are carried out with approx 2 μM ANS and approx 2 μM protein. Diffusion measurements are carried out with approx 1 mM protein dissolved in 100% D_2O with 20 µL of a 1% dioxane in D_2O solution added. The aromatic and methyl regions of the spectrum should be free of peaks from buffers or small molecule impurities as these can distort the measured diffusion coefficient. Bulk hydrogen-exchange experiments require a solution of approx 1 mM protein dissolved in 100% D_2O. The amide and aromatic region of the spectrum should be free of peaks from buffers or small molecule impurities as these will distort the calibration of peak intensities. Residue-specific hydrogen-exchange experiments require a separate sample for each time-point; samples will contain approx 1 mM protein dissolved in 100% D_2O. The urea titration HSQC experiments can be carried out with a single sample of ^{15}N-labeled protein. The sample is made up initially at a protein concentration of approx 1 mM in 95% H_2O/5% D_2O, and an HSQC spectrum is collected. Solid urea is weighed and then added to the solution to give a final urea concentration of 1 M; for a 500 µL NMR sample 30 mg of urea must be added to increase the urea concentration from 0 to 1 M. The pH of the sample is adjusted, and an HSQC spectrum is collected. This procedure is repeated for each molar increment in the urea concentration. It should be noted that the volume of the NMR solution increases as urea is added. This means that larger quantities of urea must be added for each molar increment in the urea concentration and the protein concentration will decrease during the titration.

3. Methods

3.1. Identification of Molten-Globule Species

A broad, poorly dispersed NMR spectrum is one characteristic of a molten-globule state of a protein (**Fig. 1A**). However, broadening of NMR lines can also arise from the aggregation of protein molecules. There are several spectroscopic tests for identifying molten globules, and some of these are described in the following sections.

3.1.1. Circular Dichroism (CD) Spectroscopy

CD spectra measured in the far-UV (190 to 250 nm) region provide information about α-helical and β-sheet secondary structure in a protein; strong negative ellipticity in the region of 205–225 nm is observed for both types of structure. CD spectra measured in the near-UV (250–320 nm) region provide information about the environment of aromatic sidechains in a protein. Native proteins, which usually contain significant amounts of secondary structure and a well-defined hydrophobic core containing aromatic residues, give rise to significant CD signals in both the far- and near-UV regions. Unfolded proteins, which lack secondary and tertiary structure, do not give significant signals in either the far- or near-UV CD spectra. Partially folded molten globules usually contain a significant amount of native-like α-helix or β-sheet secondary structure, and therefore, the far-UV CD spectra of these species still show a significant amount of signal. On the other hand, these species lack a well-defined packing of sidechains to form a structured core. This lack of a rigid chiral environment for aromatic sidechains leads to a significant attenuation or complete loss of the near-UV CD signal. Far- and near-UV CD spectra of human α-LA in its native, unfolded and molten-globule states are compared in **Fig. 3A,B**. The strong negative ellipticity at 208 and 222 nm in the far-UV CD spectra of both states indicates a significant amount of α-helical secondary structure. The near-UV CD spectrum of the molten globule displays signals that they are very weak compared to the native state. Therefore, the observation of significant signal in the far-UV CD spectrum combined with the loss of the near-UV CD spectrum is a characteristic feature of molten-globule states that can be used for their identification.

3.1.2. ANS Fluorescence

The hydrophobic dye ANS is also commonly used to study molten-globule states *(18)*. In aqueous solution ANS displays a weak fluorescence signal. In the presence of a partially folded protein with exposed hydrophobic surfaces, such as a molten globule, the fluorescence of ANS is enhanced significantly, and the emission maximum is blue shifted *(18)*. The fluorescence enhancements observed in the presence of native or fully unfolded proteins are much smaller. The fluorescence spectra of the hydrophobic dye ANS alone and in the presence of human α-LA under native, molten globule and unfolded conditions are compared in **Fig. 3C**. The α-LA molten globule at pH 2.0 shows an approx 40-fold fluorescence enhancement relative to ANS alone and a blue shift of the emission maximum from 504 to 473 nm. The native and unfolded states of α-LA show a marked decrease in the fluorescence enhancement (only approx three- to fourfold relative to ANS alone), and the observed emission maximum of 512 nm is close to that of free ANS. Thus, the observation of enhanced fluorescence of ANS and a blue shift of the emission maximum are another characteristic feature of molten-globule states of proteins.

3.2. Measurement of the Hydrodynamic Radius

One of the characteristic features of molten-globule states of proteins is that they are compact. In general, molten-globule states of proteins are only 10–30% expanded compared to their native state *(1,2)*. Unfolded proteins are substantially more expanded. One method of identifying a protein as a molten globule rather than an aggregated unfolded state is to measure the hydrodynamic radius (R_h) of the protein. These measurements can be carried out using a number of experimental techniques, including dynamic light scattering and small-angle X-ray scattering *(19,20)*. If a broad NMR spectrum is observed then it may be important to be able to carry out measurements directly on the NMR sample to distinguish between a molten globule and an aggregated sample. This can be done using pulsed-field-gradient NMR techniques, which are sensitive to the diffusion properties of the protein; this method is described in the following section *(21–24)*.

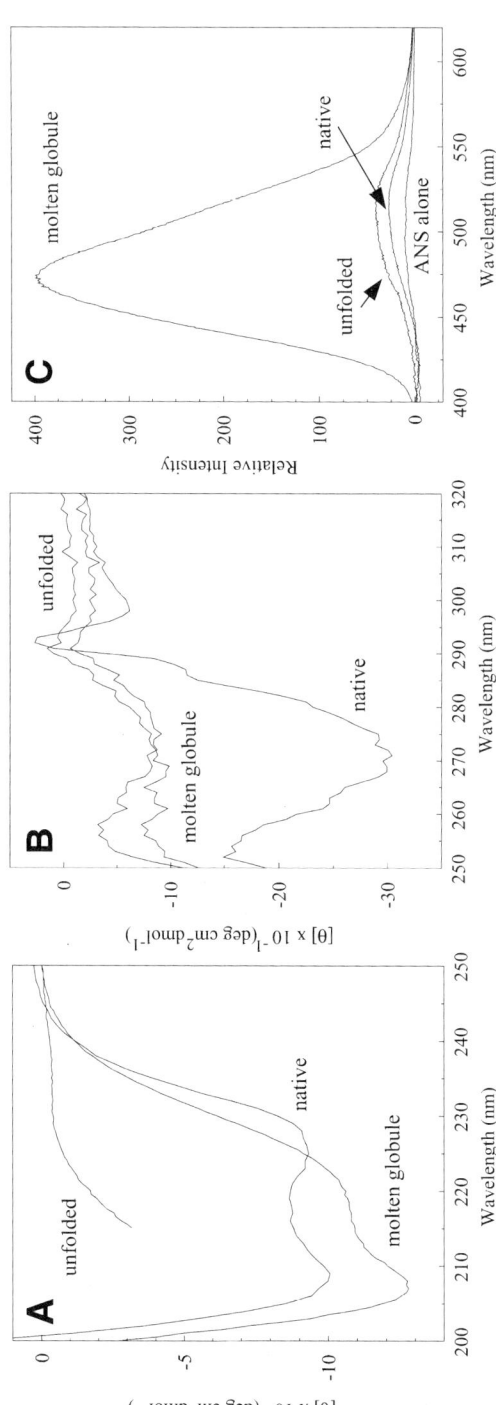

Fig. 3. (**A**) Far-UV and (**B**) near-UV CD spectra of human α-LA in its native (pH 6.3, Ca²⁺, 20°C), molten globule (pH 2.0, 20°C), and unfolded (10 *M* urea, pH 2.0, 20°C) states. The protein concentration was 50 μ*M*. The strong negative ellipticity at 208 and 222 nm in (**A**) is characteristic of α-helical structure. The far-UV CD spectrum of a β-sheet protein would show a strong negative peak at approx 215 nm. (**C**) Fluorescence emission spectra of 2 μ*M* ANS alone at pH 2.0 and 20°C, and of 2 μ*M* ANS in the presence of 2 μ*M* human α-LA in its native (pH 6.3, Ca²⁺, 20°C), molten globule (pH 2.0, 20°C), and unfolded (10 *M* urea, pH 2.0, 20°C) states. An excitation wavelength of 370 nm was used.

239

3.2.1. Pulsed-Field-Gradient NMR Measurement of the Hydrodynamic Radius

The Stokes–Einstein equation shows that the hydrodynamic radius (R_h) of a protein is inversely proportional to the diffusion coefficient (D) of the protein. However, the diffusion coefficient is also affected by temperature and solvent viscosity, and this can make measurements of absolute values of the diffusion coefficient difficult. This problem can be overcome if the diffusion coefficient of a protein is measured relative to that of a small molecule of known hydrodynamic radius present in the same solution *(23,25)*. In this case, the hydrodynamic radius of the protein (R_h^{prot}) is directly proportional to the hydrodynamic radius of the reference compound (R_h^{ref})

$$R_h^{prot} = (D_{ref}/D_{prot}) \times (R_h^{ref}),$$

where D_{prot} and D_{ref} are the measured apparent diffusion coefficients of the protein and reference compound.

Information about the diffusion coefficients of a protein and a reference compound can be obtained using the pulse-gradient stimulated echo longitudinal encode–decode (PG-SLED) sequence (**Fig. 4A**) *(21–23)*. A series of 20 spectra is collected with the strength of the diffusion gradient varying between 5 and 100% of the maximum value. The length of the diffusion gradient (δ) and the stimulated echo (τ) are adjusted so that the decay in protein signal observed from 5 to 100% gradient strength is between 80 and 90%. 1,4-dioxane has been found to be a useful reference compound; it is found not to interact with proteins and has a known hydrodynamic radius of 2.12 Å *(23,24)*. PG-SLED spectra obtained for human α-LA at pH 2.0 are shown in **Fig. 4B,C**. It can be seen that the peak belonging to dioxane decays much more rapidly than the α-LA peaks as a result of a much higher diffusion coefficient. The diffusion coefficients can be determined from a plot of integrated peak intensity as a function of the gradient strength (**Fig. 4D**). For dioxane the singlet at approx 3.74 ppm is integrated. For the protein the entire aromatic or methyl region of the spectrum is integrated. It is important that the integrated region of the protein spectrum is free from peaks arising from small molecule impurities or buffers as these will distort the measured diffusion coefficient. The decay of signal intensity follows Gaussian behavior. Typical plots for α-LA and dioxane are shown in **Fig. 4D**. The observed ratio of the dioxane and protein diffusion coefficients is 9.7. If this is multiplied by the hydrodynamic radius of dioxane, a value of 20.6 Å is obtained for the hydrodynamic radius of the α-LA molten globule *(12,23)*.

A comparison of the hydrodynamic radii of native, unfolded and partially folded species is useful for the identification of a molten globule. A level of compaction, C, can be defined as

$$C = (R_h^D - R_h)/(R_h^D - R_h^N),$$

where R_h^D and R_h^N are the hydrodynamic radii of the denatured and native states, and R_h is that observed for the partially folded species. Measurements on the molten globules of apomyoglobin, cytochrome c and α-LA have shown compaction ratios of 0.74, 0.86, and 0.86, respectively *(24)*. These values are consistent with a compact molten globule that is only slightly expanded relative to the native state. An aggregated protein would be expected to give a hydrodynamic radius significantly larger than the native state and possibly also larger than the denatured state. Therefore, measurements of hydrodynamics radius can be used to identify molten globules and to distinguish these from aggregated states of the protein.

3.3. Hydrogen–Deuterium Exchange Methods for Identifying Secondary Structure

If a protein is dissolved in D_2O, then backbone amide protons will exchange with deuterons. The intrinsic kinetics of the exchange process depend on pH, temperature, and the neighboring sidechains *(26–28)*. Hydrogen exchange is both acid and base catalysed with a pH minimum at approx 3.0. Neighboring side-chain groups can affect intrinsic exchange rates by up to two orders of magnitude *(28)*. If an amide is involved in a hydrogen bond or deeply buried in the interior of the protein, then the rate of exchange may be slowed down significantly compared to the predicted intrinsic rate. The level of protection of an amide from hydrogen–deuterium exchange, the protection factor (PF), is measured as the ratio of the intrinsic exchange rate (k_{intr}) to the observed exchange rate (k_{obs}), $PF = k_{intr}/k_{obs}$. In native proteins protection factors of 10^5–10^7 are not uncommon in regions of hydrogen-bonded secondary structure. In completely unfolded proteins the observed rates of exchange are those expected from the intrinsic rates. Molten globules contain significant levels of secondary structure, and the hydrogen bonds present in these can give rise to protection from hydrogen exchange. In the following sections, the bulk hydrogen-exchange method, which can indicate the presence of protected amides, is described first, followed by two-dimensional (2D) NMR methods, which allow the specific residues involved in hydrogen-bonded secondary structure to be identified.

3.3.1. Bulk Hydrogen-Exchange Experiments

Although extreme line broadening often precludes direct NMR study of the molten globule, information about bulk hydrogen-exchange rates can be obtained directly from a series of one-dimensional (1D) NMR spectra collected for the molten globule in D_2O (**Fig. 5A**). Such measurements can be used to determine if the observed rate of exchange is that expected from the intrinsic

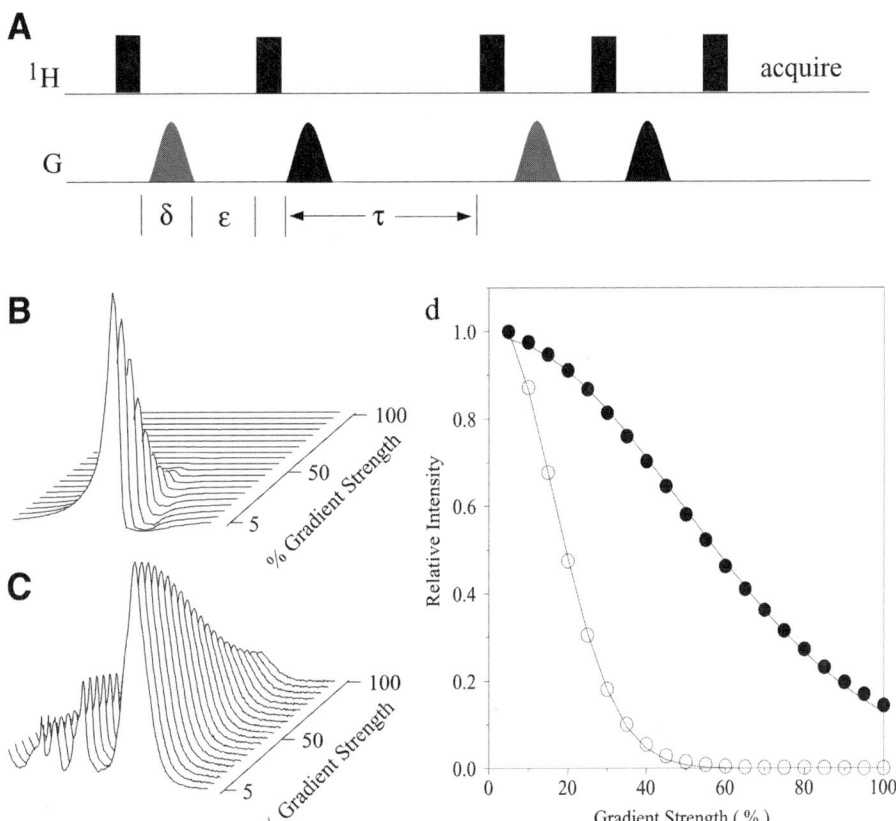

Fig. 4. **(A)** The PG-SLED sequence used to measure hydrodynamic radii. The strengths of the diffusion labeling gradients (shown in *gray*) were varied from 5–100%, whereas the crush gradients (shown in *black*) were kept at 100%. The length of the diffusion gradient (δ) and the stimulated echo (τ) are adjusted to give an 80–90% decrease in protein signal between 5 and 100% gradient strength. Details of phase cycling and other experimental aspects are described in **ref. 23**. **(B–C)** PG–SLED spectra obtained for human α-LA at pH 2.0 and 20°C showing the decay of the dioxane **(B)** and α-LA methyl **(C)** signals as the gradient strength is increased. **(D)** Plot of the integrated spectral intensity as a function of gradient strength for the α-LA methyl signals (•) and for the dioxane signal ({o}). The *solid lines* represent fits of a Gaussian decay function to the experimental data ($I(g) = I_0 \exp(-Dg^2)$, where $I(g)$ is the observed intensity at each gradient strength, g, and D is the apparent diffusion coefficient).

rates predicted from the amino acid sequence *(28)* or if a significant number of amides are protected from exchange. A comparison of the observed exchange kinetics with those predicted from random coil models will indicate the number of amides in the molten-globule state that are protected from solvent as a result of stable secondary structure.

Quantitative bulk hydrogen-exchange measurements are achieved by integration of the envelope of amide and aromatic resonances (approx 6–10 ppm) in ^{1}H NMR spectra acquired immediately after dissolution of lyophilized protein in D_2O under conditions where the molten globule is formed; in the case of human α-LA this is done at pH 2.0 and 20°C (**Fig. 5A**). The NMR spectrometer can be shimmed and tuned, and all experimental parameters adjusted prior to the beginning of the exchange experiment using a similar protein sample; this ensures that the exchange experiment can be started as quickly as possible after dissolution of the protein. A series of 1D spectra is collected as a function of time over a period of up to approx 24 h. At the end of this time the sample is heated to approx 50°C for approx 1 h to exchange all remaining amides, and the sample is then cooled to 20°C; this can be done in the spectrometer to minimize changes to the shim settings. A reference 1D spectrum is then collected. Only the resonances of nonexchangeable aromatic protons remain in the downfield region in this reference spectrum. The integrated intensity, between 6 and 10 ppm, in this spectrum is used to determine the intensity per proton on the basis of the known number of nonexchangeable aromatic protons. This reference spectrum is subtracted from each of the hydrogen-exchange spectra. The integrated intensity of each of these difference spectra is used to obtain the number of amide protons present as a function of time *(29)*. A plot of bulk hydrogen exchange observed for human α-LA at pH 2.0 is shown in **Fig. 5B**. The observed exchange kinetics are significantly slower than those expected from the intrinsic rates predicted from the amino acid sequence. This indicates that there is significant protection of amides in the α-LA molten globule resulting from the presence of hydrogen-bonded secondary structure.

3.3.2. Residue-Specific Hydrogen-Exchange Experiments

If the bulk exchange experiments demonstrate slower than expected hydrogen-exchange rates for the molten globule, then further experiments can be conducted to identify the individual amino acid residues with protected amides. However, the extreme line broadening observed in the NMR spectrum of the molten globule means that these amides cannot be identified directly from the NMR spectrum of the molten globule. Instead, the narrow lines and good resolution observed in the native state spectrum of the protein can be exploited. Hydrogen exchange is allowed to proceed under conditions that favor the molten globule. Aliquots are taken at various times following dissolution in D_2O. Suitable times can be

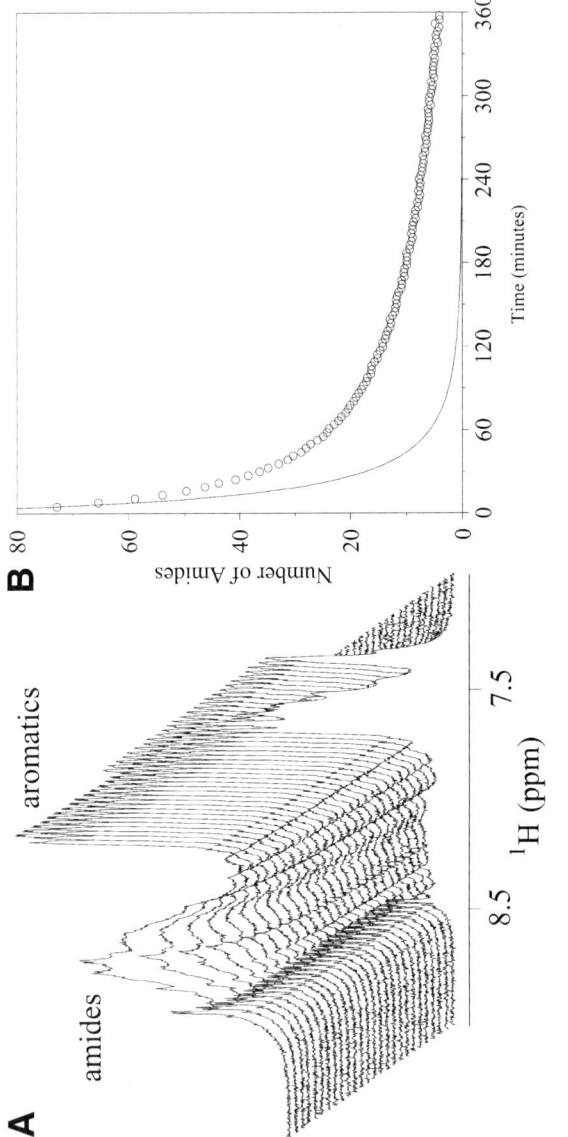

Fig. 5A,B.

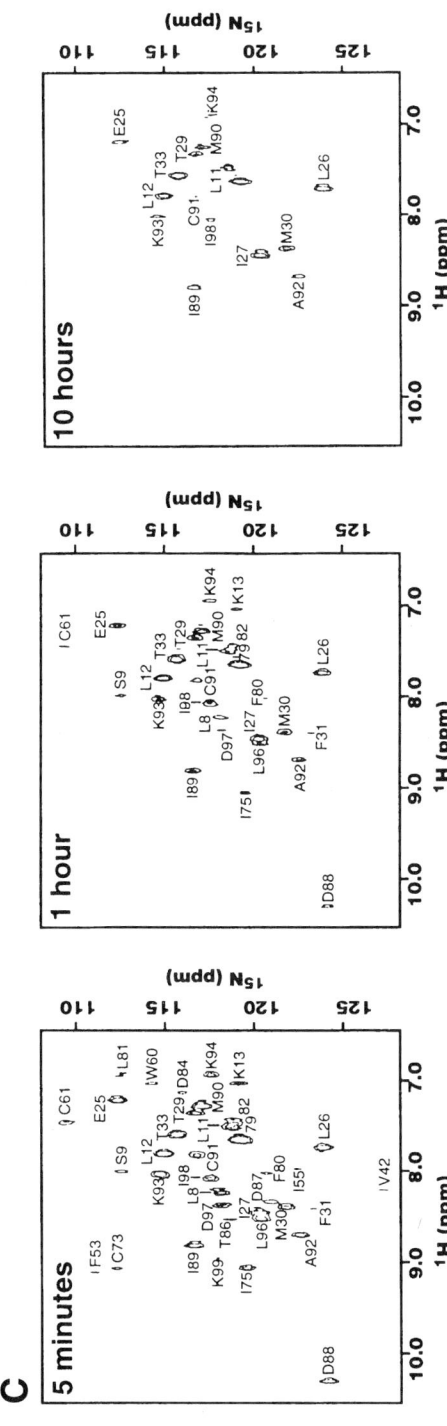

Fig. 5. (**A**) A series of 1D ¹H NMR spectra collected for human α-LA at pH 2.0, 20°C, after dissolution in D₂O. The intensity of amide peaks decreases with time, whereas that of the aromatic peaks remains constant. (**B**) Time dependence of the total number of protected amides (O) in human α-LA at pH 2.0, 20°C. The *solid line* shows the predicted exchange data for a random coil polypeptide with the amino acid sequence of human α-LA. The experimentally observed hydrogen-exchange rates are significantly slower than those predicted from the random coil model indicating significant protection of amides. (**C**) HSQC spectra of native α-LA after exchange in the molten-globule state, pH 2.0, for various periods of time. (Reprinted from **ref. 31**, with permission from Elsevier.)

245

identified from the bulk hydrogen-exchange data. The protein is then returned to its native state where further exchange is very slow because of high protection factors *(8,13,30)*. In the case of α-LA, this is done by raising the pH *(13,31)*. For apomyoglobin the native state is reconstituted by the addition of heme and an increase in pH *(8)*. Under native conditions a well-resolved NMR spectrum is obtained allowing residue-specific exchange rates to be measured using either 2D homonuclear (COSY or TOCSY) or heteronuclear (^{15}N–^{1}H HSQC) experiments. The measured rates reflect the hydrogen-exchange protection resulting from hydrogen bonds present in the molten-globule state of the protein. This method requires substantial amounts of protein as a separate sample is required for each time-point. Because the samples may vary slightly in protein concentration, it is important to collect a reference 1D ^{1}H NMR spectrum for each sample for the purposes of scaling peak intensities. This method is only suitable for identifying amides that are protected in both the molten globule and native states.

Amides that are protected in the molten globule of human α-LA have been identified using this approach *(31)*. The protein was dissolved at pH 2.0 in D_2O and aliquots taken at specific time-points were frozen to stop the exchange. Samples were then freeze-dried and redissolved in pH 6.3 buffer containing Ca^{2+}. HSQC spectra were collected to identify the protected amides. Examples of these spectra are shown in **Fig. 5C**. Amides located in the native A-, B-, and C-helices are protected from hydrogen exchange in α-LA at pH 2 indicating that these helices are also formed in the molten globule. The observed protection factors (up to approx 500) in the α-LA molten globule are substantially lower than those observed in the native state (up to approx 10^5) *(31)*.

3.4. Probing the Stability of the Molten Globule Using HSQC Spectroscopy

^{15}N-edited 2D and three-dimensional (3D) NMR spectroscopy is a powerful method for obtaining detailed structural information about proteins in solution. However, these methods can only be applied if the protein gives relatively sharp peaks in ^{15}N–^{1}H HSQC spectra *(32)*. NMR resonances of residues located in compact, folded regions of molten-globule states are often extremely broad because of conformational fluctuations on a millisecond-to-microsecond timescale that result from local side chain disorder. The HSQC spectrum of human α-LA at pH 2.0 (**Fig. 6**) contains sharp peaks from only three backbone amides *(17)*. With so few peaks observed in the HSQC spectrum it is not possible to apply ^{15}N-edited 3D methods to obtain structural information. The helical content of the α-LA molten globule at pH 2.0, as measured by far-UV CD ellipticity at 222 nm, decreases as the concentration of the denaturant urea is increased *(17)*. A 2D NMR approach using ^{15}N–^{1}H HSQC spectra collected with increasing concentrations of urea has been used for the study of the α-LA molten globule *(12,17)*; a similar approach has been used previously to study truncated variants of staphylococcal nuclease *(33,34)*.

The number of resolved peaks observed in ^{15}N–^{1}H HSQC spectra of α-LA increases as the concentration of urea is increased at 20°C (**Fig. 6**). In 3 *M* urea, 30 peaks are observed in the HSQC spectrum of α-LA. This increases to 44 peaks in 6 *M* urea and to 77 peaks in 10 *M* urea. If the temperature is increased to 50°C, then 92 of the 122 residues in α-LA give rise to detectable peaks in the HSQC spectrum in 10 *M* urea. Peaks for all residues are observed in 8 *M* guanidine HCl at 50°C *(17)*. When a significant number of peaks are observed in HSQC spectra ^{15}N-edited 3D methods can be applied to obtain specific assignments for these peaks.

The assignment of the spectrum of human α-LA in urea is feasible because of the relatively large range of ^{15}N chemical shifts observed in partially folded and unfolded proteins. This arises from the large spread of random coil ^{15}N chemical shifts for the amino acids and from the large sequence-dependent effects of up to 4.5 ppm observed on these random coil shifts *(35)*. It can be seen from **Fig. 6** that peaks arising from Gly, Ser, and Thr residues are separated from the peaks arising from other amino acids. Within the cluster of Thr and Ser peaks there is sufficient resolution arising from sequence-dependent shifts to give well-resolved peaks.

Strips from a ^{15}N-edited 3D nuclear Overhauser effect spectroscopy (NOESY)–HSQC *(32,36)* spectrum obtained for human α-LA at pH 2.0 in 5 *M* urea are shown in **Fig. 7**; intraresidue and sequential nuclear Overhauser effects (NOEs) are indicated. Numerous observations can be made from 3D NOE data collected at a range of urea concentrations and temperatures *(17)*. First, the chemical shifts of aliphatic resonances are close to values expected for unstructured peptides and are largely independent of denaturant concentrations. Second, the pattern of NOESY cross-peaks at various urea concentrations is characteristic of a highly unfolded polypeptide chain *(37)*. Both strong sequential H^N–H^N and H^α–H^N NOEs, reflecting a random distribution between α and β conformational space, are observed for all residues irrespective of their structure in the native protein *(37)*. Finally, no medium-range H^α–H^N(i,i+3) or (i,i+4) NOEs, characteristic of helical structure, are observed. These three observations suggest that the peaks observed in the HSQC spectra as urea is added correspond to residues in unfolded regions of the polypeptide chain.

The observation of a large number of sequential H^N–H^N NOEs is particularly useful for assignment as the pair of H^N giving rise to these NOEs can be assigned unambiguously in ^{15}N-edited NOESY spectra. If poor ^{1}H^N chemical-shift dispersion is a problem, then assignment can be aided using the H^N–H^N NOEs observed in the ^{15}N-edited HSQC–NOESY–HSQC experiment *(32,38)*;

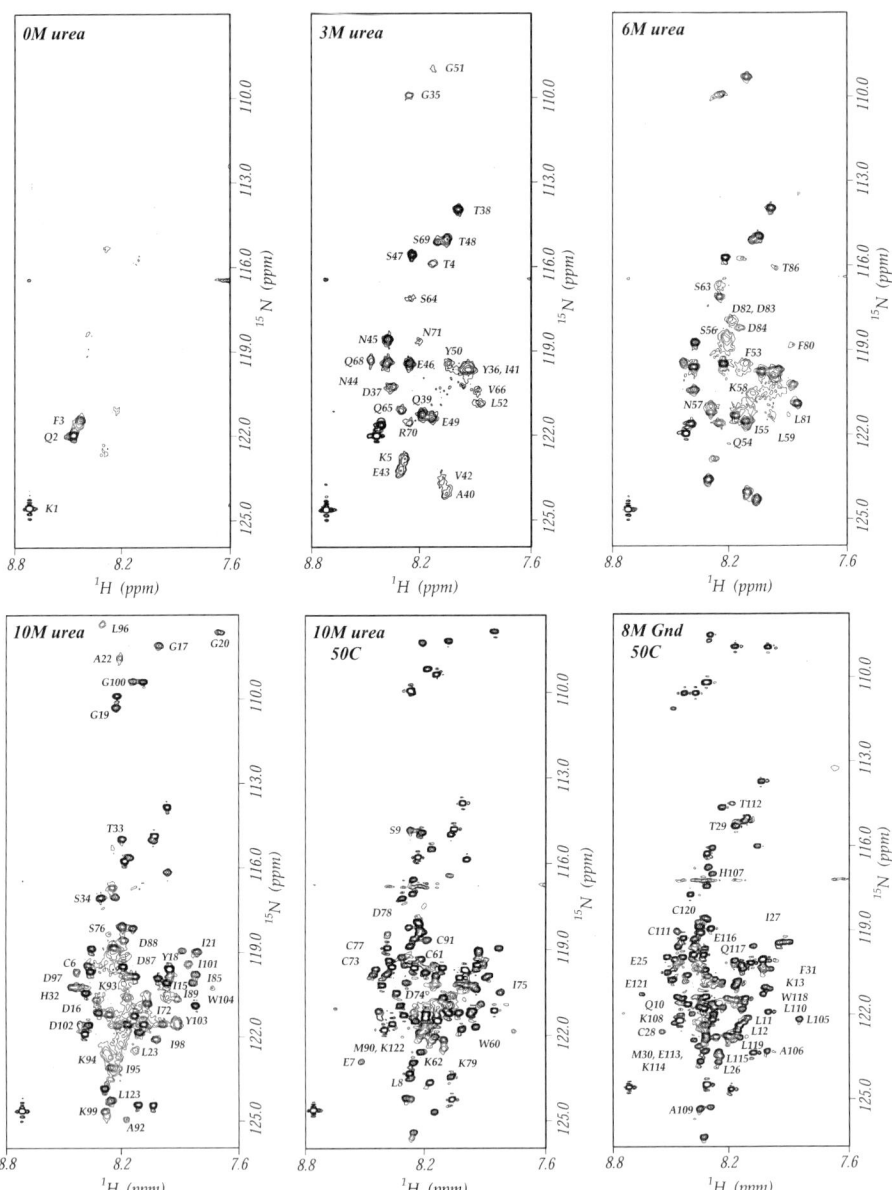

Fig. 6. Progressive appearance of unfolded NMR resonances in HSQC spectra as the human α-LA molten globule, pH 2.0, is unfolded in different concentrations of urea and guanidine HCl. Unless otherwise indicated, spectra were collected at 20°C. Peaks are labeled with their residue assignments as they become visible in the spectra. The peaks corresponding to A22 and L96 are folded in the ^{15}N dimension. (Reprinted from **ref. *17*** with permission from Nature Publishing Group.)

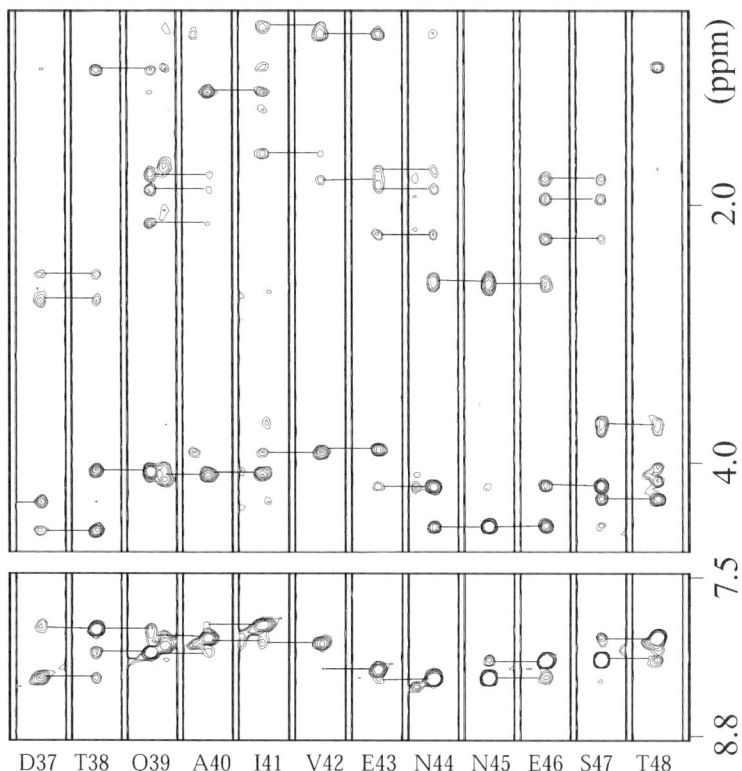

Fig. 7. Strip plot from a ^{15}N-edited 3D NOESY–HSQC spectrum of human α-LA in 5 *M* urea at pH 2.0 and 20°C. Intraresidue and sequential NOEs used for the assignment of residues 37 to 48 are indicated. The chemical shifts observed for H^α and side chain resonances are close to random coil values; this facilitates the identification of amino acid type.

in this spectrum NOEs can be observed between amides with similar $^1H^N$ chemical shifts if they have different ^{15}N shifts. Sequential assignments can be confirmed by the observation of $H^\alpha–H^N(i,i+1)$ and $H^\beta–H^N(i,i+1)$ NOEs *(39)*. Information about residue type can be determined from ^{15}N-edited 3D TOCSY–HSQC spectra *(32,36)* and from the chemical shifts of peaks giving rise to intraresidue and sequential NOEs in the aliphatic region of the 3D NOESY–HSQC spectra.

The unfolding of a native protein to its denatured state generally is a cooperative two-state transition. If this process is monitored by NMR, then the HSQC peaks arising from the native state will decrease in intensity, and a set of HSQC peaks corresponding to the unfolded state will increase in intensity as the concentration of denaturant is increased. All peaks increase or decrease

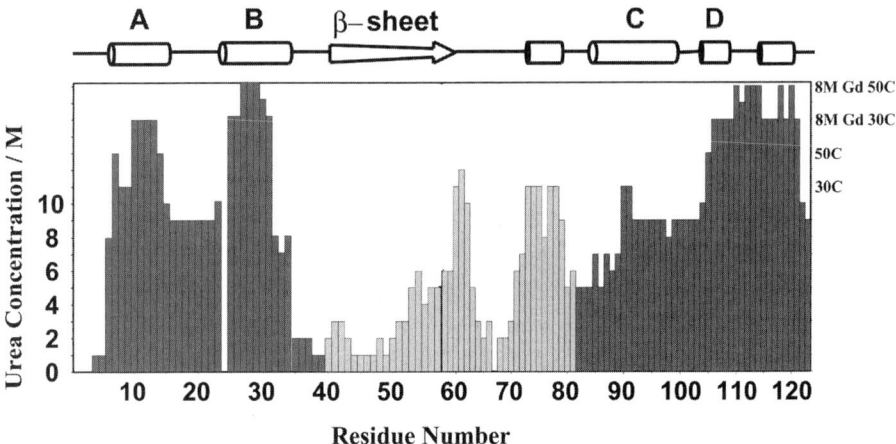

Fig. 8. Unfolding behavior of human α-LA at pH 2.0. The *bars* indicate the urea concentration at which an HSQC peak is first observed. The scale on the right indicates the more destabilizing conditions required to unfold parts of the α-LA molten globule, 10 M urea at 30, 40, and 50°C, and 8 M guanidine HCl at 20, 30, 40, and 50°C. Residues located in the α- and β-domains are shown in *dark and light gray*, respectively. The secondary structure found in native α-LA is summarized at the top of the figure *(12)*.

in intensity in the same way as a result of the cooperativity of the process. This behavior is not observed for the α-LA molten globule at pH 2.0. The stepwise appearance of HSQC peaks with increasing concentrations of the denaturant urea is consistent with noncooperative unfolding. NMR resonances of residues located in compact, folded regions of the low pH molten-globule state are extremely broad because of conformational fluctuations on a millisecond-to-microsecond time-scale. By exposing the partially folded protein to increasingly destabilizing conditions, the progressive unfolding of different regions of the structure can be followed by the appearance of sharp, well-resolved peaks in the HSQC spectrum that correspond to highly dynamic, unfolded parts of the protein. With residue-specific assignments it is possible to analyze the pattern of unfolding observed to gain insights into the stability of different regions of the molten globule (**Fig. 8**). The absence of all but three peaks in 0 M urea indicates that both the α- and β-domains are at least partially collapsed in the α-LA molten globule. Resonances appearing at the lowest concentrations of denaturant correspond to the N-terminal residues and to the β-domain of native α-LA. Previous studies have indicated that the β-domain has a lower propensity for a native fold in the molten globule *(15,16)*. The residues most resistant to denaturant are clustered together in the α-domain of the native structure of α-LA indicating that the molten globule has a highly stable native-like core. Within the α-domain, residues from the C-helix are found to be less resistant to unfolding than residues from the other helices. Residues located in the A-, B-, D-, and C-terminal 3_{10} helices form a stable core that is extremely resistant to unfolding by denaturants *(17)*.

3.5. Direct Multidimensional NMR Studies of Molten-Globule States

An increase in temperature from 20 to 50°C leads to a dramatic sharpening of peaks in the HSQC spectrum of human α-LA at pH 2.0, allowing the direct observation of peaks from the molten-globule state *(40)*. The sharpening of peaks can be attributed to modifications of the complex dynamic properties of the molten globule. A similar observation has been exploited for the detailed NMR characterization of the molten globule of apomyoglobin at pH 4.0 and 50°C *(41,42)*. Information about the location of secondary structure in the apomyoglobin molten globule has been obtained from the analysis of $^{13}C^{\alpha}$, $^{13}C^{\beta}$, and $^{13}C'$ chemical shifts in $^{13}C–^{15}N$-labeled protein. Variations in backbone dynamics of the apomyoglobin molten globule have been identified from $^{1}H–^{15}N$ NOE experiments. The observed sharpening of peaks at higher temperature provides a promising future approach for the direct characterization of molten globules. However, this is complicated by the tendency of many molten globules to aggregate at high temperature at the concentrations required for 2D and 3D NMR. The availability of higher field spectrometers equipped with cryoprobes in the future will improve the sensitivity of NMR experiments and permit the collection of data for more dilute samples that have less tendency to aggregate.

Acknowledgments

The author would like to thank the many people who have been collaborators in the study of the α-lactalbumin molten globule in Oxford; these are C. M. Dobson, L. Greene, J. A. Jones, P. S. Kim, Z.-Y. Peng, T. Pertinhez, C. Quezada, S. Ramboarina, B. A. Schulman, and R. Wijesinha-Bettoni.

References

1. Arai, M. and Kuwajima, K. (2000) Role of the molten globule state in protein folding. *Adv. Protein Chem.* **53,** 209–282.
2. Ptitsyn, O. B. (1995) Molten globule and protein folding. *Adv. Protein Chem.* **47,** 83–229.
3. Ohgushi, M. and Wada, A. (1983) Molten-globule state—a compact form of globular proteins with mobile sidechains. *FEBS Lett.* **164,** 21–24.
4. Dolgikh, D. A., Abaturov, L. V., Brazhnikov, E. V., Lebedev, I. O., Chirgadze, I. N., and Ptitsyn, O. B. (1983) Acid form of carbonic anhydrase—molten globule with a secondary structure. *Dokl. Akad. Nauk., SSSR* **272,** 1481–1484.
5. Arai, M. and Kuwajima, K. (1996) Rapid formation of a molten globule intermediate in refolding of α-lactalbumin. *Fold. Des.* **1,** 275–287.

6. Forge, V., Wijesinha, R. T., Balbach, J., Brew, K., Robinson, C.V., Redfield, C., et al. (1999) Rapid collapse and slow structural reorganisation during the refolding of bovine α-lactalbumin. *J. Mol. Biol.* **288,** 673–688.

7. Arai, M., Ito, K., Inobe, T., Nakao, M., Maki, K., Kamagata, K., et al. (2002) Fast compaction of α-lactalbumin during folding studied by stopped-flow X-ray scattering. *J. Mol. Biol.* **321,** 121–132.

8. Hughson, F. M., Wright, P. E., and Baldwin, R. L. (1990) Structural characterization of a partly folded apomyoglobin intermediate. *Science* **249,** 1544–1548.

9. Jennings, P. A. and Wright, P. E. (1993) Formation of a molten globule intermediate early in the kinetic folding pathway of apomyoglobin. *Science* **262,** 892–896.

10. Acharya, K. R., Ren, J. S., Stuart, D. I., Phillips, D. C., and Fenna, R. E. (1991) Crystal structure of human α-lactalbumin at 1.7 Å resolution. *J. Mol. Biol.* **221,** 571–581.

11. Dolgikh, D. A., Abaturov, L. V., Bolotina, I. A., Brazhnikov, E. V., Bychkova, V. E., Bushuev, V. N., et al. (1985) Compact state of a protein molecule with pronounced small-scale mobility: bovine α-lactalbumin. *Eur. Biophys. J.* **13,** 109–121.

12. Redfield, C., Schulman, B. A., Milhollen, M. A., Kim, P. S., and Dobson, C. M. (1999) α-lactalbumin forms a compact molten globule in the absence of disulfide bonds. *Nat. Struct. Biol.* **6,** 948–952.

13. Baum, J., Dobson, C. M., Evans, P. A., and Hanley, C. (1989) Characterization of a partly folded protein by NMR methods: studies on the molten globule state of guinea pig α-lactalbumin. *Biochemistry* **28,** 7–13.

14. Dolgikh, D. A., Gilmanshin, R. I., Brazhnikov, E. V., Bychkova, V. E., Semisotnov, G. V., Venyaminov, S.Yu., et al. (1981) α-lactalbumin: compact state with fluctuating tertiary structure? *FEBS Lett.* **136,** 311–315.

15. Wu, L. C., Peng, Z.-Y., and Kim, P. S. (1995) Bipartite structure of the α-lactalbumin molten globule. *Nat. Struct. Biol.* **2,** 281–286.

16. Wu, L. C. and Kim, P. S. (1998) A specific hydrophobic core in the α-lactalbumin molten globule. *J. Mol. Biol.* **280,** 175–182.

17. Schulman, B. A., Kim, P. S., Dobson, C. M., and Redfield, C. (1997) A residue-specific NMR view of the non-cooperative unfolding of a molten globule. *Nat. Struct. Biol.* **4,** 630–634.

18. Semisotnov, G. V., Rodionova, N. A., Razgulyaev, O. J., Uversky, V. N., Gripas, A. F., and Gilmanshin, R. I. (1991) Study of the "molten globule" intermediate state in protein folding by a hydrophobic fluorescent probe. *Biopolymers* **31,** 119–128.

19. Berne, B. J. and Pecora, R. (1976) *Dynamic Light Scattering with Applications to Chemistry, Biology and Physics.* Wiley, New York.

20. Lattman, E. E. (1994) Small-angle X-ray scattering studies of protein-folding. *Curr. Opin. Struct. Biol.* **4,** 87–92.

21. Stejskal, E. O. and Tanner, J. E. (1965) Spin diffusion measurements: spin echoes in the presence of a time-dependent field gradient. *J. Chem. Phys.* **42,** 288–292.

22. Gibbs, S. J. and Johnson, C. S., Jr. (1991) A PFG NMR experiment for accurate diffusion and flow studies in the presence of eddy currents. *J. Magn. Reson.* **93,** 395–402.

23. Jones, J. A., Wilkins, D. K., Smith, L. J., and Dobson, C. M. (1997) Characterisation of protein unfolding by NMR diffusion measurements. *J. Biomol. NMR* **10**, 199–203.

24. Wilkins, D. K., Grimshaw, S. B., Receveur, V., Dobson, C. M., Jones, J. A., and Smith, L. J. (1999) Hydrodynamic radii of native and denatured proteins measured by pulse field gradient NMR techniques. *Biochemistry* **38**, 16,424–16,431.

25. Chen, L., Wu, D. H., and Johnson, C. S., Jr. (1995) Determination of the binding isotherm and size of the bovine serum albumin-sodium dodecyl-sulfate complex by diffusion-ordered 2D NMR. *J. Phys. Chem.* **99**, 828–834.

26. Englander, S. W. and Kallenbach, N. R. (1984) Hydrogen exchange and structural dynamics of proteins and nucleic acids. *Q. Rev. Biophys.* **16**, 521–655.

27. Woodward, C. K., Simon, I., and Tuchsen, E. (1982) Hydrogen exchange and the dynamic structure of proteins. *Mol. Cell. Biochem.* **48**, 135–160.

28. Bai, Y., Milne, J. S., Mayne, L., and Englander, S. W. (1993) Primary structure effects on peptide group hydrogen exchange. *Proteins* **17**, 75–86.

29. Wijesinha-Bettoni, R., Dobson, C. M., and Redfield, C. (2001) Comparison of the denaturant-induced unfolding of the bovine and human α-lactalbumin molten globules. *J. Mol. Biol.* **312**, 261–273.

30. Jeng, M. -F., Englander, S. W., Elove, G. A., Wand, A. J., and Roder, H. (1990) Structural description of acid-denatured cytochrome c by hydrogen exchange and 2D NMR. *Biochemistry* **29**, 10,433–10,437.

31. Schulman, B. A., Redfield, C., Peng, Z.-Y., Dobson, C. M., and Kim, P. S. (1995) Different subdomains are most protected from hydrogen exchange in the molten globule and native states of human α-lactalbumin. *J. Mol. Biol.* **253**, 651–657.

32. Kay, L. E., Keifer, P., and Saarinen, T. (1992) Pure absorption gradient enhanced heteronuclear single quantum correlation spectroscopy with improved sensitivity. *J. Am. Chem. Soc.* **114**, 10,663–10,665.

33. Wang, Y. and Shortle, D. (1995) The equilibrium folding pathway of staphylococcal nuclease: indentification of the most stable chain-chain interactions by NMR and CD spectroscopy. *Biochemistry* **34**, 15,895–15,905.

34. Wang, Y. and Shortle, D. (1996) A dynamic bundle of four adjacent hydrophobic segments in the denatured state of staphylococcal nuclease. *Protein Sci.* **5**, 1898–1906.

35. Braun, D., Wider, G., and Wüthrich, K. (1994) Sequence-corrected [15] N "random coil" chemical shifts. *J. Am. Chem. Soc.* **116**, 8466–8469.

36. Marion, D., Driscoll, P. C., Kay, L. E., Wingfield, P. T., Bax, A., Gronenborn, A. M., et al. (1989) Overcoming the overlap problem in the assignment of 1H NMR spectra of larger proteins by use of three-dimensional heteronuclear 1H-15N Hartmann-Hahn multiple quantum coherence and nuclear Overhauser multiple quantum coherence spectroscopy: application to interleukin 1 β. *Biochemistry* **28**, 6150–6156.

37. Fiebig, K. M., Schwalbe, H., Buck, M., Smith, L. J., and Dobson, C. M. (1996) Toward a description of the conformation of denatured states of proteins: comparison of a random coil model with NMR measurements. *J. Phys. Chem.* **100**, 2661–2666.

38. Frenkiel, T., Bauer, C., Carr, M. D., Birdsall, B., and Feeney, J. (1990) HMQC-NOESY-HMQC: a three-dimensional NMR experiment which allows detection of nuclear Overhauser effects between protons with overlapping signals. *J. Magn. Reson.* **90,** 420–425.

39. Wüthrich, K. (1986) *NMR of Proteins and Nucleic Acids.* Wiley, New York.

40. Ramboarina, S. and Redfield, C. (2003) Structural characterisation of the human α-lactalbumin molten globule at high temperature. *J. Mol. Biol.* **330,** 1177–1188.

41. Eliezar, D., Jennings, P. A., Dyson, H. J., and Wright, P. E. (1997) Populating the equilibrium molten globule state of apomyoglobin under suitable conditions for structural characterization by NMR. *FEBS Lett.* **417,** 92–96.

42. Eliezer, D., Yao, J., Dyson, J. H., and Wright, P. E. (1998) Structural and dynamic characterization of partially folded states of apomyoglobin and implications for protein folding. *Nat. Struct. Biol.* **5,** 148–155.

14

Structure Determination of Protein Complexes by NMR

Daniel Nietlispach, Helen R. Mott, Katherine M. Stott, Peter R. Nielsen, Abarna Thiru, and Ernest D. Laue

Summary

This chapter describes nuclear magnetic resonance (NMR) methods that can be used to determine the structures of protein complexes. Many of these techniques are also applicable to other systems (e.g., protein–nucleic acid complexes). In the first section, we discuss methodologies for optimizing the sample conditions for the study of complexes. This is followed by a description of the methods that can be used to map interfaces when a full structure determination of the complex is not appropriate or not possible. We then describe experimental approaches for resonance assignment in complexes, these are essentially the same as those for isolated proteins. **Subheading 6.** describes the different types of so-called X-filtered NMR experiments that have been devised to separate and selectively observe either inter- or intramolecular structural information. These filtered NMR experiments are then exploited in the experimental strategies for structure determination of either protein complexes or homodimeric proteins. This is followed by a description of the calculation of their structures. Finally, we present case studies from three projects carried out in our laboratory, where we successfully used the methods presented in this chapter.

Key Words: NMR spectroscopy; protein structure; protein–protein interactions; homodimer; isotopic labeling.

1. Introduction

As the structures of more proteins and domains are solved in structural genomics projects, the study of macromolecular complexes is becoming the focus of structural biology groups. Studying protein interactions at the molecular level is crucial to the understanding of many biological processes. The study of protein–protein interactions also has important applications for understanding the molecular basis of human disease and for drug design. Nuclear magnetic resonance (NMR) spectroscopy is particularly well-suited to the investigation

From: *Methods in Molecular Biology, vol. 278: Protein NMR Techniques*
Edited by: A. K. Downing © Humana Press Inc., Totowa, NJ

of the interactions between proteins and other molecules at the atomic level in solution. Such interactions may be of variable strength and sample various time-scales. Although it is possible to obtain a high-resolution structure of a complex of up to about 40 kDa by NMR, it is also now feasible to map and study details, for example, of the interface, in much larger complexes. The majority of structural studies of molecular assemblies have been cases where the dissociation constant of the complex is at least submicromolar; a situation referred to as tight or strong binding. Under such conditions, a full structure determination of the complex is feasible. For weaker binding cases, structural information can be obtained for a ligand in the bound state, and the ligand is then modeled onto the surface of the known protein structure.

2. Preliminary Investigations

Before starting data collection, it is necessary to find the optimal set of sample conditions, particularly salt concentration, pH, and temperature. If this is not done at the beginning of the project, and problems are later identified that require a change in conditions, then a great deal of time can be wasted in both reacquiring the data and reassigning the NMR spectra. Protein complexes are generally large, and it is important to identify conditions under which spectra of sufficient signal-to-noise can be obtained. This is particularly the case because some key experiments used to study protein interactions have very low sensitivity, for example, ^{13}C-filtered, ^{13}C-edited nuclear Overhauser effect spectroscopy (NOESY), and doubly rejected experiments.

2.1. Solubility and Stability

One challenge when working with complexes is to find a set of sample conditions in which both the interaction partners and the complex are stable. Methods that screen for conditions that stabilize a protein in solution are discussed elsewhere *(1,2)*. Briefly, efficient screens can be set up to assess the solubility and structural stability of a protein in solution. Small amounts of the protein are transferred to a number of conditions, using either dialysis or hanging drop techniques, and checked for precipitation after 1–2 d. A suggested approach is first to screen for optimal pH and buffer, then change the conditions around the best combination by adding salts and other additives that do not degrade the performance of NMR experiments. Finally, if necessary, the effect of detergents and osmolytes can be investigated.

2.2. Strength of the Interaction

The choice of the experimental NMR approach depends on the strength of the interaction, which is typically measured as the K_d and determined by titration. The population of the complex is followed as a function of the ratio between the interacting partners. This can be done using any one of numerous biophysical techniques. For example, during a titration, the change in fluorescence from a chromophore in either ligand or protein can be fitted to the appropriate model *(3)*. The chromophore may be intrinsic to the molecule(s), such as the amino acid tryptophan, or it may be attached covalently (e.g., through a surface cysteine residue) or noncovalently (e.g., a modified adenosine triphosphate analog). Usually, fluorescence can be followed at concentrations of a few micromolar, making it a suitable technique for studying K_ds in this region. Surface plasmon resonance is increasingly used *(4)*. Typically, one protein is attached to the chip via a His- or glutathione-*S*-transferase tag, and it allows the measurement of both the on- and off-rates. Analytical ultracentrifugation can also be used to monitor complex formation; here detection is by absorption spectroscopy or refractometric methods *(5)*. Isothermal titration calorimetry will give more information on the interaction, such as stoichiometry, ΔH, and ΔS, but it usually requires higher concentrations of protein *(6)*. K_ds can also be measured using NMR by monitoring changes in chemical shift or relative intensity of free and bound forms. It should be noted that the K_d depends critically on the conditions, for example, ionic strength, pH, and temperature.

2.3. Identifying the Interaction Region

If the entire complex is prohibitively large for study by NMR spectroscopy, then it may be possible to replace one component with a much smaller fragment or domain that corresponds to the interacting region. In any case, it is advantageous to define the interaction region to ensure that the minimal fragments of each partner protein are studied without the inclusion of long, unbound, and flexible regions—these lead to artifacts and problems of overlap in NMR spectra. One means of addressing this is to perform limited proteolysis on the complex (reviewed in **ref. 7**). This method uses low levels of a nonspecific protease (e.g., subtilisin) to digest unstructured regions of a protein complex, leaving structured/interacting regions intact. Lower amounts of protease are often needed (e.g., 0.005–0.025% w/w subtilisin) than when delimiting the domain structure of a protein (e.g., 0.5–2.0% w/w subtilisin). Specific proteases, such as trypsin, are useful because it is possible to predict their cleavage sites, facilitating the identification of the resulting peptide fragments by mass spectrometry. However, these sites may not be coincident with the edge of the intermolecular interface. Subtilisin has lower substrate specificity and, therefore, permits a more accurate definition of the interface, although the identification of fragments by mass spectrometry is slightly more difficult.

First, small-scale trials are employed to analyze the extent of proteolysis under varying conditions. Proteolysis of the complex is compared to that of the individual components to identify fragments that are protected in the complex.

Typically, the ratio of protease to complex is varied and a time-course for the reaction is obtained at a constant temperature (between 5 and 30°C) over several hours. The reaction is stopped by adding protease inhibitors and transferring to ice, and the time-course is analyzed by sodium dodecyl sulfate-polyacrylamide gel electrophoresis (SDS-PAGE). The reaction is then repeated on a larger scale under conditions that result in stable protection of the protein fragment(s). The complex may then be purified from unbound peptide fragments using gel filtration chromatography and the protected fragments identified by a combination of mass spectrometry and N-terminal sequencing.

The fragments/domains identified in this manner will be good candidates for an optimized complex. However, it is important to check that the strength of the interaction remains the same as that of the full-length species. As limited proteolysis is a measure of the accessibility of the protease to the polypeptide backbone, it is possible that some residues on either side of the interacting region may be protected from proteolysis solely through their proximity to the binding site. For example, some residues on either side of the HP1 interaction region of chromatin assembly factor 1 (CAF-1) are protected during limited proteolysis of the complex, although ^{15}N relaxation data obtained on CAF-1 in the complex demonstrate that they are unstructured (Thiru et al., unpublished data). A more accurate assessment of the interacting region of a protein or peptide may then be obtained by chemical shift mapping (*see* **Subheading 3.1.**).

2.4. Optimizing the Sample Conditions for the Complex

Chemical or conformational exchange processes can take place on a wide range of time-scales and may affect line shapes, relaxation rates, and chemical shifts of resonances. Depending on the strength of the interaction, NMR signals observed in a complex may be in one of three exchange regimes. Strong binding refers to the situation where exchange occurs more slowly than the change in chemical shift between the free and bound form, whereas weak binding involves faster exchange. The intermediate exchange case, called coalescence, occurs when exchange rates and chemical shift differences between the free and bound species are of comparable magnitude leading to broad or vanishing signals.

The exchange regime can be determined by titrating the unlabeled compound into a solution of ^{15}N-labeled protein and monitoring the chemical shifts in a ^{15}N heteronuclear single-quantum correlation (HSQC) experiment. Under conditions of very fast exchange associated with weak binding, or a small change in chemical shift, a single peak at the weighted average chemical shift is observed, whereas in slow exchange separate signals are observed for the free and bound forms.

Weaker interactions, with K_ds of the order of 20–100 μM or more, typically give rise to signals that show intermediate to fast exchange behavior. In these cases, it is important to optimize the conditions during a titration so that intermediate exchange is avoided. This can be done by increasing the K_d—by changing the chemical composition of the solution, by changing the temperature to influence exchange rates, or by manipulating the relative populations of the different species—by adding the unlabeled species in excess so that the observed species is saturated.

The dependence of the chemical shift changes on the ligand concentration can be used to estimate the dissociation constant of the complex. Assuming that the on-rate of complex formation is diffusion limited, the largest observed shift change will provide an estimate of the lower limit of the dissociation rate constant. For example, if resonances that experience a chemical shift change of <50 Hz are in fast exchange (i.e., only one peak is observed throughout the titration), and the diffusion-limited on-rate is 10^7–10^9 $M^{-1}s^{-1}$, then the lower limit for the $K_d = K_{off}/K_{on}$ is in the range of 50 nM to 5 μM *(8)*.

When looking at strong interactions with low off-rates (typical when the K_d is in the nM range), it should be possible to adjust the sample conditions so that the majority of the resonances in the binding region are in slow exchange. It is advantageous to purify the complex using gel filtration after mixing the two components. This ensures that neither component is present in excess in the NMR sample. However, care needs to be taken that the complex separates during gel filtration from the component in excess. Alternatively, if a small excess of one component is required, it may be added back after gel filtration.

3. Determination of the Binding Interface

There are several situations where it is either inappropriate or not possible to determine a high-resolution structure of a complex. Such situations may include weak binding, high-throughput screening applications, or studies of very large complexes. NMR is uniquely suited to identify changes associated with these types of interaction at the atomic level.

3.1. Chemical Shift Mapping

As mentioned previously, in cases of fast exchange associated with weak binding or a small change in chemical shift, a single peak at the weighted average chemical shift is observed. Because chemical shifts are very sensitive to variations in the local electronic environment, small changes in ^{1}HN and ^{15}N shifts on titration of a weakly binding ligand into a ^{15}N-labeled protein can be used to map the binding interface. In cases where the assignments of the free protein are available, they can be readily transferred to the complex by tracking the changes that occur during the titration. Resonances from residues that do not participate in the interaction do not change chemical shift (except where affected by minor conformational rearrangements), whereas those that shift identify residues that are directly involved in the binding interface. Because of the high sensitivity of the ^{15}N HSQC experiment, these investigations can be done even at micromolar concentrations and are frequently used in drug-discovery-related ligand-binding studies, such as structure–activity relationships (SARs) by NMR and related techniques *(9)*.

The mapping of transient interactions by NMR can be very useful, as the crystallization of such complexes often is impossible. When mapping the interface, many of the indirect effects on chemical shift can be screened out by considering the degree of solvent exposure of resonances that experience changes. In the absence of a major conformational change, residues that are buried are unlikely to be directly contacting the ligand. If the structure of the protein is known, then the solvent accessibility can be calculated using programs such as NACCESS *(10)*. This information will complement the experimental results obtained from H/D exchange experiments in which the reduction of exchange rates indicates protection of amide protons because of exclusion of solvent water in the interface.

In cases of moderately fast exchange, a significant exchange contribution to the signal line width can be present. This effect is likely to be largest for residues in the interface, and so comparison of transverse relaxation rates with average R_2 values in the protein can help to determine the interaction region. Similarly, this applies to situations of moderately slow exchange, where free and bound forms can still be observed separately. This is especially useful when studying the formation of large complexes where, because of low sensitivity, only the free form of the smaller component is often observable.

In a strong binding situation, the majority of the observed resonances in the binding interface are in slow exchange between those in the free and bound forms so that, at the NMR sample concentrations employed, the signals appear at the chemical shifts of the complex.

The stoichiometry of the interaction can be determined during the course of a ^{15}N HSQC titration experiment. In the slow exchange situation, transfer of existing assignments from the free form can be difficult as the interaction may lead to large changes in chemical shift. A complete reassignment may thus be required.

3.2. Transferred Nuclear Overhauser Effects (NOEs)

The conformation of a ligand that is weakly bound to a protein of interest can be determined using the transferred NOE technique *(11,12)*. This allows the detection of intraligand NOEs generated in the bound state by transferring them by chemical exchange to the free ligand, where they are finally observed. In these experiments the ligand usually is present in large excess, improving the sensitivity of the experiment and facilitating the observation of NOE signals.

The resultant NOEs are relatively larger than those generated in the free ligand and have the same sign as the diagonal peaks, as a consequence of the slow molecular tumbling of the protein–ligand complex. A similar principle applies to the observation of effects originating from cross-correlated relaxation mechanisms active on the ligand in the bound state, which are subsequently transferred through chemical exchange onto the free form *(13)*. An advantage of this method is that cross-correlation rates become bigger as the size of the complex increases.

3.3. Saturation Transfer

In the case of slow exchange, the mapping of the interface is more involved. Often, peaks are broadened because of intermediate exchange, resulting in broad peaks for both the free and bound forms, and the determination of the binding interface can be difficult. Here, interaction surfaces can also be mapped using a technique known as cross-saturation, in which ^{1}H resonances on the unlabeled side of the complex are irradiated, and the saturation transferred to spins on the labeled component, which is then detected by, for example, a ^{15}N HSQC. The saturation is transferred by a combination of direct and relayed effects, thus correlating any two spins that are connected by an efficient cross-relaxation pathway. This technique is complementary to chemical shift mapping, and a joint analysis of both often increases the reliability with which the surface can be characterized because chemical shift perturbation may overestimate the extent of the interface. The main requirements for cross-saturation are a window in the spectrum that contains resonances from one component of the complex, only, and that these resonances are in or near to the interface. An alternative strategy is to deuterate one side of the complex to a high level, thus leaving the entire ^{1}H spectral width amenable to irradiation (*see* **Subheading 7.4.**) *(14)*. Deuteration has the added benefit of reducing spin diffusion on the reporter side of the complex.

3.4. Docking Approaches

If the structures of the interacting partners are known and do not change appreciably on complex formation, then chemical shift mapping data for each partner can be used in combination with the complementary shapes of the molecules to model the structure of the complex. For example, the chemical shift changes from ^{15}N HSQC experiments can be used to generate ambiguous intermolecular restraints *(15)*. These are then used in a structure calculation, starting from random orientations of the two molecules, that involves rigid body energy minimization, then semirigid torsion angle dynamical simulated annealing, where the sidechains of the interface residues are allowed to move to optimize packing. Cluster analysis is then performed to find the best model.

A similar approach has been described, which also includes residual dipolar coupling (RDC) data to provide additional information about the relative orientation of the two proteins *(16)*. In addition, a radius of gyration energy term is used to prevent expansion at the interface. If unambiguous intermolecular distance restraints are available, then they can easily be incorporated into either of these two methods.

3.5. Chemical Information at the Active Site

NMR spectroscopy is a particularly useful method for studying changes in chemical bonding at a protein interface. This may have applications in understanding enzyme mechanisms as well as in drug design. Residue-specific isotope labeling allows studies in relatively large proteins and complements the information available from an X-ray structure. Chemical shifts in NMR spectroscopy are very sensitive to the local chemical environment and can be used to monitor and determine the ionization and tautomeric states of protein sidechains and ligands. In addition, water-NOE or water-rotating frame Overhauser effect (ROE) experiments can be used to identify the presence of bound water molecules. Understanding the ionization state of acidic or basic groups at the active site is of crucial importance both for designing ligands that bind with high specificity and for understanding the activity of an enzyme where a titratable group is thought to donate or accept a proton.

To study the ionization and tautomeric state of histidine residues, one can correlate the chemical shifts of the nonexchangeable ring protons and carbons as a function of pH via the two-bond coupling to the ring nitrogens *(17,18)*. For example, the tautomeric states of all eight histidines (three bound to an active-site zinc, three to a second zinc, and two others) have been determined in studies of a matrix metalloprotease complexed with various inhibitors *(19)*. The ionization state of other titratable sidechain residues, such as those of glutamate and aspartate residues, can also be determined. Similar to the histidine sidechain resonances, the ^{13}CO and $^{13}C\beta/^{13}C\gamma$ chemical shifts of glutamic and aspartic acid sidechains are sensitive to the protonation state of the carboxylic acid. A heteronuclear experiment that correlates CO, $C\beta/C\gamma$, and $H\beta/H\gamma$ chemical shifts can thus be used to monitor a pH titration of aspartic/glutamic acids. This method has been used to determine the ionization state of the active-site aspartate in HIV protease *(20)* and the active-site aspartate in thioredoxin in the absence and presence of its target peptide from NFκB *(21,22)*.

Water molecules often play a key role in stabilizing proteins and protein complexes through hydrogen bonding. The water molecules located in the active site of proteins are of particular interest as they may be involved in catalysis or ligand recognition. Bound water molecules can be detected via NOEs between protein resonances and water *(23–25)*. The ^{15}N water–NOE experiments involve only the amide protons and are therefore applicable to both deuterated and nondeuterated proteins, whereas the ^{13}C water–NOE experiment requires the presence of aliphatic protons. The design of selective inhibitors of HIV protease made use of a unique water at the active site that is not found in other aspartyl proteases. A cyclic urea inhibitor was designed to selectively displace this bound water, and studies of bound waters in HIV protease clearly showed that it was displaced by the cyclic urea inhibitor *(26)*.

4. Homo-Oligomers

Structural studies of homo-oligomers require the production of additional samples where the symmetrical subunits of the complex are labeled differently. For homodimers, this generally involves preparation of a sample where a ^{13}C,^{15}N-labeled subunit is bound to unlabeled material. In cases where the monomer–dimer exchange is fast enough, such a sample can be obtained through simple mixing of equivalent amounts of the two appropriate protein components and will result in a statistical distribution of labeled, labeled–unlabeled, and unlabeled dimers (in a 1:2:1 ratio). To establish the efficiency and time-scale of exchange, initial small-scale tests are recommended where the swapping of the subunits is monitored through mixing equivalent amounts of proteins with and without an affinity tag. In cases where the exchange has progressed to equilibrium, purification by an affinity column followed by SDS-PAGE will reveal two bands in the ratio 2:1, with the affinity-tagged band being the more intense. Exchange that is too slow can be accelerated by an increase in temperature. Where exchange is not observed, even after several days, denaturation of the protein, mixing of the unfolded monomers, and subsequent refolding can produce the required sample.

All these procedures lead to a mixture of oligomers with different labeling patterns. To increase the sensitivity of experiments that reveal intermolecular NOEs, it is advantageous to prepare a sample where only one specific oligomer type (labeled–unlabeled) is present. This has the added benefit of removing the need to suppress signals from the other components. Also, for higher order homo-oligomers, the complexity of the problem increases with the number of symmetrical subunits so that the preparation of a sample in which only one of the subunits is labeled proves invaluable—for example, a sample where one protein is ^{13}C,^{15}N-labeled with all of the other units being completely deuterated. This can be achieved via mixing of labeled and unlabeled protein in the appropriate ratio, where the labeled protein is from a substantially longer construct or is attached to an affinity tag. The required mixed oligomer can then be separated by chromatography and the extra residues removed by enzymatic cleavage *(27)*. Provided the subunit exchange is slow enough, a single complex species will be stable for the duration of the NMR measurements. It should be noted that this is only likely to be the case for exchange-stable complexes that have been made by the unfolding/refolding method.

5. Resonance Assignment

For the most part, resonance assignment in the individual components of a complex can progress in a similar manner to the approaches established for single-chain proteins. The strategy employed depends on the size of the molecular complex being investigated. Labeling with ^{15}N and ^{13}C generally is used, and for sensitivity reasons, ^{2}H labeling may become necessary once the molecular weight of the complex exceeds 30 kDa. As discussed later, ^{2}H labeling can also aid the unambiguous separation of inter- from intramolecular distance restraints in homo-oligomeric complexes.

Complexes consisting of different proteins are best investigated by labeling one of the subunits at a time. For a complete assignment of a heterodimer, this will therefore require preparation of two samples, each being used for the assignment of the appropriately ^{13}C,^{15}N-labeled protein from J-correlation experiments. Compared to a single-chain protein of comparable size to the complex, spectral overlap will be less of a problem because fewer peaks per spectrum are present. On the other hand, faster transverse relaxation with larger complexes will cause comparable line width and sensitivity problems because these are predominantly determined by the size and shape of the macromolecular assembly if all the signals are in slow exchange. If some of the observed resonances are not entirely in the slow-exchange regime, their transverse relaxation rates can contain a substantial exchange contribution to the line width, leading to degradation in spectral quality. For complexes of a protein interacting with a peptide, labeling of the latter may not be possible, or only ^{15}N-labeling may be achievable. Under these conditions, more specialized experiments are required, which are described in the next section.

Triple-resonance experiments are recorded to allow sequential assignment of the backbone. Typically these may include HNCO, HNCA, HN(CO)CA, HNCACB, CBCANH, HN(CO)CACB, and CBCA(CO)NH experiments, which are predominantly run as three-dimensional (3D) experiments, recording the chemical shifts of ^{1}HN, ^{13}C, and ^{15}N. Each of these experiments exists in several versions and may include some or all of the following: sensitivity enhancement, water manipulation, residue-type selection, different types of evolution periods, and sequential connectivity selection. Appropriately modified versions are also available to optimize performance when used on fully or fractionally deuterated proteins. Once sequential connectivity of the backbone is established, sidechain correlations can be obtained from HC(CCO)NH, CC(CO)NH, and HB(CB)HA(CA)(CO)NH experiments, which support the traditional assignment procedures relying on 3D HCCH–total correlation spectroscopy (TOCSY) and HCCH–correlational spectroscopy (COSY) experiments. All experiments need to be repeated on each $^{13}C,^{15}N$-labeled sample. The amount of spectrometer time required to solve the structure of a complex therefore is at least twice that for a single species.

A complete assignment of backbone and side-chain resonances is required for subsequent interpretation of NOE contacts as distance restraints. Further information can be incorporated from J-coupling constant-derived torsion angle restraints, angular projection restraints obtained from cross-correlated relaxation rate constants, and RDCs measured in oriented media. Typically, however, 3D ^{13}C- and ^{15}N-separated NOESY experiments still contribute the overwhelming majority of the restraints employed in structure calculations and need to be recorded for each labeled subunit of the complex.

6. Separation of Inter- From Intramolecular Information

In the study of protein complexes, the main complication arises from the two types of NOEs that are observed: those within the subunits and those between the subunits of a complex. Furthermore, where interactions between a protein and a peptide are studied, unambiguous assignments of the latter can only be obtained through experiments that suppress the signals of the protein. Typically, all this information is obtained from isotope-filtered experiments in combination with appropriate labeling strategies, where one component is labeled and the other not. The isotope-filtered NOE experiments are recorded in addition to the conventional 3D ^{13}C- or ^{15}N-separated NOESY experiments, and comparison with the latter allows separation of inter- from intramolecular contacts. The filtered experiments are relatively insensitive, so when the complex is above a certain size, they yield little useful information. In such cases, the situation can be improved by using alternative labeling strategies, such as site-specific protonation in a fully deuterated background. Here, we describe experiments employed for fully protonated protein complexes, in which one partner is uniformly $^{15}N,^{13}C$-labeled and the other is unlabeled.

6.1. X-Filter Experiments

Most NMR spectra are recorded so as to select signals of protons based on the existence or absence of large, one-bond scalar couplings to ^{15}N or ^{13}C. Pulse sequences that remove coherences of those protons attached to ^{15}N or ^{13}C are called *filtered or rejected*, whereas those that select for proton coherences attached to ^{15}N and ^{13}C are called *edited or separated*. Filtering can be achieved with either a low-pass J-filter, a method often called *purging*, or using a spin-echo, also known as a half-filter. Both methods can be accommodated in J-correlation and NOE type experiments—for example, X-filtered or doubly rejected 2D NOESY and TOCSY experiments and can be employed on H_2O samples.

The original idea behind filtering methods comes from the work of Ernst and coworkers who introduced the concept of the low-pass filter to allow the suppression of coherences on the basis of J-coupling evolution *(28)*. Several different implementations of this approach have since been published *(29–31)*. As shown in **Fig. 1A**, in-phase proton coherence evolves during Δ_1 under J-coupling into antiphase coherence with respect to its directly attached ^{13}C or ^{15}N spin. At this point, a 90° purge pulse on the heteronucleus, X, generates unobservable multiple quantum coherence, which is removed by phase cycling of the purge pulse or by application of a pulsed-field gradient. Protons attached to ^{12}C and ^{14}N, on the other hand, are not affected by the isotope-filter. To improve the level of suppression, a second purge pulse can be applied after a further delay, Δ_2, as shown in **Fig. 1A**, without an increase in the length of the filter element. The half-filter element, shown in **Fig. 1B**, is based on a spin-echo period $(2\Delta_1)$ tuned to $1/^1J$ and involves the combination of two separate experiments *(32–38)*. In the first scan, a 180° pulse is applied to spin X (e.g., ^{13}C or ^{15}N) (**Fig. 1B**, *position a*) together with a ^{1}H refocusing pulse, resulting in the overall evolution of ^{1}H attached to ^{13}C by π and thus a sign change. In the subsequent scan, the 180° X pulse is omitted (**Fig. 1B**, *position b*) or applied far off-resonance. The transients are then stored in separate data sets and subtraction of the two experiments results in the selection of ^{1}H attached to X, whereas addition will reveal all the other protons.

High suppression factors are required in these experiments, but both isotope-filtering methods suffer from the effects of 180° pulse imperfections and large variations of $^1J_{C,H}$ (from 120–150 to 160–230 Hz for aliphatic and aromatic CH moieties), making filtered methods less efficient than editing techniques. Although a repeat of the basic filter elements, with the delay tuned to a slightly different value of ^1J, improves the level of suppression, this also significantly reduces the overall sensitivity of the experiment. Therefore, in practice, the application of more than one of the basic filter elements shown in **Fig. 1A,B** is not recommended because it results in substantial signal attenuation resulting from the additional transverse period of approx $1/J$ per frequency dimension. The overall performance of the pulse-sequence elements shown in **Fig. 1A,B** can be improved dramatically through replacement of the square 180° X pulses by adiabatic frequency-swept pulses, with their intrinsically superior *RF* properties. The most important advantage of using adiabatic pulses stems from the fact that, in proteins, the size of ^1J has a linear relationship with the ^{13}C chemical shift, δ. By exploiting this linear dependence, the timing of the adiabatic frequency sweep can be synchronized to generate maximal amounts of antiphase coherence at the end of Δ_1 (**Fig. 1C**) where this coherence is purged. Equally good suppression levels can be obtained for all protons attached to X, independent of the size of their ^1J-coupling constant. The resulting purge scheme is

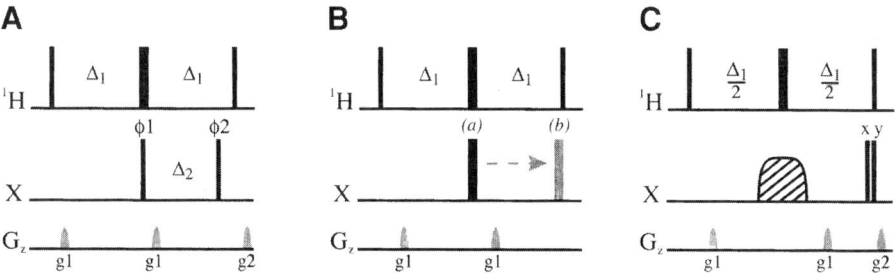

Fig. 1. Pulse scheme elements for filtered experiments: **(A)** Low-pass (or purge) J-filter element that rejects ^{1}H magnetization attached to X *(28,29)*. **(B)** Half-filter module with alternating experiments recorded with the 180° X pulse applied in position *(a)* or *(b)*, respectively, or with an alternating 180° on- and off-resonance pulse applied in position *(a)*. The two obtained data sets are stored separately and are combined as described in the text and in **Table 1**. Alternatively, both scans can be applied with pulses in position *(a)* using a 90×, 90×, and 90×, 90−× composite pulse, respectively *(33)*. **(C)** Improved purge scheme employing an adiabatic sweep 180° pulse to compensate for variations in the size of ^1J *(39)*. The frequency sweep is synchronized according to the ^1J vs δ proportionality.

In all three schemes the delays $^{\Delta}$i are set to 1/2J, whereas ensuring that $^{\Delta}$1 > $^{\Delta}$2 (scheme a) to account for variations in ^1J (e.g.,, 120 and 140 Hz). In scheme (c) $^{\Delta}$1 is adjusted with respect to the smallest $^1J_{C,X}$-coupling present and the corresponding X nucleus is inverted in the center of the sweep to allow complete evolution into antiphase.

depicted in **Fig. 1C**. With the application of two such modules in sequence, suppression factors of 100- to 140-fold can be achieved, and the total transverse time amounts to approx $1/J$. The purge scheme can efficiently be used in a 3D ^{13}C,^{15}N isotope-filtered ^{13}C-separated NOESY experiment *(39)*. This experiment exclusively yields NOEs across the binding interface, from protons in the unlabeled moiety to ^{13}C-attached protons of the ^{13}C,^{15}N-labeled component. The dispersion gained by the presence of a ^{13}C dimension makes this a very powerful experiment that gives excellent levels of artifact suppression. The synchronized adiabatic sweep can also be used in the half-filter experiments to achieve uniform inversion for all ^{1}H–^{13}C spin pairs with different values of ^1J.

In practice, high levels of labeling can be difficult to obtain, limiting the amount of filtering that can be achieved in purge-type experiments. In half-filter experiments, because of the level of ^{13}C natural abundance in the unlabeled moiety, the intramolecular spectrum will always be attenuated by approx 2% in the spectrum where the 180° X pulses are applied. Therefore, this spectrum needs to be scaled by a factor of 1.02 before subtraction *(36)*. In a similar way, using even larger correction factors compensates for incomplete ^{13}C, ^{15}N-labeling, eliminating the intramolecular peaks from the difference spectrum. In the case of a homodimer, in the half-filtered spectrum, with the 180° X pulses applied on-resonance, intermonomer contacts cancel each other out provided the ratio of labeled and unlabeled protein is exactly 1:1. This provides an independent means to identify interfacial contacts.

6.2. J-Resolved Experiments

J-resolved experiments allow the separation of the effects of chemical shift and coupling constant evolution into individual dimensions, as originally proposed by Ernst and coworkers (40). Based on this idea, a J-resolved approach has been suggested in combination with a ^{13}C-separated NOESY, resulting in a four-dimensional (4D) experiment that allows the measurement of inter- and intramolecular NOEs from a single spectrum (41). The separation of inter- and intramolecular peaks relies on modulation with respect to $^{1}J_{C,H}$ and is achieved in a constant-time (CT) experiment. This provides signals with pure phase and allows improvement of resolution using linear prediction or maximum entropy data processing. Based on the original J-resolved approach, we have employed two modified, homonuclear 3D experiments that allow separation of peptide and protein signals according to the values of the $^{1}J_{C,H}$ and $^{1}J_{N,H}$ coupling constants of the starting proton (Nietlispach et al., unpublished data). As with the original experiment, the J resolution is obtained in a CT manner and the ^{1}H chemical shift in *F1* is encoded in a semi-CT fashion. Following the homonuclear mixing period, a water suppression scheme is applied to allow the experiment to be recorded on water samples (42). The general pulse scheme for these new 3D NOESY and TOCSY experiments is shown in **Fig. 2**. The result is a 3D J-resolved NOESY spectrum, where the central, $^{1}J = 0$ plane contains both intrapeptide and peptide-to-protein NOEs. All the intraprotein peaks and intermolecular return peaks from the protein to the peptide appear in planes at the positions of their corresponding ^{1}J values. In the central plane, the intermolecular NOE signals can be separated from intrapeptide peaks because they are coupled in the acquisition dimension and their return cross-peaks can be observed in the appropriate plane in the ^{1}J dimension (*see* **Fig. 3**). Conversely, intrapeptide signals have the return peak in the same $J = 0$ plane. This allows cross checking to discriminate between real peaks and spurious signals that originate from pulse imperfections. The corresponding 3D J-resolved TOCSY experiment uniquely shows intrapeptide correlations in the central plane, whereas the protein signals are dispersed in the third dimension according to the value of the coupling constant of their starting ^{1}H–^{13}C or ^{1}H–^{15}N spin pair.

6.3. Spin Labeling Strategies

Distance restraints can also be derived from the paramagnetic broadening of ^{15}N HSQC resonances by site-specific labeling of selected amino acids using

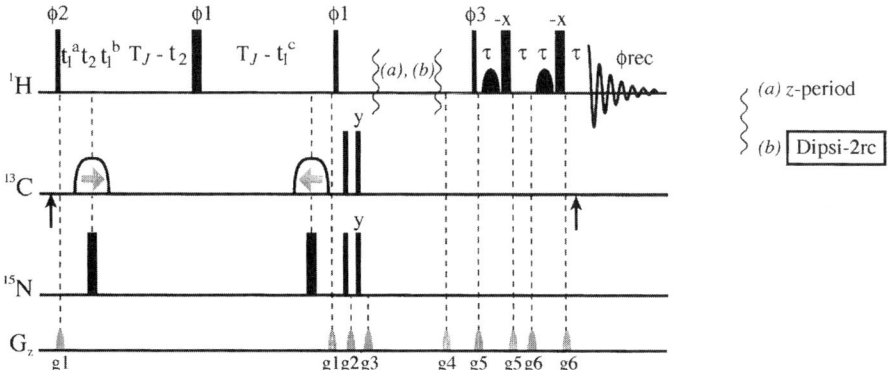

Fig. 2. Pulse schemes for $^{13}C,^{15}N$ J-resolved 3D NOESY or TOCSY experiments (Nietlispach et al., unpublished). The separation according to J (t_2) is obtained in a CT manner, whereas the ^1H shift evolution in t_1 is implemented in a semi-CT fashion (*41,77*). Incoherent (A), or DIPSI-2RC mixing (*78*) (b), follows the return of the magnetization to the z-axis. A DPFGSE (*42*) scheme provides adequate water suppression to allow measurements in H_2O samples.

spin labels. Most commonly, this is achieved by chemical ligation of a paramagnetic probe (usually a nitroxide free radical) to a free cysteine via a thiol ester linkage. R_1 and R_2 relaxation rates are first obtained for the paramagnetic (nitroxide) form, after which the probe is reduced to the diamagnetic (hydroxylamine) form by the addition of two equivalents of solid L-ascorbic acid, and the experiments are repeated. The paramagnetic rate enhancements can be converted into distances (*43–45*). Site-specific labeling of cysteine is only possible if the protein or peptide contains a single free cysteine. Often, this is achieved by site-directed mutagenesis to produce variants, each containing a single cysteine at a defined location. Nitroxide spin labels produce detectable relaxation rate enhancements up to a distance of approx 20 Å, and, therefore, several different mutants will be required to study a large protein (*46–48*). Another use of paramagnetic probes has been described, in which the labeling of one side of a symmetric head-to-head protein dimer causes a breaking of the symmetry, thus enabling intra- and intermolecular NOEs to be distinguished (*49*).

7. Experimental and Labeling Strategies

It is of considerable advantage to record all the homonuclear NMR spectra at the highest possible magnetic field available to obtain the best resolution. The experimental procedures for the doubly rejected pulse sequences are very demanding, particularly when performed on H_2O samples, and require very high spectrometer and temperature stability. Depending on the K_d, the sample conditions may have to be adjusted to provide an excess of the compound that

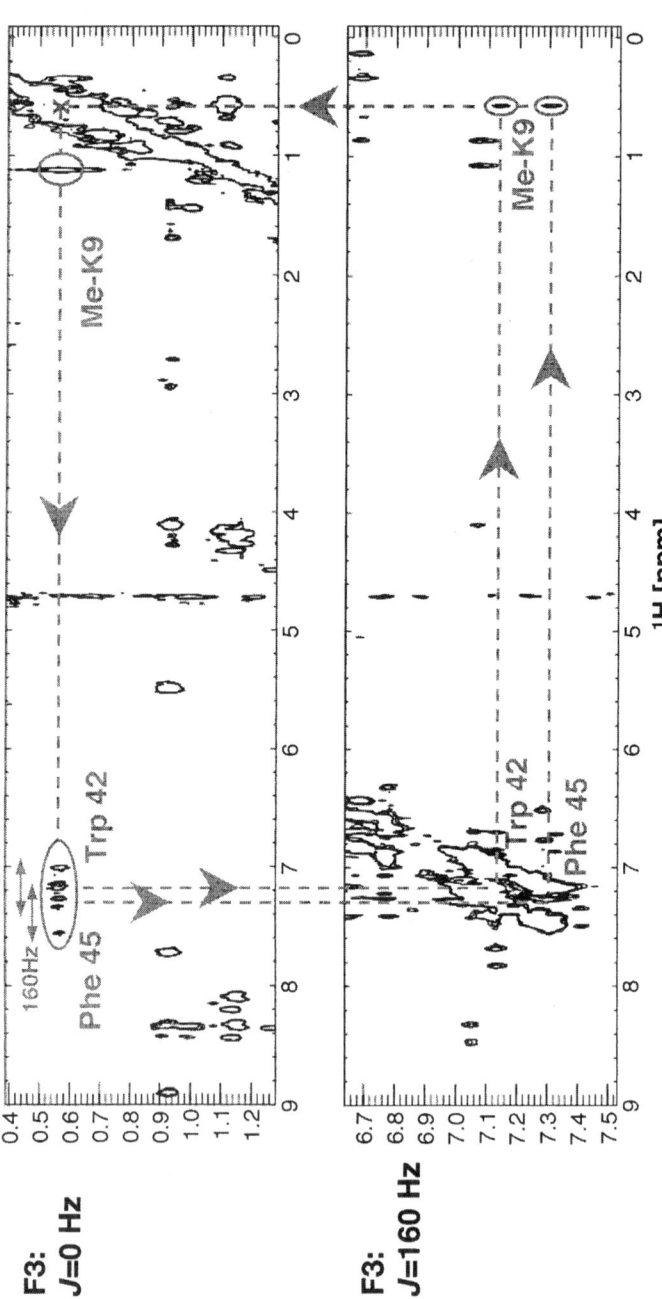

Fig. 3. 2D planes from a 3D J-resolved NOESY experiment, recorded using the sequence of Fig. 2, on a 0.8 m*M* water sample of the ^{13}C, ^{15}N-labeled N-terminal chromodomain of HP1β (**74**) bound to a lysine-methylated histone H3 peptide. The F3 plane at J = 0 Hz contains cross-peaks from one of the methyl groups of methylated Lys-9 to the aromatic residues Phe-45 and Trp-42, which form part of the binding pocket. No decoupling is applied during the acquisition so that the protein peaks appear as doublets in this dimension, facilitating the discrimination of peptide–peptide from peptide–protein cross-peaks. The corresponding return peaks from the aromatic residues of the protein to the methyl group are found in the F3 plane at the value of $^{1}J_{C_{ar},H}/2$. The observation of both related pathways in different planes allows elimination of spurious signals and reduces overlap.

is not being detected by NMR. Thus, to prevent problems associated with an excess of peptide—for example, during the isotope-filtered experiments—the sample conditions may need to be adjusted by the addition of extra protein. This procedure may have to be reversed by adding more peptide when moving on to the protein-detected experiments.

7.1. Protein–Peptide Complexes

In some cases, protein interaction regions are relatively small (10–20 amino acids), and one component of a complex can be substituted by a peptide. Because of expression or stability problems, it may be difficult to express labeled peptides in *Escherichia coli*, and they must be chemically synthesized, resulting in a sample of an unlabeled peptide bound to a ^{13}C,^{15}N-labeled protein. The conventional ^{13}C- and ^{15}N-separated 3D NOESY experiments contain both inter- and intramolecular distance information. Through direct comparison with a 3D ^{13}C,^{15}N isotope-filtered ^{13}C-separated NOESY *(39)* intra- and intermolecular contacts can readily be distinguished. Although sidechain and backbone assignments of the protein can be obtained using the conventional triple-resonance experiments mentioned above, assignments within the peptide moiety have to be obtained from J-resolved 3D or isotope-filtered 2D TOCSY, NOESY, and from double-quantum filtered (DQF)–COSY experiments *(29)*. High levels of ^{13}C- and ^{15}N-labeling in the protein are highly desirable, as any incomplete isotope labeling will lead to the appearance of unwanted protein signals in these spectra. Also, it is advisable to minimize the length of the peptide to reduce the appearance of undesirable spectral features related to the presence of highly flexible residues at the termini (**Subheading 2.3.**).

Because of the nature of the through-bond experiments, only one of the two dimensions needs to be isotope-filtered to uniquely reveal intrapeptide resonances, whereas for the 2D NOESY experiment both dimensions need to be isotope-filtered. Compared to a conventional 2D NOESY experiment, the additional periods where proton coherence is in the transverse plane amount to 2/*J* (14–16 ms), leading to a significant loss in sensitivity for these crucial experiments—this loss occurs regardless of the filtering method employed. In practice, purging approaches seem to produce cleaner spectra than the half-filter experiments *(50)*, which is of importance for observation of signals close to the diagonal. The purging experiments also allow more efficient water suppression, which is important when experiments are being recorded on H$_2$O samples.

For the 180°-based X-half-filtered method, applied to both *F1* and *F2*, each of the filters consists of the pulse element shown in **Fig. 1B** *(33,37,51)*. For a stoichiometric protein–peptide complex with the protein ^{13}C,^{15}N-labeled, four interleaved experiments (A–D) are recorded with simultaneous ^{13}C,^{15}N 180°

Table 1
Acquisition and Processing of 2D ^{13}C,^{15}N [F1,F2] Double Half-Filter NOESY Experiments *(33)*

Experiment	A	B	C	D	Dataset	NOEs observed
F1 180° (^{13}C, ^{15}N)	on	off	on	off		
F2 180° (^{13}C, ^{15}N)	on	off	off	on		
^{1}H-^{13}C/^{15}N to ^{1}H-^{13}C/^{15}N	(+)	(+)	(−)	(−)	1 = A + B − C − D	Intra-protein
^{1}H-^{12}C/^{14}N to ^{1}H-^{12}C/^{14}N	(+)	(+)	(+)	(+)	2 = A + B + C + D	Intra-peptide
^{1}H-^{12}C/^{14}N to ^{1}H-^{13}C/^{15}N	(−)	(+)	(+)	(−)	3 = −A + B + C − D	Peptide to protein
^{1}H-^{13}C/^{15}N to ^{1}H-^{12}C/^{14}N	(−)	(+)	(−)	(+)	4 = −A + B − C + D	Protein to peptide

pulses in *F1* and *F2* applied either on- or off-resonance (on, off), as shown in **Table 1**. The experiments are stored separately and combined, as shown, to reveal different subsets of cross-peaks. For each experiment the signs of the observed cross-peaks are indicated with a (+) or (−). Data set combinations 3 and 4 reveal intermolecular connections, and the diagonal signals are suppressed. For the unique observation of intermolecular NOEs—for example, in the case of a homodimer—recording of experiments A and B is sufficient.

The reduced sensitivity of isotope-filtered experiments tends to limit the size of complexes that can be investigated to <35 kDa. For larger systems, intrapeptide NOEs can be obtained with high sensitivity from a 2D NOESY spectrum recorded on a peptide sample bound to a perdeuterated protein. Using deuteration levels >98% this approach is applicable to complexes >35 kDa. If necessary, the protein amide signals can be removed through purging with a ^{15}N-filter.

7.2. Protein–Protein Complexes

Often, the interaction region is not limited to a small peptide sequence, and in these cases, the complex of two entire proteins or protein domains has to be studied. At least two samples of the complex are required, each of which contains one of the components labeled and the other component unlabeled. The full range of multidimensional experiments are recorded on each sample, enabling complete resonance assignment and measurement of scalar and RDC constants in both proteins. RDCs can prove particularly useful to define the orientation of the two domains relative to each other *(52)*.

Once the backbone resonance assignments are complete, 3D ^{13}C- and ^{15}N-separated NOESY spectra are recorded on each component of the complex. Through comparison of these, intramolecular and intermolecular contacts can be extracted without having to rely entirely on the less sensitive isotope-filtered experiments. As both proteins can be labeled differentially, a larger range of experiments is possible to resolve remaining ambiguities. More recent approaches include using samples where one protein is ^{15}N-labeled with perdeuterated (>98%) sidechains, whereas the other protein is ^{13}C-labeled. These samples can be used to record 3D ^{1}H/^{15}N–^{1}H/^{13}C NOESY experiments, which exclusively contain intermolecular information. Complementary information can be obtained from a 3D NOESY ^{15}N HSQC of a ^{15}N/^{2}H-(sidechain) labeled protein bound to an unlabeled protein *(53)*. This approach has the advantage that the amide protons are detected with high sensitivity because of the deuterated environment, and the magnetization originating from the aliphatic and aromatic protons relaxes more slowly because of the absence of ^{13}C. These spectra have much sharper lines and can be recorded with high resolution as NOESY–[^{1}H,^{15}N] TROSY type experiments *(54)*, making them useful for structural studies of complexes greater than 35 kDa. This ^{15}N-separated approach can offer advantages compared to ^{13}C-edited spectra, although backbone-to-sidechain NOEs are likely to be less abundant than sidechain–sidechain contacts in the interface, unless an intermolecular β-sheet is formed. Deuteration leads to efficient reduction of spin diffusion so that it is possible to detect NOE contacts beyond 5 Å at mixing times up to 250 ms *(55)*. The observation of interamide proton distances of up to 7 Å have been reported in highly deuterated samples with longer mixing times of 0.6 to 1 s *(56)*.

7.3. Symmetric Oligomers

A more specialized case of protein complexes arises in homo-oligomeric protein assemblies. In a symmetric dimer, the distinction between inter- and intramolecular NOE contacts relies primarily on the preparation of a mixed-labeled sample, consisting of equivalent amounts of ^{13}C, ^{15}N-labeled and unlabeled monomers. For some proteins the monomer–dimer equilibrium is sufficiently fast to produce a mixed-labeled sample within a few days, whereas others require harsher conditions, such as high temperatures or mixing under denaturing conditions followed by refolding. Intermolecular NOEs can then be observed with the purging or half-filter experiments described earlier in this chapter. Unfortunately, in symmetric dimers, aliphatic intermolecular side-chain NOE contacts often fall close to the diagonal, and these regions are prone to artifacts resulting from incomplete suppression in the isotope-filtered experiments. By contrast, NOEs between the sidechain protons and the backbone amides can be observed easily in a 3D ^{15}N-separated NOESY recorded on a mixed dimer of ^{2}H/^{15}N and unlabeled subunits because the cross-peaks are far away from the diagonal *(57,58)*.

The higher the number of subunits, the more complicated the problem of differentiation between intramonomer and intermonomer contacts because each intermonomer NOE can potentially consist of several contributions (*see*, for example, **refs.** *59* and *60*). Intramolecular contacts can be obtained by labeling one of the subunits with ^{13}C, whereas all the other components are highly deuterated *(61)*. The NOESY spectra recorded on this sample can be compared with NOESY spectra of the complex with all subunits ^{13}C,^{15}N-labeled to assign the intrasubunit restraints. Nevertheless, some of the distances obtained may also contain contributions from intermonomer interactions, thus reducing the number of fully unambiguous NOEs. Structure calculation methods need to be used that, through the introduction of symmetry restraints, force ambiguous NOE distance restraints to consist of contributions that satisfy the correct symmetry *(62–65)*.

7.4. Large Protein–Protein Complexes

Identification of the binding site in very large complexes can be achieved through a cross-saturation experiment in which one of the components is perdeuterated, combined with a more sensitive TROSY pulse scheme *(14)*. As discussed in **Subheading 3.3.**, selective saturation of the aliphatic resonances of the unlabeled species leads to spin diffusion within the unlabeled protein and eventually transfers to the contacting ^{1}HN of the ^{15}N/^{2}H-labeled protein across the binding interface. The amount of saturation transfer through spin diffusion can be monitored by comparison of the reduced intensities with a reference TROSY spectrum. This more recently developed methodology complements techniques that monitor chemical shift changes on complex formation. In addition, the results are often less ambiguous because the intensity changes are directly related to the true distance of the amide protons from the binding interface. In chemical shift mapping experiments the spectral changes may be a consequence of direct molecular interaction, but they could also be the result of an induced secondary conformational rearrangement.

For molecular complexes >50 kDa, chemical shift mapping methods rely on extensive use of deuteration in combination with experimental methods that benefit from constructive relaxation interference to reduce line widths (TROSY) and improve the coherence transfer efficiency (cross-correlated relaxation-induced polarization transfer and cross-correlated relaxation-enhanced polarization transfer) *(66,67)*. Structures of large protein assemblies whose formation is accompanied by only minor conformational changes of the individual components may be modeled by the characterization of the isolated subunits, which are then subsequently oriented relative to each other based on information obtained from extensive measurements of RDC constants.

8. Structure Calculation of Protein Complexes

At the time of writing, it is not possible to use automated structure calculation routines such as ARIA *(68)* for calculating the structures of protein complexes. Our approach thus far has been either to run ARIA with the input data modified accordingly or to use CNS *(69)* directly. In most cases, the systems being investigated contain numerous ambiguous NOE restraints, so it is imperative to use a structure calculation protocol that can handle such restraints. The structures produced will be improved if the ambiguity of the restraints can be decreased by filtering out the ambiguous possibilities that do not contribute to the overall peak intensity. This methodology is covered elsewhere in the book so will not be dealt with here (*see* Chapters 17 and 18). Structure calculation of complexes using other programs such as DYANA *(70)* will also not be discussed here, although the principles of the methods are the same.

8.1. Geometrical Constraints

When dealing with complexes, it is usually best to assign a separate segment identifier (or chain identifier) to each chain. This has the advantage that the separate chains can then be selected more readily when analyzing the final NOE tables. In the case of a complex made up of two different components with the same residue numbers, it is suggested to number one chain from, for example, 501 rather than 1. The consequence of having the same numbering system for both components is that segment names would have to be used explicitly in the NOE tables. This may not be compatible with the NMR assignment/analysis package that is used to generate the NOE tables.

As with all structure calculations, the protein sequences are used to generate topology files containing lists of bonds, angles, and dihedrals present in the protein chain. These can be generated separately for each chain in the complex, although if using ARIA it is more convenient to put them into the same file, as less editing of the structure calculation scripts is required. In this case, each segment (i.e., chain) is defined separately, from a different sequence file.

In programs such as CNS, it is common to start structure calculations from a template coordinate set, which is an extended chain with ideal geometry. It is important that the two chains do not lie on top of each other at the start of the calculation, otherwise their relative orientation may bias the structure calculation because some ambiguous intermolecular NOEs may be satisfied in these starting structures. In addition, if the two chains are lying on top of each other, the van der Waals energies will be unfavorable from the beginning of the calculation.

8.2. Experimental Restraints

When working with complexes, the NOE tables will be generated from experiments recorded on several samples, with different labeling schemes. It is essential that the referencing of the different data sets is consistent, so that an ambiguous NOE includes the correct assignment in its list of possibilities.

For the most part, the NOE tables generated for structure calculation of complexes are very similar to those used for structure calculation of free proteins. The ambiguous NOE tables must be edited so that the list of possibilities for the destination ^{1}H only includes contributions from the labeled component, whereas the source ^{1}H includes contributions from the labeled and unlabeled species. There will be some cross-peaks in the NOESY ^{13}C HSQC that also appear in the 3D ^{13}C,^{15}N-filtered ^{13}C-separated NOESY spectrum. These should be treated in the same way as the other NOESY ^{13}C HSQC cross-peaks because they may contain intra- as well as intermolecular contributions to the intensity. NOEs involving Ser, Thr, and Tyr −OH and Cys −SH will also appear in the ^{13}C,^{15}N-filtered ^{13}C-separated NOESY spectrum because the ^{1}H are not directly attached to ^{13}C or ^{15}N. Although these ^{1}H are often not observable, these residue types should be carefully checked in the NOESY ^{13}C HSQC and NOESY ^{15}N HSQC experiments for any evidence for cross-peaks arising from these moieties.

The NOE tables generated from the 3D ^{13}C-separated, ^{13}C,^{15}N-filtered NOESY spectrum should only contain intermolecular contributions. As these experiments can contain considerable amounts of breakthrough from flexible residues and also show intramolecular signals because of incomplete labeling of the protein, these peaks must not be included during peak picking. Possible candidates for intermolecular NOEs will lack a return peak in the ^{15}N- or ^{13}C-separated NOESY spectrum and/or can be validated by screening for the return peak in the appropriate plane of the 3D J-resolved NOESY spectrum. Breakthrough originating from intense peaks can be partially identified by comparison with these peaks in the ^{13}C-separated NOESY spectrum. In cases where a F1/F2 ^{13}C-rejected NOESY experiment is used to assign the peptide moiety, both source and destination ^{1}H will be from the peptide (i.e., the unlabeled component). The doubly filtered experiments are also prone to artifacts, so the NOEs in the doubly rejected experiment should be unambiguously assigned as far as possible. In the case of relatively short peptides there are few long-range NOEs, as such peptides are often linearly arranged along the protein, making few intramolecular contacts.

8.3. Structure Calculation

Once the starting structures have been set up correctly and the data is appropriately filtered, the structure calculations are essentially the same as those for single chains. It may be necessary to calculate an initial set of structures where only the unambiguously assigned intermolecular NOE restraints are used. This is required in cases where the ambiguity of the system is too high to allow convergence—for example, if there is a high degree of chemical shift degeneracy and few unambiguous NOEs. If the structure of one of the components is known in advance, using it as a starting structure for the calculation may also circumvent this problem.

9. Structure Calculation of Symmetric Oligomers

The approach to structure calculation of symmetric oligomers is slightly different because these systems must be treated as a complex where the components have identical chemical shifts.

9.1. Distance Restraints

It is almost certain that in structure calculations of any symmetric oligomer the majority of the NOEs will not be assigned to be inter- or intramonomer. The NOE restraints will thus be ambiguous; even if they are assigned to a particular pair of atoms, they cannot be assigned to specific segments. This means that each NOE restraint implicitly has n contributions to its intensity, where n is the number of monomers in the oligomer.

9.2. Symmetry Restraints

To maintain the symmetry between the monomers in the oligomer during the calculation, symmetry restraints must be applied. There are two types of restraints necessary to maintain symmetry; the noncrystallographic symmetry (NCS) restraints, which keep the monomers superimposable, and the distance symmetry restraints, which maintain the symmetry of the oligomers with respect to each other *(62,71,72)*.

The weighting that is used on the NCS restraints may depend on the system that is being studied. In general, it is best to use the smallest NCS weighting that maintains the symmetry. The weighting may vary between 0.1 and 10.0 in our experience, with a weighting of 2.0 usually being sufficient to maintain the symmetry.

The distance symmetry restraints are specified as pairs of distances that should be the same between the sets of monomers, that is, the distance from x in monomer A to y' in monomer B should be the same as that between x' in monomer B and y in monomer A. In the case of a homodimer, the symmetry table is trivial to generate because there is only one possible symmetry group. The residues chosen for the distance comparisons are systematic, and for a homodimer they are as follows: the $A(1)$ to $B(n)$ distance is equivalent to $B(1)$ to $A(n)$; $A(2)$ to $B(n-1)$ is equivalent to $B(2)$ to $A(n-1)$, and so forth, where n is the last residue in the chain. If there are an odd number of residues, then the central residue is excluded. The symmetry tables for higher order oligomers are more complicated and depend on the symmetry group. If this cannot be defined by other means, the structure calculations would have to be run with all possible symmetry groups, and the one that best fits the experimental data would be selected.

9.3. Starting Structures

The starting structures used for homo-oligomer calculations may have to be varied if the symmetry is not known in advance. One possibility is to initiate calculations with all the chains centered on the same origin, that is, they are on top of each other. There may be convergence problems with such an obviously incorrect starting point, and in this case, some experimentation will be required to improve the convergence. The multiple starting structures are generated by duplicating a single set of coordinates that are then assigned names, A, B, C, and so forth. After this, the sets of coordinates can be translated and rotated so that they are aligned head-to-head or head-to-tail, and so on, in the various symmetry groups. For homodimers there is only one possible symmetry group, but the structure calculations of higher order oligomers show better convergence if the starting structures are in the correct symmetry group.

10. Case Studies

10.1. The HP1β Chromodomain in Complex With a Histone H3 Peptide

The structure of the mouse chromodomain protein HP1β complexed with a peptide corresponding to the N-terminal tail of histone H3 methylated at the sidechain position of Lys-9 was determined employing the methods described above. Because of the posttranslational modification of the Lys sidechain, it was not straightforward to make the peptide recombinantly. Therefore, the strategy adopted involved working on samples containing $^{13}C,^{15}N$-labeled chromodomain mixed with unlabeled peptide made by chemical synthesis.

Many of the resonances in the ^{15}N HSQC were in the intermediate-to-fast exchange regime so that the peptide had to be added in a 1.2-fold excess to record satisfactory protein spectra. The protein was assigned using standard heteronuclear correlation experiments recorded on a $^{13}C,^{15}N$-labeled sample. ^{15}N- and $^{13}C,^{15}N$-labeled samples were also used to record 3D ^{13}C- and ^{15}N-separated NOESY and $^{13}C,^{15}N$-filtered, ^{13}C-edited NOESY experiments. The peptide resonances were assigned by analysis of a set of 2D $^{13}C,^{15}N$ doubly rejected DQF COSY, TOCSY, and NOESY spectra. For these filtered experiments, the protein to peptide ratio was readjusted to reduce the appearance of signals from the free ligand. This prevented the observation of peptide chemical shifts at positions averaged according to their distribution between free and bound forms. A sample containing a 1:1 ratio of the two components was used because, in this case, 97% of the peptide was bound. Even so, many observed peptide resonances still contained a significant exchange contribution, reducing the sensitivity of the corresponding X-filtered experiments. Under these circumstances, information from the more sensitive 2D ^{1}H,^{1}H NOESY proved invaluable and was obtained on a sample where the protein was ^{13}C,^{15}N-labeled. As no heteronuclear decoupling was applied, intrachromodomain signals appear as doublets in *F1* and *F2*, intermolecular signals as doublets in *F1* or *F2* (identifying the protein dimension as the one where the splitting occurs), and intrapeptide signals are single peaks.

The methyl groups on the Nε of Lys-9 were not readily assigned from these spectra. In an attempt to identify them, a 3D J-separated NOESY spectrum was recorded (**Subheading 6.2.**). The signals in this spectrum are split in the third dimension according to the value of the one bond ^{1}H–^{13}C and ^{1}H–^{15}N couplings of the starting nucleus. Therefore, the center plane shows intrapeptide and intermolecular correlations, whereas *F1* planes at, for example, 60 Hz, 80 Hz, and 45 Hz show NOE connectivities from aliphatic, aromatic, or amide protons, respectively. A Lys-9 methyl candidate resonance was identified in this spectrum, but the chemical shift was significantly different from that in the free peptide (1.2 vs 2.9 ppm). To confirm the assignment, a sample was prepared where all the NH_2 groups of the peptide (N-terminus, Lys-4, and Lys-9 sidechains) were ^{13}C-methylated by reductive alkylation using ^{13}C formaldehyde followed by borohydride reduction *(73)*. A ^{13}C HSQC and a ^{13}C-edited ^{1}H homonuclear NOESY spectrum confirmed the assignment of the methyl groups on Nε and also showed that only methylated Lys-9 was essential for the peptide interaction, whereas methylated Lys-4 had no effect on binding. **Figure 3** shows the 3D J-resolved NOESY region of the HP1β/histone H3 peptide complex, which helped to define the hydrophobic methylated Lys-9 binding pocket (defined by the aromatic residues Tyr-21, Trp-42, and Phe-45). The intermolecular cross-peaks from the methyl group of the peptide to the aromatic residues of the protein are confirmed by the presence of the return cross-peak in the plane corresponding to the value of the $^1J_{C,H}$-coupling of the phenyl ring. The binding pocket around Lys-9 is shown in **Fig. 4** *(74)*.

10.2. The Cdc42/PAK Complex

When both binding partners can be produced recombinantly, the assignment strategy is less prone to the problems described in **Subheading 10.1.** This was the case for the complex between the small G protein Cdc42 and a fragment of its downstream effector, the p21-activated kinase (PAK). Cdc42 contains two regions (switches I and II) that undergo intermediate chemical or conformational exchange in the free form of the protein and whose resonances only

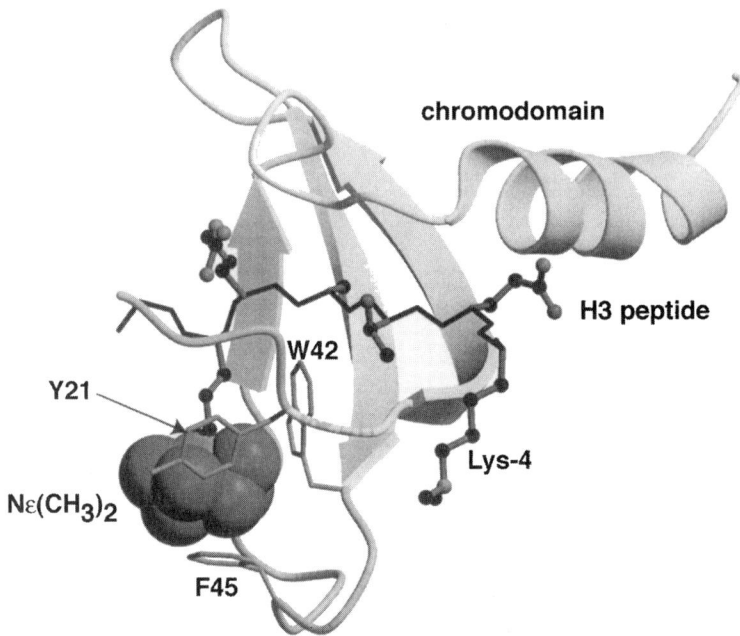

Fig. 4. The structure of the HP1β chromodomain/histone H3 peptide complex. Interaction with the methylated Lys-9 leads to structuring of the *N*-terminal region of the protein and the formation of an intermolecular β-sheet. The Nε (CH$_3$)$_2$ group of the modified Lys-9 is recognized by an aromatic box composed of Tyr-21, Trp-42, and Phe-45.

become visible when bound to the PAK fragment. PAK is largely unstructured in its free form, but it interacts with Cdc42 burying approx 2500 Å^2. Consequently, on binding, dramatic changes were observed in the ^{15}N HSQC of both PAK and Cdc42. The rotational correlation time of the complex was 14 ns, which allowed several backbone triple-resonance experiments to be performed on Cdc42. However, because of the low sensitivity of the HNCACB experiment, most ^{13}C^β resonances were missing. Based only on ^{13}C^α shifts, the sequential assignment of the protein could not be completed. At this point, instead of producing a deuterated sample, which would have allowed the recording of more sensitive spectra, we decided to use the chemical shift information of the previously assigned free Cdc42. Resonances that did not shift in the ^{15}N HSQC on complex formation were used as starting points in the assignment. The PAK resonances could be assigned using HNCA and CBCA(CO)NH experiments because the fragment was only 46 amino acids, and the chemical shifts were well dispersed.

NOEs in the complex were assigned using ^{13}C- and ^{15}N-separated NOESY experiments recorded on four different samples: each of them containing one of the proteins unlabeled and the other either ^{15}N- or ^{13}C,^{15}N-labeled. These samples were only just stable enough to record all of the experiments. Because Cdc42 was larger and required more experiments for the assignment, the ^{13}C,^{15}N-Cdc42/unlabeled PAK sample had degraded by the time all the data necessary for the assignment had been collected. Thus, the ^{13}C,^{15}N-labeled PAK sample was used for the 3D ^{13}C,^{15}N-filtered, ^{13}C-edited NOESY experiment, critical for the assignment of intermolecular NOEs. This had the disadvantage that the larger component in the complex (Cdc42) was only represented by ^{1}H chemical shifts, and the ambiguity level was thus higher. In an ideal situation, with stable samples and ample spectrometer time, this experiment should be recorded on each of the ^{13}C,^{15}N-labeled samples so that a larger number of unambiguous NOEs can be assigned. In the PAK complex, we were able to pick 62 intermolecular NOEs in this experiment, of which approximately half could be assigned. The final ARIA structure calculations used a total of 4000 NOE restraints, of which 173 contained only intermolecular contributions, and the structures had a backbone root-mean-square deviation of 0.94 Å over all nonterminal residues (**Fig. 5**) *(75)*.

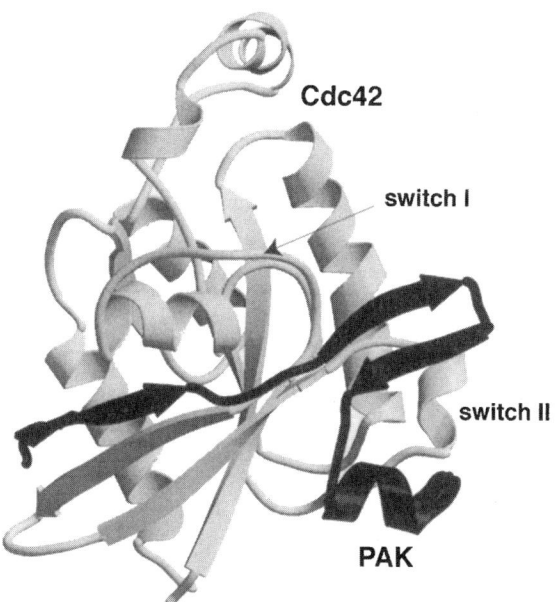

Fig. 5. The structure of the small G protein Cdc42 complexed with a fragment of the p21 activated kinase PAK. The interaction with the effector fixes the two switch regions of Cdc42 in a single conformation so that they become visible in the spectra.

10.3. The HP1β Chromo Shadow Domain in Complex With CAF-1 Peptide

The complex of the HP1β chromo shadow domain (HP1C) and CAF-1 adopts an unusual stoichiometry, in which the homodimeric HP1C binds to a single CAF-1 peptide *(76)*. Consequently, the assignment and structure calculation of the complex resembled the assignment of a heterotrimeric complex but also involved problems commonly associated with studies of homodimers. The availability of recombinant CAF-1 peptide enabled production of samples of labeled HP1C in complex with unlabeled CAF-1 peptide and unlabeled HP1C bound to labeled peptide.

The assignment of NMR spectra recorded on the ^{15}N- or ^{15}N,^{13}C-labeled HP1C in complex with CAF-1 was nontrivial because of the partial loss of chemical shift degeneracy between the two monomers caused by removal of the C_2-symmetry of the free homodimer on peptide binding. For example, on addition of peptide, only approx 25% of the HP1C resonances in the ^{15}N HSQC spectrum of labeled protein remain unaffected, whereas the remainder split into two and shift, to varying extents, from their positions in the spectrum of the free protein. Therefore, it was necessary to assign resonances to a specific monomer as well, as to a given residue in the sequence. For residues near the binding site, the differences in chemical shift between the free and bound protein were so large that the backbone and sidechain assignment was performed *ab initio*, and was equivalent to the assignment of a heterodimer. It was thus possible to assign blocks of sequential residues corresponding to equivalent regions in the two monomers. Resonances that maintained chemical shift degeneracy on peptide binding were assigned to both monomers. Finally, the analysis of long-range NOEs (e.g., those across the β-sheet) enabled the identification of noncontiguous blocks of residues that belonged within the same monomer, permitting monomer assignment.

The intra-HP1C NOEs in the complex could be divided into intramonomer, intermonomer, and comonomer (containing both an intra- and an intermonomer contribution). Symmetry restraints (discussed in **Subheading 9.2.**) were incorporated into the structure calculations to include information about the symmetry of HP1C that was inherent in the NMR spectra. Distance symmetry restraints were placed on the $^{13}C^{\alpha}$ atoms of residues that showed identical chemical shifts between the two monomers. NCS restraints were placed on all $^{13}C^{\alpha}$ atoms; however, the weight on the restraints was varied to reflect the extent to which chemical shift degeneracy was maintained between the two monomers, ranging from 2.0 (complete chemical shift degeneracy between equivalent residues in the two monomers) to 0.1 (both backbone and sidechain nuclei have different chemical shifts in the two monomers). The use of symmetry restraints greatly

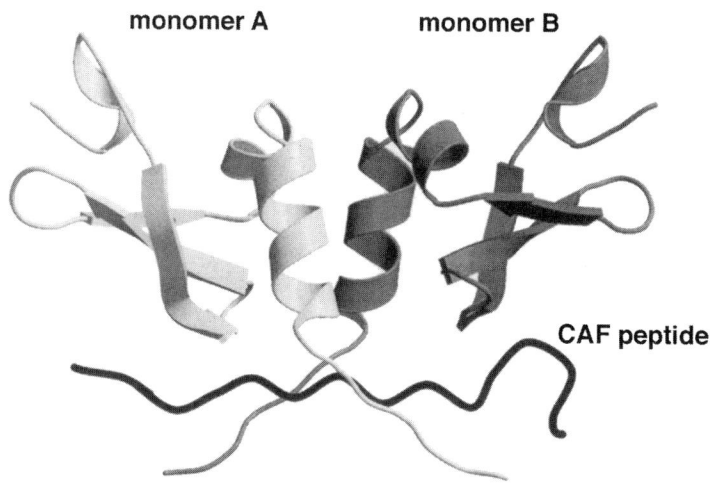

Fig. 6. The structure of the HP1β chromo shadow domain dimer complexed with a 29-residue CAF-1 peptide. The peptide binds between the C-termini of the two monomers, leading to a partial removal of the chemical shift degeneracy.

improved the overall convergence and the definition of the dimer interface. The final ARIA structure calculations used 5000 NOE-derived distance restraints, of which 195 contained only intermolecular contributions. The structures had a backbone RMSD of 0.65 Å over the structured regions of the protein and peptide (**Fig. 6**).

References

1. Lepre, C. A. and Moore, J. M. (1998) Microdrop screening: a rapid method to optimize solvent conditions for NMR spectroscopy of proteins. *J. Biomol. NMR* **12,** 493–499.
2. Bagby, S., Tong, K. I., and Ikura, M. (2001) Optimization of protein solubility and stability for protein nuclear magnetic resonance. *Methods Enzymol.* **339,** 20–41.
3. Eftink, M. R. (1997) Fluorescence methods for studying equilibrium macro-molecule–ligand interactions. *Methods Enzymol.* **278,** 221–257.
4. Lofas, S. and Johnsson, B. (1990) A novel hydrogel matrix on gold surfaces in surface-plasmon resonance sensors for fast and efficient covalent immobilization of ligands. *J. Chem. Soc. Chem. Commun.* **21,** 1526–1528.
5. Laue, T. M. and Stafford, W. F. (1999) Modern applications of analytical ultracentrifugation. *Annu. Rev. Biophys. Biomol. Struct.* **28,** 75–100.
6. Leavitt, S. and Freire, E. (2001) Direct measurement of protein binding energetics by isothermal titration calorimetry. *Curr. Opin. Struct. Biol.* **11,** 560–566.

7. Hubbard, S. J. (1998) The structural aspects of limited proteolysis of native proteins. *Biochim. Biophys. Acta.* **1382,** 191–206.

8. Emerson, S. D., Madison, V. S., Palermo, R. E., Waugh, D. S., Scheffler, J. E., Tsao, K. L., et al. (1995) Solution structure of the Ras-binding domain of C-Raf-1 and identification of its Ras interaction surface. *Biochemistry* **34,** 6911–6918.

9. Shuker, S. B., Hajduk, P. J., Meadows, R. P., and Fesik, S. W. (1996) Discovering high-affinity ligands for proteins: SAR by NMR. *Science* **274,** 1531–1534.

10. Hubbard, S. J. and Thornton, J. M. (1993) *NACCESS.* Department of Biochemistry and Molecular Biology, University College London, London.

11. Clore, G. M. and Gronenborn, A. M. (1982) Theory and applications of the transferred nuclear Overhauser effect to the study of the conformations of small ligands bound to proteins. *J. Magn. Reson.* **48,** 402–417.

12. Clore, G. M. and Gronenborn, A. M. (1983) Theory of the time-dependent transferred nuclear Overhauser effect—applications to structural-analysis of ligand protein complexes in solution. *J. Magn. Reson.* **53,** 423–442.

13. Blommers, M. J. J., Stark, W., Jones, C. E., Head, D., Owen, C. E., and Jahnke, W. (1999) Transferred cross-correlated relaxation complements transferred NOE: structure of an IL-4R-derived peptide bound to STAT-6. *J. Am. Chem. Soc.* **121,** 1949–1953.

14. Takahashi, H., Nakanishi, T., Kami, K., Arata, Y., and Shimada, I. (2000) A novel NMR method for determining the interfaces of large protein–protein complexes. *Nat. Struct. Biol.* **7,** 220–223.

15. Dominguez, C., Boelens, R., and Bonvin, A. (2003) HADDOCK: a protein–protein docking approach based on biochemical or biophysical information. *J. Am. Chem. Soc.* **125,** 1731–1737.

16. Clore, G. M. and Schwieters, C. D. (2003) Docking of protein–protein complexes on the basis of highly ambiguous intermolecular distance restraints derived from H-1(N)/N-15 chemical shift mapping and backbone N-15-H-1 residual dipolar couplings using conjoined rigid body/torsion angle dynamics. *J. Am. Chem. Soc.* **125,** 2902–2912.

17. Markley, J. L. (1975) Observation of histidine residues in proteins by nuclear magnetic resonance spectroscopy. *Acc. Chem. Res.* **8,** 70–80.

18. Pelton, J. G., Torchia, D. A., Meadow, N. D., and Roseman, S. (1993) Tautomeric states of the active-site histidines of phosphorylated and unphosphorylated III(Glc), a signal-transducing protein from *Escherichia-coli*, using 2-dimensional heteronuclear NMR techniques. *Protein Sci.* **2,** 543–558.

19. Gooley, P. R., Johnson, B. A., Marcy, A. I., Cuca, G. C., Salowe, S. P., Hagmann, W. K., et al. (1993) Secondary structure and zinc ligation of human recombinant short-form stromelysin by multidimensional heteronuclear NMR. *Biochemistry* **32,** 13,098–13,108.

20. Yamazaki, T., Nicholson, L. K., Torchia, D. A., Wingfield, P., Stahl, S. J., Kaufman, J. D., et al. (1994) NMR and X-Ray evidence that the HIV protease catalytic aspartyl groups are protonated in the complex formed by the protease

and a nonpeptide cyclic urea-based inhibitor. *J. Am. Chem. Soc.* **116,** 10,791–10,792.

21. Qin, J., Clore, G. M., and Gronenborn, A. M. (1996) Ionization equilibria for side-chain carboxyl groups in oxidized and reduced human thioredoxin and in the complex with its target peptide from the transcription factor NF kappa B. *Biochemistry* **35,** 7–13.

22. Jeng, M. F. and Dyson, H. J. (1996) Direct measurement of the aspartic acid 26 pK(a) for reduced *Escherichia coli* thioredoxin by C-13 NMR. *Biochemistry* **35,** 1–6.

23. Otting, G., Liepinsh, E., and Wuthrich, K. (1991) Protein hydration in aqueous-solution. *Science* **254,** 974–980.

24. Grzesiek, S. and Bax, A. (1993) Measurement of amide proton-exchange rates and NOEs with water in C-13/N-15-enriched calcineurin-B. *J. Biomol. NMR* **3,** 627–638.

25. Clore, G. M., Bax, A., Omichinski, J. G., and Gronenborn, A. M. (1994) Localization of bound water in the solution structure of a complex of the erythroid transcription factor GATA-1 with DNA. *Structure* **2,** 89–94.

26. Grzesiek, S., Bax, A., Nicholson, L. K., Yamazaki, T., Wingfield, P., Stahl, S. J., et al. (1994) NMR evidence for the displacement of a conserved interior water molecule in HIV protease by a nonpeptide cyclic urea-based inhibitor. *J. Am. Chem. Soc.* **116,** 1581, 1582.

27. Jasanoff, A., Wagner, G., and Wiley, D. C. (1998) Structure of a trimeric domain of the MHC class II-associated chaperonin and targeting protein Ii. *EMBO J.* **17,** 6812–6818.

28. Kogler, H., Sorensen, O. W., Bodenhausen, G., and Ernst, R. R. (1983) Low-pass J-filters—suppression of neighbor peaks in heteronuclear relayed correlation spectra. *J. Magn. Reson.* **55,** 157–163.

29. Ikura, M. and Bax, A. (1992) Isotope-filtered 2D NMR of a protein peptide complex—study of a skeletal-muscle myosin light chain kinase fragment bound to calmodulin. *J. Am. Chem. Soc.* **114,** 2433–2440.

30. Lee, W., Revington, M. J., Arrowsmith, C., and Kay, L. E. (1994) A pulsed-field gradient isotope-filtered 3D C-13 HMQC-NOESY experiment for extracting intermolecular NOE contacts in molecular-complexes. *FEBS Lett.* **350,** 87–90.

31. Vuister, G. W., Kim, S. J., Wu, C., and Bax, A. (1994) 2D and 3D NMR-study of phenylalanine residues in proteins by reverse isotopic labeling. *J. Am. Chem. Soc.* **116,** 9206–9210.

32. Otting, G., Senn, H., Wagner, G., and Wuthrich, K. (1986) Editing of 2D H-1-NMR spectra using X half-filters—combined use with residue-selective N-15 labeling of proteins. *J. Magn. Reson.* **70,** 500–505.

33. Otting, G. and Wuthrich, K. (1989) Extended heteronuclear editing of 2D H-1-NMR spectra of isotope-labeled proteins, using the X(omega-1, omega-2) double half filter. *J. Magn. Reson.* **85,** 586–594.

34. Otting, G. and Wuthrich, K. (1990) Heteronuclear filters in 2-dimensional [H-1, H-1] NMR- spectroscopy—combined use with isotope labeling for studies of macromolecular conformation and intermolecular interactions. *Q. Rev. Biophys.* **23,** 39–96.

35. Wider, G., Weber, C., and Wuthrich, K. (1991) Proton proton Overhauser effects of receptor-bound cyclosporine-A observed with the use of a heteronuclear-resolved half-filter experiment. *J. Am. Chem. Soc.* **113,** 4676–4678.

36. Folmer, R. H. A., Hilbers, C. W., Konings, R. N. H., and Hallenga, K. (1995) A C-13 double-filtered NOESY with strongly reduced artifacts and improved sensitivity. *J. Biomol. NMR* **5,** 427–432.

37. Folkers, P. J. M., Folmer, R. H. A., Konings, R. N. H., and Hilbers, C. W. (1993) Overcoming the ambiguity problem encountered in the analysis of nuclear Overhauser magnetic-resonance spectra of symmetrical dimer proteins. *J. Am. Chem. Soc.* **115,** 3798, 3799.

38. Burgering, M., Boelens, R., and Kaptein, R. (1993) Observation of intersubunit NOEs in a dimeric P22 Mnt repressor mutant by a time-shared [N-15,C-13] double half-filter technique. *J. Biomol. NMR* **3,** 709–714.

39. Zwahlen, C., Legault, P., Vincent, S. J. F., Greenblatt, J., Konrat, R., and Kay, L. E. (1997) Methods for measurement of intermolecular NOEs by multinuclear NMR spectroscopy: application to a bacteriophage lambda N-peptide/boxB RNA complex. *J. Am. Chem. Soc.* **119,** 6711–6721.

40. Mueller, L., Kumar, A., and Ernest, R. R. (1975) Two-dimensional carbon-13 NMR spectroscopy. *J. Chem. Phys.* **63,** 5490, 5491.

41. Melacini, G. (2000) Separation of intra- and intermolecular NOEs through simultaneous editing and J-compensated filtering: a 4D quadrature-free constant-time J-resolved approach. *J. Am. Chem. Soc.* **122,** 9735–9738.

42. Hwang, T. L. and Shaka, A. J. (1995) Water suppression that works—excitation sculpting using arbitrary wave-forms and pulsed-field gradients. *J. Magn. Reson. Ser. A* **112,** 275–279.

43. Solomon, I. and Bloembergen, N. (1956) Nuclear magnetic interactions in the HF molecule. *J. Chem. Phys.* **25,** 261–266.

44. Krugh, T. R. (1976) in *Spin Labeling: Theory and Applications* (Berliner, L. J., ed.). Academic Press, New York, pp. 339–372.

45. Kosen, P. A. (1989) Spin labeling of proteins. *Methods Enzymol.* **177,** 86–121.

46. Gillespie, J. R. and Shortle, D. (1997) Characterization of long-range structure in the denatured state of staphylococcal nuclease, 1: paramagnetic relaxation enhancement by nitroxide spin labels. *J. Mol. Biol.* **268,** 158–169.

47. Gillespie, J. R. and Shortle, D. (1997) Characterization of long-range structure in the denatured state of staphylococcal nuclease ,2: distance restraints from paramagnetic relaxation and calculation of an ensemble of structures. *J. Mol. Biol.* **268,** 170–184.

48. Battiste, J. L. and Wagner, G. (2000) Utilization of site-directed spin labeling and high-resolution heteronuclear nuclear magnetic resonance for global fold determination of large proteins with limited nuclear Overhauser effect data. *Biochemistry* **39,** 5355–5365.

49. Gaponenko, V., Altieri, A. S., Li, J., and Byrd, R. A. (2002) Breaking symmetry in the structure determination of (large) symmetric protein dimers. *J. Biomol. NMR* **24,** 143–148.

50. Ogura, K., Terasawa, H., and Inagaki, F. (1996) An improved double-tuned and iso-tope-filtered pulse scheme based on a pulsed field gradient and a wide-band inversion shaped pulse. *J. Biomol. NMR* **8,** 492–498.

51. Wider, G., Weber, C., Traber, R., Widmer, H., and Wuthrich, K. (1990) Use of a double-half-filter in 2-dimensional 1H nuclear-magnetic-resonance studies of receptor-bound cyclosporine. *J. Am. Chem. Soc.* **112,** 9015, 9016.

52. Skrynnikov, N. R., Goto, N. K., Yang, D. W., Choy, W. Y., Tolman, J. R., Mueller, G.A., et al. (2000) Orienting domains in proteins using dipolar couplings measured by liquid-state NMR: differences in solution and crystal forms of maltodextrin binding protein loaded with beta-cyclodextrin. *J. Mol. Biol.* **295,** 1265–1273.

53. Walters, K. J., Matsuo, H., and Wagner, G. (1997) A simple method to distinguish intermonomer nuclear Overhauser effects in homodimeric proteins with C-2 symmetry. *J. Am. Chem. Soc.* **119,** 5958, 5959.

54. Xia, Y. L., Sze, K. H., and Zhu, G. (2000) Transverse relaxation optimized 3D and 4D N-15/N-15 separated NOESY experiments of N-15 labeled proteins. *J. Biomol. NMR* **18,** 261–268.

55. Venters, R. A., Metzler, W. J., Spicer, L. D., Mueller, L., and Farmer, B. T. (1995) Use of H-1(N)-H-1(N) NOEs to determine protein global folds in perdeuterated proteins. *J. Am. Chem. Soc.* **117,** 9592, 9593.

56. Mal, T. K., Matthews, S. J., Kovacs, H., Campbell, I. D., and Boyd, J. (1998) Some NMR experiments and a structure determination employing a {N-15,H-2} enriched protein. *J. Biomol. NMR* **12,** 259–276.

57. Walters, K. J., Dayie, K. T., Reece, R. J., Ptashne, M., and Wagner, G. (1997) Structure and mobility of the PUT3 dimer. *Nat. Struct. Biol.* **4,** 744–750.

58. Ferentz, A. E., Opperman, T., Walker, G. C., and Wagner, G. (1997) Dimerization of the UmuD' protein in solution and its implications for regulation of SOS muta-genesis. *Nat. Struct. Biol.* **4,** 979–983.

59. Caffrey, M., Cai, M. L., Kaufman, J., Stahl, S. J., Wingfield, P. T., Covell, D. G., et al. (1998) Three-dimensional solution structure of the 44 kDa ectodomain of SIV gp41. *EMBO J.* **17,** 4572–4584.

60. Nooren, I. M. A., Kaptein, R., Sauer, R. T., and Boelens, R. (1999) The tetrameri-zation domain of the Mnt repressor consists of two right-handed coiled coils. *Nat. Struct. Biol.* **6,** 755–759.

61. Jasanoff, A. (1998) An asymmetric deuterium labeling strategy to identify inter-protomer and intraprotomer NOEs in oligomeric proteins. *J. Biomol. NMR* **12,** 299–306.

62. Nilges, M. (1993) A calculation strategy for the structure determination of sym-metrical dimers by H-1-NMR. *Proteins* **17,** 297–309.

63. O'Donoghue, S. I., Junius, F. K., and King, G. F. (1993) Determination of the structure of symmetrical coiled-coil proteins from NMR dta—application of the leucine-zipper proteins Jun and Gcn4. *Protein Eng.* **6,** 557–564.

64. O'Donoghue, S. I., Chang, X. Q., Abseher, R., Nilges, M., and Led, J. J. (2000) Unraveling the symmetry ambiguity in a hexamer: calculation of the R-6 human insulin structure. *J. Biomol. NMR* **16,** 93–108.

65. Nilges, M. (1995) Calculation of protein structures with ambiguous distance restraints–automated assignment of ambiguous NOE crosspeaks and disulfide connectivities. *J. Mol. Biol.* **245,** 645–660.

66. Pervushin, K., Riek, R., Wider, G., and Wuthrich, K. (1997) Attenuated T-2 relaxation by mutual cancellation of dipole- dipole coupling and chemical shift anisotropy indicates an avenue to NMR structures of very large biological macromolecules in solution. *Proc. Natl. Acad. Sci. USA* **94,** 12,366–12,371.

67. Riek, R., Wider, G., Pervushin, K., and Wuthrich, K. (1999) Polarization transfer by cross-correlated relaxation in solution NMR with very large molecules. *Proc. Natl. Acad. Sci. USA* **96,** 4918–4923.

68. Linge, J. P., O'Donoghue, S. I., and Nilges, M. (2001) Automated assignment of ambiguous NOEs with ARIA. **339,** 71–90.

69. Brunger, A. T., Adams, P. D., Clore, G. M., DeLano, W. L., Gros, P., Grosse-Kunstleve, R.W., et al. (1998) Crystallography & NMR system: a new software suite for macromolecular structure determination. *Acta Crystallogr. D Biol. Crystallogr.* **54,** 905–921.

70. Herrmann, T., Guntert, P., and Wuthrich, K. (2002) Protein NMR structure determination with automated NOE assignment using the new software CANDID and the torsion angle dynamics algorithm DYANA. *J. Mol. Biol.* **319,** 209–227.

71. O'Donoghue, S. I., King, G. F., and Nilges, M. (1996) Calculation of symmetric multimer structures from NMR data using a priori knowledge of the monomer structure, co-monomer restraints, and interface mapping: the case of leucine zippers. *J. Biomol. NMR* **8,** 193–206.

72. Nilges, M. and O'Donoghue, S. I. (1998) Ambiguous NOEs and automated NOE assignment. *Prog. Nucl. Magn. Reson. Spectrosc.* **32,** 107–139.

73. Means, G. E. and Feeney, R. E. (1995) Reductive alkylation of proteins. *Anal. Biochem.* **224,** 1–16.

74. Nielsen, P. R., Nietlispach, D., Mott, H. R., Callaghan, J., Bannister, A., Kouzarides, et al. (2002) Structure of the HP1 chromodomain bound to histone H3 methylated at lysine 9. *Nature* **416,** 103–107.

75. Morreale, A., Venkatesan, M., Mott, H. R., Owen, D., Nietlispach, D., Lowe, P. N., et al. (2000) Structure of Cdc42 bound to the GTPase binding domain of PAK. *Nat. Struct. Biol.* **7,** 384–388.

76. Brasher, S. V., Smith, B. O., Fogh, R. H., Nietlispach, D., Thiru, A., Nielsen, P. R., et al. (2000) The structure of mouse HP1 suggests a unique mode of single peptide recognition by the shadow chrome domain dimer. *EMBO J.* **19,** 1587–1597.

77. Grzesiek, S. and Bax, A. (1993) Amino-acid type determination in the sequential assignment procedure of uniformly C-13/N-15-enriched proteins. *J. Biomol. NMR* **3,** 185–204.

78. Hwang, T. L., Kadkhodaei, M., Mohebbi, A., and Shaka, A. J. (1992) Coherent and incoherent magnetization transfer in the rotating frame. *Magn. Reson. Chem.* **30,** S24–S34.

15

NMR Studies of Protein–Nucleic Acid Interactions

Gabriele Varani, Yu Chen, and Thomas C. Leeper

Summary

Protein–DNA and protein–RNA complexes play key functional roles in every living organism. Therefore, the elucidation of their structure and dynamics is an important goal of structural and molecular biology. Nuclear magnetic resonance (NMR) studies of protein and nucleic acid complexes have common features with studies of protein–protein complexes: the interaction surfaces between the molecules must be carefully delineated, the relative orientation of the two species needs to be accurately and precisely determined, and close intermolecular contacts defined by nuclear Overhauser effects (NOEs) must be obtained. However, differences in NMR properties (e.g., chemical shifts) and biosynthetic pathways for sample productions generate important differences. Chemical shift differences between the protein and nucleic acid resonances can aid the NMR structure determination process; however, the relatively limited dispersion of the RNA ribose resonances makes the process of assigning intermolecular NOEs more difficult. The analysis of the resulting structures requires computational tools unique to nucleic acid interactions. This chapter summarizes the most important elements of the structure determination by NMR of protein–nucleic acid complexes and their analysis. The main emphasis is on recent developments (e.g., residual dipolar couplings and new Web-based analysis tools) that have facilitated NMR studies of these complexes and expanded the type of biological problems to which NMR techniques of structural elucidation can now be applied.

Key Words: RNA; DNA; proteins; nucleic acids; structure determination; structure analysis; residual dipolar couplings.

1. Introduction

Proteins and nucleic acids function as part of complexes with other proteins, nucleic acids or carbohydrates. In the past, structure determination of proteins and nucleic acids was central to structural biology; it is now the structure of their complexes, and of multimolecular assemblies, which is progressively taking center stage. The opportunity to conduct nuclear magnetic resonance

From: *Methods in Molecular Biology, vol. 278: Protein NMR Techniques*
Edited by: A. K. Downing © Humana Press Inc., Totowa, NJ

(NMR) investigations of multimolecular assemblies is beginning to be explored through the use of transverse relaxation-optimized spectroscopy-based techniques that allow ever larger complexes to be studied; however, this area of structural biology is likely to remain dominated by the application of electron microscopy and X-ray crystallography. In contrast, NMR has an essential role in studying the structure and dynamics of protein–nucleic acids and protein–protein complexes. Here, we present NMR methods pertinent to the study of protein–RNA and protein–DNA complexes. The emphasis is on aspects of the technology that still represent significant critical points in the structure-determination process.

This contribution is organized following the common path of a structure-determination project: sample preparation, mapping interaction surfaces, collection of constraints, structure calculation, analysis. RNA preparation is still a somewhat specialized task discussed in the first section. New methods to map interaction surfaces and obtain qualitative information on protein–nucleic acid complexes are presented second; these experiments represent not only important steps on the way to a complete structure determination but also valuable sources of biological insight when structures cannot be obtained because of the size of the complex or the dynamics of the interface. The sections on data collection and structure calculation review methods specific to the study of protein–nucleic acid complexes. Computational tools to analyze protein–nucleic acid complexes are presented last.

2. RNA Synthesis and Preparation

Two strategies are available for preparing large quantities of RNA of defined sequence and high purity for NMR studies: in vitro enzymatic transcription using RNA polymerases *(1–3)* and phosphoramidite-based chemical synthesis. In vitro enzymatic synthesis is much more efficient and cost effective, and most RNA molecules used in NMR structural studies have been prepared by this method using synthetic DNA templates or linearized plasmids. However, enzymatic synthesis has some serious drawbacks. First of all, certain sequences transcribe more efficiently than others, and some RNA sequences simply cannot be prepared at all. Furthermore, functionalities within RNA are often modified in living organisms; modified nucleotides can only be incorporated by chemical synthesis *(4,5)*. Finally, the incorporation of isotopic labels at specific residues is very problematic. The synthetic approach is more flexible, but its applications are limited by high cost and the difficult synthesis of large RNAs.

2.1. Enzymatic RNA Synthesis Using Bacteriophage RNA Polymerases

Milligram amounts of RNA are produced with transcription volumes of 5–10 mL using T7 RNA polymerase. Although transcription off linearized plasmid

DNA generally provides improved yields and increased flexibility in the choice of RNA sequence, the use of synthetic DNA templates is remarkably simple and robust. Template synthesis on a 0.1 μM scale easily produces enough material to prepare tens of milligrams of RNA. The amount of RNA containing additional nucleotides beyond the desired sequence can be reduced significantly by incorporating 2′-O-CH$_3$ in the two nucleotides preceding the end of the template *(6)*. If increased homogeneity is required, cotranscription of the desired RNA with flanking sequences that fold into ribozymes *(3,7)* is possible. For some transcripts, however, self-cleavage at ribozyme sites is inhibited by the formation of stable secondary structures.

The essential components for in vitro transcription using phage polymerases are monovalent and divalent ions (K$^+$ and Mg^{2+}), dithiothreitol to stabilize the enzyme, detergents (e.g., Triton X-100), polyethylene glycol, pyrophosphatase (this enzyme is used to hydrolyze pyrophosphates that inhibit the polymerase reaction, but it is not strictly required), polycations (spermidine) and the substrate nucleotide-triphosphates (NTPs), in addition to the polymerase itself and the DNA template. T7 RNA polymerase can be obtained commercially but is expensive (Epicentre, Madison, WI: 10,000 U in 50 μL, $110; New England Biolabs, Beverly, MA: 10,000 U in 200 μL, $110; Fermentas, Hanover, MD: 10,000 U in 100 μL, $200; Promega, Madison, WI: 10,000 U in 125 μL, $265). The polymerase can be overexpressed in *Escherichia coli* and purified to very high levels of homogeneity. All other materials are available commercially at small cost, except uridine 5′-triphosphate, which can be expensive. The optimal concentration of each component and optimal reaction time (as a rule of thumb, 4 h is usually a suitable period for transcription) are both template- and polymerase-batch dependent. It is advisable to carry out a series of transcription reactions on relatively small scales (20 μL typically) to optimize reaction conditions before moving on to large-scale synthesis. Strategies for optimizing other components in the reaction mixture have been discussed thoroughly in the literature *(2)*. Reaction time should be optimized separately for each template and polymerase batch.

Besides the quality of the polymerase, the transcriptional efficiency is determined primarily by the choice of the template sequence. Although no general rules exist for predicting which sequence will transcribe well, the polymerases require a G at the 5′ end of the transcript and the best templates for transcription include purine-rich sequences near the 5′ end, such as GGG, GAG, or GGA *(1,2)*. The template DNA is composed of an 18-basepair promoter region whose sequence is critical for recognition by the polymerase and therefore for efficient synthesis. The RNA sequence is encoded by a complementary template region that can be either single or double stranded; this choice can affect the transcription efficiency, and a comparison during the optimization stage is suggested if yields are unsatisfactory. If a single-stranded DNA template is used, then the primer corresponding to the promoter region is common to all RNAs. In many cases, transcription off a linearized plasmid or supercoiled template will improve the RNA yields, but at the expense of the efforts required to clone and purify the plasmid DNA templates.

2.2. RNA Synthesis by Phosphoramidite Chemistry

The application of phosphoramidite chemistry to RNA synthesis is more difficult than for DNA because of the requirement to protect the 2'-hydroxyl functionality during chain elongation to prevent 5'-2' linkage from occurring in place of the correct 5'-3' connection. The presence of a bulky group alongside the 3'-hydroxyl reactive functionality leads to reduced coupling efficiency and much slower reaction rates than for the DNA case, and deprotection of the 2'-hydroxyl functionality after synthesis requires an additional step. The introduction of new protecting groups *(8)* and activating agents in the coupling reaction has decreased coupling time *(9,10)* and increased yields *(11)*. There are now very good commercial sources for RNA; for example, Dharmacon, which introduced a new protecting group for further increased yields and ease of decoupling *(12,13)*. It is now both time and cost effective, with the exception of specialized applications, to simply purchase the RNA from companies that specialize in RNA synthesis.

2.3. Isotopic Labeling

Isotopic labeling with ^{13}C resolves overlapped resonances in multidimensional spectra—a more severe problem for RNA compared to DNA or proteins of comparable size because the sugar proton resonances are spread over a very narrow chemical shift range (but the corresponding carbons are dispersed). In addition, spectra of complexes can be edited, based on the presence or absence of isotopic labels in either component, to allow their selective observation in the complex and the unambiguous identification of intermolecular nuclear Overhauser effect (NOE) interactions. Finally, new constraints for structure determination (heteronuclear scalar coupling constants and residual dipolar couplings [RDCs]) can be extracted from ^{13}C-edited multidimensional experiments. Because labeled nucleotides can now be purchased at a convenient price (e.g., Silantes, Cambridge Isotope Labs, and Spectra Gases; approx \$100–200 per 5 mg of ^{13}C/^{15}N NTPs; this corresponds to \$50/\$100 per milliliter of transcription), extraction of ribosomal RNA from bacterial preparations is not needed anymore. Therefore, preparation of labeled RNA is analogous to the preparation of unlabeled material, although some purification of the NTPs (e.g., ethanol precipitation, sometimes some form of chromatographic separation) is generally required, in our experience, to obtain good transcriptional yield.

The T7 RNA polymerase expression system provides two straightforward possibilities for isotopic labeling *(14,15)*: uniform enrichment at either a high (>98%) or dilute (30%) level and selective labeling by residue type. In addition, by using nucleotide triphosphate substrates within the transcription mixture that contain isotopic labels at one or more specific positions within each nucleotide, it is possible to produce RNA uniformly labeled at that position for one or all four bases *(16)*. Such samples could be especially useful for NMR relaxation studies of motion in RNA complexes because ^{13}C–^{13}C couplings produce alternate relaxation pathways that complicate the analysis of the results.

2.4. RNA Purification

RNA purification can best be performed by polyacrylamide electrophoresis on denaturing gels (**Fig. 1**), which is the method of choice to separate full-length fragments from incorrectly terminated RNA products. When short oligonucleotides are transcribed, the polymerase often fails to terminate transcription properly. Most aborted transcripts are very short (2–10 nucleotides) and can be easily separated from full-length products. However, transcripts are also produced with 1 or 2 additional nucleotides at the 3'-end and single nucleotide resolution is required to separate transcripts of the correct 'N' length from aborts ('N − 1') or add-ons ('N + 1', 'N + 2'). Nucleotide resolution can only be achieved when the gels are not too heavily loaded; consequently, several gels are required to avoid overloading of the matrix. When RNAs are prepared with self-cleaving ribozymes *(7)*, very clean 3' ends are produced lacking either add-ons or shorter transcripts, making RNA purification straightforward. The desired RNA product has only to be separated from the self-cleaved ribozyme sequences, that are generally of very different length from the product and that can therefore be separated from the desired RNA without difficulties. After electrophoresis, the RNA band is excised from the gel and electroeluted from the gel strips (**Fig. 1**). Dialysis is recommended to remove impurities. Typically, dialysis is executed at varying high-ionic strength (1 *M*–0 *M*); after concentration, desalting and size-exclusion chromatography are used to remove remaining impurities.

3. Mapping Interaction Surfaces

An important first step in the determination of the structure of a protein–RNA or protein–DNA complex is to identify the interaction surface and determine in a qualitative sense at least the relative orientation of the protein with respect to the RNA or DNA. This task can be accomplished by chemical shift perturbation analysis *(17,18)*, cross-saturation *(19,20)*, hydrogen-exchange protection *(21)* and by monitoring the exclusion of broadening/shift reagents *(22)*. Orientational information can be obtained from the broadening

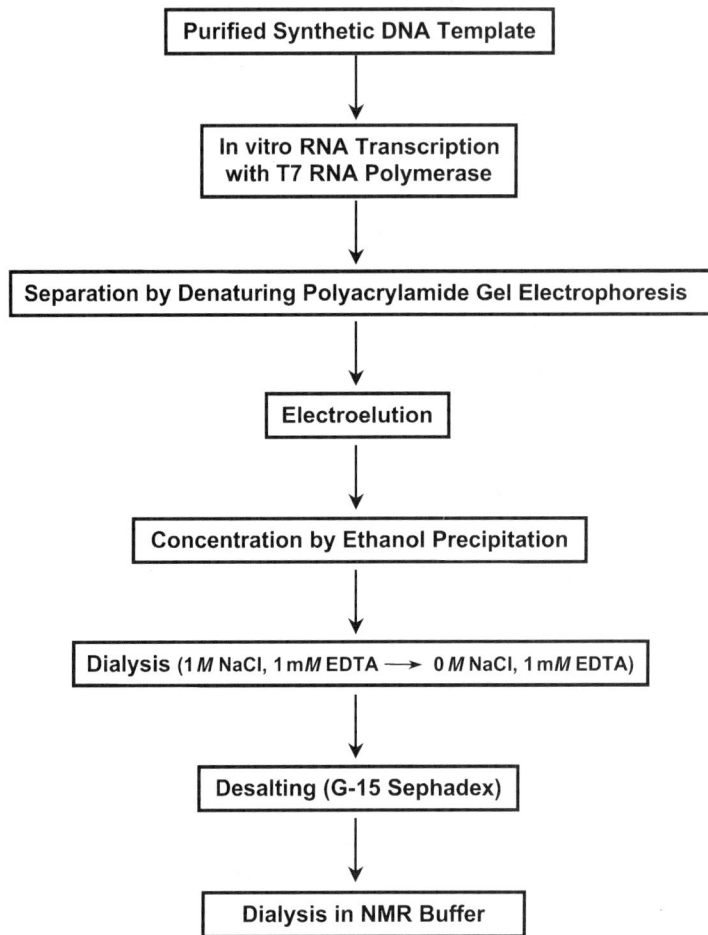

Fig. 1. Schematic protocol for the purification of RNAs prepared by in vitro transcription.

afforded by paramagnetic spin labels *(23,24)*. Accomplishing these first steps make the task of assigning resonances and intermolecular NOEs much simpler because many possible choices will not be within the distances allowed by the crude models constructed from this information. It is very wise to take these steps early in the structure-determination process.

3.1. Chemical Shift Perturbation Analysis and Cross-Saturation

The evaluation of chemical shift changes on binding is one of the simplest ways to define a binding interface. The experiments used for this task are fast

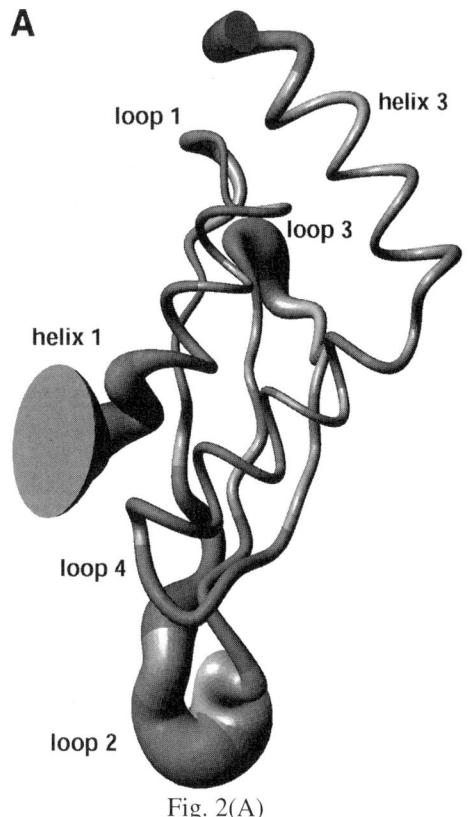

Fig. 2(A)

and can be used in pilot studies when even the presence or absence of an interaction may be unclear. A ^{15}N heteronuclear single-quantum coherence (HSQC) is first measured on the free protein, and then it is remeasured after introduction of the DNA/RNA. Those amides that have differing chemical shifts (proton, nitrogen, or both) are deemed to be near the interaction surface, which can be mapped by identifying shifted resonances on the protein structure (**Fig. 2A**) or its sequence (**Fig. 2B**). This effect is sometimes expressed quantitatively by a weighted average that can be used to rank the shifted amides to lie outside, on the edge, or well within the interaction surface *(25)*. This weighted average over the three nuclei (amide proton ^{1}H, amide nitrogen ^{15}N, ^{13}C$_\alpha$) measured in ^{15}N and ^{13}C HSQC spectra of a protein engaged in a protein–nucleic acid complex is as follows *(26)*.

$$\Delta\delta_{av} = \{[(\Delta\delta^1 HN/3.5)^2 + (\Delta\delta^{15}N/30)^2 + (\Delta\delta^{13}C\alpha/20)^2]/3\}^{1/2}$$

This equation weights the relative contributions of the three nuclei by their relative chemical shift dispersion. Each individual term, such as $\Delta\delta^{15}$N, represents

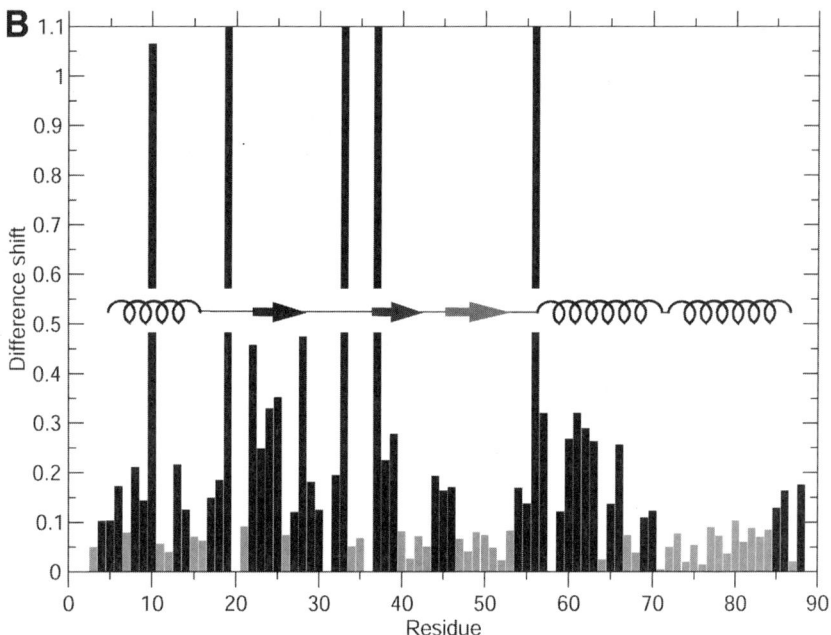

Fig. 2. **(A)** Structure of the yeast RNA processing factor Rnt1p protein structure determined in the presence of its cognate RNA; darker spots identify residues whose amides chemical shifts change significantly in the presence of the RNA; the thickness is proportional to the local rmsd. **(B)** Changes in amide chemical shift are plotted vs residue number on the protein sequence. The RNA binding surface is identified as comprising helix α1 and the loops connecting the β strands.

the change in chemical shift of that nucleus when in complex relative to the free protein.

There are two drawbacks of this approach. First, the chemical shifts of resonances far from the interaction surface occasionally can be perturbed as a result of conformational changes within the monitored species. Second, chemical shift changes in nucleic acids are much smaller than those for proteins; as a consequence, it is very difficult to define unambiguously interaction surfaces by using a comparable approach. Thus, chemical shift perturbation data must be evaluated with caution and can only be useful for coarsely defining the interaction surfaces.

A complementary technique to overcome this limitation is based on cross-saturation *(19,27)*. A saturation pulse selective for one of the partners (e.g., RNA/DNA NH resonances that are very well separated from protein signals) is applied during the preparation period of defined and incremented duration and a typical two-dimensional (2D) spectrum, such as a ^{15}N HSQC is then collected for the binding partners. Resonances spatially close to the saturated resonance will have reduced intensity, as can easily be visualized in differenct spectra. By applying shaped pulses that saturate only a relatively narrow spectral region of about 1–2 ppm, the RNA can be saturated and cross-saturation effects (NOEs) can be observed in the protein amide ^{15}N HSQC. By incrementing the saturation pulse duration, the rate and lag time associated with the cross-saturation can be monitored, thus quantitating the distance between saturated and observed nuclei, and also distinguishing cases where cross saturation is the result of relayed transfer (i.e., spin diffusion). Occasionally, some residues near the interface may not be detectable; yet, because careful buildup curves are measured, few false positives are observed. This is in stark contrast to chemical shift perturbation where spurious changes resulting from conformation exchange can sometimes be misleading.

3.2. Hydrogen-Exchange Protection and the Exclusion of Broadening Agent

These two techniques use the same basic principle, that is, to map accessibility and protection by a ligand. In the case of hydrogen exchange, free protein is exchanged from H_2O solvent to D_2O: amide resonances that exchange slowly are deemed buried within the protein core. The same experiment is repeated while monitoring protein amide exchange rates in the presence of the binding partner (21). Changes in protection are because of the additional burial of the interaction surface by the ligand; thus, protected residues are marked as existing within the interaction area. The exclusion of the broadening/shift reagent method is based on the same principle but uses compounds such as gadolidium-ethylenediaminetetraacetic acid or lanthanides that can alter the chemical shift or relaxation properties of amides and thus delineate buried surface (22). Both hydrogen-exchange protection and broadening/shift reagents are preferred to simple chemical shift perturbation because they exclusively report changes in molecular interface. Reverse effects such as the unexpected exposure in the presence of the RNA/DNA can also be revealing by indicating a conformational change that exposes a previously buried residue.

3.3. Paramagnetic Spin Labels

Proximity between proteins and nucleic acids can also be detected by using paramagnetic spin labels, such as nitroxides, that broaden NMR lines (23). A paramagnetic center induces relaxation of spin 1/2 nuclei to an extent inversely proportional to the sixth power of the electron–nucleus distance, in analogy to the NOE. However, the broadening occurs over a much greater distance than typical NOE measurements because the electron dipole is approx 2000 times larger than the proton dipole. Systematic studies with protein samples have demonstrated that distances of at least 15 Å can be detected by measuring the dipolar relaxation induced by the paramagnetic species on neighboring protons *(28–30)*.

To execute these experiments, a spin label must be covalently attached to one species to identify residues in the other species that are affected by the electron dipole. For the study of protein–nucleic acid complexes, it is most convenient to label the RNA/DNA, which can be synthesized with a specific chemically modified nucleotide—for example, a nucleotide with a 2′ amino group instead of a 2′ hydroxyl, or with 4-thio-uracyl. For 2′-linked nitroxide RNA, the modified RNA is reacted with an isocyanate derivative of 4-amino-TEMPO to conjugate a nitroxide group to the 2′ position of the modified nucleotide *(31)*. Spin labeling of 4-thio-Uracyl-modified RNA can be done similarly, except that the labeling reaction uses 3-(2-iodoacetamideoproxyl) as the proxyl source *(23)*. The spin-labeled RNA is then purified by electrophoresis to remove degraded species.

The work would then proceed as follows: (a) ^{15}N HSQC spectra are collected on the complex in the presence of the spin-labeled RNA; (b) reducing agents, such as sodium ascorbate or sodium hydrosulfite, are added to the NMR sample and another ^{15}N HSQC is collected. The paramagnetic nitroxide group is reduced into a hydroxylamine group, which is diamagnetic, and any spin-label-specific increase in relaxation disappears. (c) Resonances present at near-normal intensity in the spectra of reduced complex but completely absent from the spectra in the presence of the spin label are deemed to be at close range to the spin label, that is, less than 15 Å. Resonances that are only partially broadened and have only a partial reduction in intensity when in the presence of the spin label are deemed to be at intermediate range, 12–20 Å.

Qualitative distance restraints obtained as described in the previous paragraph are too crude to be used effectively in structure calculations but are often sufficient to orient the species with respect to one another if structures of the individual components are available. An example of the application of this technique is provided in **Fig. 3**, where resonances within yeast Rnt1p protein in proximity to a spin label on its cognate RNA stem loop are easily identified. In principle, it should be possible to measure precise electron–proton distances by measuring the electron–proton dipolar contribution to T_2 relaxation *(28,29)*, but great care must be exercised. The paramagnetic group is attached through a flexible two-atom linker. As a result, the spin label is relatively free to move and its precise position cannot be established *a priori*. Furthermore, if transient sample aggregation occurs, even for a small fraction of the time, or if unreacted spin label is present in the solution, then electron–proton relaxation can be mistakenly observed in exposed hydrophobic patches. In our experience, it is difficult to remove all unreacted spin label, even after extensive dialysis. To

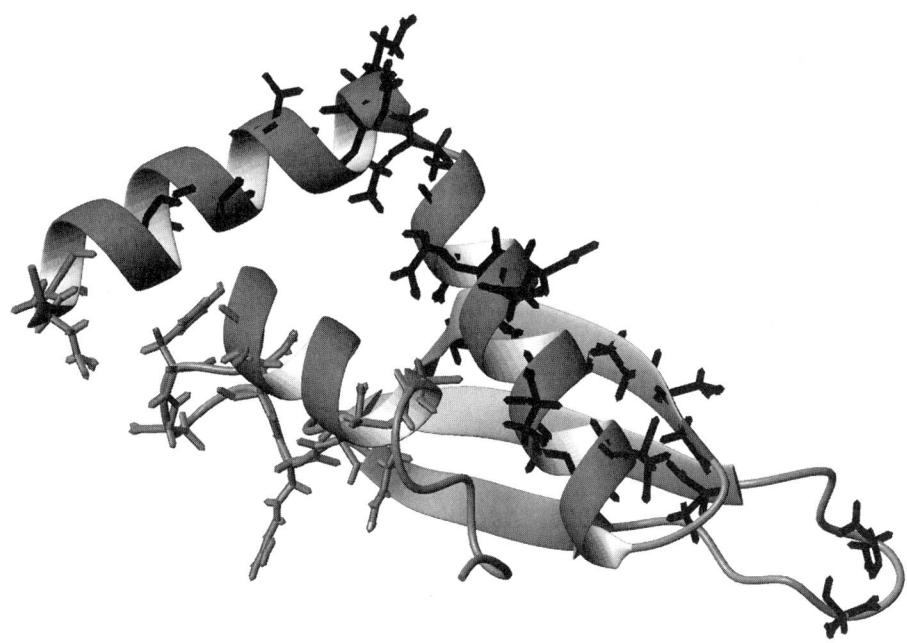

Fig. 3. An RNA tagged with nitroxide spin label was used to identify residues within Rnt1p protein close to the site of labeling (*gray sidechains*).

reduce the problem, labeling reactions can be conducted under conditions of slight excess of protein or nucleic acid over spin-label, then verify completion of the reaction using mass spectrometry.

4. Structure Determination

Only aspects specific to protein–nucleic acid complexes are described here. NMR data collection and analysis for the protein or DNA/RNA components in isolation can be executed as described in other sections of this book.

4.1. NMR Data Collection

Of particular importance to the generation of high-resolution structure of macromolecular complexes is, of course, the detection of intermolecular NOE interactions, which must be reliably distinguished from intramolecular NOEs. Protein–RNA/DNA complexes are somewhat, but not significantly, easier to study than protein–protein complexes because some nucleic acid and protein resonances can be found in distinct spectral regions (**Fig. 4**). For example, nucleic acid imino protons resonate from 10 to 15 ppm, where there are few protein resonances, and few RNA protons have chemical shifts below 3 ppm

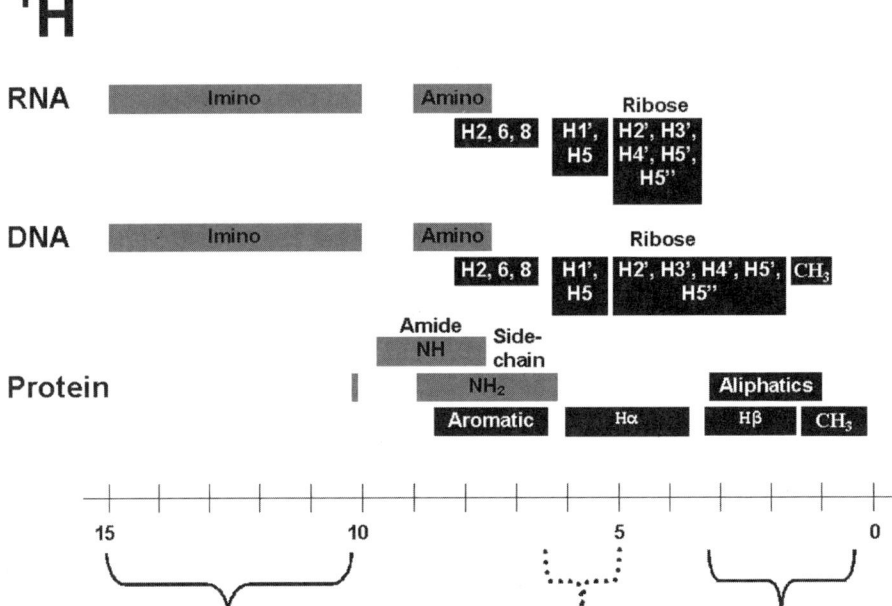

Fig. 4. Protein and nucleic acid proton chemical shift distribution. Protons that readily exchange with solvent and that are only observed in water samples are shown as *gray bars*, whereas protons observable in D_2O are shown in *black*. Regions where nonoverlapping chemical shifts, denoted by *brackets*, are present, are particularly useful for selective excitation when using selective cross-saturation as a surface delineation tool.

(though DNA H2′/H2″ and -CH$_3$ groups do). Furthermore, protons from protein and nucleic acids within overlapping regions are often attached to different heteronuclei. For example, protein amides and DNA/RNA base resonances overlap, yet these two sets of protons are obviously attached to nitrogen and carbon, respectively. However, most successful studies of intermolecular complexes rely on preparing samples where one component (e.g., the protein or the RNA) is uniformly $^{13}C/^{15}N$ labeled and the other is unlabeled; NOE-filtered experiments then provide unambiguous identification of intermolecular NOE interactions in a complex.

The basic technique used to distinguish intra- and intermolecular NOEs is the so-called half-X filter *(32,33)*. Half-filter experiments generate 2D proton–proton NOE spectra in which one dimension is filtered to observe only protons attached to ^{13}C (or ^{15}N), and the other only protons attached to ^{12}C (or ^{14}N). By labeling only the protein or RNA with ^{13}C (and/or ^{15}N) and leaving the other molecule unlabeled, these experiments can yield intermolecular NOEs quickly and unambiguously. Limitations originate from artifacts resulting from both instrumental imperfections and to the spread in coupling constants used to execute the heteronuclear filter. Despite several proposals to improve the performance of the filters *(34,35)*, artifacts still severely affect their performance. A particularly effective solution to improve the performance of heteronuclear filters relies on adiabatically swept pulses to reduce artifacts resulting from imperfect tuning of the filter element with respect to the heteronuclear ^{13}C–^1H coupling constants *(35)*.

Protein–nucleic acid interfaces are rich in hydrogen bonds. The presence of hydrogen bonds can often be inferred based on NOE patterns, and hydrogen-bonding restraints for structure calculation can be generated from these inferred interactions. However, if these interactions are incorrectly attributed, mistakes occur. At the interface between proteins and nucleic acids, the regular patterns of NOEs that characterize protein and nucleic acid secondary structures that often lead to hydrogen-bond identification, cannot often be used. Thus, it is very useful to measure the presence of hydrogen bonds directly. HNN–COSY experiments take advantage of the small yet measurable (2–10 Hz) though bond J-coupling that develops between the donor and accepter in a hydrogen bond *(36)*. This pair can then be restrained throughout the structure calculations. This technique has been used to identify unusual basepairs in nucleic acids *(37,38)*, such as what might be recognized by proteins, but has not been extensively applied to detecting hydrogen bonds between proteins and nucleic acids with perhaps one exception *(39)*.

4.2. Residual Dipolar Coupling Refinement

Determination of the structure of DNA and RNA molecules by NMR using constraints derived from analysis of NOE and J couplings is more difficult than for proteins *(40)*; this difficulty is inevitably reflected in studies of their complexes. The relative low abundance of protons and reduced chemical dispersion result in fewer NOE constraints, and only a small fraction of all NOEs (seldom more than 10%) involve nucleotides separated by more than 1–2 residues in the sequence. Although isotopic labeling has significantly reduced the chemical dispersion problem and enables the determination of J-coupling constants, the problem of the nonglobular nature of nucleic acid structures remains. The RNA and DNA components of protein complexes are poorly determined in their global features, even when the structure of the protein and of the protein–RNA interface are determined to very high levels of precision *(41)*. The most promising approach to improve the global precision and accuracy of the NMR structures of nucleic acids and their complexes is undoubtedly RDCs *(42)*. Normally dipolar couplings average to zero because of motional reorientation, but the magnitude of the dipolar coupling can be tuned pretty much at will by controlling the amount of partial orientation in the sample induced by adding an external ordering component. When only a fraction (1–2%) of the population of molecules is oriented, then the residual dipolar couplings are 10–100 Hz. These observables can provide the relative orientation of NH, CH, and CC bond vectors relative to a reference axis and report on the global structure of a biological molecule.

The key step to obtain useful RDC data is the method chosen to obtain partial alignment of the molecule with respect to the magnetic field, which should be done without causing appreciable line broadening and/or aggregation. Several common media types and conditions exist: inherent alignment, strained gels, phospholipid bycelles *(42)*, phage particles, and polyethylene glycol mixed with organic alcohols, but not all samples are stable or soluble in all media. For example, phospholipid solutions contain a nuclease activity that degrades RNA samples *(43)*, and basic proteins tend to be less soluble in these media. Phospholipid bicelles are lipid disks that have the negatively charged phosphate group pointing out into solvent, whereas the nonpolar lipids lie alongside one another as a result of their hydrophobicity. These media are often used with negatively charged proteins because the proteins will be resistant to adhere directly to the bicelles; these align in magnetic fields and the protein cosolutes partially align as a result of steric collisions. Proteins that bind RNA are often positively charged and hence can adhere directly to the bicelles. Although this interaction generates alignment, it can create problems for NMR because this changes the effective correlation time of the protein and hence broadens the NMR lines *(43)*. Although there are cases where the RDCs of positive proteins have been collected in the presence of lipid bicelles *(43,44)*, generally this medium is not optimal for protein–nucleic acid complexes. An additional drawback of the bicelles is that they are ineffective at high-ionic strength and temperatures outside of 22–50°C. The preferred orienting media for studies of nucleic acids is represented by filamentous phage particles *(45–47)*. These long rod-like particles align in magnetic fields under various temperature and ionic strengths. Care should be used with these media because phages sometimes fail to align at the low pH and ionic strength that are typically used for studies of protein–nucleic acid complexes as a result of aggregation. Furthermore, if mishandled, phages can become heterogonous in length and will not align. Nevertheless, they are very popular media for RDC studies because they are inexpensive to prepare and can be stored for long periods of time as a pellet that can be dissolved in the appropriate buffer of interest. Polyethylene glycol and organic alcohols have become popular recently as aligning media *(48)*. Their stability at various ionic strengths, buffer conditions, and pH values makes them an attractive alternative to bicelles and filamentous phage. Although these media are particularly sensitive to temperature, and the range of ratios that permit generation of the laminar phase is small, they nicely compliment the other media.

In principle, residual dipolar couplings are easily measured by determining the coupling constants for one-bond spin systems, such as $^1H/^{15}N$ or $^1H/^{13}C$, by turning off decoupling during either proton or heteronuclear chemical shift measurement. The apparent coupling constant difference is the RDC. When measuring $^1H/^{15}N$ dipolar couplings in proteins the increased dispersion and larger chemical shift ranges of the ^{15}N dimension are usually used in preference to the proton dimension. In contrast, when measuring $^1H/^{13}C$ residual dipolar couplings in RNA or DNA, most often, the couplings are measured in the proton dimension where the resolution is greatest. The doubling of the peaks in these coupled spectra can lead to resonance overlap that is not present in the decoupled spectra. Spin-state selective methods for collecting the individual downfield and upfield components separately alleviate these problems *(49,50)*. This so-called in-phase antiphase HSQC experiment has been applied successfully to nucleic acids as well *(51)*.

RDCs can be included in structure calculations of protein–nucleic acid complexes in two ways. The first method uses RDCs to refine both local structure and global structure simultaneously *(43,51,52)*. Basically, RDCs are applied as constraints throughout the molecule(s) of interest alongside more typical NOE, hydrogen bonding, and dihedral angle constraints. To implement these constraints, the magnitude (D_a) and asymmetry (rhombicity R) of the alignment tensor for the molecule must be estimated. This is easier for globular proteins: The distribution of the RDCs when plotted as a histogram mimics a solid-state NMR powder pattern *(53)*. Analysis of this distribution can provide reliable initial values for D_a and R, and refinement of these values by optimization of the fit to the RDCs can be achieved by a grid search. This strategy is effective because the bond vectors associated with the measured RDCs nearly exhaustively sample all possible directions in space. However, for nucleic acids, the distribution of measurable RDCs only covers a small range of all possible directions. Consequently, alternative methods for estimating D_a and R are required *(52)* based on simultaneously fitting NOE and residual dipolar coupling data and systematically varying the values of these parameters until best-fit values are found.

The second approach to the incorporation of RDCs to structure calculations fits regions of well-defined local structure to the RDC data, thus solely defining the global conformation *(46,54–56)*. This approach is particularly advantageous for multidomain proteins, nucleic acids, or protein–nucleic acid complexes. Often, it is possible to define the structures of the individual molecules (or domains) accurately and precisely, but the number of intermolecular (or interdomain) restraints may be relatively small. Consequently, application of RDCs as a long-range refinement tool becomes quite useful. This strategy effectively uses the rigid structures of the individual molecules to reduce the number of solutions that can satisfy the RDC data. Furthermore, because all of the bonds within a given rigid body will have the same D_a and R, this method can also circumvent situations where the initial estimations of these parameters are unreliable.

5. Structure Calculations

Structure determination by NMR requires efficient and reliable computational methods to convert experimental constraints into coordinates. Unlike proteins, where a consensus emerged, nucleic acid structures have been calculated using a variety of methods until recently. Complexes present additional problems, however, because of the induced-fit nature of the interaction between the components *(41,57)*. Methods that rely on predetermining protein and RNA structures and docking them against each other using intermolecular NOE constraints suffer from some bias because of the conformational changes that take place in these interfaces and the fact that the RNA conformation may be underdetermined without intermolecular distance constraints. Simulated annealing protocols developed for determining protein structures start from completely randomized coordinates to obtain a global fold of the protein and then refine its conformation using the NMR constraints, and even relatively simple adaptation of the method allows it to work well in the RNA environment *(40,58,59)*, and even with RNA–protein complexes *(60,61)*. Determining the structure of the protein and RNA in the complex simultaneously, rather than predetermining the RNA and protein structures and then docking them, allows all components of a complex to fold together under the influence of the experimental information to avoid bias in the conformational search toward the free forms.

Structure calculations based on restrained molecular dynamics use both experimental restraints and nonexperimental information introduced in the form of a semiempirical force field provided by software packages such as CNS *(62)*. The force field includes all information needed to define protein and DNA/RNA stereochemistry and should guarantee reasonable bond lengths and bond angles, base planarity, and proper chirality. In addition, other terms describe nonbonded interactions such as van der Waals and electrostatics contacts and the conformation of dihedral angles. NMR constraints (distance and dihedral angle) are introduced in the form of additional "pseudoenergy" terms into the force field defined by the software package. An effective protocol for nucleic acid–protein structure calculation using X-PLOR or CNS divides the overall protocol in three steps *(40,58)*: (a) During a first *global fold* phase, only NOE and hydrogen-bonding-distance constraints are active together with energy terms designed to reproduce the covalent structure. Repulsive van der Waals interactions are inactive initially to allow atoms to pass each other more easily and facilitate sampling of conformational space and improve convergence, but repulsive terms are slowly turned on during the simulation. It is helpful at this stage to avoid using lower bounds on distance constraints; setting the lower distance bounds to 0 Å *(63)* (rather than to the 1.8 Å sum of the radii of two protons) facilitates sampling of conformational space and also improves convergence. At the end of the restrained molecular dynamics step, the system is slowly cooled to generate a *folded* structure. (b) Dihedral angle constraints are then slowly introduced during a second restrained molecular dynamics stage, followed once again by slow cooling of the system to a *refined* structure. (c) At this stage, a final energy-minimization step is carried out during which non-bonded energy terms (such as electrostatics and attractive van der Waals) can be introduced. It is important not to introduce the nonbonding terms too early in the calculation to avoid potentially biasing the outcome. More recent methods that are equally or more effective include torsion angle refinement within DYANA or CNS *(64,65)*; although the principles of the methods are the same, torsion angle refinements allow greater convergence and obviate the need for the multistep refinement protocol that appears to be needed if minimization is conducted in Cartesian coordinates.

6. Computational Tools to Analyze and Display Protein–Nucleic Acid Interactions

Once structure is solved, it is necessary to analyze its structure–activity relationship and sequence-dependent intermolecular interactions. There are now several recently developed and easy-to-use computational tools to analyze and schematically display protein–RNA (DNA) interactions. They are available for free from websites for researchers at academic and non-profit-making institutions.

> *HBPLUS* (http://www.biochem.ucl.ac.uk/~mcdonald/hbplus/home.html). BPLUS calculates hydrogen bonds by reading a Protein Data Bank (PDB) file and extracting the geometries of all hydrogen-bond interactions (bond angles and length). For a protein–nucleic acid complex, those interactions include hydrogen bonds within protein, nucleic acid, and at the protein and nucleic acid interface. The program is suitable for analysis of PDB structures obtained by both NMR and X-ray methods. For X-ray structures, the program can automatically calculate hydrogen atom positions and suggest optimal conformations for Asn, Gln, and His sidechains.
>
> *NACCESS* (http://wolf.bms.umist.ac.uk/naccess/). NACCESS is a stand-alone program that calculates the accessible surface area of a molecule from a PDB format file. The program rolls a probe of given radius on the surface of the molecule, and the

Chain D

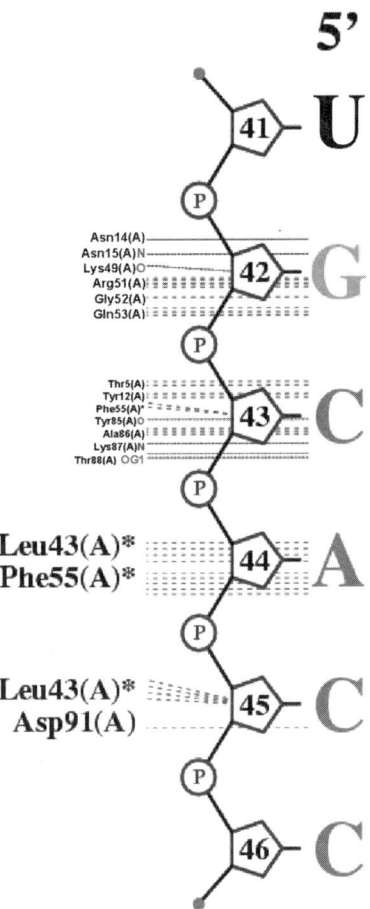

Key

③	Backbone-sugar and base-number	⎯⎯ Hydrogen bond to RNA
℗	Phosphate group	⋯⋯ Nonbonded contact to RNA (<3.35 Å)
		88 ⓦ Water molecule and number
*	Residue/water on plot more than once	

Fig. 5. Schematic representation of interactions between U1A protein and PIE RNA (pdb file 1DZ5) generated by NUCPLOT.

path traced out by its center is the accessible surface. Depending on the chosen radii, the program can calculate the atomic and residue accessibilities for both proteins and nucleic acids. Typically, the probe has the same radius as water (1.4 Å), and hence, the surface described often is referred to as the solvent accessible surface.

NUCPLOT (http://bioinfo.mbb.yale.edu/~nick/nucplot.html). NUCPLOT generates schematic diagrams of protein–nucleic acid interactions. The input is a standard PDB file and the output is a PostScript file that provides a simple representation of the hydrogen bonds and hydrophobic contacts between protein and nucleic acids. By default, the program uses results calculated by the HBPLUS program. An example of such a plot is provided in **Fig. 5**.

Entangle (http://www-bioc.rice.edu/~shamoo/project/entangle/entangle.htm). Entangle is a JAVA program. Compared with the UNIX-based programs listed previously, *Entangle* can be installed on a PC. It reads a PDB file containing a protein–nucleic acid complex and gives a list of interactions that occur at the interface, including hydrogen bonds, electrostatic, hydrophobic, van der Waals, and stacking interactions between protein and nucleic acids. The interactions can be visualized in a simple 3D mode.

Protein–Nucleic Acid Interaction Server (http://www.biochem.ucl.ac.uk/ bsm/ DNA/server/). This Web-based server is a tool to analyze the protein interface of any protein–nucleic acid complex. The coordinates of a complex can be submitted in PDB format, and tables describing the nature of the interface are output. The output includes information about accessible surface area, hydrogen bonds, gap volume index, bridging water molecules, list of interface residues, and so forth.

7. Concluding Remarks

There are now effective NMR methods to determine the structures of protein–nucleic acid complexes, but significant obstacles remain. Effective methods to observe longer range distances would be of immense value because they would very significantly improve the accuracy of the structures by increasing the number of nonlocal interactions. In addition to being short-ranged, NOE interactions can also be quenched by various dynamic processes occurring at intermolecular interfaces, particularly when binding constants are not strong (less than nm), as is the case with many regulatory systems. Furthermore, RNA- and DNA-binding proteins often interact, through groups located on protein sidechains, with the nucleic acid phosphodiester backbone, which is not always possible to assign because of spectral overlap. Even the incorporation of qualitative distance constraints could have a significant impact on structure determination. Technical developments in this area would be of significant value to structural biologists.

Acknowledgments

Work in our laboratory on RNA-protein recognition is supported by the NIH (GM 1RO1 64440).

References

1. Milligan, J. F., Groebe, D. R., Witherell, G. W., and Uhlenbeck, O. C. (1987) Oligoribonucletide synthesis using T7 RNA polymerase and synthetic DNA templates. *Nucleic Acids Res.* **15,** 8783–8789.
2. Milligan, J. F. and Uhlenbeck, O. C. (1989) Synthesis of small RNAs using T7 RNA polymerase. *Methods Enzymol.* **180,** 51–62.
3. Price, S. R., Oubridge, C., Varani, G., and Nagai, K. (1998) Preparation of RNA-protein complexes for X-ray crystallography and NMR. In *RNA–Protein Interaction: A Practical Approach* (Smith, C. W. R., ed.). Oxford University Press, Oxford, pp. 37–74.
4. Koshlap, K. M., Guenther, R., Sochacka, E., Malkiewicz, A., and Agris, P. F. (1999) A distinctive RNA fold: the solution structure of an analogue of the yeast tRNA^{Phe} TΨC domain. *Biochemistry* **38,** 8647–8656.
5. Sengupta, R., Vainauskas, S., Yaruan, C., et al. (2000) Modified construct of the tRNA TΨC domain to probe substrate conformational requirements of the m^1A_{58} and m^5U_{54} tRNA methyltransferase. *Nucleic Acids Res.* **28,** 1374–1380.
6. Kao, C., Zheng, M., and Rudisser, S. (1999) A simple and efficient method to reduce nontemplated nucleotide addition at the 3′ ends of RNAs transcribed by T7 RNA polymerase. *RNA* **5,** 1268–1272.
7. Price, S. R., Ito, N., Oubridge, C., Avis, J. M., and Nagai, K. (1995) Crystallization of RNA-protein complexes I: a novel method for the large-scale preparation of RNA suitable for crystallographic studies. *J. Mol. Biol.* **249,** 398–408.
8. Whoriskey, S. K., Usman, N., and Szostak, J. W. (1995) Total chemical synthesis of a ribozyme derived from a group I intron. *Proc. Nat. Acad. Sci. USA* **92,** 2465–2469.
9. Wincott, F., DiRenzo, A., Shaffer, C., Grimm, S., Tracz, D., Workman, C., et al. (1995) Synthesis, deprotection, analysis and purification of RNA and ribozymes. *Nucleic Acids Res.* **23,** 2677–2684.
10. Sproat, B., Colonna, F., Mullah, B., Tsou, D., Andrus, A., Hampel, A., et al. (1995) An efficient method for the isolation and purification of oligoribonucleotides. *Nucleos. Nucleot.* **14,** 255–273.
11. Schmidt, S., Grenfell, R. L., Fogg, J., Smith, T. V., Grasby, A., Mersmann, K., et al. (1996) Solid phase synthesis of oligoribonucleotides containing site-specific modifications. In *Innovation and Perspectives in Solid Phase Synthesis and Combinatorial Libraries 1996 (Proceedings of the 4th International Symposium, Edinburgh).* (Epton, R. ed.), Mayflower, Birmingham, UK, pp. 11–18.
12. Scaringe, S. A. (2000) Advanced 5′-silyl-2′-orthoester approach to RNA oligonucleotide synthesis. *Methods Enzymol.* **317,** 3–18.
13. Scaringe, S. A. (2001) RNA oligonucleotide synthesis via 5′-silyl-2′-orthoester chemistry. *Methods* **23,** 206–217.
14. Batey, R. T., Inada, M., Kujawinski, E., Puglisi, J. D., and Williamson, J. R. (1992) Preparation of isotopically labeled ribonucleotides for multidimensional NMR spectroscopy of RNA. *Nucleic Acids Res.* **20,** 4515–4523.

15. Batey, R. T., Battiste, J. L., and Williamson, J. R. (1995) Preparation of isotopically enriched RNAs for heteronuclear NMR. *Methods Enzymol.* **261,** 300–322.
16. Santalucia, J. J., Shen, L. X., Cai, Z., Lewis, H., and Tinoco, I. J. (1995) Synthesis and NMR of RNA with selective isotopic enrichment in the bases. *Nucleic Acids Res.* **23,** 4913–4921.
17. Görlach, M., Wittekind, M., Beckman, R. A., Mueller, L., and Dreyfuss, G. (1992) Interaction of the RNA-binding domain of the hnRNP C proteins with RNA. *EMBO J.* **11,** 3289–3295.
18. Howe, P. W. A., Nagai, K., Neuhaus, D., and Varani, G. (1994) NMR studies of U1 snRNA recognition by the N-terminal RNP domain of the human U1A protein. *EMBO J.* **13,** 3873–3881.
19. Ramos, A., Kelly, G., Hollingworth, D., Pastore, A., and Frenkiel, T. (2000) Mapping the interfaces of protein-nucleic acid complexes using cross-saturation. *J. Am. Chem. Soc.* **122,** 11,311–11,314.
20. Lane, A. N., Kelly, G., Ramos, A., and Frenkiel, T. A. (2001) Determining binding sites in protein-nucleic acid complexes by cross-saturation. *J. Biomol. NMR* **21,** 127–139.
21. Kalodimos, C. G., Boelens, R., and Kaptein, R. (2002) A residue-specific view of the association and dissociation pathway in protein–DNA recognition. *Nat. Struct. Biol.* **9,** 193–197.
22. Arumugam, S., Hemme, C. L., Yoshida, N., Suzuki, K., Nagase, H., Berjanskii, M., et al. (1998) TIMP-1 contact sites and perturbations of stromelysin 1 mapped by NMR and a paramagnetic surface probe. *Biochemistry* **37,** 9650–9657.
23. Ramos, A. and Varani, G. (1998) A new method to detect long-range protein-RNA contacts: NMR detection of electron-proton relaxation induced by nitroxide spin-labeled RNA, *J. Am. Chem. Soc.* **120,** 10,992–10,993.
24. Gopalan, V., Kuhne, H., Biswas, R., Li, H., Brudvig, G. W., and S. Altman (1999) Mapping RNA-protein interactions in ribonuclease P from *Escherichia coli* using electron paramagnetic resonance spectroscopy. *Biochemistry* **38,** 1705–1714.
25. Zuiderweg, E. R. P. (2002) Mapping protein–protein interactions in solution by NMR spectroscopy. *Biochemistry* **41,** 1–7.
26. Pellecchia, M., Montgomery, D. L., Stevens, S. Y., Vander Koii, C. W., Feng, H. P., Gierasch, L. M., et al. (2000) Structural insights into substrate binding by the molecular chaperone DnaK. *Nat. Struct. Biol.* **7,** 298–303.
27. Takahashi, H., Nakanishi, T., Kami, K., Arata, Y., and Shimada, I. (2000) A novel NMR method for determining the interfaces of large protein–protein complexes. *Nat. Struct. Biol.* **7,** 220–223.
28. Gillespie, J. R. and Shortle, D. (1997) Characterization of long-range structure in the denatured state of staphyloccocal nuclease, I: paramagnetic relaxation enhancement by nitroxide spin labels. *J. Mol. Biol.* **268,** 158–169.
29. Battiste, J. L. and Wagner, G. (2000) Utilization of site-directed spin labeling and high resolution heteronuclear NMR for global fold determination of large proteins with limited NOE data. *Biochemistry* **39,** 5355–5365.

30. Gochin, M. (2000) A high resolution structure of a DNA-chromomycin-Co(II) complex determined from pseudocontact shifts in nuclear magnetic resonance. *Structure* **8**, 441–452.

31. Edwards, T. E., Okonogi, T. M., Robinson, B. H., and Sigurdsson, S. T. (2001) Site-specific incorporation of nitroxide spin labels into internal sites of the TAR RNA. Structure-dependent dynamics of RNA by EPR spectroscopy. *J. Am. Chem. Soc.* **123**, 1527, 1528.

32. Otting, G. and Wüthrich, K. (1989) Extended heteronuclear editing of 2D ^{1}H NMR spectra of isotope-labeled proteins, using the X(ω_1,ω_2) double filters. *J. Magn. Reson.* **85**, 586–594.

33. Otting, G. and Wüthrich, K. (1990) Heteronuclear filters in two-dimensional [1H-1H]-NMR spectroscopy: combined use with isotope labeling for studies of macromolecular conformation and intermolecular interactions. *Q. Rev. Biophys.* **23**, 39–96.

34. Gemmecker, G., Olejniczak, E. T., and Fesik, S. W. (1992) An improved method for selectively observing protons attached to ^{12}C in the presence of ^{1}H-^{13}C spin pairs. *J. Magn. Reson.* **96**, 199–204.

35. Zwahlen, C., Legault, P., Vincent, S. J. F., Greenblatt, J., Konrat, R., and Kay, L. E. (1997) Methods for measurement of intermolecular NOEs by multinuclear NMR spectroscopy: application to a bacteriophage λ N-peptide/*boxB* RNA complex. *J. Am. Chem. Soc.* **119**, 6711–6721.

36. Dingley, A. J. and Grzesiek, S. (1998) Direct observation of hydrogen bonds in nucleic acid base pairs by internucleotide ^{2}J$_{NN}$ couplings. *J. Am. Chem. Soc.* **120**, 8293–8297.

37. Hennig, M. and Williamson, J. R. (2000) Detection of N-H··N hydrogen bonding in RNA via scalar couplings in the absence of observable imino proton resonances. *Nucleic Acids Res.* **28**, 1585–1593.

38. Wohnert, J., Dingley, A. J., Stoldt, M., Gorlach, M., Grzesiek, S., and Brown, L. R. (1999) Direct identification of NH . . . N hydrogen bonds in noncanonical base pairs of RNA by NMR spectroscopy. *Nucleic Acids Res.* **27**, 3104–3110.

39. Liu, A., Majumdar, A., Jiang, F., Chernichenko, N., Skripkin, E., and Patel, D. J. (2000) NMR detection of intermolecular NH-N hydrogen bonds in the human T cell leukemia virus-1 Rex peptide-RNA aptamer complex. *J. Am. Chem. Soc.* **122**, 11,226–11,227.

40. Allain, F. H.-T. and Varani, G. (1997) How accurately and precisely can RNA structure be determined by NMR? *J. Mol. Biol.* **267**, 338–351.

41. Allain, F. H.-T., Gubser, C. C., Howe, P. W. A., Nagai, K., Neuhaus, D., and Varani, G. (1996) Specificity of ribonucleoprotein interaction determined by RNA folding during complex formation. *Nature* **380**, 646–650.

42. Tjandra, N. and Bax, A. (1997) Direct measurement of distances and angles in biomolecules by NMR in a dilute liquid crystalline medium. *Science* **278**, 1111–1114.

43. Bayer, P., Varani, L., and Varani, G. (1999) Refinement of the structure of protein-RNA complexes by residual dipolar coupling analysis. *J. Biomol. NMR* **14**, 149–155.

44. Wu, B., Arumugam, S., Gao, G., Lee, I., Semenchenko, V., Huang, W., et al. (2000) NMR structure of tissue inhibitor of metalloproteinases-1 implicates localized induced fit in recognition of matrix metalloproteinases. *J. Mol. Biol.* **295,** 257–268.

45. Hansen, M. R., Mueller, L., and Pardi, A. (1998) Tunable alignment of macro-molecules by filamentous phage yields dipolar coupling interactions. *Nat. Struct. Biol.* **5,** 1065–1074.

46. Bondensgaard, K., Mollova, E. T., and Pardi, A. (2002) The global conformation of the hammerhead ribozyme determined using residual dipolar couplings. *Biochemistry* **41,** 11,532–11,542.

47. Mollova, E. T., Hansen, M. R., and Pardi, A. (2000) Global structure of RNA determined with residual dipolar couplings. *J. Am. Chem. Soc.* **122,** 11,561–11,562.

48. Ruckert, M. and Otting, G. (2000) Alignment of biological macromolecules in novel nonionic liquid crystalline media for NMR experiments. *J. Am. Chem. Soc.* **122,** 7793–7797.

49. Andersson, P., Nordstrand, K., Sunnerhagen, M., Liepinsh, E., Turovskis, I., and Otting, G. (1998) Heteronuclear correlation experiments for the determination of one-bond coupling constants. *J. Biomol. NMR* **11,** 445–450.

50. Ottiger, M., Delaglio, F., and Bax, A. (1998) Measurement of J and dipolar cou-plings from simplified two-dimensional NMR spectra. *J. Magn. Reson.* **131,** 373–378.

51. Leeper, T. C., Martin, M. B., Kim, H., Cox, S., Semenchenkio, V., Schmidt, F. J., and Van Doren, S. R. (2002) Structure of the UGAGAU hexaloop that braces *Bacillus* RNase P for action. *Nat. Struct. Biol.* **9,** 397–403.

52. Warren, J. J. and Moore, P. B. (2001) Application of dipolar couplings data to the refinement of the solution structure of the sarcin-ricin loop RNA. *J. Biomol. NMR* **20,** 311–323.

53. Clore, G. M., Gronenborn, A. M., and Bax, A. (1998) A robust method for deter-mining the magnitude of the fully asymmetric alignment tensor of oriented macromolecules in the absence of structural information. *J. Magn. Reson.* **133,** 216–221.

54. Al-Hashimi, H. M., Gorin, A., Majumdar, A., and Patel, D. J. (2001) Alignment of the HTLV-1 Rex peptide bound to its target RNA aptamer from magnetic field-induced residual dipolar couplings and intermolecular hydrogen Bonds. *J. Am. Chem. Soc.* **123,** 3179, 3180.

55. Al-Hashimi, H. M., Gosser, Y., Gorin, A., Hu, W., Majumdar, A., and Patel, D. J. (2002) Concerted motions in HIV-1 TAR RNA may allow access to bound state conformations: RNA dynamics from NMR residual dipolar couplings. *J. Mol. Biol.* **315,** 95–102.

56. Al-Hashimi, H. M., Pitt, S. W., Majumdar, A., Xu, W., and Patel, D. J. (2003) Mg^{2+}-induced variations in the conformation and dynamics of HIV-1 TAR RNA probed using NMR residual dipolar couplings. *J. Mol. Biol.* **329,** 867–873.

57. Mao, H., White, S. A., and Williamson, J. R. (1999) A novel loop–loop recognition motif in the yeast ribosomal protein L30 autoregulatory RNA complex. *Nat. Struct. Biol.* **6,** 1139–1147.

58. Wimberly, B., Varani, G., and Tinoco, I., Jr. (1993) The conformation of loop E of eukaryotic 5S ribosomal RNA. *Biochemistry* **32,** 1078–1087.

59. Allain, F. H.-T. and Varani, G. (1995) Structure of the P1 helix from group I self splicing introns. *J. Mol. Biol.* **250,** 333–353.

60. Howe, P. W. A., Allain, F. H.-T., Varani, G., and Neuhaus, D. (1998) Determination of the NMR structure of the complex between U1A protein and its RNA polyadenylation inhibition element. *J. Biomol. NMR* **11,** 59–84.

61. Varani, L., Spillantini, M. G., Goedert, M., and Varani, G. (2000) Structural basis for recognition of the RNA major groove in the tau exon 10 splicing regulatory element by aminoglycoside antibiotics. *Nucleic Acids Res.* **28,** 710–719.

62. Brünger, A. T., Adams, P. D., Clore, G. M., Delano, W. L., Gros, P., Grosse-Kunstleve, R. W., et al. (1998) Crystallography and NMR system (CNS): a new software suite for macromolecular structure determination. *Acta Cryst.* **D54,** 905–921.

63. Gubser, C. C. and Varani, G. (1996) Structure of the polyadenylation regulatory element of the human U1A pre-mRNA 3′-untranslated region and interaction with the U1A protein. *Biochemistry* **35,** 2253–2267.

64. Güntert, P., Mumenthaler, C., and Wüthrich, K. (1997) Torsion angle dynamics for NMR structure calculation with the new program DYANA. *J. Mol. Biol.* **273,** 283–298.

65. Stein, E. G., Rice, L. M., and Brunger, A. T. (1997) Torsion-angle molecular dynamics as a new efficient tool for NMR structure calculation. *J. Magn. Reson.* **124,** 154–164.

16

Using NMRView to Visualize and Analyze the NMR Spectra of Macromolecules

Bruce A. Johnson

Summary

Nuclear magnetic resonance (NMR) experiments on macromolecules can generate a tremendous amount of data that must be analyzed and correlated to generate conclusions about the structure and dynamics of the molecular system. NMRView is a computer program that is designed to be useful in visualizing and analyzing these data. NMRView works with various types of NMR datasets and can have multiple datasets and display windows opened simultaneously. Virtually all actions of the program can be controlled through the Tcl scripting language, and new graphical user interface components can be added with the Tk toolkit. NMR spectral peaks can be analyzed and assigned. A suite of tools exists within NMRView for assigning the data from triple-resonance experiments.

Key Words: NMR; computer; software; assignment; chemical shift; NOE; spectrum.

1. Introduction

If nuclear magnetic resonance (NMR) spectroscopy were as simple as ultraviolet, infrared, or most other spectroscopies the life of an NMR spectroscopist would be much less interesting. Fortunately, the physics of NMR is such that the spectroscopist can perform an almost limitless variety of experiments. Different experiments can be used to tease out different sorts of information from the molecular system that can ultimately be used to understand both the structure and dynamics of a molecular system. A downside of this flexibility is the tremendous amount of data that may need to be interpreted. A series of experiments on a reasonably complicated molecule such as a protein or nucleic acid can generate literally gigabytes of data. Furthermore, to interpret these data requires correlating the information from the multitude of different experiments. NMRView exists to help the spectroscopist manage and use this vast

From: *Methods in Molecular Biology, vol. 278: Protein NMR Techniques*
Edited by: A. K. Downing © Humana Press Inc., Totowa, NJ

quantity of information *(1)*. Other excellent programs, with similar goals, exist *(2–6)*. Why use NMRView instead of these?

1.1. Useful

NMRView is developed by and for NMR spectroscopists. The features are those needed by practicing NMR spectroscopists in their day-to-day work. Of course that does not mean that the features we need coincide exactly with features that every other NMR spectroscopist needs. On the other hand, many of the features we have added are likely to be of use to others. We do listen to requests for new features, and we try to incorporate them as time allows.

1.2. Scriptable

One of the first features requested after the basic version of NMRView was completed was the ability to automate frequently used operations. The logical way to accomplish this was by incorporating the ability to write scripts: lists of instructions for NMRView to carry out. The best decision made in the development of NMRView was to not develop a new scripting language. Instead, NMRView incorporates the scripting language Tool Command Language (Tcl) into NMRView *(7,8)*.

With Tcl, NMRView gained not just the ability to automate simple tasks, but also the ability to write complex programs that could be used to carry out novel analyses. Thus, the end user was to a large extent freed from the limitations placed on NMRView by the author's limited time. He or she could extend it in entirely new directions.

Starting with version 3.0, NMRView incorporated not just the Tcl language, but its amazing companion, Tk. Tk is a library of Tcl commands that can be used to create graphical user interfaces (GUIs). With this version of NMRView, the entire GUI is created using Tk. Thus, not only can end users write new analysis techniques or simple scripts to automate actions, they can also add a GUI to these extensions.

Although the compiled NMRView executables are still developed solely by the author, a worldwide community of spectroscopists now contributes to the Tcl scripts, such as the enhanced chemical shift index (CSI) facility *(9)*. In 2000, a group of NMRView developers gathered in Utrecht, The Netherlands, to bring together various extensions to NMRView and unite them into what became version 5.0.

1.3. Free

Finally, NMRView is free software, available to anyone who wishes to use it. This is one way that the developers can contribute to the overall growth of the technique of NMR.

2. Obtaining NMRView

The standard method for obtaining the most up-to-date version of NMRView is by downloading it from the website www.onemoonscientific.com. The distribution format consists of two files. One is a tar archive containing all the platform-independent files, such as Tcl scripts, documentation, and icons. The other is a platform-dependent executable. At the time this chapter was written, executables were available for Sun computers with the Solaris OS (operating system), SGI computers with the IRIX OS, IBM computers with the AIX OS, Intel processor computers with the Linux OS, and Macintosh computers with the MacOS X OS.

A recent event in the history of NMRView has been the development of a version written in the Java programming language (http://www.java.com). The promise of Java being a portable language is realized quite well in NMRViewJ. The identical Java bytecode file executes successfully on Windows NT and XP, Macintosh MacOS X, SGI Irix, Linux, Solaris, and probably other Java implementations. In the case of NMRViewJ, the end user need only download a single file. This file in the Java "jar" format contains all the resource files (Tcl, documentation, etc.) and the Java binary file. A key feature of the Java version of NMRView is the ability for end users to extend it not only with new Tcl scripts, but also with new Java classes.

2.1. Installation

2.1.1. Installing NMRView (C Version for UNIX/Linux)

After downloading NMRView follow these steps: First, create a directory in which to store the NMRView libraries (mkdir /home/username/nmrview5). Copy the NMRView tar file to this new directory (cp nmrview5.2.1.lib.tar.Z /home/username/nmrview5) and then change to the nmrview directory (cd /home/username/nmrview5). Now, uncompress the files (gunzip nmrview5.2.1.lib.tar.gz) and extract them from the archive (tar xvf nmrview5.2.1.lib.tar). Finally, uncompress the executable (gunzip nmrview5.2.1.linux.gz), make it executable (chmod +x nmrview5.2.1.linux) and copy it to a useful directory (cp nmrview5.2.1.linux /home/username/nmrview). The exact location you install NMRView in is not important, but it is crucial that three environment variables are set correctly. The best way to ensure that they are set correctly, and do not conflict with the use of Tcl/Tk with other applications, is to place the necessary commands in a single file (an example file "nv5" comes with the distribution):

```
#!/bin/csh
setenv NMRVIEW5HOME /home/username/nmrview
setenv TCL_LIBRARY ${NMRVIEW5HOME}/tc18.3
setenv TK_LIBRARY ${NMRVIEW5HOME}/tc18.3
${NMRVIEW5HOME}/nmrview5.2.1.linux — -s ${HOME}/nmrview/startup.tcl
```

Once this file is made executable (chmod +x nv5) it can be used to start NMRView. The "-s" flag is used to specify an optional startup script. Do not include this flag and the following file path if you are not using such a script. You can, however, use a startup script loaded in this way to customize NMRView as you wish.

2.1.2. Installing NMRView (Java Version)

Before installing NMRViewJ you should ensure that you have an appropriate version of Java installed. Issuing the command "java -version" will provide the necessary information. If java is installed and in your "path," then you will get a response similar to the following:

> java version "1.4.1_01"
> Java(TM) 2 Runtime Environment, Standard Edition (build 1.4.1_01-39)
> Java HotSpot(TM) Client VM (build 1.4.1_01-14, mixed mode)

At the time of this writing Java version 1.4.1 or later is the preferred version. Version 1.4.1 is available for at least Linux, Solaris, Windows NT and XP, and IRIX, and it comes preinstalled on MacOS X. If the java command does not run, then you should check to be sure if Java is installed on your system and install it if it is not. Java is available for download from www.javasoft.com (for Solaris, Linux, and Windows) and from the appropriate hardware vendor for other platforms. When downloading Java, it is preferable to get the software development kit, as the simpler Java Runtime Environment does not include the "jar" program needed for installation of NMRViewJ.

Installation of NMRViewJ is similar to the protocol just described: Create a directory and copy the archive (cp nmrviewj5.2.1.jar /home/username/nmrview) file to it. Now, use the Java "jar" tool to uncompress and extract the contents (jar xvf nmrviewj5.2.1.jar; tip: this example only works if the "jar" command is in your "path").

As with NMRView, it is best to start NMRViewJ with a script file that sets up the appropriate environment:

> setenv NVJHOME /usr/local/NMRView/java
> setenv CLASSPATH ${NVJHOME}/swank.jar:${NVJHOME}/tcljava.jar:
> ${NVJHOME}/jacl.jar:${NVJHOME}/nmrview.jar:${NVJHOME}/colt.jar:
> ${NVJHOME}/xml.jar:${NVJHOME}/svggen.jar:${NVJHOME}/cluster.jar:
> ${NVJHOME}/mrk_swank.jar
> java -ms48m -mx128m tcl.lang.NvShell

(Tip: The second "setenv" command should be a single line in the file, and the correct version of the "java" command needs to be in your path.)

The -ms48m and -mx128m parameters let Java know the preferred and maximum memory (48 and 128 megabytes in this example) amounts to use for Java. Set these to larger values to take advantage of more memory if your computer has it.

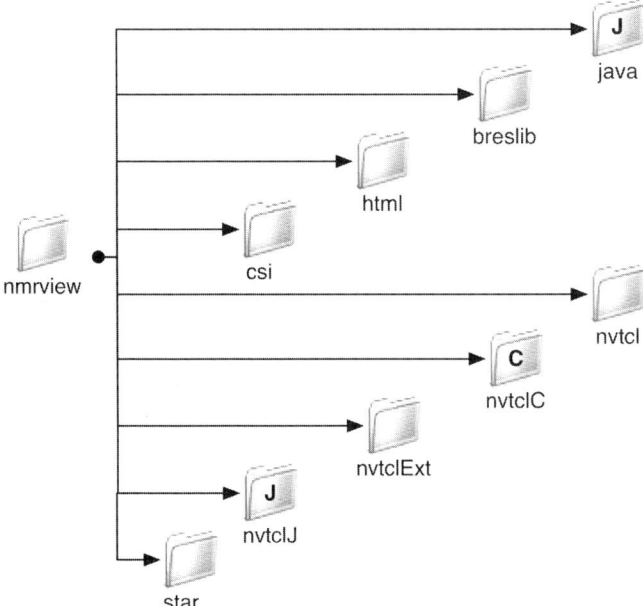

Fig. 1. The folders/directories associated with NMRView. All the "nvtcl" folders contain Tcl scripts used to create the GUI and provide NMR analysis routines. Contents of the other folders are as follows: html—documentation, CSI—parameters for chemical-shift index calculations, star—NMR Star saveframe definitions, breslib—topology files for protein and nucleic acid residues, java—Java jar files containing NMRViewJ. Folders labeled with a "J" are used only by the Java version and the one labeled with a "C" is used only by the C version.

Again, make this file executable (chmod +x nvj). The format of the nvj file on computers running a version of Windows is somewhat different than this UNIX shell script. An example "nvj.bat" file is available with the NMRViewJ distribution.

After extracting the contents of the tar or jar archives you will have a set of directories as illustrated in **Fig. 1**. The various nvtclX directories contain the Tcl scripts that create the NMRView interface and provide various analysis tools. The majority of the scripts, those in the nvtcl directory, are common to both the C and Java versions of NMRView.

By installing the NMRView libraries from NMRViewJ's "jar" file you get all the same files that come with the C version. By adding the appropriate binary executable of NMRView to the installation you can interchangeably use both the C and Java versions of NMRView. All data files can be read by both versions of the program.

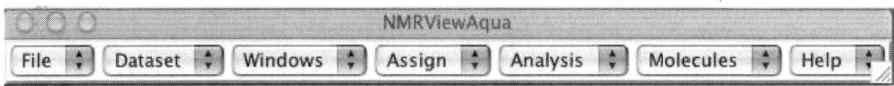

Fig. 2. The NMRView control window. GUI commands are accessible via the series of menus on this menu bar.

2.2. Starting NMRView

Simply type the name of the scripts to start them (including the full path if the script location is not in your path (/home/username/nmrview/nv5 or /home/username/nmrview/nvj). NMRView should start up and display the main control window and a console window. The control window (**Fig. 2**) provides access to NMRView features via the GUI, whereas the console window (**Fig. 3**) provides access to NMRView features via keyboard commands.

3. Data Sets

The most common use of NMRView involves visualizing and analyzing NMR spectra. Many users of NMRView will use other programs to transform the original time domain data into the frequency domain. Because there is no standard data format for NMR spectra that is in common use, NMRView has been designed to be very flexible in its ability to read spectra. **Table 1** lists the file formats that NMRView can read directly without the user needing to specify any parameters. Other file formats can be read by creating an NMRView parameter file that contains a description of the layout of the file.

3.1. Data Set Formats

The discussion of NMR file formats does not make for very exciting reading. On the other hand many NMR spectroscopists have experienced the frustration of having collected perfectly good spectra but not being able to visualize them

```
000                          TkCon 1.6 Main
>Main< (brucejohnson) 8 % nv_peak lists
hsqc hnco cbcaconh hncacb chsqc cconhtocsy hnha hcchcosy
>Main< (brucejohnson) 9 % nv_peak n hcchcosy
357
>Main< (brucejohnson) 10 % nv_peak elem int hnco.4
1.47000
>Main< (brucejohnson) 11 % source test.tcl
>Main< (brucejohnson) 12 %
```

Fig. 3. The NMRView console. This window provides the command line interface to NMRView. Shown are examples of executing several NMRView commands and reading in ("sourcing") a file of Tcl commands.

Table 1
File Formats for Several Common NMR Processing Programs

Program	*Format*	*File header*[a]	*Block header*[b]	*Parameters*[c]
NmrPipe	Serial	2048	0	File header
Felix	Sub-matrix	2048	0	File header
VNMR	Serial × 2	32 / 28	24	File header & Param file
XWINNMR	Sub-matrix	0	0	Param file
NMRView	Sub-matrix	2048	0	File header & Param file

[a]The number of bytes of data in the file header, a section of parameters at the beginning of the file that describe various attributes of the file.

[b]The number of bytes of data in the block header, a section of parameters at the beginning of the block of data that describe various attributes of the subsequent data block.

[c]The file location of the parameters associated with the data file. Parameters include such experiment-specific items as sweep-width and spectrometer frequencies as well as parameters that describe the layout of data in the file, such as the block sizes for files in a sub-matrix format.

because of confusion over the data formats. In light of this perhaps the reader will bear with us as we summarize some of the features of the more common NMR file formats. As indicated in **Table 1**, files containing NMR data are commonly found in two different formats. In the serial format the experimental progression of data points (all the points in one row, followed by all the points in the next row) corresponds exactly to the order of data points in the actual data file (**Fig. 4**). In the submatrix format a portion (the block size) of a row is followed by further portions of subsequent rows before the remaining data values of the row are found (**Fig. 4**).

Of the two, the submatrix format provides for a layout of data on the disk that is much more efficient for NMRView to access. This is particularly true if the data are to be displayed in different orientations or if multiple data sets are to be analyzed simultaneously. Instead of accessing files with a serial format directly, it is generally better to convert them to the more efficient submatrix format.

3.1.1. Working With Felix Files

The data file format for Felix files *(10)* is the submatrix format, which is therefore already in an optimal format. NMRView can read these files directly. As noted in **Table 1** the parameters describing the layout of the Felix file are contained within the file itself. NMRView can extract these parameters allowing it to properly read the data without any additional instructions. Furthermore, when copying the Felix format file to another directory it is not necessary to

Serial

```
57 58 59 60 61 62 63 64

49 50 51 52 53 54 55 56

41 42 43 44 45 46 47 48

33 34 35 36 37 38 39 40

25 26 27 28 29 30 31 32

17 18 19 20 21 22 23 24

 9 10 11 12 13 14 15 16

 1  2  3  4  5  6  7  8
```

Submatrix

```
45 46 47 48 | 61 62 63 64

41 42 43 44 | 57 58 59 60

37 38 39 40 | 53 54 55 56

33 34 35 36 | 49 50 51 52

13 14 15 16 | 29 30 31 32

 9 10 11 12 | 25 26 27 28

 5  6  7  8 | 21 22 23 24

 1  2  3  4 | 17 18 19 20
```

Fig. 4. Serial and submatrix data formats illustrated for a 2D matrix consisting of 64 data points with 8 data points on each axis. The submatrix file is logically divided into 4 blocks, each containing 16 points. The position (row and column) of each number corresponds to frequencies for each axis. The value of each number is the index of that data point in the actual data file (as it exists on the disk drive or other medium).

copy any additional files with it. Note, that NMRView will only read Felix files that are in a real, rather than complex, data format.

3.1.2. Working With NMRPipe Files

The data file format for nmrPipe (11) files is the serial format. It is possible, by creating an appropriate parameter file, for NMRView to read these files directly. However, it is preferable to convert them to the native NMRView format if they are to be used for more than a brief examination. Because of NMRPipe's elegant use of pipes, the user does not need to actually create an NMRPipe file and then convert it.

The processing scheme in NMRPipe is implemented by having a series of instances of the NMRPipe program executing at the same time. Each instance carries out a single processing step, for example, the Fourier transform. The output of one instance is directed to the input of the next instance by the use of UNIX pipes. At the beginning of the pipeline of programs is a program for reading data files, and at the end of the pipeline is a program for writing files. To produce a file in the format used by NMRView the last program in the pipeline should be pipe2xyz executing with the "-nv" flag.

var2pipe - in fid I nmrPipe -SB I nmrPipe -FT I pipe2xyz -nv > NMRViewFile

By executing the processing scheme in this manner the data are moved directly, via the pipeline, from the original FID to the NMRView format output file. Parameters are stored in the NMRPipe file and are correctly carried all the way through the processing into the NMRView file. Because of this, it is useful to ensure that the referencing parameters are set up correctly at the head of the pipeline.

3.1.3. Working With VNMR Files

The data file format for VNMR *(12)* phase files is the serial format. As with NMRPipe files NMRView can read these files directly, but it is preferable to convert them to a submatrix format such as that used by NMRView. VNMR stores parameters in both file and block headers and in an associated text-format parameter file. One-dimensional (1D) and two-dimensional (2D) files can be converted to NMRView format using NMRView itself.

To convert a VNMR file to the NMRView file format choose the "Convert Varian Phasefile" item from the "Dataset" menu. The program will prompt you for input and output files and then convert the file. It is very important to note that Varian phasefiles can contain "holes" where the actual data have not been written to the disk file. To ensure that this is not the case, draw the entire file (full plot limits) and then issue a flush command within VNMR.

3.1.4. Working With XWIN-NMR Files

The data file format for processed XWIN–NMR *(13)* files (1r, 2rr, etc.) is the submatrix format. NMRView can read these files directly. At present, NMRViewJ can directly extract the necessary parameters to open these files, whereas with NMRView it is necessary to set up of a parameter file describing the content (*see* **Subheading 3.2.**).

3.2. Adding File Formats

Every spectral data file that NMRView opens can have associated with it a parameter file, the name of which is the base name of the data file with a ".par" extension (e.g., the par file for noesy.nv is noesy.par). This file can be used to store different reference parameters (which will override those contained in the actual data file) and store parameters that define the layout of data within the data file. **Figure 5** shows a sample parameter file for a three-dimensional (3D) file.

Opening a data file in a format that NMRView does not automatically recognize can be accomplished by creating a parameter file defining the layout of the content in the data file. The most important lines in the parameter file are the first two, which specify the header and dimensionality of the data file. Some examples:

A 512×512 file with 64×64 submatrix blocks and no header (typical of a Bruker file):

```
header 0 0
dim 2 512 512 64 64
```

```
header 2048 0                        #  2048 bytes at the head of the file,
                                            0 bytes before each block
dim 3 512 256 128 32 16 8            #  3 dimensional, size 512 x 256 x 128,
                                         with blocks of size 32x16x8
posneg 0                             #  display both positive and negative contours
lvl 0.04                             #  default contour level is 0.04
scale 1e+06                          #  scale all data points by dividing by 1.0x106
rdims 3                              #  all three dimensions are frequency
                                         dimensions (rather than arrays of other
                                         parameters
datatype 0                           #  the data is real (not complex)
byteorder 0                          #  data is BigEndian
sw 1 4000.0000                       #  sweepwidth in dimension 1 is 4000.0
sf 1 598.8557                        #  spectrometer frequency in dimension 1 is 598.8557
ref 1 4.8061 513.0000                #  point 513.0 of dimension 1 corresponds
                                         to 4.8061 ppm
label 1 1HN                          #  label dimension 1 with "1HN"
sw 2 2112.1184                       #  following same as above, for dimensions 2 and 3
sf 2 150.6075
ref 2 178.0583 129.0000
label 2 13CO
sw 3 2200.0000
sf 3 60.6883
ref 3 121.1320 65.0000
label 3 15N
```

Fig. 5. This is an example of an NMRView parameter file. The # characters and comments after them are not allowed in an actual parameter file.

A 2048 × 2048 file in a serial format with a 2048 block file header (typical of an NMRPipe file):

header 2048 0
dim 2 2048 2048 2048 1

3.3. Loading Data Sets

To load a data set into NMRView, choose the "Open Dataset" item from the Datasets menu. From the dialog that appears you can choose one or more data sets to open. Remember, NMRView has no arbitrary limits on the number of data sets open at once, so go ahead and open all the data sets you need. As a

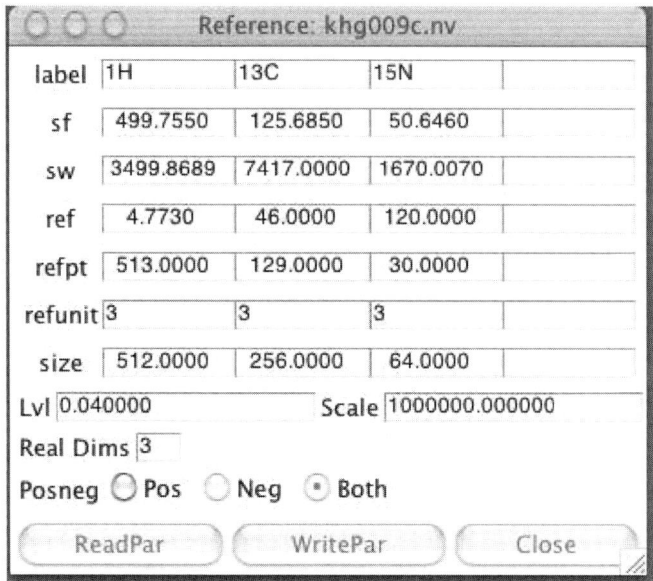

Fig. 6. The GUI for interacting with reference parameters for data sets. Clicking the "WritePar" button will create an NMRView parameter file (**Fig. 5**).

convenience, you can also choose "Open and Draw Dataset" from the same menu. This will automatically create a spectrum display window and draw the selected data set in it. You can also type the "nv_dataset open datasetName" command in the console or add one or more of these commands to a Tcl script file to automate opening data sets.

3.4. Changing Reference Information

NMRView uses a conservative approach in changing the reference information of a data file. The original data file is never modified; instead, changes to the referencing are made persistent by creating the above-mentioned parameter file. A dialog panel (**Fig. 6**) exists for adjusting all the reference information and then creating the parameter file. It is important to remember that if a data file is copied from one disk location to another the parameter file needs to be copied to the new location as well.

4. Visualizing Spectra

4.1. Overview

NMRView has few limits on the user's ability to visualize NMR spectra. As with the number of data sets one can open, there are no arbitrary limits on the

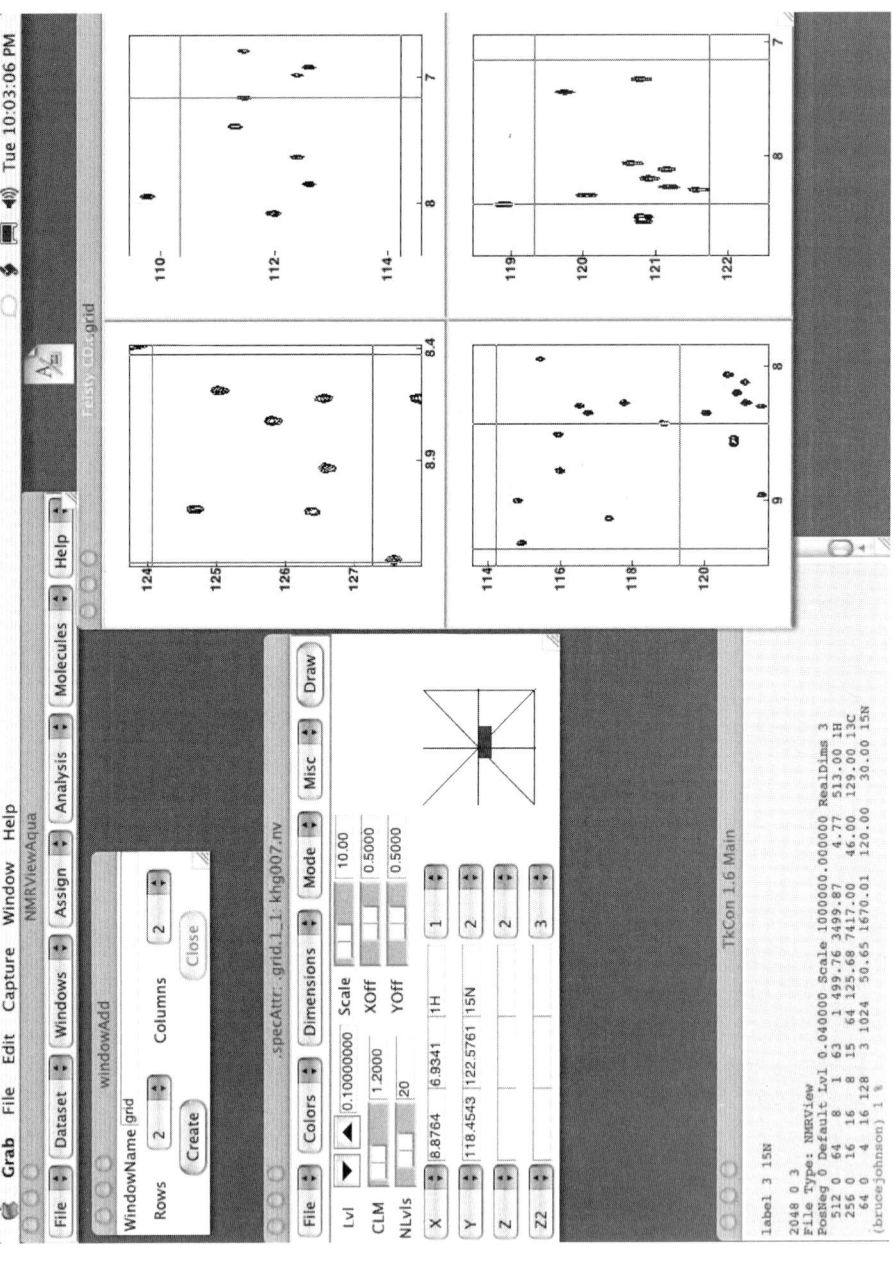

Fig. 7. A screen image showing an NMRView session. Visible are the dialogs for creating new spectral display windows and altering the spectral display attributes as well as a window containing a 2×2 grid of spectra.

number of spectral display windows that can be open at once. Likewise they can be displayed with arbitrary locations and sizes on the desktop. Large numbers of windows in arbitrary locations can sometimes be more confusing than useful so NMRView also provides a variety of ways to group related windows. Although only 1D and 2D displays of spectra are possible, the choice of which data set dimensions correspond to which display axes is up to the user. Different views, including variations on such parameters as data set dimensions; plot regions; and contour levels of the same spectrum, can be displayed simultaneously in different display windows. A hallmark of NMRView is the fact that no matter how the spectra are displayed in different windows the positions of the crosshairs in each window are correlated with the positions in every other window in an automatic and intuitive way.

4.2. Creating Spectral Display Windows

New spectral display windows can be created from the "Add" item of the Windows menu. Selecting this menu item will bring up a dialog box in which the name for the window to be created is specified. Tip: Window names should start with a lower case letter. By default a single spectrum is created, but by specifying values for the row and/or column entry in this dialog a grid of windows can be created (**Fig. 7**). In this way, a group of windows containing related spectra can be created and kept together in a single "top-level" window.

4.3. Cursor Correlation

The crosshairs in different windows automatically track each other in what is generally an appropriate manner (**Fig. 8**). No commands are required to start correlated crosshair tracking. Crosshair correlation is dependent on the label given to each axis of the spectrum during the referencing process. The crosshair in each window only tracks the motion of the moved crosshair if the plot limits of the window overlap the position of the moved crosshair. Crosshair tracking can be disabled in a window by changing the window's axis label(s).

An important implication of the above mechanism is that it is important to use a consistent scheme for labeling the various dimensions of experiments. Also, multidimensional data sets should always have unique labels for the different dimensions. A similar mechanism exists for displaying peaks on spectra, so the need for consistent labeling of spectra, and unique labeling within a data set, are also necessary for the proper rendering of peak displays.

4.4. Window Commands

A large variety of commands exist for creating and manipulating spectral display windows. These commands consist of the command "nv_win" followed by an option, and zero or more arguments. For example: "nv win dataset

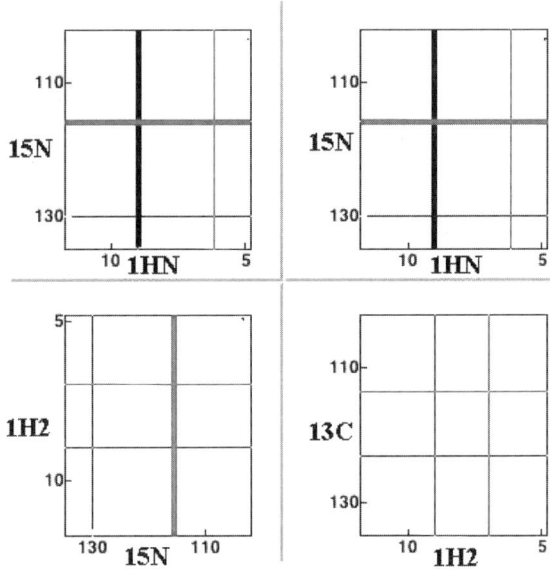

Fig. 8. Schematic illustration of cursor correlation. Moving the thick vertical cursor in the top left window will cause a correlated movement of the thick vertical cursor in the top right window. Moving the thick horizontal cursor in the top left window will cause a correlated movement of the thick horizontal cursor in the top right window, and the thick vertical cursor in the bottom left window. This pattern of correlated movement is determined by the labeling of the two display axes in each window.

datasetName" assigns a specific data set (which must already be opened) to the window, "nv_win ppm × 3.0 5.0" sets the *x*-axis plot limits from 3.0 to 5.0 ppm, and "nv_win draw" will redraw the spectrum with the current parameters.

4.5. Preserving the State

An NMRView user may spend considerable time creating and configuring a group of display windows. To avoid having to recreate this display in the case of a computer malfunction or when starting up a new session of work, it is useful to save the current state of the window parameters. The user will find the "Save State" item in the File menu. Selecting this entry will create a text file named ".nmrview" in the current directory. This file contains all the parameters describing the NMRView display environment, including window names, sizes, locations, and display attributes, as well as which data sets were opened in the session. When restarting NMRView, simply choose the "Load State" item in the File menu, and the previous files will be loaded and the display environment recreated. At present this file can only be named ".nmrview." To maintain different states for different projects the user should run NMRView from within different directories. (Tip: Examining the ".nmrview" file with a text editor can be educational.)

5. Working With Spectral Peaks

Although it is possible to analyze NMR data by directly decomposing the raw data into lists of parameters, such as frequencies and line widths *(14–16)*, most users still rely on Fourier transformation of the FID and then identifying peak positions in the transformed data. Accordingly, NMRView provides many tools for locating and analyzing these spectral peaks.

5.1. Peak-Pick Interface

The spectral region and intensity threshold used for peak picking corresponds to the region currently displayed. The cursors can also be used to limit the peak-pick region to a subset of the display region. **Figure 9** shows the peak picking interface.

5.2. Peak-Pick Algorithm

The algorithm used for locating peaks is quite simple but relatively robust and rapid. Peaks are considered points of local maxima (that is, any point that has a higher intensity than all adjacent points). When NMRView locates peaks, it also performs the following steps: identify the peak bounds (the width of the peak at the level of the intensity threshold); estimate the half-height peak width; determine whether the peak is on the edge of the spectrum or adjacent to other peaks; calculate the center position by interpolating the intensities of the adjacent data points. Locating peak positions does not, however, include estimating the peak volume. This must be done as a separate operation. Note: When NMRView picks peaks it reports the number of "good" and "bad" peaks found. Bad peaks are not really ones that have a tendency to misbehave, rather they have some exceptional condition: narrow or wider line widths than certain limits, being adjacent to another peak, or on the edge of the spectrum.

5.3. Peak Filtering and Analysis

The simple nature of the NMRView peak picker can lead to the inclusion of peaks that correspond to noise or artifacts. Instead of eliminating these during the initial identification of peak positions, NMRView provides various routines for filtering out these peaks. Peaks can be deleted interactively via the GUI, and routines are available for deleting peaks that are thinner, wider, or less intense than specified criteria or are on the diagonal. Furthermore, simple Tcl scripts can be written by users to filter out peaks based on other criteria.

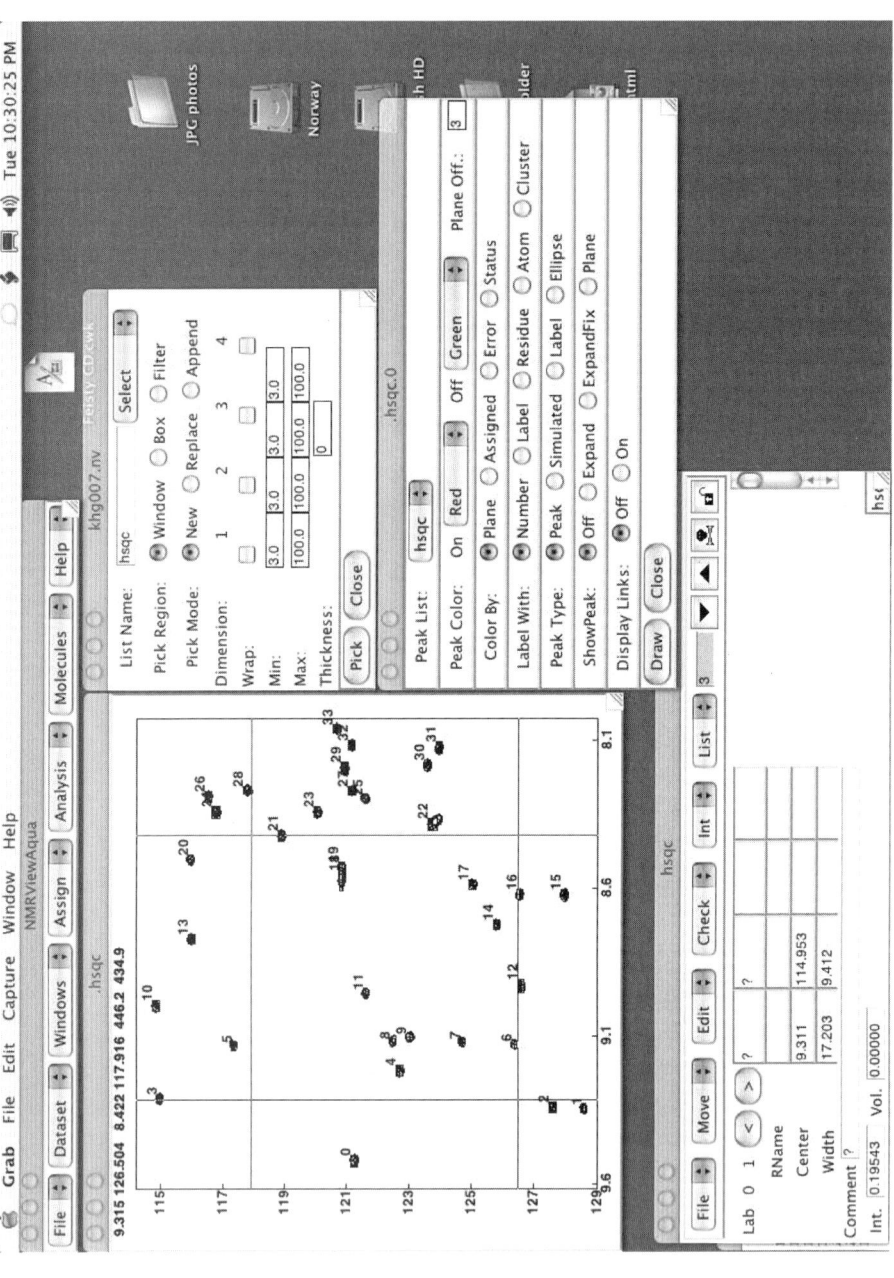

Fig. 9. A screen image showing an NMRView session with various peak tools in use. Visible are the peak-pick dialog, the peak display attributes dialog, and the peak analysis dialog. The spectrum is displayed with peak boxes and their corresponding indexes visible.

328

```
proc Nv_PeakCutLow {plist threshold} {
    set ncut 0
    foreachpeak iPeak $plist {
        set lvl [eval nv_peak elem int $plist.$iPeak]
        if {[ expr {abs($lvl)<$threshold} ] } {
            eval nv_peak delete $plist.$iPeak
            incr ncut
        }
    }
    puts "Cut $ncut of $n peaks"
}
```

Fig. 10. An example of a Tcl script for filtering out NMR peaks. This script marks for deletion all peaks whose intensity is below a specified threshold.

Figure 10 is an example of a simple Tcl procedure that deletes any peaks in a specified list whose intensity is below a specified threshold. The user could invoke this script by typing in the NMRView console "Nv_PeakCutLow noesy 0.4." Two commands in the script are worth emphasizing. The "foreachpeak" command loops over all the peaks in a specified list (setting the iPeak variable to the peak number of each in turn). By using this command the user does not need to worry about the number of peaks in the list or whether there are gaps in the peak numbering. The "nv_peak elem" command provides read-and-write access to all the parameters that characterize a particular peak. In this example, the "nv_peak" command is used to obtain the peak intensity, but all other elements of each peak, such as chemical shift, width, volume, assignment label, and comments, are accessible as well. Together, these two commands can form the basis for scripts that perform a wide variety of operations on peak lists.

5.4. Examining Peaks

The positions of peaks are displayed in the spectral display windows as boxes whose position corresponds to the peak center and whose outline corresponds to the position where the peak intensity falls below the contour threshold used when the peak was picked. The peak display is charactcrized by a variety of attributes: color, label type, and whether 3D peaks whose centers are not exactly on the display plane should be displayed. The color used for each peak can be dynamically determined by a specified peak property, such as assignment status, peak plane, or a user settable status flag. The choice of these attributes can be specified in the "Peak Attributes" dialog.

NMRView provides a very powerful and relatively rapid method for interactively examining peaks. One or more windows can be set (on the "Peak Attributes" dialog) to have their spectral display region adjusted whenever the choice of peak displayed in the "Peak Analysis" dialog changes. The controlled windows have their display region changed so that they are centered on the chemical shift of the corresponding peak. When setting this up it is a good idea to first set the "ShowPeaks" attribute to the Expand mode. With this setting, the width of the display region will be set to a multiple of five times that of the width of the peak's bounding box. Step through a few peaks in the "Peak Analysis" dialog until you find a peak that has a fairly typical intensity for the particular peak list. Now change the "Show Peaks" to the ExpandFixed mode. With this setting, the display will center on the peak position, but the display regions width will remain constant. Now, as you step through the individual peaks in the list, their relative widths will be visually obvious.

This facility for peak examination is very useful in the analysis of biomolecular NMR spectra. Although it is possible to automate the elimination of artifactual peaks as described earlier (**Subheading 5.3.**), and perhaps a more sophisticated model *(17)* should be included, an interactive method can also be valuable. Actually looking at the shape and distribution of peaks can be very important for discovering artifacts in spectra that may lead to insights into new NMR experiments, and for recognizing extra peaks or missing peaks that indicate conformational dynamics, multiple species in solution, or ligand-binding effects.

Using this method for quickly examining peaks has been very useful for cleaning up peak lists prior to using the lists in relaxation analysis or assignment using the CBCA tool described below. It has also been very useful in comparing peaks of spectra collected under different conditions—for example, comparing spectra of proteins collected in the presence of different ligands *(18)*, or for comparing wild-type and mutant forms of proteins.

5.5. Measuring Peak Volumes

The peak-picking routines of NMRView do not determine the volumes of the peaks. Instead, so that the peak picking itself can be as fast as possible, a subsequent analysis routine is used. The peak volume routine is initiated by choosing either "Get Volumes" or "Get eVolumes" from the Int menu of the Peak Analysis interface panel. Both methods will calculate the sum of intensities in the data set of a footprint region calculated from the peak attributes. By default, the intensities are obtained with the data set from which the peak list was originally picked, and the data set must be open within NMRView prior to using the

```
proc getVolumes {plist datasets} {

    global Nv_Value

    set result ""

    foreachpeak iPeak $plist {

        set volumes $iPeak

        foreach dataset $datasets {

            eval nv_peak analysis $dataset $plist.$iPeak

            lappend volumes $Nv_Value(evolume)

        }

        lappend result $volumes

    }

}

set plist relax1

set datasets "relax1.nv relax2.nv relax3.nv"

set result [getVolumes $plist $datasets]
```

Fig. 11. An example of a Tcl script for analyzing NMR peaks. This script calls a routine which measures the volume of an NMR peak and returns a list of the peak indexes and their corresponding volumes.

command. The footprint used by the first method is a rectangular region corresponding to the peak bounds identified when the peak was picked (**Subheading 5.2.**), whereas the latter (and preferred) method uses an elliptical region whose width is a certain "optimal" multiple of the peak width. The elliptical region minimizes the contribution of noise and neighboring peaks to the calculated peak intensity.

As with all NMRView operations, peak volumes can be obtained using custom scripts. For example, the script shown in **Fig. 11** could be used to obtain volumes (or intensities) from a whole series of data sets with regions corresponding to the peaks in a single peak list, such as in the analysis of relaxation data.

5.6. Identifying Peaks

Analysis of the peaks in the NMR spectra is obviously much more useful if one has the assignment of which atoms in the molecule (or molecules) give rise to which peaks in the spectrum. **Subheading 7.** focuses on some tools in

NMRView for the initial assignment of peaks from triple-resonance (^{13}C, ^{15}N, ^{1}H) spectra and **Subheading 5.7.** focuses on the more specific problem of identifying the pairs of hydrogen atoms that give rise to NOE peaks. Here, we describe some tools that are of general utility in working with the identification of NMR peaks.

Every peak list stores within itself several parameters that facilitate peak searching and matching. The peak template parameter is a list of dimension name and tolerance pairs that are used in searching peak lists for peaks with certain chemical shift values. The search template can have fewer dimensions than the peak list. So, for example, a 3D HNCO experiment could have its template set with "nv_peak template hnco ^{1}HN 0.1 ^{15}N 0.3". Searching this peak list with two chemical shifts, "nv_peak find hnco 8.3 117.0," would find any peaks whose ^{1}HN chemical shift is between 8.2 and 8.4 ppm, and whose ^{15}N chemical shift is between 116.7 and 117.3 ppm. Because the ^{13}C dimension is not specified in the template, the ^{13}C chemical shift is ignored in the search.

The peak-list's pattern and tolerance parameters are used when a search is done to find atoms whose assigned chemical shifts are consistent with those of a given peak. The peak pattern of an individual dimension has the format "sequenceNumber.atomType." For example, a peak pattern could be set for a peak list named "hnco" with the command: "nv_peak pattern hnco i.hn i.n i-1.c." This implies that any atom assignments for the first dimension should be of atom type "hn," for the second dimension should be of type "n," and the third dimension of type "c." Because the first and second dimensions share the same symbolic sequence number (which can be either an "i" or a "j,"), the atom assignments for these two dimensions should always have the same sequence number. The third dimensions atom assignment should have a residue that is one less than the residue assignment for the first and second dimensions. The tolerance value indicates the allowable deviation between the peaks chemical shifts and the chemical shift of any atoms assigned to the peak. For example, "nv_peak tolerance hnco 0.1 0.3 0.4," would set the tolerance for the first dimension to 0.1, the second dimension to 0.3, and the third to 0.4. The peak pattern and tolerance values can also be set within the "Peak Reference" panel. Note: The peak pattern and tolerance values are not enforced for peaks in the present versions of NMRView. They exist as storage for parameters that can be used by Tcl scripts for the analysis and assignment of peaks.

The pattern and tolerance parameters are central to the use of the "Peak Identification Dialog." This dialog (**Fig. 12**), brought up from the Edit → Identify menu of the Peak Analysis panel, can be used to interactively assign atom identifiers to individual peaks for a molecule with existing chemical shift assignments. If this dialog is open, then each time you step to a new peak in the Peak Analysis panel this PeakID panel will be updated with a list of the atoms.

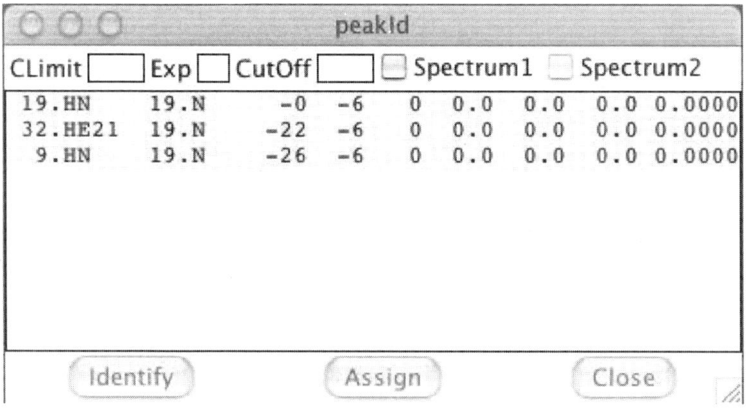

	peakId						
CLimit ☐ Exp ☐ CutOff ☐		☐ Spectrum1 ☐ Spectrum2					
19.HN	19.N	-0	-6	0	0.0	0.0	0.0 0.0000
32.HE21	19.N	-22	-6	0	0.0	0.0	0.0 0.0000
9.HN	19.N	-26	-6	0	0.0	0.0	0.0 0.0000

Identify   Assign   Close

Fig. 12. The Peak Id dialog. Three possible assignments are listed for this 2D peak. The two numbers following the pair of assignments are the deviation of the atom chemical shifts from the peak chemical shift (expressed as a percentage of a specified tolerance). The remaining numbers would provide information about the distance between protons if coordinates for the molecule were available. Had a peak pattern been specified that restricted assignments to atoms from the same residue (i.hn i.n), then only the first of the three assignments would be listed.

The atoms listed are those atoms for which their chemical shifts are within the peak list's tolerance of the peaks' chemical shift values, and for which their atom names are consistent with the peak list's pattern. For each dimension of the peak list a score is given that measures the deviation between the atom's chemical shift and the peak's chemical shift (for the corresponding dimension). The scores are normalized as a percentage of the tolerance (ranging from -100 to $+100$). If 3D coordinates of the atoms are available, and two of the peak dimensions correspond to hydrogen atoms, then additional information is given about the distance between the pair of hydrogen atoms.

If the Spectrum1 or Spectrum2 check boxes are activated, then the region of the spectrum (of the data set assigned to the peak list) around the peak can be displayed in the panels on the right side of the PeakId panel. Double-clicking on an entry in the atoms list will refresh the display of the spectrum and draw the crosshair cursor at the position of the chemical shifts of the atoms in that entry.

The NMRView users should themselves synthesize the information from the chemical shift deviations, hydrogen pair distances, and spectral display to reach a conclusion about which, if any, of the atom entries are the correct assignment for the peak. Selecting the entry and then clicking the "Assign" button will update the assignment labels for the current peak with the names of the atoms in that entry.

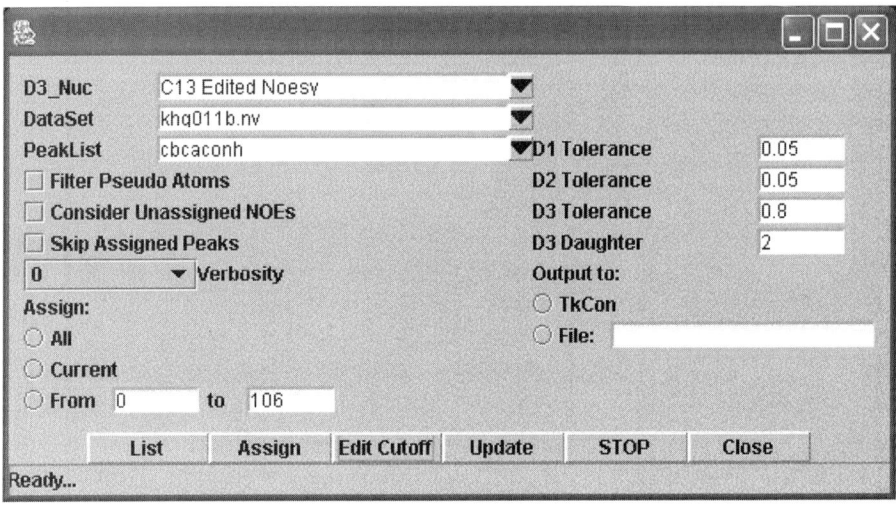

Fig. 13. The NOE assignment dialog provides an alternative to the Peak Id dialog. It provides access to routines that can automatically assign NOEs based on both the chemical shift assignments of the molecule and coordinates from trial structures.

At any given point in a project it may not be possible to assign a unique set of atoms to a particular peak. Instead, because of overlap of chemical shifts there may be an ambiguous group of atoms that could give rise to the peak. NMRView supports the explicit assignment of an ambiguous set of atoms to a particular peak. This can be done within the PeakId panel by selecting more than one row of atom sets in the list and clicking the assign button. The entire group of atoms will be stored with the peak. These assignments can be viewed within the Peak Analysis panel by using the arrows adjacent to the peak label entry.

5.7. NOE Analysis

The aforementioned PeakId panel can be used to interactively assign NOE peaks, but for this specific task an enhanced version of the PeakId panel and underlying Tcl script is also available. This "NOE Analysis Tool" (**Fig. 13**) allows for the control of more parameters to define the criteria for peak assignments and can automatically assign peaks without user intervention. The basic use of the tool is summarized next.

If one has an initial set of structures, they should be loaded prior to use of the tool. Next, set up the experiment type (D3_Nuc Type), the data set, and the peak list to be used. The D1, D2, and D3 values establish the tolerances to be used when searching for atoms whose chemical shifts are consistent with those of the peaks. Next, set the D3 Daughter dimension. This is the dimension that is correlated with D3 in the 3D spectrum of interest. For example, in a ^{15}N-edited nuclear Overhauser effect spectroscopy (NOESY) peak list with dimensions HN, H, and N, the D3 daughter dimension would be the dimension number of the HN dimension. In a ^{13}C-edited NOESY peak list with dimensions H, HC, C, the D3 daughter dimension would be the dimension number of the HN dimension.

Only potential NOEs with distances that satisfy the distance cutoff are considered. The user may enter any expression, where the distance returned by nv_idpeak is given by $d($i). This affords flexibility in searching. For example, to search for long-range NOEs one could enter: (($d($i)> = 5) && ($d($i)< = 12)). Additional filtering may be achieved by selecting the "Use Nilges . . ." option. The lower the coefficient, the more stringent the filter will be. Select "Filter Pseudo Atoms" to convert atoms returned by nv_idpeak into the corresponding pseudoatoms when appropriate. Set "Consider Unassigned NOEs" if you want to consider NOEs to currently unassigned protons. Note that at present, this mode is very slow. Select "Skip Assigned Peaks" to operate only on peaks that are unassigned in at least one dimension.

Having set all the above parameters you can now use the tool to generate assignments for the peaks. Go to the peak of interest manually or with keyboard shortcut g (i.e., g 456). Hit the "List" button to list possible assignments for the selected peak(s). Hit the "Assign" button to list and enter possible assignments for the selected peak(s).

5.8. Displaying Strips

It is often useful to display a set of NMR peaks adjacent to each other, with their positional order on one axis determined by a parameter other than their chemical shift. For example, a series of peaks could be displayed in order of their sequential assignment. The conventional method for this is to display a series of strips of spectra. NMRView provides an interactive strip viewer for setting up and viewing strips (**Fig. 14**).

The Strip Plot GUI automatically generates strip plots based on the chemical shifts of a series of specified peaks. Several means are provided for specifying the peaks, and the peaks can be conveniently reordered. The list box on the right side of the Strip Plot GUI contains the peakID used to generate the plot limits for each window that composes the strip plot, along with their assignments along the X and Z axes of the strip display, if any. One spectral window in the plot will be created for each peak in the list. However, there is a maximum number of windows that can be displayed in the Strip Plot. If the number of peaks exceeds the maximum number of windows, then the scrollbar at the bottom of the Strip Plot can be used to adjust the range of peaks that are displayed.

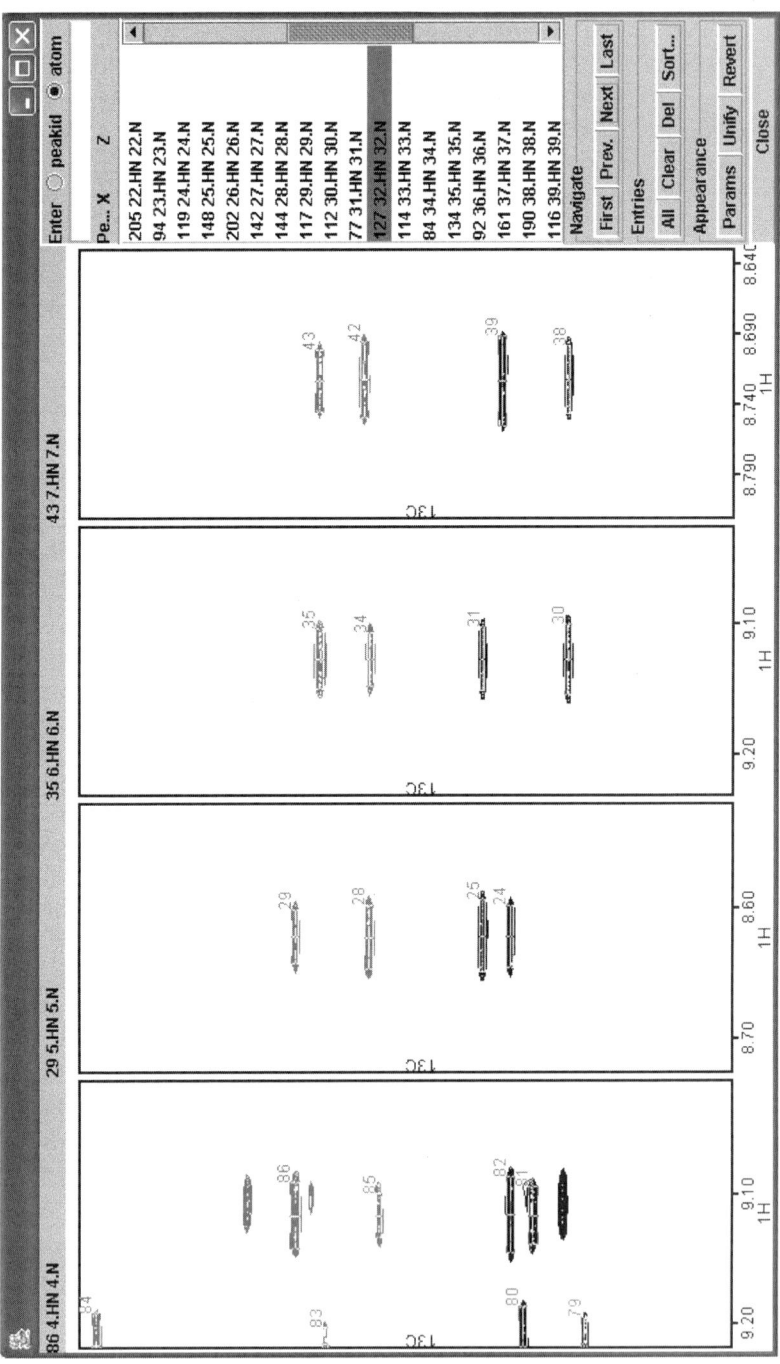

Fig. 14. The strips dialog allows the user to visualize a series of narrow regions of an NMR data set. The regions can be chosen manually or automatically based on either peaks or atoms. The dialog allows the user to sort and/or interactively reorder the strips.

336

6. Using STAR Files

6.1. Overview

NMRView can export and import peak lists, assignments, and polymer sequences from and to simple tabular files. Many users rely on this method for maintaining a persistent copy of their data between NMRView sessions. Import and export of peak lists is invoked via the File menu of the Peak Analysis panel (Read Peaks and Write Peaks), assignments through the File menu of the Atom Assignments panel (Read PPM and Write PPM), and sequences through the Molecule → Read_Topology → Sequence menu item on the main control panel.

Although supported, the above protocol is not the most appropriate method for maintaining a persistent copy of the data. The various read/write methods were actually designed as a means for transferring data between NMRView and other programs. Using an external program for automated assignments, for example, could be done by exporting the peak list from NMRView and then using programs such as AWK, Tcl, or PERL to translate the list to the native format of the external program.

The preferred method for persistent data storage in NMRView is to write and read files in the NMR-STAR format (http://www.bmrb.wisc.edu). This format was chosen for NMRView because it is platform independent and conforms to a standard. Because it is the format developed and used by the BioMagResBank (BMRB) for archival of NMR data, the end results of users' NMR assignment process is already in the format necessary for uploading to the BMRB. Hopefully, this feature will encourage users to upload their data. Finally, NMRView is one of the few programs that can actually read and display BMRB NMR-STAR files. This means it is possible to download files from the BMRB and load them directly into NMRView. The large quantity of data available from the BMRB is thus available to NMR users for aiding in the assignment of homologous proteins or statistical analysis of chemical shift assignments of proteins.

6.2. STAR Editor

Data stored in the NMR-STAR format is structured with related data stored in "SaveFrames." The NMRView Star Editor allows the user to view and edit existing Saveframes and add new Saveframes (**Fig. 15**). The dictionary of elements contained within each type of Saveframe is known by NMRView so it is easy to add information in a manner consistent with the NMR-STAR standard. Within the STAR Editor choose "Add Entry" from the File menu and select the desired Saveframe type from the menu. You will then be prompted to enter a unique name for the new Saveframe.

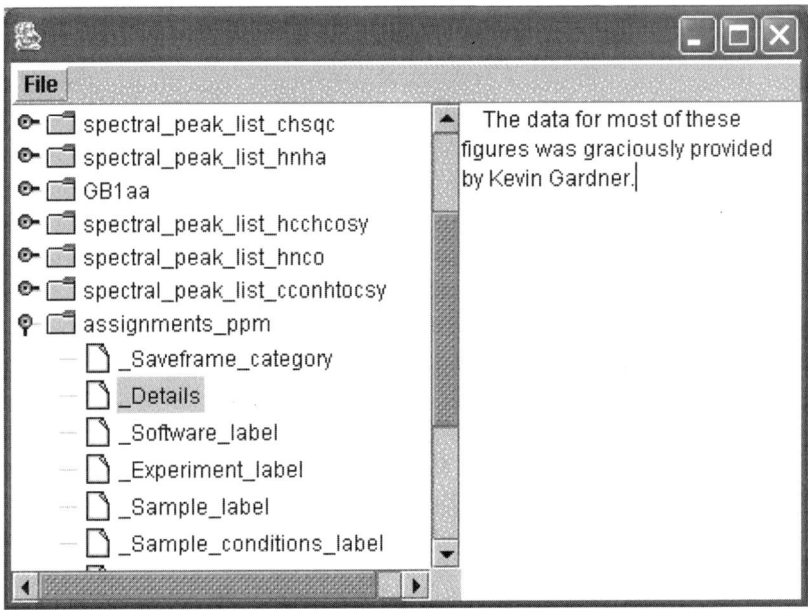

Fig. 15. The STAR editor provides a tree-like view of the data in an NMR STAR file. New saveframes can be edited and the elements of saveframes can be edited.

6.3. Retrieving STAR Files

At the time this chapter was written, the BMRB contained over 2500 entries. These entries can be downloaded within NMRView by choosing the "Fetch STAR file" entry from the main File menu. In the dialog that appears, enter the accession number (e.g., 5709) and click the "Fetch" button. The selected entry will be downloaded to a file on your computer's disk and loaded into NMRView. You can then use the STAR Editor within NMRView to view the contents. Before downloading another entry you should type "StarReset" in the NMRView console window. This will clear out the existing entry. (Tip: If your are behind a firewall you will need to enter a proxy name and port number to access the BMRB.)

7. Analysis of Triple-Resonance Experiments

7.1. Overview

The assignment of the ^{1}H, ^{13}C, and ^{15}N chemical shifts forms an essential step of many protein NMR studies. NMRView provides the CBCA tool for aiding in this process (**Fig. 16**). In particular, this tool is appropriate for 3D experiments giving CB and CA correlations to the HN and N resonances. It could also be used, with or without customization, for analyzing other sequential

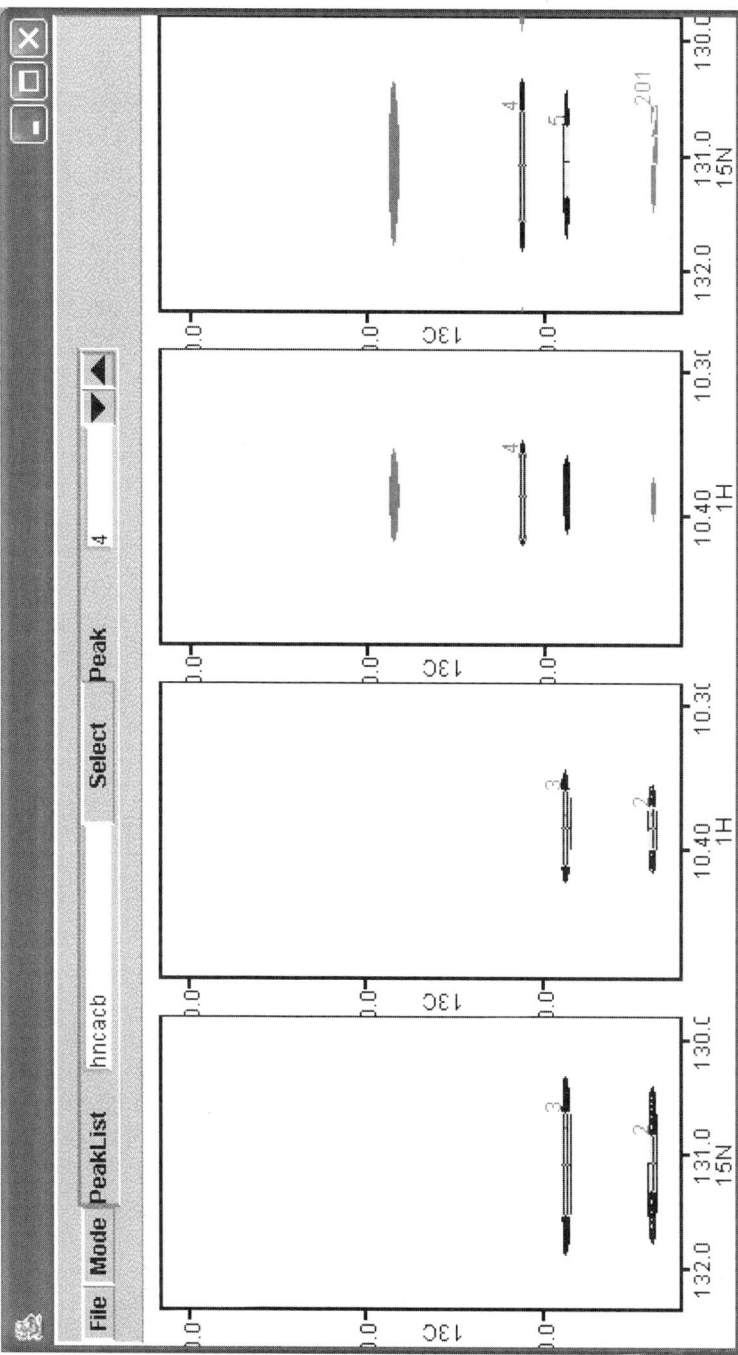

Fig. 16. The CBCA dialog provides access to manual and automated tools for assigning multidimensional NMR spectra (typically 3D, triple-resonance experiments). In the mode illustrated, two orthogonal views of both the CBCACONH (left two panes) and the HNCACB (right two panes) experiments are being used to visualize and verify the peaks that will be used in subsequent steps.

339

assignment experiments. The guiding philosophy of this tool is to provide substantial automation to the process without precluding the user from using his or her insight to facilitate the process. Accordingly, use of the tool involves a series of automated and manual steps. At each manual step the user can confirm and alter the results of the previous automated step. Confirmation of each step can involve both the assessment of quantitative scores and visual analysis of the spectral regions giving rise to the sequential connectivity.

Although the CBCA tool provides a variety of automation routines for different steps of the analysis, the design of the tool is such that other automation routines could be used as well. Indeed, many other computer algorithms exist for sequential assignment using heteronuclear NMR data *(19–28)*. With a little effort in Tcl programming many of these could be easily plugged into the following cycles of automation and visualization.

7.2. Interface

In the graphical user interface for the CBCA analysis system, a control bar appears across the top of the window, and a set of spectra are displayed below. Up to four rows of spectra can be displayed. The controls and spectra that are displayed depend on what mode the system is in, and what types of data sets were collected. The CBCA window is divided into a grid, with four columns and a number of rows that depend on the type of experimental data that were collected. Typically, there is a single row for each type of carbon that is uniquely selected. For example, if you have collected experiments in which CA and CB carbons are detected in separate experiments you will have two rows: one for the $C\alpha$ carbons and one for the $C\beta$ carbons. On the other hand, where both types of carbons are collected in a single experiment you will have a single row. In some modes a separate control panel with one or more list boxes will appear as well.

7.3. Analysis Steps

The analysis proceeds in a series of steps, with both automated and manual components. The overall procedure is illustrated in a flowchart (**Fig. 17**) and summarized as follows:

1. Pick peaks in relevant spectra.
2. Set up CBCA analysis parameters.
3. Clean and verify peaks.
4. Filter peaks to eliminate artifacts.
5. Cluster peaks based on common HN and N frequencies.
6. Verify and adjust peak clusters.
7. Link clusters based on common CB and CA frequencies.
8. Edit and verify matched clusters.
9. Form fragments from overlapping clusters.
10. Assign fragments to the protein sequence.

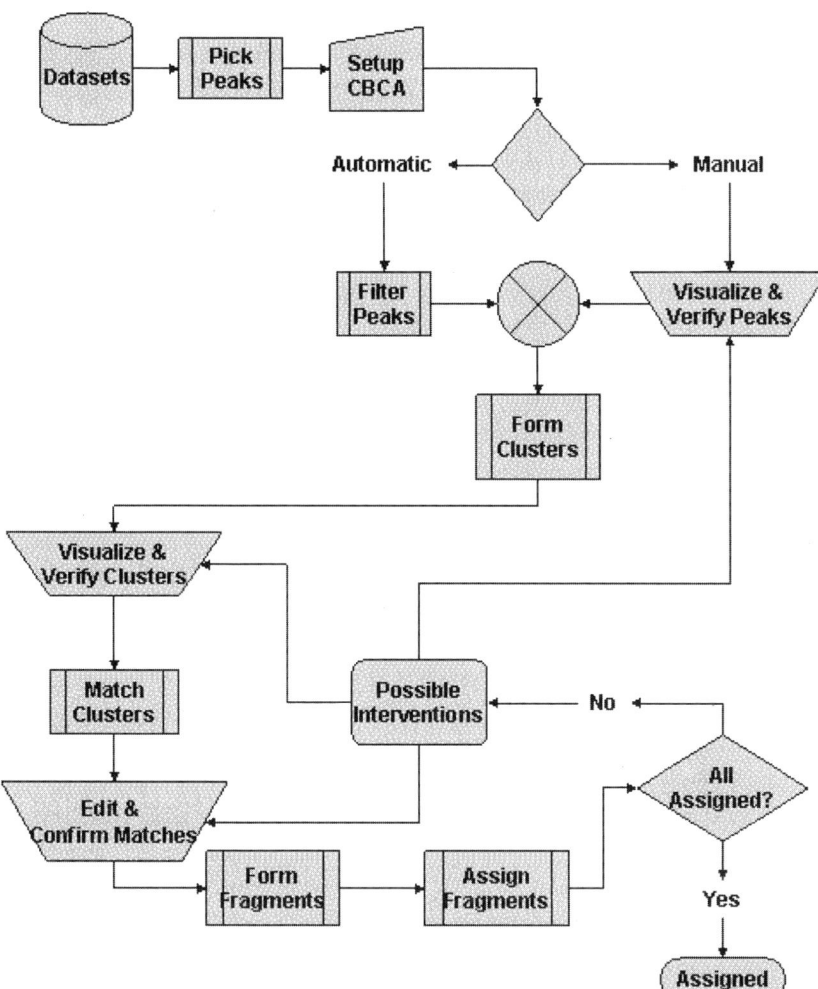

Fig. 17. A flowchart illustrating the steps involved in using NMRView to assign NMR experiments using the CBCA dialogs. Rectangular boxes represent automated steps and trapezoidal boxes represent steps involving user-driven visualization and verification. In practice, with high-quality data sets the manual steps could be skipped for a fully automated procedure. Also, each of the automated routines could be substituted with external routines that use a different algorithm than that provided in NMRView.

7.4. Pick-Peaks in Relevant Spectra

First, use the standard NMRView peak-picking tools to peak-pick your spectra. If the protein under study is deuterated, then the set of experimental data is likely to include at least four spectra (HN(CO)CACB, HN(CO)CA, HN(CA)CB, HNCA), whereas with nondeuterated proteins, it is likely that the data set consists of only two spectra (CACB(CO)NNH, HNCACB). (Tip: Spend some time at this stage to choose reasonable peak-pick thresholds and picking regions. All subsequent steps depend on the quality of the peak picking done at this stage. Also, having chosen a reasonable contour level for display and peak picking of the spectra, it is a good idea to assign this level to the data set in the Dataset Manage dialog. If a contour level is set for a data set, then whenever the data set is displayed in subsequent steps this contour level will be used automatically.)

7.5. Set Up CBCA Analysis Parameters

The next step is to set up various parameters that determine which spectra and peak lists will be displayed. Click on the Prefs button to open a Preferences dialog. Because different users label spectra in different manners, you need to first specify what labels are used for your data sets. The proton, nitrogen, and carbon dimensions should each have a unique label, but these labels should be the same for each of the spectra. Enter the labels for the three dimensions exactly as they are used in the spectra. The Tolerances fields will be filled out later. Click on the radio button ("0," "1," "2," or "3") to display the parameters for the row that you want to control. If you want this row to be displayed, then select the "Show row" check button. By default, the spectrum range used for the y axis (typically the carbon dimension) is the full range. If you want to limit it to a certain chemical shift range, then unselect the "Full" check button and enter the desired upper and lower bounds in the two entry boxes. Now enter the file name and peak lists that will be displayed on this row. Clicking on the select buttons will allow you to choose the data sets and peak lists from a menu (the data sets and peak lists must already be opened). Repeat this process for each row that you want to activate (or deactivate). Values for MatchLists and MatchScripts are set later. (Tip: It is a good idea to use the "Save" item in the CBCA analysis window's File menu on a regular basis. This will save your setup settings and intermediate assignment data to a file for subsequent retrieval.)

7.6. Clean and Verify Peaks

In this mode you can choose a single peak list, typically the most sensitive one (2D heteronuclear single-quantum coherence (HSQC) spectroscopy or 3D HNCO), to use to step through the spectra and verify the picked peaks (**Fig. 16**). Go into the "Examine Peak" mode. Select the desired peak list with the "Select" button. You can then step through the peak list with the up/down buttons, or jump to a specific peak by entering it in the entry box and hitting the Enter key. As you move to each peak the spectral displays will be updated to show the region corresponding to the peak. The y axis of all spectra will be the full carbon axis. The x axis of the outer two spectra will be a range of the nitrogen dimension chosen to bracket the peaks ^{15}N chemical shift. The z axis (the chosen planes) will be at the ^{1}H chemical shift of the peak. The inner two spectra have the ^{1}H and ^{15}N axes switched. The two spectra on the left correspond to the interresidue experiments; the two spectra on the right correspond to the intraresidue experiments. The mouse bindings are set up so that when you point to a peak with the crosshair and double-click the left mouse button it sets the status flag of the peak to a value of 1. By stepping through all of the peaks, one can click on all the peaks that "look good" to confirm them. Then a script (cutncf) can be used to delete all peaks that have not been confirmed in this way (obviously you probably want to save a copy of the database before deleting them). This may or may not be the best protocol, but it seems to work reasonably well. The opposite protocol would also be possible—to delete all peaks that do not look good. "Looking good" is taken to mean that several peaks in the different spectra line up with each other in both the HN and N views, unlike noise peaks, which should be uncorrelated in the different spectra. It should be apparent that a simple script could be written to do this as well, but it is perhaps therapeutic or educational to look at actual peaks occasionally.

7.7. Filter Peaks to Eliminate Artifacts

The 3D peak lists can be filtered for peaks that have no equivalents (common ^{1}H and ^{15}N shifts) in some reference peak list (e.g., a good 2D HSQC or 3D HNCO) These lone peaks are most likely noise peaks and should be deleted before the cluster formation. Change the mode to filter peaks. The ^{1}H and ^{15}N N shifts of the picked peaks in the filtered spectrum will be compared to those of the reference spectrum using tolerances specified in the preference window: Select the desired reference peak list with the "Select" button, and hit the "Filter Peaks" button. The peaks will be truly deleted (not just marked for deletion), so please save data to a new STAR file first. (Tip: Both manual peak verification (7.5) and automatic filtering can be done; *see* **Fig. 17**.)

7.8. Cluster Peaks Based on Common HN and N Frequencies

Next, switch the CBCA mode to "Link Peaks." Enter values for the tolerances for the intralist comparisons in the preference window. Then click Link; this will form links within each peak list. Then repeat this step, perhaps with slightly larger tolerances, for the interlist link. The process of linking the peaks is done by executing a 2D cluster algorithm on the ^{1}HN and ^{15}N frequencies of the peaks. The results are stored as so-called links between the peaks. The link attributes of peaks are automatically stored in the database. However, this step of the analysis also creates a Tcl array variable named "Clust." This array contains additional information that will be supplemented at later stages of the analysis. It is not saved to the database, so you should periodically save it to a text file. This can be done by choosing the "Save" entry from the "File" menu of the CBCA panel. This will save the Clust array and several other relevant arrays to files. Selecting the "Restore" entry will reread these arrays.

7.9. Verify and Adjust Peak Clusters

This is an essential step in the CBCA analysis; without clean clusters the sequential assignment will fail. Switch the CBCA mode to "Edit Clusters (HN-C)." A new control panel will appear (**Fig. 18**). You will see in the left list box a list of all peaks linked together in that cluster. The region of the spectra corresponding to the cluster HN and N frequencies will also appear. If you click on an entry in the list, then it will move from the left to right list (and vice versa). Use this if you can resolve a set of peaks into two clusters. Just move all the ones that should be split out to the other box, and then click on Relink. This action will remove the links between all the peaks in the two list boxes and then separately form new links between the peaks in the left and right boxes. Remember that the Cluster is not saved in the database so you need to manually use the Save command to save it. Do this periodically to save your work. In this "Verify Cluster" mode mouse bindings are set up so that pointing to a peak with the mouse cursor and typing the "a" key will add that peak to the "left" list. Typing the "s" key will switch the peak to the other list. After switching a peak you must hit "Relink" to reform the clusters. Hitting the "Neighbor" button will search through the clusters to find the cluster nearest (by HN and N) frequencies to the current cluster. Then you can easily move peaks back and forth between the clusters if the automatic clustering algorithm did not separate them correctly. If more than two clusters partly overlap, then you will need to go back and forth between the clusters a couple of times to resolve it. Rather than looking at proton–carbon and nitrogen–carbon planes, the "Edit Clusters (HN-N)" mode can be used to inspect the clusters in proton–nitrogen planes. First set the number of rows to the proper value and hit "update." Then go through the clusters in a similar manner as above; use the up/down arrows on the control bar to step through the clusters in the Clust array (or type in a cluster number and hit return).

You can use the cluster editor to look at and change clusters without resorting to command line functions. After linking the peaks, each cluster is a data structure containing the average HN and N shifts and a list of peak identifiers for each of the experiment types. During the course of linking the clusters into fragments, more information is stored in the Clust array, such as the preceding and succeeding clusters' assignment, and so forth.

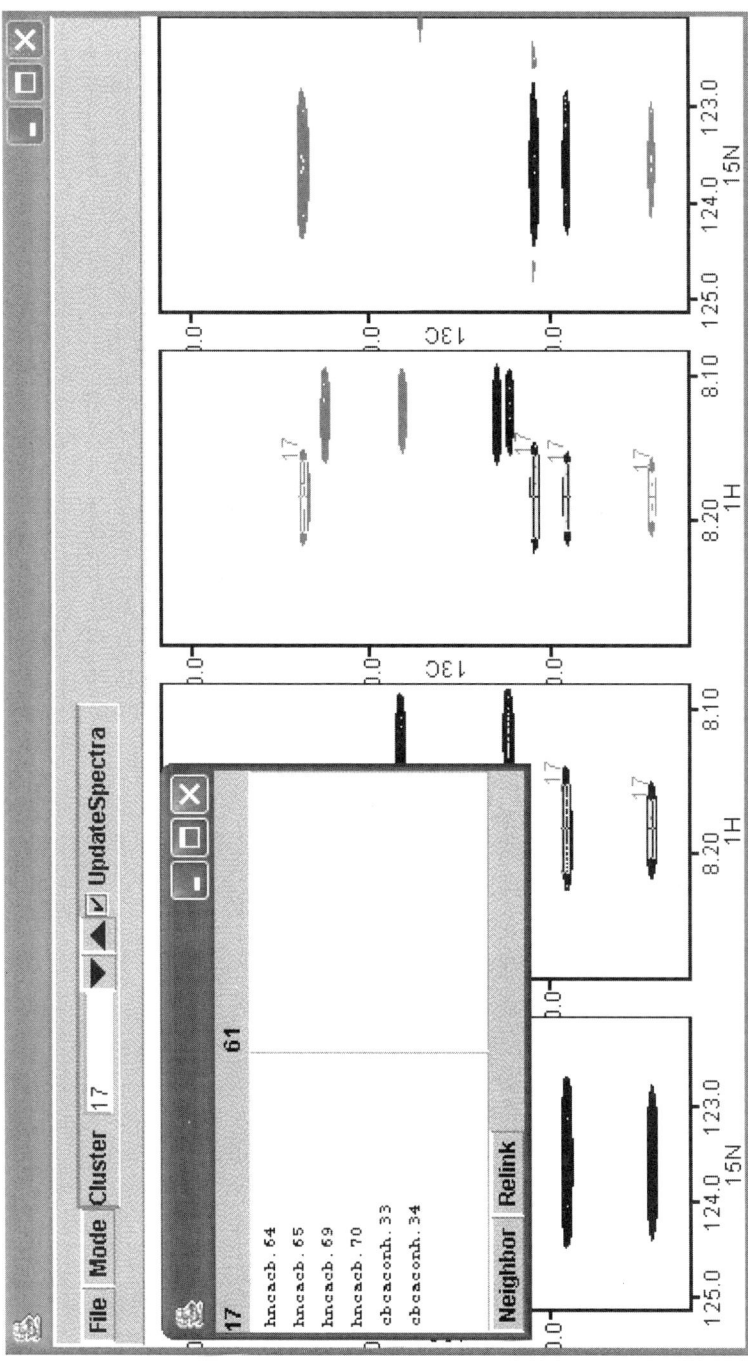

Fig. 18. The CBCA dialog in the "Edit Clusters" mode. Here, the user can interactively verify the clustering of a set of peaks that have a similar pair of chemical shifts of the HN and N atoms. Clusters can be split, merged and supplemented with additional peaks.

7.10. Link Clusters Based on Common CA and CB Frequencies

The next stage involves finding clusters that are likely to correspond to sequentially adjacent residues. A given cluster should share a set of carbon chemical shifts with the cluster corresponding to the next residue in the sequence. For example, the CA carbon chemical shift from a peak in the HNCA experiment should be the same as the CA carbon shift of the peak in the HN(CO)CA experiment that originates from the subsequent amide.

Switching to "MatchCluster" mode provides the controls to initiate a global comparison of all possible pairs of clusters. Prior to starting the search, enter a tolerance for the 13C comparison in the preference window (a value of about 0.6 is reasonable) and select the proper matching algorithm: Use MatchCBCA4 in case four experiments were collected with CBCA information (HNCA, HN(CO)CA, HN(CA)CB, and HN(CO)CACB) and MatchCBCA2 in case two experiments with CBCA information were collected. (Tip: Modification of these scripts can be done to accommodate other experiments that might be used.)

The cluster matching is initiated by clicking on the "Match" button. The progress of the matching will be displayed, and the quit button can be used to interrupt the process. Do not forget to save the CBCA after matching is complete.

7.11. Edit and Verify Matched Clusters

With high-quality, well-resolved spectra, one can expect that at this stage a high percentage of correct cluster matches will have been formed. For these spectra, it would be possible to skip this next step and use an automatic procedure to accept the cluster matches. However, with many real-world proteins, spectral artifacts, chemical shift overlap, and missing peaks can make it beneficial for the user to intervene in this process. The next phase of the CBCA tool allows the user to examine cluster matches, and based on his or her assessment of both the quantitative score information and the graphical alignment of the peaks, to confirm the validity of the automatically generated matches (**Fig. 19**).

To perform this assessment, switch the CBCA mode to "Edit Matches." In this mode, the middle two spectral windows will display the regions corresponding to the inter- and intraresidue peaks of the currently selected cluster. An additional control panel, in a separate window, will list the clusters that the "match cluster" algorithm has found as likely neighbors to the currently selected cluster. The left half of the control panel shows a list of candidate clusters for the preceding residue, and the right half of the control panel shows a list of candidate clusters for the succeeding residue.

The fields on each line in the list box represent the following information:

1. The number of the matching cluster.
2. The score of the match (higher better, max is 3.0).

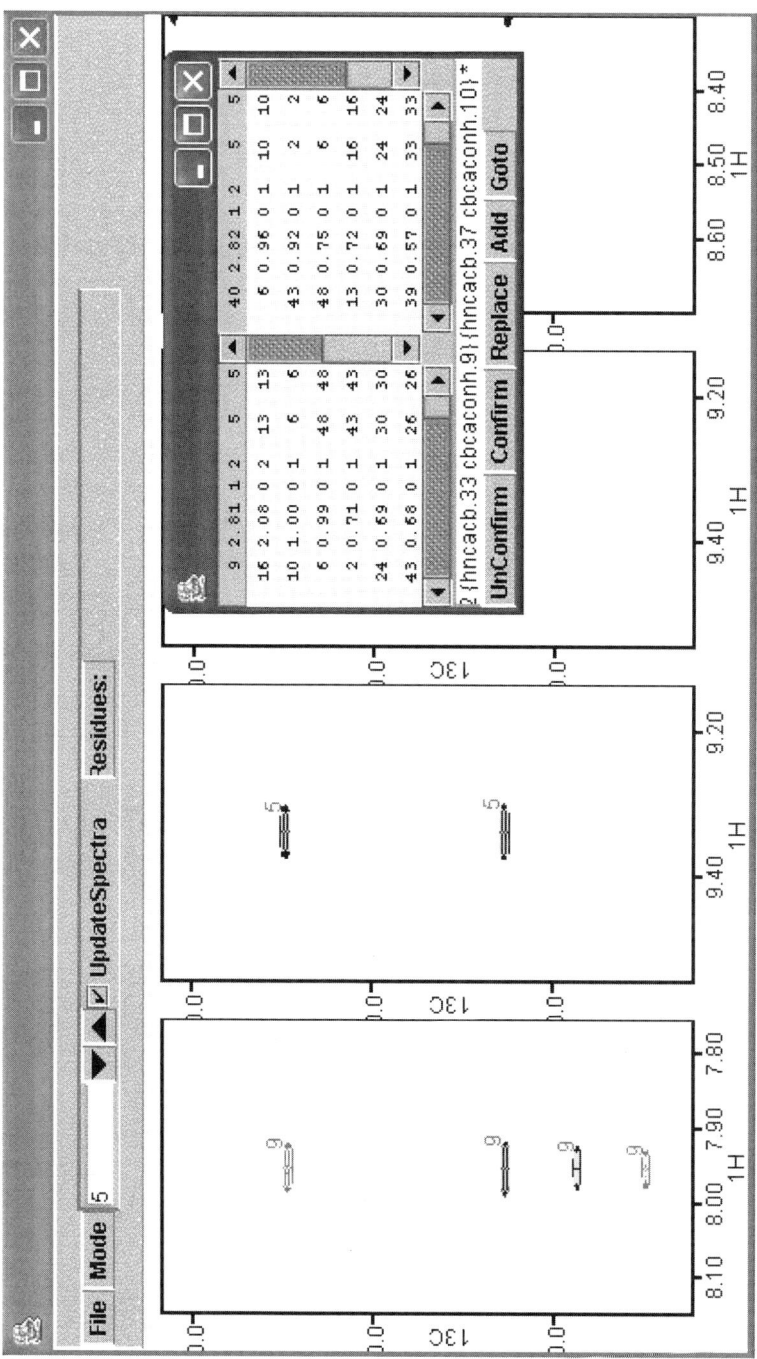

Fig. 19. The CBCA dialog in the "Edit Matches" mode. Here, the user can interactively verify the matching of clusters that were done based on carbon chemical shifts of peaks in the clusters. The alignment score and visual alignment of peaks can be used to verify the automatic matching.

347

3. "1" if match confirmed manually, "0" if not.
4. Number of carbons matching (1 or 2).
5. The cluster that the matching cluster matches.
6. The cluster that the matching cluster has been confirmed as matching (or "–1" if it is unconfirmed).

For a good match, one expects the fifth field to show the current cluster; if cluster 16 is succeeding 27, then cluster 27 should be preceding 16.

If the user clicks on any of the clusters in either the left or right list, then the leftmost or rightmost spectrum will display the region corresponding to the selected cluster. Now, it is possible to visually compare the alignment of peaks from the two clusters in the adjacent pair of windows.

After assessing the quantitative information in the list box entries and the alignment of peaks in the spectral displays, it may be clear that one of the preceding or succeeding clusters is clearly the correct match for the current cluster. If this is so, then by selecting that cluster in the list and clicking on the Confirm button you mark that cluster pair as suitable for including in fragments formed in the subsequent analysis step. The left (preceding) and right (succeeding) cluster matches have to be confirmed individually. If you change your mind, then select the cluster in the list box and click on UnConfirm. Step through all the clusters this way. If the cluster matching process reveals that some clusters still need to be cleaned up, then you can always go back into the "Edit Cluster" mode and do so.

7.12. Form Fragments From Overlapping Clusters

In the context of the CBCA assignment tool, a *fragment* is a series of overlapping clusters that correspond to a stretch of adjacent residues in the protein sequence. Fragments are built automatically. The process begins with the first cluster. If this cluster has a preceding cluster that matches it, and that match has been marked as confirmed in the previous analysis step, then that cluster will be added to the fragment. The search is then repeated for each cluster that is added until a cluster is reached that does not have a confirmed match for a preceding cluster. The same process is now repeated, beginning at the initial cluster, and searching for succeeding clusters with confirmed matches. When this search is finished a fragment has been completed. Now, the entire fragment generation process is started again, beginning with the next cluster that has not been assigned to a fragment.

The fragment assembly process is initiated after selecting the "Assemble Fragment" mode. Click on the "Create Fragment" button to begin the process.

7.13. Assign Fragments to the Protein Sequence

The final step of the analysis is to match each fragment to a specific stretch of amino acids in the protein sequence. The matching algorithm used is similar

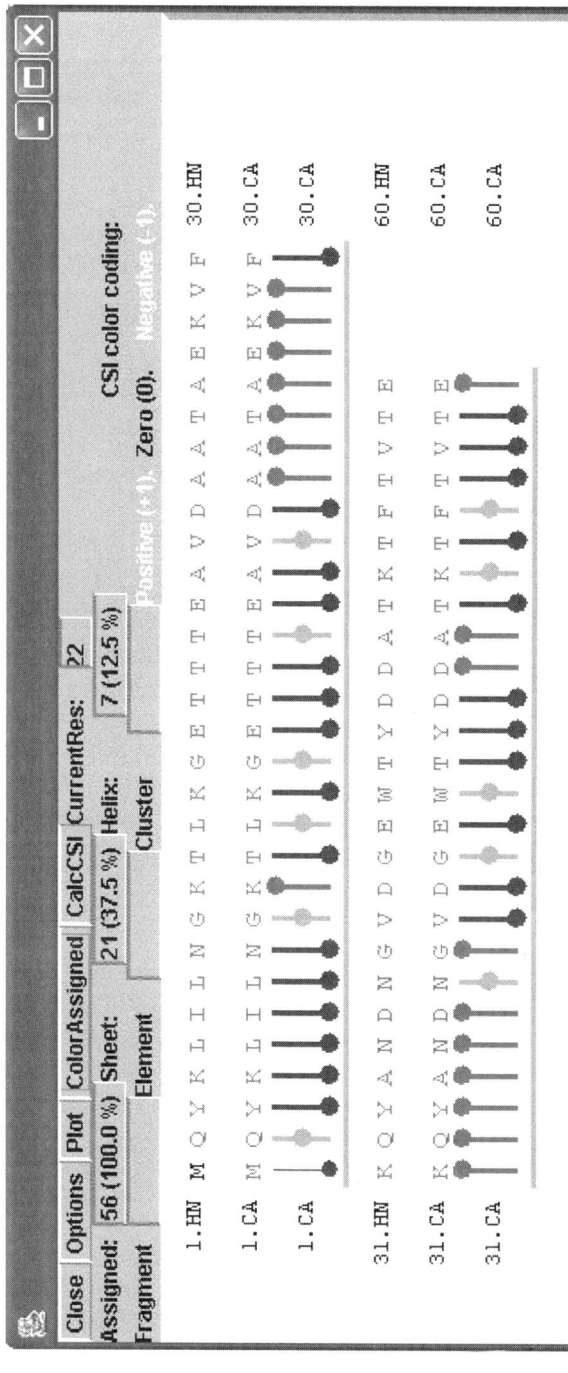

Fig. 20. The sequence editor being used to visualize the results of the assignment process illustrated in **Fig. 17**. Assigned atoms can be colored and chemical shift indexes displayed. The fragments (stretches of linked clusters) and the clusters that give rise to the assignments can be displayed. This information can be used to aid in resolving breaks in the assignment process (box labeled "Possible Interventions" in **Fig. 17**).

349

to others previously used *(29)*. The matching of fragments to sequence specific locations is dependent on a score that measures how well the chemical shifts of the peaks that form the clusters in the fragment match the expected values for the types of amino acids present in that location. The default values are listed in the "Fragment Preferences" dialog, which can be displayed by clicking the obvious button in the main CBCA Preferences dialog. The match also takes into account the sign of the intensity of the peaks in the clusters. The expected value for the sign can be set along with the chemical shift ranges in the "Fragment Preferences" dialog.

The fragment-matching process is initiated by clicking on the button labeled "Align Fragments in Sequence." The results of the fragment assignment can be observed by looking at the NMRView "SeqDisplay" window and coloring the residue symbols based on whether they have been assigned (**Fig. 20**).

8. Conclusion

There are many other tools within NMRView that could not be described in the scope of this chapter. Among these are the Relaxation Analysis tool, the Chemical Shift Analysis tool, and various tools for vector manipulation. Further details can be found in the manual that comes with NMRView, or at the NMRView website www.nmrview.com.

Acknowledgments

I would like to thank the many users who have contributed bug fixes, new analysis scripts, answers to users on the NMRView mailing list, and feedback that has helped make NMRView the program that it is. Special thanks go to Kevin Gardner for allowing me to use his NMR data to illustrate this chapter.

References

1. Johnson, B. A. and Blevins, R. A. (1994) NMRView: a computer program for the visualization and analysis of NMR data. *J. Biomol. NMR* **4,** 603–614.
2. Bartels, C., Xia, T. H., Billeter, M., Güntert, P., and Wüthrich, K. (1995) The program XEASY for computer-supported NMR spectral analysis of biological macromolecules. *J. Biomol. NMR* **6,** 1–10.
3. Goddard, T. D. and Kneller, D. G. (2001) Sparky 3. University of California, San Francisco.
4. Hoch, J. C. (1985) Rowland NMR Toolkit. Rowland Institute, Harvard University, Cambridge, MA.
5. Pons, J. L., Malliavin, T. E., and Delsuc, M. A. (1996) Gifa V 4: a complete package for NMR data set processing. *J. Biomol. NMR* **8,** 445–452.
6. Kjaer, M., Andersen, K. V., and Poulsen, F. M. (1994) Automated and semiautomated analysis of homo- and heteronuclear multidimensional nuclear magnetic resonance spectra of proteins: the program Pronto. *Methods Enzymol.* **239,** 288–307.

7. Ousterhout, J. (1994) *Tcl and the Tk Toolkit,* Addison, Boston.
8. Welch, B. B. (1999) *Practical Programming in Tcl and Tk.* 3rd. ed. Prentice Hall, Upper Saddle River, NJ.
9. Schwarzinger, S., Kroon, G. J. A., Foss, T. R., Wright, P. E., and Dyson, H. J. (2000) Random coil chemical shifts in acidic 8 *M* urea: implementation of random coil shift data in NMRView. *J. Biomol. NMR* **18,** 43–48.
10. Felix. (2002) Accelrys, San Diego.
11. Delaglio, F., Grzesiek, S., Vuister, G. W., Zhu, G., Pfeifer, J., and Bax, A. (1995) NMRPipe: a multidimensional spectral processing system based on UNIX pipes. *J. Biomol. NMR* **6,** 277–293.
12. VNMR. (2003) Varian, Inc., Palo Alto, CA.
13. XWIN-NMR. (2004) Bruker Biospin GmbH, Rheinstetten, Germany.
14. Chylla, R. A., Volkmann, B. F., and Markley, J. L. (1998) Practical model fitting approaches to the direct extraction of NMR parameters simultaneously from all dimensions of multidimensional NMR spectra. *J. Biomol. NMR* **12,** 277–297.
15. Hoch, J. C. and Stern, A. S. (1996) *NMR Data Processing,* Wiley, New York.
16. Chen, J., De Angelis, A. A., Mandelshtam, V. A., and Shaka, A. J. (2003) Progress on the two-dimensional filter diagonalization method. An efficient doubling scheme for two-dimensional constant-time NMR. *J. Magn. Reson.* **162,** 74–89.
17. Koradi, R., Billeter, M., Engeli, M., Güntert, P., and Wüthrich, K. (1998) Automated peak picking and peak integration in macromolecular NMR spectra using AUTOPSY. *J. Magn. Reson.* **135,** 288–297.
18. Johnson, B. A., Wilson, E. M., Li, Y., Moller, D. E., Smith, R. G., and Zhou, G. (2000) Ligand-induced stabilization of PPARgamma monitored by NMR spectroscopy: implications for nuclear receptor activation. *J. Mol. Biol.* **298,** 187–194.
19. Atreya, H. S., Sahu, S. C., Chary, K. V. R., and Govil, G. (2000) A tracked approach for automated NMR assignments in proteins (TATAPRO). *J. Biomol. NMR* **17,** 125–136.
20. Croft, D., Kemmink, J., Neidig, K.-P., and Oschkinat, H. (1997) Tools for the automated assignment of high-resolution three-dimensional protein NMR spectra based on pattern recognition techniques. *J. Biomol. NMR* **10,** 207–219.
21. Hitchens, K. T., Lukin, J. A., Zhan, Y., McCallum, S. A., and Rule, G. S. (2003) MONTE: an automated Monte Carlo based approach to nuclear magnetic resonance assignment of proteins. *J. Biomol. NMR* **25,** 1–9.
22. Hyberts, S. G. and Wagner, G. (2003) IBIS—a tool for automated sequential assignment of protein spectra from triple resonance experiments. *J. Biomol. NMR* **26,** 335–344.
23. Leutner, M., Gschwind, R. M., Liermann, J., Schwarz, C., Gemmecker, G., and Kessler, H. (1998) Automated backbone assignment of labeled proteins using the threshold accepting algorithm. *J. Biomol. NMR* **11,** 31–43.
24. Lukin, J. A., Gove, A. P., Talukdar, S. N., and Ho, C. (1997) Automated probabilistic method for assigning backbone resonances of (^{13}C,^{15}N)-labeled proteins. *J. Biomol. NMR* **9,** 151–166.

25. Malmodin, D., Papavoine, C. H. M., and Billeter, M. (2003) Fully automated sequence-specific resonance assignments of hetero-nuclear protein spectra. *J. Biomol. NMR* **27,** 69–79.

26. Moseley, H. N. B. and Montelione, G. T. (1999) Automated analysis of NMR assignments and structures for proteins. *Curr. Opin. Struct. Biol.* **9,** 635–642.

27. Mumenthaler, C., Güntert, P., Braun, W., and Wüthrich, K. (1997) Automated combined assignment of NOESY spectra and three-dimensional protein structure determination. *J. Biomol. NMR* **10,** 351–362.

28. Zimmerman, D. E., Kulikowski, C. A., Huang, Y. P., Feng, W. Q., Tashiro, M., Shimotakahara, S., et al. (1997) Automated analysis of protein NMR assignments using methods from artificial intelligence. *J. Mol. Biol.* **269,** 592–610.

29. Friedrichs, M. S., Mueller, L., and Wittekind, M. (1994) An automated procedure for the assignment of protein (HN)-H-1,N-15, C-13(alpha), H-1(alpha), C-13(beta) and H-1(beta) resonances. *J. Biomol. NMR* **4,** 703–726.

17

Automated NMR Structure Calculation With CYANA

Peter Güntert

Summary

This chapter gives an introduction to automated nuclear magnetic resonance (NMR) structure calculation with the program CYANA. Given a sufficiently complete list of assigned chemical shifts and one or several lists of cross-peak positions and columes from two-, three-, or four-dimensional nuclear Overhauser effect spectroscopy (NOESY) spectra, the assignment of the NOESY cross-peaks and the three-dimensional structure of the protein in solution can be calculated automatically with CYANA.

Key Words: Protein structure; NMR structure determination; conformational constraints; automated structure determination; automated assignment; NOESY assignment; CYANA program; network anchoring; constraint combination; torsion angle dynamics.

1. Introduction

Until recently, nuclear magnetic resonance (NMR) protein structure determination has remained a laborious undertaking that occupied a trained spectroscopist over several months for each new protein structure. It was recognized that many of the time-consuming interactive steps carried out by an expert during the process of spectral analysis could be accomplished by automated, computational approaches *(1)*. Today, automated methods for NMR structure determination are playing a more and more prominent role and are superseding the conventional manual approaches to solving three-dimensional (3D) protein structures in solution.

In *de novo* 3D structure determinations of proteins in solution by NMR spectroscopy, the key conformational data are upper distance limits derived from nuclear Overhauser effects (NOEs) *(2–4)*. To extract distance constraints from a nuclear Overhauser effect spectroscopy (NOESY) spectrum, its cross-peaks have to be assigned—that is, the pairs of interacting hydrogen atoms have

From: *Methods in Molecular Biology, vol. 278: Protein NMR Techniques*
Edited by: A. K. Downing © Humana Press Inc., Totowa, NJ

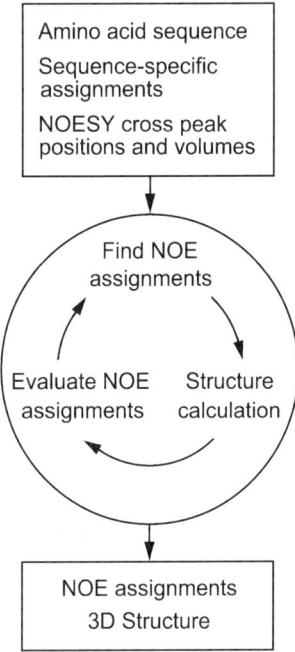

Fig. 1. General scheme of automated combined NOESY assignment and structure calculation.

to be identified (*see* **Note 1** for a summary of conventions used in this chapter). The NOESY assignment is based on previously determined chemical shift values that result from chemical shift assignment (*see* **Note 2**). Because of the limited accuracy of chemical shift values and peak positions, many NOESY cross-peaks cannot be attributed to a single unique spin pair but have an ambiguous NOE assignment composed of multiple spin pairs (*see* **Note 3**). Obtaining a comprehensive set of distance constraints from a NOESY spectrum is by no means straightforward under these conditions but becomes an iterative process in which preliminary structures, calculated from limited numbers of distance constraints, serve to reduce the ambiguity of cross-peak assignments. In addition to the problem of resonance and peak overlap, considerable difficulties may arise from spectral artifacts and noise, and from the absence of expected signals because of fast relaxation or conformational exchange. These inevitable shortcomings of NMR data collection are the main reason why until recently laborious interactive procedures have dominated 3D protein structure determinations. The automated NOESY assignment method CANDID *(5)* implemented in CYANA follows the same general scheme but does not require manual intervention during the assignment/structure calculation cycles (**Fig. 1**).

Two main obstacles have to be overcome by an automated approach starting without any prior knowledge of the structure: First, the number of cross-peaks with unique assignment based on chemical shifts is, as just pointed out, in general not sufficient to define the fold of the protein. Therefore, the automated method must have the ability to make use also of NOESY cross-peaks that cannot yet be assigned unambiguously. Second, the automated program must be able to substitute the intuitive decisions of an experienced spectroscopist in dealing with the imperfections of experimental NMR data by automated devices that can cope with the erroneously picked or inaccurately positioned peaks and with the incompleteness of the chemical shift assignment of typical experimental data sets. If used sensibly, automated NOESY assignment with CYANA has no disadvantage compared to the conventional, interactive approach but is a lot faster and more objective. With CYANA, the evaluation of NOESY spectra is no longer the time-limiting step in protein structure determination by NMR.

2. Materials

1. Computer with a UNIX -based operating system (e.g., Linux, Silicon Graphics IRIX, Compaq Alpha OSF1, IBM AIX, or MacOS X). The time-consuming structure calculations are most efficiently performed on a cluster of Linux computers using the message passing interface (MPI) for interprocess communication *(6)*, or on a shared-memory multiprocessor system. A minimum of 256 megabytes of memory per processor is recommended.
2. CYANA software package for structure calculation using torsion angle dynamics, automated NOESY assignment, and structure analysis.
3. MOLMOL software package *(7)* for molecular graphics and structure analysis.
4. Input file with the amino acid sequence.
5. One or several input files with lists of cross-peaks from 2D [^{1}H,^{1}H]–NOESY, 3D or 4D ^{13}C- or ^{15}N-resolved [^{1}H,^{1}H] NOESY spectra. The input NOESY peak lists can be prepared either using interactive spectrum analysis programs such as XEASY *(8)*, NMRView *(9)*, ANSIG *(10,11)*, or by automated peak-picking methods such as AUTOPSY *(12)* or ATNOS *(13)*, which permit starting the NOE assignment and structure calculation process directly from the NOESY spectra. The peak lists must give the positions and volumes of the NOESY cross-peaks, but initial assignments are not required for the NOESY cross-peaks.
6. Input file(s) with ^{1}H and, if available, ^{13}C and ^{15}N chemical shifts in the format of the program XEASY *(8)* or of the BioMagResBank *(14)*.
7. Optional: Previously assigned NOE upper distance constraints or other previously assigned conformational constraints. These will not be touched during automated NOE assignment but will be used for the CYANA structure calculation.

3. Methods

In this section, automated structure calculation with CYANA is described. Automated NOESY assignment is described in **Subheading 3.1.**, and structure calculation by torsion angle dynamics-driven simulated annealing in **Subheading 3.2.** The effect of incomplete chemical shift assignment is discussed in **Subheading 3.4.**, quality control of structure calculations using automated NOESY assignment in **Subheading 3.5.**, and strategies to overcome problems with insufficient input data in **Subheading 3.5.** The approach presented here has been used successfully in many NMR structure determinations of proteins with hitherto unknown structure (*see* **Note 4**).

3.1. Automated NOESY Assignment

In the program, CYANA automated NOESY assignment is performed by the CANDID algorithm *(5)* that combines features from NOAH *(15,16)* and ARIA *(17–20)*, such as the use of 3D structure-based filters and ambiguous distance constraints, with the new concepts of network anchoring and constraint combination that enable an efficient and reliable search for the correct fold already in the initial cycle of *de novo* NMR structure determinations.

3.1.1. Overview of the CANDID Algorithm for Automated NOE Assignment

In CYANA, the automated CANDID method *(5)* proceeds in iterative cycles of ambiguous NOE assignment followed by structure calculation using torsion angle dynamics (**Fig. 1**):

1. *Read experimental input data.* Amino acid sequence, chemical shift list from sequence-specific resonance assignment, list of NOESY cross-peak positions and volumes, and, optionally, conformational constraints from other sources for use in addition to the input from automated NOE assignment.

2. *Create initial assignment list.* For each NOESY cross-peak, one or several initial assignments are determined based on chemical shift agreement within a user-defined tolerance range.

3. *Rank initial assignments.* For each individual NOESY cross-peak the initial assignments are weighted with respect to several criteria, and initial assignments with low overall score are discarded. The filtering criteria include the agreement between the values of the chemical shift list and the peak position, self-consistency within the entire NOE network (*see* "network anchoring" in **Subheading 3.1.3.**), and, if available (i.e., in cycles 2, 3, . . .), the compatibility with the 3D structure from the preceding cycle (**Fig. 2**). The assessment of self-consistency also includes a check for the presence of symmetry-related cross-peaks.

4. *Calibrate distance constraints.* From the NOESY peak volumes or intensities upper distance bounds are derived for the corresponding, ambiguous or unambiguous distance constraints.

5. *Eliminate spurious NOESY cross-peaks.* Only those cross-peaks are retained that have at least one assignment with a network-anchoring score above a user-defined threshold and that are compatible with the intermediate 3D protein structure generated in the preceding cycle (cycles 2, 3, . . .).

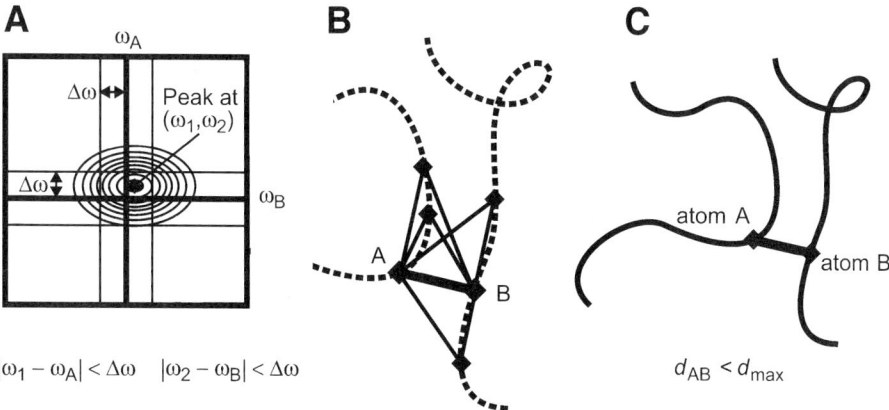

$$|\omega_1 - \omega_A| < \Delta\omega \qquad |\omega_2 - \omega_B| < \Delta\omega \qquad\qquad d_{AB} < d_{max}$$

Fig. 2. Three conditions that must be fulfilled by a valid assignment of a NOESY cross-peak to two protons A and B in the CANDID-automated NOESY assignment algorithm *(5)*: **(A)** agreement between chemical shifts and the peak position, **(B)** network anchoring, and **(C)** spatial proximity in a (preliminary) structure.

6. *Constraint combination.* In cycles 1 and 2 groups of 2 or 4, *a priori* unrelated long-range distance constraints are combined into new virtual distance constraints that each carry the assignments from two of the original constraints (*see* **Subheading 3.2.2.**).

7. *Structure calculation.* Using torsion angle dynamics (*see* **Subheading 3.1.4.**) a 3D structure of the protein is calculated that is added to the input for the following cycle. Distance constraints from NOEs with multiple assignments and those resulting from constraint combination are introduced as ambiguous distance constraints into the structure calculation. Return to **step 1**.

Between subsequent cycles, information is transferred exclusively through the intermediary 3D structures, in that the molecular structure obtained in a given cycle is used to guide the NOE assignment in the following cycle (**Fig. 1**). Otherwise, the same input data is used for all cycles. For each cross-peak, the retained assignments are interpreted in the form of an upper distance limit derived from the cross-peak volume. Thereby, a conventional distance constraint is obtained for cross-peaks with a single retained assignment. Otherwise an ambiguous distance constraint is generated that embodies several assignments (*see* **Subheading 3.1.2.**). Cross-peaks with a poor score are temporarily discarded. To reduce deleterious effects on the resulting structure from erroneous distance constraints that may pass this filtering step, long-range distance constraints are incorporated into "combined distance constraints" (*see* "constraint combination" in **Subheading 3.1.4.**). The distance constraints are then included in the input for the structure calculation with the CYANA torsion angle dynamics algorithm. An automated structure calculation typically

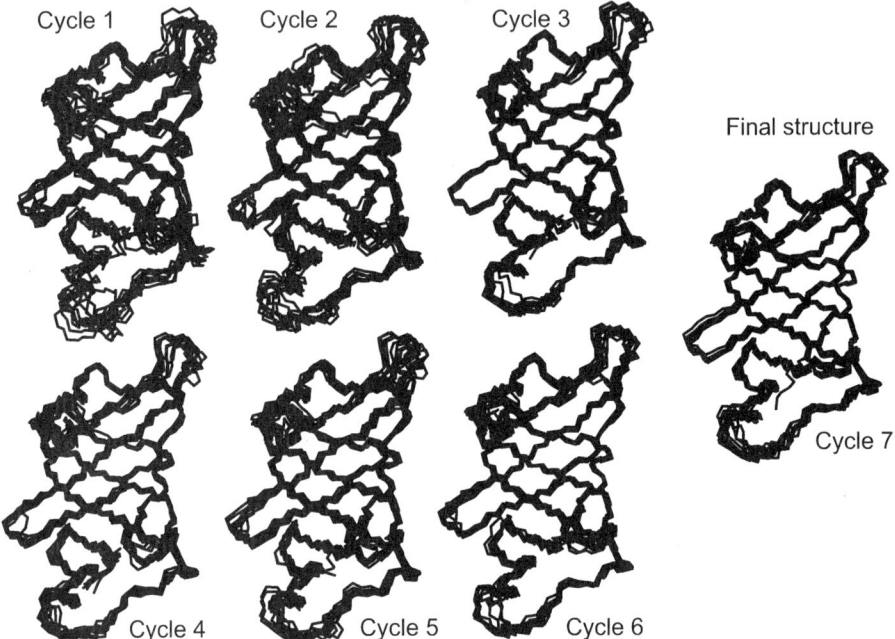

Fig. 3. Structures of the heme chaperone CcmE *(37)* obtained with the program CYANA in seven consecutive cycles of combined automated NOESY assignment with CANDID *(5)* and structure calculation with torsion angle dynamics. The backbones of the 10 conformers with lowest target function value in each cycle were drawn with the program MOLMOL *(7)*.

involves seven cycles (**Fig. 3**). In the first cycle, the structure-independent NOE self-consistency check (*see* "network anchoring" in **Subheading 3.1.3.**) has a dominant impact because structure-based criteria cannot be applied yet. The second and subsequent cycles differ from the first cycle by the use of additional selection criteria for cross-peaks and NOE assignments that exploit the protein 3D structure from the preceding cycle. Because the precision of the structure determination normally improves with each subsequent cycle (**Fig. 3**), the criteria for accepting assignments and distance constraints are tightened in more advanced cycles of the structure calculation. The output from a CANDID cycle includes a listing of NOESY cross-peak assignments, a list of comments about individual assignment decisions that can help to recognize potential artifacts in the input data, and a 3D structure in the form of a bundle of conformers. In the final cycle, an additional filtering step ensures that all NOEs have either unique assignments to a single pair of hydrogen atoms, or are eliminated from the input for the structure calculation. This allows for the direct use of the NOE distance constraints in subsequent refinement and analysis programs that do not handle ambiguous distance constraints.

3.1.2. Ambiguous Distance Constraints

Ambiguous distance constraints *(21,22)* provide a very important concept for handling ambiguities in the initial, chemical-shift-based NOESY cross-peak assignments. Prior to the introduction of ambiguous distance constraints, in general only unambiguously assigned NOEs could be used as distance constraints in the structure calculation. Because the majority of NOEs cannot be assigned unambiguously from chemical shift information alone, this lack of a general way to incorporate ambiguous data into the structure calculation considerably hampered the performance of early automatic NOESY assignment algorithms *(15)*.

When using ambiguous distance constraints, each NOESY cross-peak is treated as the superposition of the signals from each of its multiple assignments, using relative weights proportional to the inverse sixth power of the corresponding interatomic distance. A NOESY cross-peak with a unique assignment possibility gives rise to an upper-bound b on the distance $d(\alpha,\beta)$ between two hydrogen atoms, α and β. A NOESY cross-peak with $n > 1$ assignment possibilities can be seen as the superposition of n degenerate signals and interpreted as an ambiguous distance constraint, $\bar{d} \leq b$, with

$$\bar{d} = \left(\sum_{k=1}^{n} d_k^{-6} \right)^{-1/6}.$$

Each of the distances $d_k = d(\alpha_k, \beta_k)$ in the sum corresponds to one assignment possibility to a pair of hydrogen atoms, α_k and β_k. Because the "r^{-6}-summed distance" $\bar{d}$ is always shorter than any of the individual distances d_k, an ambiguous distance constraint is never falsified by including incorrect assignment possibilities, as long as the correct assignment is present.

3.1.3. Network Anchoring

Network anchoring *(5)* exploits the observation that the correctly assigned constraints form a self-consistent subset in any network of distance constraints that is sufficiently dense for the determination of a protein 3D structure. Network anchoring thus evaluates the self-consistency of NOE assignments independent of knowledge on the 3D protein structure, and in this way it compensates for the absence of 3D structural information at the outset of a *de novo* structure determination (**Fig. 2**). The requirement that each NOE assignment must be embedded in the network of all other assignments makes network anchoring a sensitive approach for detecting erroneous, "lonely" constraints that might artificially constrain unstructured parts of the protein. Such artifact constraints would not lead to systematic constraint violations during the structure calculation and, therefore, can not be eliminated by 3D structure-based peak filters. The network-anchoring score $N_{\alpha\beta}$ for a given initial assignment of a NOESY cross-peak to an atom pair (α,β) is calculated by searching all atoms γ in the same or in the neighboring residues of either α *or* β that are connected simultaneously to both atoms α *and* β. The connection may either be an initial assignment of another peak (in the same or in another peak list) or the fact that the covalent structure implies that the corresponding distance must be short enough to give rise to an observable NOE. Each such indirect path contributes to the total network-anchoring score for the assignment (α,β), an amount given by the product of the generalized volume contributions *(5)* of its two parts, $\alpha \rightarrow \gamma$ and $\gamma \rightarrow \beta$. $N_{\alpha\beta}$ has an intuitive meaning as the number of indirect connections between the atoms α and β through a third atom γ, weighted by their respective generalized volume contributions. The calculation of the network-anchoring score is recursive in the sense that its calculation for a given peak requires the knowledge of the generalized volume contributions from other peaks, which in turn involves the corresponding network-anchored assignment contributions. Therefore, the calculation of these quantities is iterated three times, or until convergence. Note that the peaks from all peak lists contribute simultaneously to the network-anchored assignment.

3.1.4. Constraint Combination

Spurious distance constraints may arise in NMR protein structure determinations from misinterpretation of noise and spectral artifacts. This situation is particularly critical at the outset of a structure determination, before a preliminary structure is available for 3D structure-based filtering of constraint assignments. Constraint combination *(5)* aims at minimizing the impact of such imperfections on the resulting structure at the expense of a temporary loss of information. Constraint combination is applied in the first two cycles. It consists of generating distance constraints with combined assignments from different, in general unrelated, cross-peaks (**Fig. 4**). The basic property of ambiguous distance constraints, namely that the constraint will be fulfilled by the correct structure whenever at least one of its assignments is correct, regardless of the presence of additional, erroneous assignments, implies that such combined constraints have a lower probability of being erroneous than the corresponding original constraints (provided that less than half of the original constraints are erroneous; *see* **Note 5**). Two modes of constraint combination are provided in CYANA *(5)*: "2 $\rightarrow$ 1" combination of all assignments of two long-range peaks each into a single constraint, and "4 $\rightarrow$ 4" pairwise combination of the assignments of four long-range peaks into four constraints. Let *A, B, C, D* denote the sets of assignments of four peaks. Then, 2 $\rightarrow$ 1 combination replaces two constraints with assignment sets *A* and *B*, respectively, by a single ambiguous constraint with

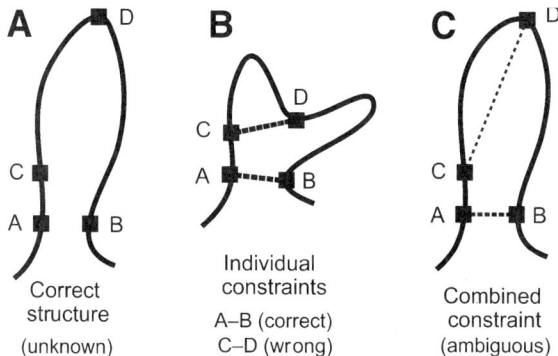

Fig. 4. Schematic illustration of the effect of constraint combination *(6)* in the case of two distance constraints, a correct one connecting atoms A and B, and an incorrect one between atoms C and D. A structure calculation that uses these two constraints as individual constraints that have to be satisfied simultaneously will, instead of finding the correct structure **(A)**, result in a distorted conformation **(B)**, whereas a combined constraint that will be fulfilled already if one of the two distances is sufficiently short leads to an almost undistorted solution **(C)**.

assignment set $A \cup B$, the union of sets A and B (**Fig. 4**). The $4 \rightarrow 4$ pairwise combination replaces four constraints with assignments $A, B, C,$ and D by four combined ambiguous constraints with assignment sets $A \cup B, A \cup C, A \cup D,$ and $B \cup C,$ respectively. In both cases, constraint combination is applied only to the long-range peaks, that is, the peaks with all assignments to pairs of atoms separated by five or more residues in the sequence, because in case of error their effect on the global fold of a protein is more pronounced than that of erroneous short- and medium-range constraints. The number of long-range constraints is halved by the $2 \rightarrow 1$ combination but stays constant on $4 \rightarrow 4$ pairwise combination. Therefore, the latter approach preserves more of the original structural information. Furthermore, it can take into account that certain peaks and their assignments are more reliable than others because the peaks with assignment sets A, B, C, D are used 3, 2, 2, and 1 times, respectively, to form combined constraints. To this end, the long-range peaks are sorted according to their total residue-wise network-anchoring score *(5)*, and the $4 \rightarrow 4$ combination is performed by selecting the assignments A, B, C, D from the first, second, third, and fourth quarter of the sorted list, respectively. The upper distance bound b for a combined constraint is formed from the two upper distance bounds b_1 and b_2 of the original constraints either as the r^{-6}-sum, $b = \left(b_1^{-6} + b_2^{-6}\right)^{-1/6}$, or as the maximum, $b = \max(b_1, b_2)$. The first choice minimizes the loss of information if two already correct constraints are combined, whereas the second choice avoids the introduction of too small an upper bound if a correct and an erroneous constraint are combined.

3.2. Structure Calculation

The calculation of the 3D structure forms a cornerstone of the NMR method for protein structure determination. Because of the complexity of the problem—a protein typically consists of more than 1000 atoms, which are restrained by a similar number of experimentally determined constraints in conjunction with stereochemical and steric conditions—it is in neither feasible to do an exhaustive search of allowed conformations nor to find solutions by interactive model building. Therefore, the calculation of the 3D structure is formulated in CYANA as a minimization problem for a target function that measures the agreement between a structure and the given set of constraints.

3.2.1. Target Function

The CYANA target function (23,24) is defined such that it is zero if and only if all experimental distance constraints and torsion angle constraints are fulfilled and all nonbonded atom pairs satisfy a check for the absence of steric overlap. A conformation that satisfies the constraints more closely than another one will lead to a lower target function value. The exact definition of the CYANA target function is:

$$V = \sum_{c=u,l,v} w_c \sum_{(\alpha,\beta)\in I_c} \left(d_{\alpha\beta} - b_{\alpha\beta}\right)^2 + w_a \sum_{i\in I_a} \left[1 - \frac{1}{2}\left(\frac{\Delta_i}{\Gamma_i}\right)^2\right]\Delta_i^2$$

Upper and lower bounds, $b_{\alpha\beta}$, on distances $d_{\alpha\beta}$ between two atoms α and β, and constraints on individual torsion angles θ_i in the form of allowed intervals $[\theta_i^{min}, \theta_i^{max}]$ are considered. I_u, I_l, and I_v are the sets of atom pairs (α,β) with upper, lower, or van der Waals distance bounds, respectively, and I_a is the set of restrained torsion angles. w_u, w_l, w_v, and w_a are weighting factors for the different types of constraints. $\Gamma_i = \pi - (\theta_i^{max} - \theta_i^{min})/2$ denotes the half width of the forbidden range of torsion angle values, and Δ_i is the size of the torsion angle constraint violation.

3.2.2. Torsion Angle Dynamics

The minimization algorithm in CYANA is based on the idea of simulated annealing (25) by molecular dynamics simulation (see **Note 6**) in torsion angle space. The distinctive feature of molecular dynamics simulation when compared to the straightforward minimization of a target function is the presence of kinetic energy that allows overcoming barriers of the potential surface, which reduces greatly the problem of becoming trapped in local minima. Torsion angle dynamics, that is, molecular dynamics simulation using torsion angles instead of Cartesian coordinates as degrees of freedom *(24,26–34)*, provides at present the most efficient way to calculate NMR structures of biological macromolecules. The only degrees of freedom are the torsion angles, that is, rotations about single bonds, such that the conformation of the molecule is uniquely specified by the values of all torsion angles. Covalent bonds that are incompatible with a tree structure because they would introduce closed flexible rings, for example, disulphide bridges, are treated, as in Cartesian space dynamics, by distance constraints. The efficiency of the torsion angle dynamics algorithm *(30)* implemented in the program CYANA, and previously in DYANA *(24)*, is high because of the fact that it requires a computational effort that increases only linearly with the system size (*see* **Note 7**). In contrast, the computation time for "naïve" approaches to torsion angle dynamics rises with the third power of the system size (e.g., **ref. 29**), which renders these algorithms unsuitable for use with macromolecules. With the fast torsion angle dynamics algorithm in CYANA, the advantages of torsion angle dynamics, especially the much longer integration time steps that can be used, are effective for molecules of all sizes.

3.2.3. Simulated Annealing

The potential energy landscape of a protein is complex and studded with many local minima, even in the presence of experimental constraints and when using the simplified target function of **Subheading 3.2.1.** Because the temperature, that is, kinetic energy, determines the maximal height of energy barriers that can be overcome in a molecular dynamics trajectory, the temperature schedule is important for the success and efficiency of a simulated annealing calculation. Elaborated protocols have been devised for structure calculations using molecular dynamics in Cartesian space *(35,36)*. In addition to the temperature, other parameters, such as force constants and repulsive core radii, are varied in these schedules, which may involve several stages of heating and cooling. The fast exploration of conformation space with torsion angle dynamics allows for much simpler schedules. The standard simulated annealing protocol in the program CYANA *(24)* starts from a conformation with all torsion angles treated as independent, uniformly distributed random variables and consists of five stages:

1. *Initial minimization.* A short minimization to reduce high-energy interactions that could otherwise disturb the torsion angle dynamics algorithm: 100 conjugate gradient minimization steps are performed, including only distance constraints between atoms up to 3 residues apart along the sequence, followed by a further 100 minimization steps including all constraints. For efficiency, until **step 4**, all hydrogen atoms are excluded from the check for steric overlap, and the repulsive core radii of heavy atoms with covalently bound hydrogens are increased by

0.15Å with respect to their standard values. The weights in the target function of **Subheading 3.2.1.** are set to 1 for user-defined upper and lower distance bounds, to 0.5 for steric lower distance bounds, and to 5Å^2 for torsion angle constraints.

2. *High-temperature phase.* A torsion angle dynamics calculation at constant high temperature: One-fifth of all *N* torsion angle dynamics steps are performed at a constant high reference temperature of, typically, 10,000 K. The time step is initialized to 2 fs (= 2×10^{-15} s). The list of van der Waals lower distance bounds is updated every 50 steps using a cutoff of 4.2Å for the interatomic distance throughout all torsion angle dynamics phases.

3. *Slow cooling.* Torsion angle dynamics calculation with slow cooling close to zero temperature: The remaining 4*N*/5 torsion angle dynamics steps are performed during which the reference value for the temperature approaches zero according to a fourth-power law.

4. *Low-temperature phase with individual hydrogen atoms.* Incorporation of all hydrogen atoms into the check for steric overlap: After resetting the repulsive core radii to their standard values, and increasing the weighting factor for steric constraints to two, 100 conjugate gradient minimization steps are performed, followed by 200 torsion angle dynamics steps at zero reference temperature.

5. *Final minimization.* A final minimization consisting of 1000 conjugate gradient steps.

With the CYANA torsion angle dynamics algorithm it is possible to efficiently calculate protein structures on the basis of NMR data. Even for a system as complex as a protein, the program CYANA can execute thousands of torsion angle dynamics steps within minutes of computation time. For instance, the computation time for the calculation of one conformer of the 136-residue heme chaperone protein CcmE on the basis of 2453 NOE upper distance bounds and 56 torsion angle constraints *(37)* using 10,000 torsion angle dynamics steps on a single processor is below 1 min on up-to-date hardware:

Linux PC, Pentium IV, 3.06 GHz:	29 s
Linux PC, Pentium IV, 1.8 GHz:	42 s
Compaq Alpha server GS 320:	23 s
Silicon Graphics, R16000, 700 MHz:	39 s
Silicon Graphics, R12000, 400 MHz:	59 s

Furthermore, because an NMR structure calculation always involves the computation of a group of conformers, it is highly efficient and straightforward with CYANA to run calculations of multiple conformers in parallel. Nearly ideal speedup, that is, an overall computation time almost inversely proportional to the number of processors, can be achieved with CYANA *(24)*.

3.3. Effect of Incomplete Chemical Shift Assignments

A limiting factor for the application of the automated NOE assignment algorithm CANDID is that it relies on the availability of an essentially complete list of chemical shifts from the preceding sequence-specific resonance assignment. At present, chemical shift assignment remains largely the domain of semiautomated and interactive methods, in spite of promising attempts toward automation *(1)*. Experience shows that in general the majority of the chemical shifts can be assigned readily, whereas others pose difficulties that may require a disproportionate amount of the spectroscopist's time. Hence, NMR structure determination would be speeded up significantly if NOE assignment and structure calculation could be based on incomplete lists of assigned chemical shifts, or if chemical shift assignments could be completed during the structure calculation (*see* **Note 8**), provided that the reliability and robustness of the NMR method for protein structure determination is not compromised.

It has been shown *(38)* that for reliable automated NOESY, assignment with the CANDID algorithm around 90% completeness of the chemical shift assignment is necessary (*see* **Note 9**). In certain cases, the lack of a small number of "essential" chemical shifts can lead to a significant deviation of the structure. On the other hand, in practice the algorithm might be expected to tolerate a slightly higher degree of incompleteness in the chemical shift assignments provided that most missing assignments are of "unimportant" atoms that are involved in only few NOEs. This is usually the case because the chemical shifts of protons that are involved in many NOEs, and if absent prevent the program from correctly assigning any of these NOEs, are intrinsically easier to assign than those exhibiting only a few NOEs. This effect is confirmed by the finding that the lack of aromatic chemical shifts is in general more harmful to the outcome of a structure calculation than that of a similar number of other protons because aromatic protons tend to be located in the hydrophobic core of the protein where they give rise to a higher-than-average number of NOEs. Network anchoring and constraint combination are two methods that have been designed and shown to be effective in minimizing the impact of incomplete and/or erroneous pieces of input data (*see* **Subheadings 3.1.3.** and **3.1.4.**). Chemical shift assignment-based automated NOE assignment without the safeguards of network anchoring and constraint combination is expected to be more susceptible to deleterious effects from missing chemical shift assignments and artifacts in the input data. In contrast to missing or incorrect entries in the chemical shift list, the algorithm is remarkably tolerant regarding incompleteness of the NOESY peak list (*see* **Note 10**). This suggests that it is better to strive for correctness than for ultimate completeness of the input NOESY peak lists.

3.4. Quality Control

In this section, simple criteria based on the output of CYANA are given that allow assessing the reliability of the resulting structure without cumbersome recourse to independent interactive verification of the NOESY assignments.

Final structures from an automatic algorithm that have a low root-mean-square deviation (RMSD) within the bundle of conformers but differ significantly from the "correct" reference structure are problematic because, without knowledge of an independently determined "reference" structure, they may appear at first glance as good, well-defined solutions. In a conventional structure calculation based on manual NOESY assignment, incomplete or inconsistent input data will be manifested by large RMSD and/or target function values of the final structure bundle, which will prompt the spectroscopist to correct and/or complete the input data for a next round of structure calculation. Test calculations showed that for structure calculation with automated NOE assignment, neither the RMSD value of the final structure nor the final target function value are suitable indicators to discriminate between correct and biased results *(38)*. Other criteria are needed to evaluate the outcome. On the basis of the initial experience with the CANDID algorithm, guidelines for successful CANDID runs were proposed *(5)*. These comprised six criteria that should be met simultaneously:

1. Average CYANA target function value of cycle 1 below 250 Å^2.
2. Average final CYANA target function value below 10 Å^2.
3. Less than 20% unassigned NOEs.
4. Less than 20% discarded long-range NOEs.
5. RMSD value in cycle 1 below 3 Å.
6. RMSD between the mean structures of the first and last cycle below 3 Å.

Criterion 4 refers to the percentage of NOEs discarded by the CANDID algorithm among all NOEs with assignments exclusively between atoms separated by four or more residues along the polypeptide sequence. Criteria 3 and 4 impose a limit on the number of NOEs that are not used to generate distance constraints for the final structure calculation and, thus, measure the completeness with which the picked NOE cross-peaks can be explained by the resulting structure. The validity of the original guidelines as sufficient conditions for successful CYANA runs was confirmed by the fact that all the structure calculations in a systematic study *(38)* with an RMSD bias *(39)* to the reference structure higher than 2 Å violated one or several of the six criteria. On the other hand, the same test calculations revealed a certain redundancy among the six original criteria. Provided that the input peak lists do not deliberately misinterpret the underlying NOESY spectra (to which the algorithm has no direct access), the aforementioned criteria can be replaced by just two conditions for successful structure calculation with CYANA:

1. Less than 25% of the long-range NOEs must have been discarded by the automated NOESY assignment algorithm for the final structure calculation (*see also* **Note 11**).
2. The backbone RMSD to the mean coordinates for the structure bundle of the *first* cycle must not exceed 3 Å.

The ability of the program to find a well-defined structure in the initial cycle of NOE assignment and structure calculation, as measured by the RMSD within the structure bundle in cycle 1, is an important factor that strongly influences the accuracy of the final structure. This can be understood by considering the iterative nature of automated NOESY assignment, by which each cycle except cycle 1 is dependent on the structure obtained in the preceding cycle. Using network anchoring and constraint combination, the algorithm tries to obtain a well-defined structure already in the first cycle. A low precision of the structure from cycle 1 may hinder convergence to a well-defined final structure, or, more dangerously, opens the possibility of a structural drift in later cycles toward a precise but inaccurate final structure. In practice, it is safe to apply both criteria, even though in test calculations *(38)* the percentage of discarded long-range NOEs alone would have been sufficient to detect all runs that resulted in a structure with more than 2 Å RMSD bias. In these test calculations a large dispersion in the accuracy of the final structure was reflected reliably by the percentage of discarded long-range NOEs and the RMSD in cycle 1, but it could not readily be discerned from the values of the target function after cycle 1 or 7, the RMSD at cycle 7, or, in a few cases, the percentage of unassigned NOEs.

3.5. Troubleshooting

If the output of a CYANA structure calculation based on automated NOESY assignment with CANDID does not fulfill the guidelines of **Subheading 3.4.**, then the structure will in many cases still be essentially correct but should not be accepted without further validation. Within the framework of CYANA, the recommended approach is to improve the quality of the input chemical shift and peak lists and to perform a new complete CYANA run with seven cycles, until the criteria are met. Usually, this can be achieved efficiently because the output from an unsuccessful CYANA run, even though the structure should not be trusted *per se,* clearly points out problems in the input—for example, peaks that cannot be assigned and might therefore be artifacts or indications of erroneous or missing sequence-specific assignments. To facilitate this task, the program gives for each peak informational output that includes the list of its chemical-shift-based assignment possibilities, the assignment(s) finally chosen, and the reasons why an assignment is chosen or not, or why a peak is not used at all. Even when the criteria of **Subheading 3.4.** are already met, a higher precision and local accuracy of the structure might still be achieved by further improving the input data. In principle, a *de novo* protein structure determination requires one run of CYANA with seven cycles of automated NOE assignment and structure calculation. This is realistic when almost complete chemical shift assignments and exhaustive high-quality NOESY peak lists are available. In practice, it is often more efficient to start a first CYANA calculation from an initial, slightly incomplete list of "safely identifiable" NOESY cross-peaks. The results of this first CYANA calculation can then be used to prepare an improved, more complete NOESY peak list for a second CYANA calculation. This can be done more efficiently than it would be possible *ab initio* because only peaks and regions of the protein that gave rise to problems in the first CYANA calculation need to be checked.

4. Notes

1. *Definitions.* For consistency and simplicity, the following conventions are used: An interaction between two or more atoms is manifested by a *signal* in a multidimensional spectrum. A *peak* refers to an entry in a peak list that has been derived from an experimental spectrum by *peak picking*. A peak may or may not represent a signal, and there may be signals that are not represented by a peak. *Chemical shift assignment* is the process and the result of attributing a specific chemical shift value to an atom. *Peak assignment* is the process and the result of identifying in each spectral dimension the atom(s) that are involved in the signal represented by the peak. *NOESY assignment* is peak assignment in NOESY spectra.

2. *Automated chemical shift assignment algorithms.* There have been many attempts to automate the chemical shift assignment that has to precede the collection of conformational constraints and the structure calculation. These methods have been reviewed recently *(1)*. Some automated approaches to chemical shift assignment target the question of assigning the backbone and, possibly, β chemical shifts, usually on the basis of triple-resonance experiments that delineate the protein backbone through one- and two-bond scalar couplings, whereas others are concerned with the more demanding problem of complete assignment of the amino acid sidechain chemical shifts. In most cases, these algorithms require peak lists from a specific set of NMR spectra as input and produce lists of chemical shifts of varying completeness and correctness, depending on the quality and information content of the input data, and on the capabilities of the algorithm.

3. *Ambiguity of chemical shift based NOE assignment.* A simple mathematical model of the NOESY assignment process by chemical shift matching gives insight into this problem *(16)*. It assumes a protein with n hydrogen atoms, for which complete and correct chemical shift assignments are available, and N cross-peaks picked in a 2D [^{1}H,^{1}H]–NOESY spectrum with an accuracy of the peak position of $\Delta\omega$, that is, the position of the picked peak differs from the resonance frequency of the underlying signal by no more than $\Delta\omega$ in both spectral dimensions. Under the simplifying assumption of a uniform distribution of the proton chemical shifts over a spectral width $\Delta\Omega$, the chemical shift of a given proton falls within an interval of half width $\Delta\omega$ about a given peak position with probability $p = 2\Delta\omega/\Delta\Omega$. Peaks with unique chemical shift-based assignment have in both spectral dimensions exactly 1 out of all n proton shifts inside the tolerance range $\Delta\omega$?from the peak position. Their expected number, $N^{(1)} = N(1 - p)^{2n-2} \approx Ne^{-2np} = Ne^{-4n\Delta\omega/\Delta\Omega}$,

decreases exponentially with increasing size of the protein (n) and increasing chemical shift tolerance range ($\Delta\omega$). For a typical small protein such as the *Williopsis mrakii* killer toxin (WmKT), with 88 amino acid residues, $n = 457$ proton chemical shifts and $N = 1986$ NOESY cross-peaks within a range of $\Delta\Omega = 9$ ppm *(40)*, this model predicts that only about 4% of the NOEs can be assigned unambiguously based solely on chemical shift information with an accuracy of $\Delta\omega = 0.02$ ppm— an insufficient number to calculate a preliminary 3D structure. For peak lists obtained from ^{13}C- or ^{15}N-resolved 3D [^{1}H,^{1}H]–NOESY spectra, the ambiguity in one of the proton dimensions can usually be resolved by reference to the heterospin, so that $N^{(1)} \approx Ne^{-np} = Ne^{-2n\Delta\omega/\Delta\Omega}$. Regarding assignment ambiguity, 3D NOESY spectra are thus equivalent to homonuclear NOESY spectra from a protein of half the size or with twice the accuracy in the determination of the chemical shifts and peak positions. Once available, a preliminary 3D structure may be used to resolve ambiguous NOE assignments. The ambiguity is resolved if only one out of all chemical shift-based assignment possibilities corresponds to an interatomic distance shorter than the maximal NOE-observable distance, d_{max}. Assuming that the hydrogen atoms are evenly distributed within a sphere of radius R that represents the protein, the probability q that two given hydrogen atoms are closer to each other than d_{max} can be estimated by the ratio between the volumes of two spheres with radii d_{max} and R, respectively: $q = \left(d_{max}/R\right)^3$. Using $d_{max} = 5$ Å, one obtains for WmKT, a nearly spherical protein with a radius of about 15 Å, $q \approx 4\%$. Hence, not more than 96% of the peaks with two assignment possibilities can be assigned uniquely by reference to the protein structure. Even by reference to a perfectly refined structure it is therefore impossible, on fundamental grounds, to resolve all assignment ambiguities because q will always be larger than 0.

4. *Structure determinations with automated NOE assignment by CANDID.* The automated structure calculation method described in this chapter has been evaluated in test calculations *(5,13,38)* and used for various *de novo* structure determinations, including four variants of the human prion protein *(41,42)*, the calreticulin P-domain *(43)*, two distinct forms of the pheromone-binding protein from *Bombyx mori (44,45)*, the class I human ubiquitin-conjugating enzyme 2b *(46)*, the heme chaperone CcmE *(37)* (**Fig. 3**), and the nucleotide-binding domain of Na,K-ATPase *(47)*. The NOESY assignments and the corresponding distance constraints for these *de novo* structure determinations were made automatically by the program, confining interactive work to the stage of the preparation of the input chemical shift and peak lists. These structure determinations have confirmed the viability of CYANA for automated NOESY assignment and structure calculation without prior knowledge about NOESY assignments or the 3D structure.

5. *Effect of constraint combination.* The effect of constraint combination on the expected number of erroneous distance constraints can be estimated quantitatively in the case of $2 \rightarrow 1$ combination by assuming an original data set containing N long-range peaks and a uniform probability $p \ll 1$ that a long-range peak would

lead to an erroneous constraint *(5)*. By $2 \rightarrow 1$ constraint combination, these are replaced by $N/2$ constraints that are erroneous with probability p^2. In the case of $4 \rightarrow 4$ combination, it may be assumed that the same N long-range peaks can be classified into four equally large classes with probabilities to be erroneous of αp, p, p, $(2 - \alpha)p$, respectively. The overall probability for an input constraint to be erroneous is again p. The parameter α, $0 \leq \alpha \leq 1$, expresses how much "safer" the peaks in the first class are compared to those in the two middle classes, and in the fourth, "unsafe" class. After $4 \rightarrow 4$ combination, there are still N long-range constraints but with an overall error probability of $\left(\alpha + \left(1 - \alpha^2\right)/4\right)p^2$, which is smaller than the probability p^2 obtained by simple $2 \rightarrow 1$ combination provided that the classification into more and less safe classes was successful ($\alpha < 1$). For instance, $4 \rightarrow 4$ combination will transform an input data set of 900 correct and 100 (10%) erroneous long-range cross-peaks (i.e., $N = 1000$, $p = 0.1$) that can be split into four classes with $\alpha = 0.5$ into a new set of approx 993 correct and 7 (0.7%) erroneous combined constraints. Alternatively, $2 \rightarrow 1$ combination will yield under these conditions approx 495 correct and 5 (1%) erroneous combined constraints. Unless the number of erroneous constraints is high, $4 \rightarrow 4$ combination is thus preferable over $2 \rightarrow 1$ combination in the first two CANDID cycles.

6. *Molecular dynamics simulation vs NMR structure calculation.* There is a fundamental difference between molecular simulation that has the aim of simulating the trajectory of a molecular system as realistically as possible to extract molecular quantities of interest and NMR structure calculation that is driven by experimental constraints. Classical molecular dynamics simulations *(48)* rely on a full empirical force field to ensure proper stereochemistry and are generally run at a constant temperature, close to room temperature. Substantial amounts of computation time are required because the empirical energy function includes long-range pair interactions that are time-consuming to evaluate and because conformation space is explored slowly at room temperature. When molecular dynamics algorithms are used for NMR structure calculations, however, the objective is quite different. Here, such algorithms simply provide a means to efficiently optimize a target function that takes the role of the potential energy. Details of the calculation, such as the course of a trajectory, are unimportant, as long as its end point comes close to the global minimum of the target function. Therefore, the efficiency of NMR structure calculation can be enhanced by simplification or modification of the force field and/or the algorithm that does not significantly alter the location of the global minimum (the correctly folded structure) but shortens (in terms of computation time needed) the way by which it can be reached from the start conformation. A typical "geometric" force field used in NMR structure calculation therefore retains only the most important part of the nonbonded interaction by a simple repulsive potential that replaces the Lennard–Jones and electrostatic interactions of the full empirical energy function. This short-range repulsive function can be calculated much faster and significantly facilitates large-scale conformational changes that are required during the folding process by lowering energy barriers induced by the overlap of atoms.

7. *Fast algorithm for torsion angle dynamics.* The key idea of the fast torsion angle dynamics algorithm in CYANA is to exploit the fact that a chain molecule such as a protein or nucleic acid can be represented in a natural way as a tree structure consisting of $n + 1$ rigid bodies that are connected by n rotatable bonds (**Fig. 5A**) *(28,49)*. Each rigid body is made up of one or several mass points (atoms) with fixed relative positions. The tree structure starts from a base, typically at the N-terminus of the polypeptide chain, and terminates with "leaves" at the ends of the sidechains and at the C-terminus. The angular velocity vector ω_k and the linear velocity v_k of the reference point of the rigid body k (**Fig. 5B**) are calculated recursively from the corresponding quantities of the preceding rigid body

$$\omega_k = \omega_{p(k)} + e_k \dot{\theta}_k,$$

$$v_k = v_{p(k)} - \left(\mathbf{r}_k - \mathbf{r}_{p(k)} \right) \wedge \omega_{p(k)}.$$

Denoting the vector from the reference point to the center of mass of the rigid body k by Y_k, its mass by m_k, and its inertia tensor by I_k (**Fig. 5B**), the kinetic energy can be computed in a linear loop over all rigid bodies:

$$E_{\text{kin}} = \frac{1}{2} \sum_{k=0}^{n} \left[m_k v_k^2 + \omega_k \cdot I_k \omega_k + 2 v_k \cdot \left(\omega_k \wedge m_k Y_k \right) \right].$$

The calculation of the torsional accelerations, that is, the second-time derivatives of the torsion angles, is the crucial point of a torsion angle dynamics algorithm. The equations of motion for a classical mechanical system with generalized coordinates are the Lagrange equations

$$\frac{d}{dt} \left(\frac{\partial L}{\partial \dot{\theta}_k} \right) - \frac{\partial L}{\partial \theta_k} = 0 \qquad (k = 1, \ldots, n)$$

with the Lagrange function $L = E_{\text{kin}} - E_{\text{pot}}$. They lead to equations of motion of the form $M(\theta)\ddot{\theta} + C(\theta, \dot{\theta}) = 0$. In the case of torsion angles as degrees of freedom, the mass matrix $M(\theta)$ and the n-dimensional vector $C(\theta, \dot{\theta})$ can be calculated explicitly *(28,29)*. To generate a trajectory, this linear set of n equations would have to be solved in each time step for the torsional accelerations $\ddot{\theta}$, which requires a computational effort proportional to n^3, which is prohibitively expensive for larger systems. Therefore, in CYANA the fast recursive algorithm of Jain et al. *(30)* is implemented to compute the torsional accelerations, which makes explicit use of the tree structure of the molecule to obtain $\ddot{\theta}$ with a computational effort that is only proportional to n. The mathematical details of the CYANA torsion angle dynamics algorithm are given in *(24,30)*. It suffices to note here that the torsional accelerations can be obtained by executing a series of three linear loops over all rigid bodies similar to the single one that is needed to compute the kinetic energy, E_{kin}. The integration scheme for the equations of motion in torsion angle dynamics is a variant of the "leap-frog" algorithm *(48,50)* used in Cartesian space molecular dynamics. To obtain a trajectory, the equations of motion are numerically integrated

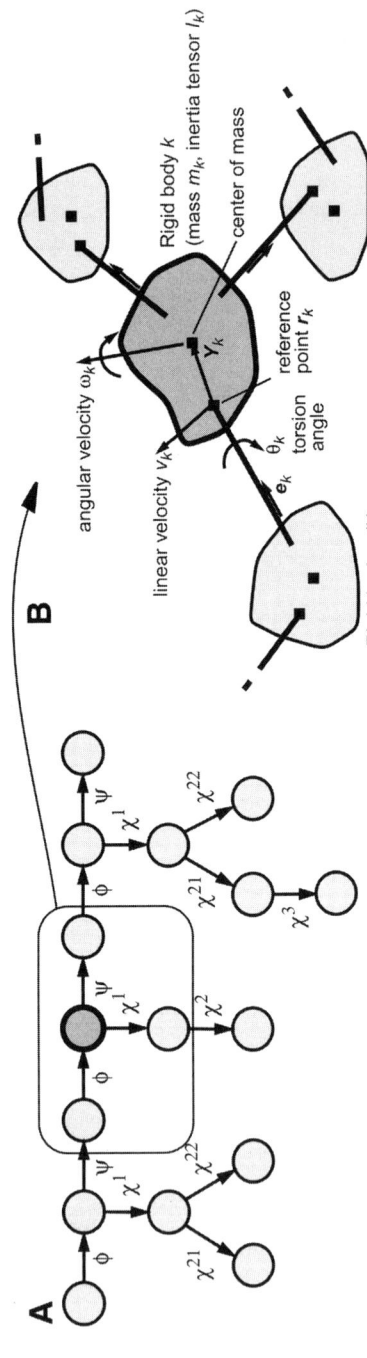

Fig. 5. (**A**) Tree structure of torsion angles for the tripeptide Val-Ser-Ile. *Circles* represent rigid units. Rotatable bonds are indicated by *arrows* that point toward the part of the structure that is rotated if the corresponding torsion angle is changed. (**B**) Excerpt from the tree structure formed by the torsion angles of a molecule, and definition of quantities required by the CYANA torsion angle dynamics algorithm.

by advancing the $i = 1, \ldots, n$ (generalized) coordinates q_i and velocities $\dot{q}_i$ that describe the system by a small but finite time step Δt:

$$\dot{q}(t + \Delta t / 2) = \dot{q}_i(t - \Delta t / 2) + \Delta t \ddot{q}_i(t) + O(\Delta t^3)$$
$$q_i(t + \Delta t) = q_i(t) + \Delta t \dot{q}_i(t + \Delta t / 2) + O(\Delta t^3)$$

The degrees of freedom, q_i, are the Cartesian coordinates of the atoms in conventional molecular dynamics simulation, or the torsion angles in CYANA. The $O(\Delta t^3)$ terms indicate that the errors with respect to the exact solution incurred by the use of a finite time step Δt are proportional to Δt^3. The time step Δt must be small enough to sample adequately the fastest motions. Because the fastest motions in conventional molecular dynamics simulation are oscillations of bond lengths and bond angles, which are "frozen" in torsion angle space, longer time steps can be used for torsion angle dynamics than for molecular dynamics in Cartesian space *(24)*. The temperature is controlled by weak coupling to an external bath *(51)*, and the length of the time step is adapted automatically based on the accuracy of energy conservation *(24)*. It could be shown that in practical applications with proteins time steps of about 100, 30, and 7 fs at low (1 K), medium (400 K), and high (10,000 K) temperatures, respectively, can be used in torsion angle dynamics calculations with CYANA *(24)*, whereas time steps in Cartesian space molecular dynamics simulation generally have to be in the range of 2 fs. The concomitant fast exploration of conformation space provides the basis for the efficient CYANA structure calculation protocol.

8. *Chemical shift assignment during NOE assignment and structure calculation.* Methods to find additional chemical shift assignments simultaneously with automated NOESY assignment and the structure calculation have been proposed and applied with some success in the case when a preliminary structure was available *(52)*: Starting from nearly complete chemical shift assignments for the backbone and for 348 sidechain protons of the 28 kDa single-chain T-cell receptor protein, the chemical shifts of 40 additional sidechain protons could be found by a combination of chemical shift prediction with the program SHIFTS *(53,54)* and NOE assignment with ARIA *(17)*. The same approach can be used with CYANA.

9. *Impact of incomplete chemical shift assignments.* The influence of incomplete chemical shift assignments on the reliability of NMR structures has been investigated using the program CYANA with input data that represents various degrees of completeness of the chemical shift assignment *(38)*. The effect of missing chemical shift assignments was assessed by randomly omitting entries from the "complete" experimental ^1H chemical shift lists that had been used for the earlier, conventional structure determinations of two proteins, the *Bombyx mori* pheromone-binding protein form A (BmPBP^A) *(44)* and the killer toxin WmKT *(40)*. Sets of structure calculations were performed with different numbers and selections of randomly omitted chemical shifts and the results compared to those obtained when using the complete experimental chemical shift list. The deviation of the structures obtained with incomplete chemical shift assignments from the reference structure was monitored by the RMSD bias, the RMSD between the mean coordinates of the

two structure bundles *(39)*. In the representative case of randomly selecting the omitted chemical shifts among all ^{1}H chemical shift assignments of BmPBP^A, the RMSD bias increased only slowly with increasing omission ratio p up to about $p = 10\%$, from where onward the RMSD bias rose abruptly, reflecting that severely distorted structures had been obtained. Higher omission ratios not only resulted in high mean values of the RMSD bias but also in pronounced variations among the individual runs at a given p value with different random selections of the omitted shifts. The CYANA target function values of the final structures were, regardless of the omission ratio, almost always in the range below 5 Å^2—that is, indicative of a structure that essentially fulfills all the input conformational constraints. The percentages of unassigned NOEs increased, and the number of distance constraints for the final cycle of structure calculation decreased almost linearly with the omission rate. The algorithm was more tolerant against the lack of chemical shifts when run with data from the uniformly ^{13}C- and ^{15}N-labeled protein BmPBP^A than with the homonuclear data for the protein WmKT, even though BmPBP^A (142 residues) is much larger than WmKT (88 residues). This is because of the availability of ^{13}C and ^{15}N chemical shifts that allow resolution of many ^{1}H chemical shift degeneracies such that the probability of accidental erroneous NOE assignments is decreased compared to the case of homonuclear data. The omission of aromatic ^{1}H chemical shift assignments in general causes more severe problems than the omission of the same number of chemical shifts chosen randomly among all assigned ^{1}H chemical shifts *(38)*. In the case of BmPBP^A the omission of all assigned aromatic chemical shifts, corresponding to 6.0% of all assigned protons, led already to 2 Å RMSD bias. In the case of WmKT, with only homonuclear data, significant deviations from the reference structure were in some cases already observed at 20% omission of the aromatic chemical shifts, which corresponds to an overall omission ratio of merely 1.6% of all assigned ^{1}H chemical shifts.

10. *Effect of incomplete NOESY peak picking.* In contrast to the effects seen under the omission of chemical shift assignments, the random omission of NOESY peaks does not cause severe problems *(38)*. Even when 50% of the NOESY peaks were omitted from the experimental input peak lists for BmPBP^A, most RMSD bias values remained in the region of 2 Å. An outlier with RMSD bias close to 4 Å shows that for BmPBP^A the algorithm starts to lose its stability at 50% NOE omission ratio. The results with the homonuclear data from WmKT showed similar patterns, albeit with a somewhat stronger dependence on the omission rate, and RMSD bias values occasionally exceeding 2 Å in runs with 30% NOESY peak omission ratio. The CYANA structure calculation protocol is thus remarkably tolerant with respect to incomplete NOESY peak picking and can tolerate the omission of up to 50% of the NOESY cross-peaks with only a moderate decrease in the precision and accuracy of the resulting structure.

11. *Alternative criterion to assess the completeness of the NOESY assignment.* The percentage of discarded long-range NOEs cannot be calculated readily outside the CYANA program because it requires knowledge of the possible assignments also for the NOESY cross-peaks that were excluded from the generation of conforma-

tional constraints. In this case, an overall percentage of unused cross-peaks of less than 15% can be used as an alternative criterion that is straightforward to evaluate from the final assigned output peak lists, in which unused cross-peaks remain unassigned. However, among these two criteria, the percentage of discarded long-range NOEs is a slightly more sensitive indicator of the accuracy of the final structure than the overall percentage of unused cross-peaks because the latter includes also peaks with short-range assignment or with no assignment possibility at all that are expected to have little distorting effect on the resulting structure.

References

1. Moseley, H. N. B. and Montelione, G. T. (1999) Automated analysis of NMR assignments and structures for proteins. *Curr. Opin. Struct. Biol.* **9,** 635–642.
2. Solomon, I. (1955) Relaxation processes in a system of two spins. *Phys. Rev.* **99,** 559–565.
3. Macura, S. and Ernst, R. R. (1980) Elucidation of cross relaxation in liquids by 2D NMR spectroscopy. *Mol. Phys.* **41,** 95–117.
4. Neuhaus, D. and Williamson, M. P. (1989) *The Nuclear Overhauser Effect in Structural and Conformational Analysis.* VCH, Weinheim, Germany.
5. Herrmann, T., Güntert, P., and Wüthrich, K. (2002) Protein NMR structure determination with automated NOE assignment using the new software CANDID and the torsion angle dynamics algorithm DYANA. *J. Mol. Biol.* **319,** 209–227.
6. Gropp, W., Lusk, E., Doss, N., and Skjellum, A. (1996) A high-performance, portable implementation of the MPI message passing interface standard. *Parallel Computing* **22,** 789–828.
7. Koradi, R., Billeter, M., and Wüthrich, K. (1996) MOLMOL: a program for display and analysis of macromolecular structures. *J. Mol. Graph.* **14,** 51–55.
8. Bartels, C., Xia, T. H., Billeter, M., Güntert, P., and Wüthrich, K. (1995) The program XEASY for computer-supported NMR-spectral analysis of biological macromolecules. *J. Biomol. NMR* **6,** 1–10.
9. Johnson, B. A. and Blevins, R. A. (1994) NMR View—a computer program for the visualization and analysis of NMR data. *J. Biomol. NMR* **4,** 603–614.
10. Kraulis, P. J. (1989) ANSIG—a program for the assignment of protein H-1 2D NMR spectra by interactive computer graphics. *J. Magn. Reson.* **24,** 627–633.
11. Helgstrand, M., Kraulis, P., Allard, P., and Härd, T. (2000) ANSIG for Windows: an interactive computer program for semiautomatic assignment of protein NMR spectra *J. Biomol. NMR* **18,** 329–336.
12. Koradi, R., Billeter, M., Engeli, M., Güntert, P., and Wüthrich, K. (1998) Toward fully automatic peak picking and integration of biomolecular NMR spectra. *J. Magn. Reson.* **135,** 288–297.
13. Herrmann, T., Güntert, P., and Wüthrich, K. (2002) Protein NMR structure determination with automated NOE-identification in the NOESY spectra using the new software ATNOS. *J. Biomol. NMR* **24,** 171–189.

14. Doreleijers, J. F., Mading, S., Maziuk, D., Sojourner, K., Yin, L., Zhu, J., Markley, J. L., et al. (2003) BioMagResBank database with sets of experimental NMR constraints corresponding to the structures of over 1400 biomolecules deposited in the Protein Data Bank. *J. Biomol. NMR* **26,** 139–146.

15. Mumenthaler, C. and Braun, W. (1995) Automated assignment of simulated and experimental NOESY spectra of proteins by feedback filtering and self-correcting distance geometry. *J. Mol. Biol.* **254,** 465–480.

16. Mumenthaler, C., Güntert, P., Braun, W., and Wüthrich, K. (1997) Automated procedure for combined assignment of NOESY spectra and three-dimensional protein structure determination. *J. Biomol. NMR* **10,** 351–362.

17. Nilges, M., Macias, M., O'Donoghue, S. I., and Oschkinat, H. (1997) Automated NOESY interpretation with ambiguous distance constraints: the refined NMR solution structure of the pleckstrin homology domain from β-spectrin. *J. Mol. Biol.* **269,** 408–4228

18. Nilges, M. and O'Donoghue, S. I. (1998) Ambiguous NOEs and automated NOE assignment. *Prog. NMR Spectrosc.* **32,** 107–139.

19. Linge, J. P., O'Donoghue, S. I., and Nilges, M. (2001) Automated assignment of ambiguous nuclear Overhauser effects with ARIA. *Methods Enzymol.* **339,** 71–90.

20. Linge, J. P., Habeck, M., Rieping, W., and Nilges, M. (2003) ARIA: automated NOE assignment and NMR structure calculation. *Bioinformatics* **19,** 315–316.

21. Nilges, M. (1993) A calculation strategy for the structure determination of symmetric dimers by ¹H NMR. *Proteins* **17,** 297–309.

22. Nilges, M. (1995) Calculation of protein structures with ambiguous distance restraints: automated assignment of ambiguous NOE crosspeaks and disulphide connectivities. *J. Mol. Biol.* **245,** 645–660.

23. Güntert, P., Braun, W., and Wüthrich, K. (1991) Efficient computation of three-dimensional protein structures in solution from nuclear magnetic resonance data using the program DIANA and the supporting programs CALIBA, HABAS and GLOMSA. *J. Mol. Biol.* **217,** 517–530.

24. Güntert, P., Mumenthaler, C., and Wüthrich, K. (1997) Torsion angle dynamics for NMR structure calculation with the new program DYANA. *J. Mol. Biol.* **273,** 283–298.

25. Kirkpatrick, S., Gelatt, C. D., Jr., and Vecchi, M. P. (1983) Optimization by simulated annealing. *Science* **220,** 671–680.

26. Katz, H., Walter, R., and Somorjay, R. L. (1979) Rotational dynamics of large molecules. *Computers Chemistry* **3,** 25–32.

27. Bae, D. S. and Haug, E. J. (1987) A recursive formulation for constrained mechanical system dynamics, part I: open loop systems. *Mech. Struct. Mech.* **15,** 359–382.

28. Mazur, A. K. and Abagyan, R. A. (1989) New methodology for computer-aided modelling of biomolecular structure and dynamics (I): non-cyclic structures. *J. Biomol. Struct. Dyn.* **4,** 815–832.

29. Mazur, A. K., Dorofeev, V. E., and Abagyan, R. A. (1991) Derivation and testing of explicit equations of motion for polymers described by internal coordinates. *J. Comp. Phys.* **92,** 261–272.

30. Jain, A., Vaidehi, N., and Rodriguez, G. (1993) A fast recursive algorithm for molecular dynamics simulation. *J. Comp. Phys.* **106,** 258–268.
31. Kneller, G. R. and Hinsen, K. (1994) Generalized Euler equations for linked rigid bodies. *Phys. Rev. E Stat. Nonlin. Soft Matter Phys.* **50,** 1559–1564.
32. Mathiowetz, A. M., Jain, A., Karasawa, N., and Goddard, W. A., III. (1994) Protein simulations using techniques suitable for large systems: the cell multipole method for nonbond interactions and the Newton-Euler inverse mass operator method for internal coordinate dynamics. *Proteins* **20,** 227–247.
33. Rice, L. M. and Brünger, A. T. (1994) Torsion angle dynamics: reduced variable conformational sampling enhances crystallographic structure refinement. *Proteins* **19,** 277–290.
34. Stein, E. G., Rice, L. M., and Brünger, A. T. (1997) Torsion-angle molecular dynamics as a new efficient tool for NMR structure calculation. *J. Magn. Reson.* **124,** 154–164.
35. Nilges, M., Clore, G. M., and Gronenborn, A. M. (1988) Determination of three-dimensional structures of proteins from interproton distance data by hybrid distance geometry-dynamical simulated annealing calculations. *FEBS Lett.* **229,** 317–324.
36. Brünger, A. T. (1992) *X-PLOR version 3.1: a system for X-ray crystallography and NMR.* Yale University Press, New Haven, CT.
37. Enggist, E., Thöny-Meyer, L., Güntert, P., and Pervushin, K. (2002) NMR structure of the heme chaperone CcmE reveals a novel functional motif. *Structure* **10,** 1551–1557.
38. Jee, J. G. and Güntert, P. (2003) Influence of the completeness of chemical shift assignments on NMR structures obtained with automated NOE assignment. *J. Struct. Funct. Genomics* 4, 179–189.
39. Güntert, P. (1998) Structure calculation of biological macromolecules from NMR data. *Q. Rev. Biophys.* **31,** 145–237.
40. Antuch, W., Güntert, P., and Wüthrich, K. (1996) Ancestral βγ-crystallin precursor structure in a yeast killer toxin, *Nat. Struct. Biol.* **3,** 662–665.
41. Calzolai, L., Lysek, D. A., Güntert, P., von Schroetter, C., Riek, R., Zahn, R., et al. (2000) NMR structures of three single-residue variants of the human prion protein. *Proc. Natl. Acad. Sci. USA* **97,** 8340–8345.
42. Zahn, R., Güntert, P., von Schroetter, C., and Wüthrich, K. (2003) NMR structure of a human prion protein with two disulfide bridges. *J. Mol. Biol.* **326,** 225–234.
43. Ellgaard, L., Riek, R., Herrmann, T., Güntert, P., Braun, D., Helenius, A., et al. (2001) NMR structure of the calreticulin P-domain. *Proc. Natl. Acad. Sci. USA* **98,** 3133–3138.
44. Horst, R., Damberger, F., Luginbühl, P., Güntert, P., Peng, G., Nikonova, L., et al. (2001) NMR structure reveals intramolecular regulation mechanism for pheromone binding and release. *Proc. Natl. Acad. Sci. USA* **98,** 14,374–14,379.
45. Lee, D., Damberger, F. D., Peng, G., Horst, R., Güntert, P., Nikonova, L., et al. (2002) NMR structure of the unliganded *Bombyx mori* pheromone-binding protein at physiological pH. *FEBS Lett.* **531,** 314–318.

46. Miura, T., Klaus, W., Ross, A., Güntert, P., and Senn, H. (2002) The NMR structure of the class I human ubiquitin-conjugating enzyme 2b. *J. Biomol. NMR* **22**, 89–92.

47. Hilge, M., Siegal, G., Vuister, G. W., Güntert, P., Gloor, S. M., and Abrahams, J. P. (2003) ATP-induced conformational changes of the nucleotide binding domain of Na,K-ATPase. *Nat. Struct. Biol.* **10**, 10–18.

48. Allen, M. P. and Tildesley, D. J. (1987) *Computer Simulation of Liquids.* Clarendon Press, Oxford, UK.

49. Abe, H., Braun, W., Noguti, T. and Go, N. (1984) Rapid calculation of first and second derivatives of conformational energy with respect to dihedral angles in proteins: general recurrent equations. *Computers Chemistry* **8**, 239–247.

50. Hockney, R. W. (1970) The potential calculation and some applications. *Meth. Comput. Phys.* **9**, 136–211.

51. Berendsen, H. J. C., Postma, J. P. M., van Gunsteren, W. F., DiNola, A., and Haak, J. R. (1984) Molecular dynamics with coupling to an external bath. *J. Chem. Phys.* **81**, 3684–3690.

52. Hare, B. J. and Wagner, G. (1999) Application of automated NOE assignment to three-dimensional structure refinement of a 28 kDa single-chain T cell receptor. *J. Biomol. NMR* **15**, 103–113.

53. Ösapay, K. and Case, D. A. (1991) A new analysis of proton chemical shifts in proteins. *J. Am. Chem. Soc.* **113**, 9436–9444.

54. Sitkoff, D. and Case, D. A. (1997) Density functional calculations of proton chemical shifts in model peptides. *J. Am. Chem. Soc.* **119**, 12,262–12,273.

18

NOE Assignment With ARIA 2.0

The Nuts and Bolts

Michael Habeck, Wolfgang Rieping, Jens P. Linge, and Michael Nilges

Summary

The assignment of nuclear Overhauser effect (NOE) resonances is the crucial step in determining the three-dimensional structure of biomolecules from nuclear magnetic resonance (NMR) data. Our program, Ambiguous Retraints for Iterative Assignment (ARIA), treats Noe assignment as an integral part of the structure determination process. This chapter briefly outlines the method and discusses how to carry out a complete structure determination project with the new version 2.0 of ARIA. Two new features greatly streamline the procedure: a new graphical user interface (GUI) and the incorporation of the data model of the Collaborative Computing Project for the NMR community (CCPN). The GUI supports the user in setting up and managing a project. The CCPN data model facilitates data exchange with a great variety of other programs. We give practical guidelines for how to use ARIA and how to analyze results.

Key Words: Automated assignment; ambiguous distance restraints; CCPN; structure calculation; XML; graphical user interface.

1. Introduction

The interpretation of spectra from nuclear magnetic resonance (NMR) experiments is a complex data analysis problem. Despite great progress toward automation, some crucial analysis steps are still only amenable to semiautomatic or manual treatment. The primary observables for structure determination are cross-relaxation rates measured in a nuclear Overhauser effect spectroscopy (NOESY) experiment. Other experiments provide complementary information that can be used in combination with the nuclear Overhauser effect (NOE)-based approach.

The problem of analyzing a NOESY spectrum is twofold. First, cross-peaks must be attributed to pairs of magnetically interacting spins (NOE assignment

From: *Methods in Molecular Biology, vol. 278: Protein NMR Techniques*
Edited by: A. K. Downing © Humana Press Inc., Totowa, NJ

problem). Second, structures must be calculated that fulfill conformational restraints derived from assigned NOEs. Regarding the latter aspect, NMR structure determination benefits from advances in simulation techniques, such as molecular dynamics, minimization, and simulated annealing. Robust methods for structure calculation thus exist, whereas methods for NOE assignment are still in the development stage.

The introduction of restraints from ambiguously assigned cross-peaks marked an important conceptual change in the treatment of degenerate NOE assignments. An ambiguous distance restraint (ADR) *(1)* combines alternative assignment possibilities in one restraint. Molecular conformations can be calculated from a list of ADRs and subsequently used to filter the assignment possibilities. The resulting iterative protocol converges to a structure ensemble and NOE assignments that are consistent. The method has been implemented and made publicly available, as in the program package Ambiguous Restraints for Iterative Assignment (ARIA) *(2–4)*.

2. Concepts

The standard procedure to determine macromolecular structures from solution NMR data is to minimize an objective function that incorporates experimental data and physical knowledge. Conformational restraints integrate experimental data; a molecular dynamics force field quantifies physical knowledge.

2.1. NOE Models

Traditionally, NOESY measurements serve as the primary source of structural information. ARIA offers two models to analyze cross-peak volumes quantitatively: the isolated spin-pair approximation and relaxation matrix analysis.

2.1.1. Isolated Spin-Pair Approximation

For short mixing times, a good approximation for relating an NOE volume, V_{ij}, to the distance, d_{ij}, of the two contributing spins is

$$V_{ij} = \alpha d_{ij}^{-6} \tag{1}$$

The scale, α, depends on properties of the system under investigation as well as on the experimental setup and cannot be measured directly. Therefore, it has to be estimated during the course of structure calculation (cf. **Subheading 4.2.1.**).

If an NOE is caused by dipolar interaction of two groups of magnetically equivalent spins, such as methyl groups and aromatic rings, **Eq. 1** must be extended to account for averaging effects. Let I and J denote two groups of spins with n_I and n_J members, respectively. We calculate the theoretical cross-peak volume as r^{-6} average over pairwise contributions:

$$V_{IJ} = \alpha n_I n_J \hat{d}_{IJ}^{-6} \text{ where } \hat{d}_{IJ}^{-6} = \frac{1}{N_I N_J} \sum_{I \times J} d_{ij}^{-6} \tag{2}$$

Introduction of the *effective* distance $\hat{d}_{IJ}$ retains the functional form of **Eq. 1**. **Equation 2** relies on a discrete slow jump model where spins I and J jump between N_I and N_J equilibrium sites, respectively *(5)*. Other models exist but quantitative differences are small *(6)*. Most commonly, r^{-6} averaging is applied to protons belonging to methyl groups or aromatic rings.

2.1.2. Relaxation Matrix Analysis

Observed NOE intensities involving two spins are affected by the presence of vicinal spins through spin diffusion. The spatial configuration of all proton spins forms a network that establishes alternative pathways for indirect magnetization transfer. Therefore, interproton distances are mostly underestimated when applying the isolated spin-pair approximation. We use relaxation matrix theory to account for indirect magnetization transfer. This formalism allows the calculation of cross-peak volumes at mixing time τ_m given the volumes at $\tau_m = 0$ and the matrix of auto- and cross-relaxation rates, R *(5)*:

$$V_{ij}(\tau_m) = \alpha V_{ij}(0)(\exp(-R\tau_m))_{ij} \tag{3}$$

2.2. Ambiguous Distance Restraints

To formulate restraints for structure calculation, distances are derived from cross-peak volumes. By means of **Eq. 1** the relation

$$D = (\alpha^{-1}V)^{-1/6} \tag{4}$$

converts the observed volume V into a target distance D. Isolated spin-pair approximation and relaxation matrix analysis are only approximate models of NOE volumes; they do not consider all mechanisms that affect relaxation rates. To account for this problem, it is common practice to introduce lower and upper *distance bounds*, L and U, by

$$L = \max(0, D - \Delta), U = D + \Delta \tag{5}$$

to restrain the distance to an interval rather than to a unique value. This reduces systematic bias because of potentially erroneous restraint distances. By default, we use a second order polynomial in the target distance to calculate the error margin $\Delta = 0.12D^2$. Other functional forms are supported. The distance bounds enter the objective function via a *flat-bottom harmonic-wall* potential with linear asymptotes *(2)*.

For degenerate resonance assignments with n_c pairs of contributing spins or groups of equivalent spins, we sum over *partial volumes* $\{V_i\}$ (**Eq. 2**), to calculate the effective distance

$$\overline{D} = \left(\alpha^{-1} \sum_{i=1}^{n_c} V_i\right)^{-1/6} \tag{6}$$

An ADR restrains the effective distance to lie between lower and upper distance bounds. Ambiguous distances have proved to provide an efficient way of expressing ambiguities at the logical level. Logical ambiguities occur during initial cross-peak assignment when compiling lists of possible contributions for every cross-peak. The ambiguity is resolved later during the course of structure calculation.

3. The ARIA Package

The program package ARIA implements the outlined ideas for structure determination from unassigned NOESY spectra. Methods for data management and data analysis are bundled in as a software library, thereby avoiding a rigid monolithic program structure and encouraging the user to modify and extend the standard protocols. This is further facilitated by choosing Python *(7)*, a plain and modern scripting language, for programming.

The ARIA core protocols derive distance restraints from a NOESY spectrum. These restraints define molecular conformations that are compatible with the observed cross-peaks. We use CNS *(8)* employing a simulated annealing strategy *(9)* to calculate structures. In principle, other structure calculation programs could replace CNS.

ARIA needs a definition of the molecular system and a NOESY spectrum accompanied by a list of chemical shift assignments. Other experimental data, such as scalar and residual dipolar couplings (RDCs), dihedral angles, hydrogen bonds, and distance restraints, can be added easily.

Two new features enhance the functionality and usability of the program: a graphical user interface (GUI) for complete project management and full support of the Collaborative Computing Project (CCPN) data model *(10)*.

3.1. The Project File

A single file contains the locations of the input data as well as all program and protocol parameters. Thus, the user can easily survey and change program settings by directly editing the project file or by using the GUI. The project file can be created during data conversion (cf. **Subheading 5.1.**), on the command line, or via the user interface. The latter option is the recommended and most convenient way for creating and modifying projects.

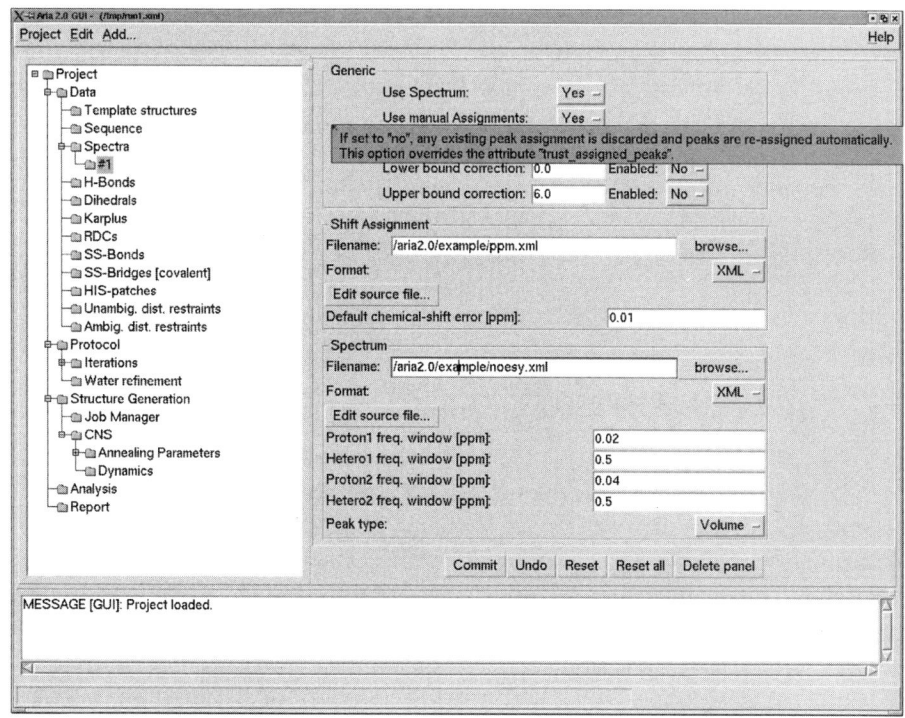

Fig. 1. ARIA 2.0 provides a new user interface for streamlined project management. The complete set of program and protocol settings can now be modified graphically.

3.2. A GUI for Project Management

To rationalize and simplify the setup of an ARIA run, version 2.0 offers a new GUI that replaces the HTML-based interface used in previous versions (*see* **Fig. 1**). All relevant program and protocol parameters can be adjusted via the GUI. Default values are given if appropriate and a pop-up help menu explains the meaning of each parameter.

3.3. New Formats for Data Representation

NMR data processing packages have introduced several proprietary data formats, each with its own advantages and drawbacks. Experience with earlier versions showed that many problems occurring at later stages of a calculation were the result of misformatted or inconsistent input files. Therefore, version 2.0 runs thorough validation checks on the input data before starting the calculation. We have developed a new data format that is based on the eXtensible Markup Language (XML) *(11)*. As of ARIA 2.0, all input data are XML compliant.

XML allows definition of portable and human readable formats for information exchange and supports document validation, thus guaranteeing consistency and integrity of the input data. XML is easy to interpret and can be manually corrected by using either an XML or a text editor. Our data format streamlines the validation process and represents experimental information as explicitly as possible. For instance, we do not support pseudoatoms or wild cards (as, e.g., the methylene group HB#). Instead, the user has to explicitly specify the corresponding group of atoms (i.e., HB2, HB3). Another example is the interpretation of chemical shift assignments of prochiral groups: In the input file, it is explicitly stated whether the assignment is stereospecific or nonstereospecific; this clarifies how an assignment is used throughout the calculation.

3.4. Data Exchange With Other Software Packages

The heterogeneity of NMR data formats hampers integrative analysis approaches. It is often very cumbersome to move data between different software packages, and this process almost always requires manual interference. The CCPN for NMR is an initiative that provides a service to NMR spectroscopists analogous to CCP4 in the X-ray community. One objective is the development of a data model for harvesting and exchange of data between different NMR processing and analysis software packages. ARIA uses the CCPN data model to store all results and analyses of a calculation in a general format. Furthermore, an existing CCPN project can be used to launch an ARIA calculation. However, we also support the more traditional approach of directly converting data generated by other programs. Besides CNS tbl-files, various data formats have become amenable via use of the CCPN program suite: Ansig *(12)*, NMRDraw *(13)*, NMRView *(14)*, Pipp *(15)*, Pronto *(16)*, CYANA *(17)*, XEasy *(18)*, Diana *(19)*, NMRStar *(20)*, and Sparky (Goddard, T.D. and Kneller, D.G. SPARKY3, University of California, San Francisco).

3.5. Complementary Experimental Information

Additional data, such as RDCs, complement NOE measurements and provide valuable information for structure calculation. We allow the integration of various data types.

3.5.1. Hydrogen Bonds

Hydrogen bonds are incorporated by restraining the distance between hydrogen donor and acceptor as well as the distance between acceptor and hydrogen.

3.5.2. J-Couplings

The Karplus curve describes the approximate functional relationship between a three-bond measured J-coupling and the involved dihedral angle. Calculated J-couplings are directly refined against observed J-couplings.

3.5.3. Residual Dipolar Couplings

ARIA offers two approaches to use residual dipolar coupling (RDC) data as restraints: the SANI statement *(21)* and the VEAN statement *(22)*. For SANI, the user has to specify the rhombicity and magnitude of the alignment tensor. VEAN uses angular restraints that must be precalculated with a separate program *(22)*.

3.5.4. Disulfide Bridges

If the connectivity of a disulfide bridge is known, the respective bond is added to the molecular topology. During torsion angle dynamics (TAD) *(23)*, unambiguous disulfide bridges cannot be treated as fixed; they are modeled as standard harmonic bonds. ARIA automatically deletes the disulfide bonds from the molecular topology during TAD and incorporates them as distance restraints while slowly increasing the weight of the harmonic potential. Unknown connectivities are treated like ambiguous distance restraints from NOE measurements *(1)*.

3.5.5. Dihedral Angle Restraints

ARIA supports the programs TALOS *(24)* and CSI *(25)*, which predict likely values of phi/psi main-chain dihedral angles given a list of chemical shift assignments. The predictions are incorporated as dihedral angle restraints using a *flat-bottom harmonic-wall* potential.

3.6. IUPAC Atom Name Nomenclature

Markley et al. *(26)* have proposed a standard for the presentation of NMR structures of proteins and nucleic acids; the standard has been approved by the International Union of Pure and Applied Chemistry (IUPAC). ARIA follows these recommendations. Owing to the inconsistent use of atom name nomenclatures in the NMR community, we want to explicitly list the most common naming problems:

1. The C-terminal carboxyl group is named O' and O''. O'' contains two apostrophes (ASCII 39), not a quotation mark (ASCII 34). The Protein Data Bank (PDB) uses O and OXT or OT1 and OT2 instead.
2. The N-terminus consists of H1, H2, and H3 (not HT1, HT2 and HT3).
3. The protein backbone amide proton is called H (instead of HN).
4. The glycine alpha protons are HA2 and HA3.
5. Pseudoatoms are not supported; r^{-6} averaging is applied to equivalent groups.

4. Program Flow

Figure 2 shows the work flow of an ARIA project. It consists of the following steps: a preparatory stage, iterative structure calculation and NOE assignment, structure refinement in explicit solvent, and final analysis of the results. We explain these steps in more detail.

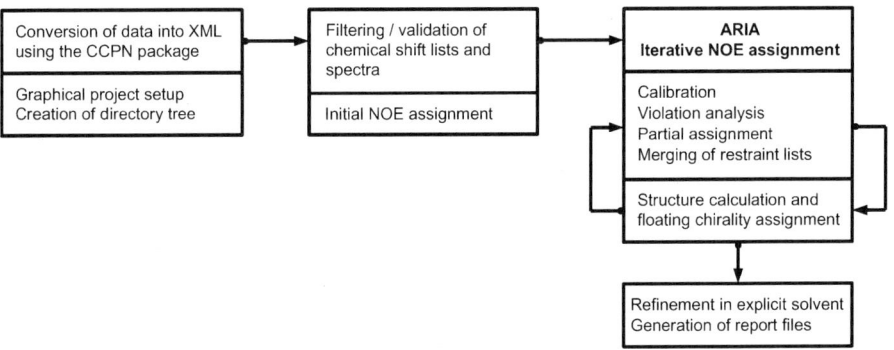

Fig. 2. ARIA integrates the following tasks: conversion of experimental data to XML, iterative structure calculation and NOE assignment, refinement in explicit solvent, and final analysis.

4.1. Preparation Stage

The following steps are performed before structure calculation. First, the data are filtered for errors and inconsistencies. The program then proceeds with creation of the molecular topology. Afterward a seed NOE assignment is derived, based on the chemical-shift lists.

4.1.1. Filtering of Data

ARIA checks the chemical shift assignments for consistency. We consider three cases: First, a unique assignment consisting of one atom and one chemical shift; second, a degenerate chemical shift assignment, where one group of equivalent atoms is assigned to exactly one chemical shift; third, an assignment of the two constituents of a prochiral group, which can have one or two chemical shifts. In the latter case, we use floating chirality assignment *(27)* to treat the resulting restraints (cf. **Subheading 4.2.2.**). To clean the peak lists, we first delete diagonal peaks that are detected by comparing the proton dimensions within the chemical shift tolerances. Second, peaks with incomplete frequency information or missing peak sizes are removed. We always use absolute values of peak sizes. Files reporting results of data cleaning are generated (cf. **Subheading 5.4.**).

4.1.2. Creating the Molecular Topology

ARIA runs the CNS script `generate.inp` to create a molecular topology file (MTF). The MTF defines the name, chemical type, charge and mass of each atom, as well as the covalent connectivity. The CNS script `geneate_template.inp` constructs the molecule in an extended conformation and writes the coordinates to a PDB file. By default, the extended conformation is randomized and used as the starting structure for calculation, but it can be replaced with any other structure. In the case of standard biopolymers, the molecular topology is created automatically. The user can define disulfide bridges and histidine protonation states in the GUI. For nonstandard residues or chemical compounds the user has to intervene manually. The script `generate.inp` for setting up the MTF can be edited and then run from ARIA. Alternatively, an MTF can be specified in the project file. Changes must also be made to the CNS topology, linkage, and parameter files.

4.1.3. Initial NOE Assignment

ARIA uses the chemical shift lists to derive possible assignments for every cross-peak. These *seed assignments* solely refer to the peak position in frequency space. A proton is assigned to all cross-peaks whose frequency windows include the proton chemical shift. Either the frequency error or the global frequency window determines the window size. In the case of heteronuclear NOESY experiments, the hetero atom attached to the proton must also match the corresponding chemical shift window. Frequency window sizes are important parameters that ought to be chosen carefully. On the one hand, narrow windows lead to potentially incomplete seed assignments. On the other hand, generous window sizes produce highly degenerate initial assignments that can cause severe convergence problems. The preferred strategy is therefore to choose the window size just large enough that the initial assignments contain the correct assignment, as far as possible. Recommended window sizes are 0.02 ppm for direct, 0.04 ppm for indirect proton dimensions, and 0.5 ppm for heteronuclear dimensions.

Furthermore, the completeness of the chemical shift list affects the accuracy of the initial assignment. Evidently, atoms with missing resonance assignment cannot be assigned to any cross-peak. In this case, automatically generated assignments are almost certainly wrong. As a rule of thumb, the completeness of a chemical shift list should not fall below 90% to achieve reasonable convergence.

ARIA provides two ways for adding manual cross-peak assignments: First, assignments contained in the original peak list are looped through to the spectrum XML file during data conversion (cf. **Subheading 5.1.**). Second, the user may edit the spectrum XML file to add manual assignments (element `assignment`). Manual assignments always enter the seed assignment list and are not modified.

4.2. Iterative Structure Calculation

ARIA detects inconsistent cross-peaks and reduces cross-peak assignment possibilities by iterating through the following scheme:

1. Analysis of the structure ensemble, calibration of spectra, and detection of inconsistent cross-peaks.

2. Creation of a restraint set with fewer assignment possibilities and calculation of a new structure ensemble.

Both stages depend on numerous user-definable parameters. The objective is to achieve mostly unambiguous cross-peak assignments after the last iteration. The parameters are adjusted such that the algorithm converges after a given number of iterations, nine by default. Obviously, the parameter values affect the performance of the algorithm and hence the quality of the produced structures. We have optimized the default parameter settings in terms of consistency and quality of the structure ensemble, not necessarily with respect to speed.

4.2.1. Analysis of the Structure Ensemble

Every cycle begins with an analysis of the previously calculated structures. All analyses relate to ensemble averages. To obtain reliable estimates, it is essential to exclude outliers. The minimization algorithm, however, cannot be expected to converge in every instance. Hence, it is important to identify nonconverged structures. We use the total energy to discriminate between converged and nonconverged conformers by selecting S structures with lowest energy, 7 out of 20 by default. In the following, all ensemble averages are calculated with respect to the S lowest energy structures.

4.2.1.1. CALIBRATION

According to **Eq. 4** calculation of distances from observed volumes involves an unknown scale or "calibration" factor, α. We use the ratio of the average of the experimental to the average of the theoretical volumes

$$\alpha = \frac{\sum_i V_i^{exp}}{\sum_i V_i^{th}} \tag{7}$$

as an estimator for the calibration constant. Theoretical volumes can be calculated using either the isolated spin-pair approximation or relaxation matrix analysis. For the latter case, test calculations demonstrated a significant improvement of quality indexes, such as Ramachandran coverage, packing quality, and a reduction of NOE restraint violations *(28)*. Partial volumes are calculated as ensemble averages. We then introduce a *cutoff distance, d_c.* Cross-peaks with theoretical volume less than $V_c = d_c^{-6}$ are considered as outliers and therefore excluded from calibration. d_c is a user-defined parameter that is set to 6.0 Å by default. We estimate a preliminary calibration factor from the remaining cross-peaks according to **Eq. 7**. Every spectrum is calibrated separately.

For the first iteration, the user may specify an initial structure ensemble for calibration. If omitted, the calibration routine is modified: We assume that the average distance, d_{avg}, of spin pairs causing a NOE is known. Using **Eq. 7**, the initial estimate of the calibration factor, α_0, is

$$\alpha_0 = n^{-1} \sum_{i=1}^{n} v_i^{exp} / d_{avg}^{-6} \qquad (8)$$

We use the calibration factor, α (respectively, α_0), to calculate a set of pre-liminary distance bounds on the basis of **Eq. 6**.

4.2.1.2. DETECTION OF INCONSISTENT RESTRAINTS BY VIOLATION ANALYSIS

To identify wrong assignments and noise peaks, the obtained restraints are subject to a violation analysis. Violation analysis relies on the hypothesis of structural consistency *(2,29)*. To assess whether a restraint follows the "general trends" imposed on the structures by the entire data set, we compare its distance bounds with the corresponding distances found in the ensemble. A restraint is "violated" if the distance found in the structure lies outside the bounds by more than a user-defined *violation tolerance, t.* To identify restraints that are system-atically violated, each structure in the ensemble is analyzed. Let L_i and U_i denote the lower and upper bound of the *i*-th restraint. We calculate the fraction, f_i, of structures violating restraint *i* according to

$$f_i := S^{-1} \sum_{j=1}^{s} \left[\theta\left(L_i - d_i^{(j)} + t\right) + \theta\left(d_i^{(j)} - U_i - t\right) \right] \qquad (9)$$

where $d_i^{(j)}$ denotes the distance found in the *j*-th structure; $\theta(\cdot)$ is the Heaviside function. We classify a restraint as violated if f_i exceeds a user-defined *violation threshold*, which is set to 0.5 by default. To calculate the final calibration factor we apply **Eq. 7** to the set of nonviolated restraints.

4.2.2. Reduction of Assignment Possibilities and Calculation of a New Ensemble

Cross-peaks are assigned in an indirect fashion by eliminating unlikely assignment possibilities. Because of the r^{-6} dependence, assignments with large distances contribute only marginally to the NOE intensity. We weight each assignment possibility by its normalized partial volume, $w_i = V_i / \sum_i^{n_c} V_i$. To reduce the number of possibilities, only the *m* largest contributions satisfying $\sum_1^m w_i \geq w_c$ are kept. w_c denotes a user-defined *ambiguity cutoff* that is set to 1.0 in the first iteration. The cutoff is gradually decreased to 0.8 to obtain almost unambiguous assignments after the last iteration. It may also be advantageous to limit the maximum number of contributions, *max_n*, to improve the quality of NMR structure ensembles *(3)* (the default value is 20).

4.2.2.1. STRUCTURE CALCULATION

Restraints involving the same set of atoms lead to a biased restraint list. They are the result of symmetry peaks or duplicated peaks from multiple spectra. Equivalent restraints produce artifact in the structures because of overrepresentation of certain distance data. Therefore, we detect nonviolated restraints with equivalent atom content. We keep the restraint with smallest distance and discard the others.

We calculate a new structure ensemble on the basis of the merged restraint set and other restraints (e.g., hydrogen bonds, dihedral angles, RDCs). For every iteration, the number of structures can be controlled with the parameter *n_structures.*

4.2.2.2. FLOATING CHIRALITY ASSIGNMENT

It is often difficult to assign the chemical shifts of the two substituents of a prochiral center stereospecifically. In proteins, these are the two methylene protons or the methyl protons of the isopropyl groups of valine or leucine. A resonance matching one of the chemical shifts in the proton dimensions potentially involves either of the two prochiral substituents. ARIA compensates this lack of information by testing both alternatives during structure calculation. For each calculated structure, the energetically preferred options are written to a file with a .float extension.

4.3. Refinement in Explicit Solvent

Because of the simplified treatment of nonbonded forces and missing solvent contacts, calculated structures often show artifacts, such as unrealistic sidechain packing and unsatisfied hydrogen bond donors or acceptors. ARIA provides a protocol for refining macromolecular structures in a shell of water or dimethyl sulfoxide molecules. For each structure of the final ensemble, we calculate a short trajectory with a full molecular dynamics force field, including electrostatic and Lennard–Jones potentials (*see* ref. *30* for details). The parameters for refinement (PARALLHDG 5.3) are consistent with the force field used for structure calculation and validation. Hence, there are no systematic differences that could influence validation results.

4.4. Analyses and Output Files

ARIA generates various output files to report analysis results. Calculated structures are stored in PDB format respecting IUPAC atom name nomenclature. For every iteration, ARIA creates the following report files:

1. noe_restraints.unambig, noe_restraints.ambig
   These files tabulate unambiguous and ambiguous restraints, respectively. Restraints discarded by the merging procedure are excluded. For every restraint, information is given on its reference cross-peak, restraint bounds, the average distance found in the ensemble, and the result of violation analysis.

2. noe_restraints.violations

   Lists all violated restraints sorted with respect to their upper-bound violations and contains the same information as no. 1.

3. noe_restraints.assignments

   Lists primarily restraint-wise assignments and gives information on whether the assignment(s) stem from fully, partially, or unassigned cross-peaks.

4. noe_restraints.merged

   Reports all restraints that have been discarded by the merging procedure.

5. noe_restraints.xml, noe_restraints.pickle

   The complete list of NOE-based distance restraints stored in XML and Python binary format for persistent object serialization. Both files are intended to serve as a database for advanced users working with their own analysis scripts.

6. report

   Summarizes analyses of the restraint lists and the structure ensemble.

4.4.1. MOLMOL

For the last iteration, MOLMOL *(31)* distance restraint files containing lower and upper bounds are written to the subdirectory */molmol/*. These can be used to visualize distance restraints from ARIA and may assist in identification of erroneous restraints.

4.4.2. Quality Checks

ARIA uses the programs WHAT IF *(32)*, PROCHECK *(33)*, and PROSA II *(34)* to evaluate the quality of both the final set of structures and the solvent-refined ensemble. For every program separate report files, quality_checks, are stored in the directories of the respective ensembles.

4.4.3. Miscellaneous Analyses

Several CNS scripts calculate restraint energies, ensemble root-mean-square deviations (RMSDs), and an average structure. Results are stored in */analysis/*.

By default, ARIA creates all analysis and report files except for Python binary data.

5. Working With ARIA

An ARIA calculation is easy to set up and proceeds largely automatically. The standard procedure is to run ARIA repeatedly and use the results to revise the input data (**Fig. 3**). We give some practical guidelines that enable beginners to start their own structure determination project.

5.1. Conversion of Input Data to XML

Prior to structure calculation, all input data must be converted to XML. Besides formats and filenames of the raw data (sequence, chemical shift lists and spectra), data conversion requires some additional information that must be

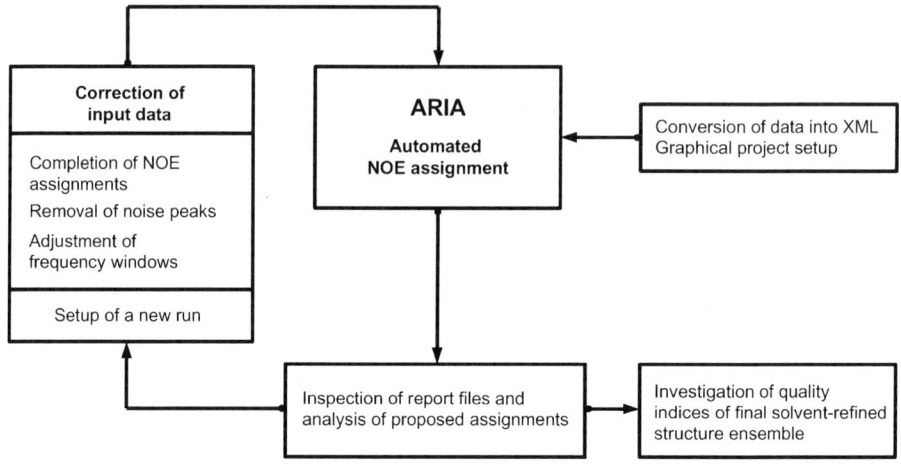

Fig. 3. A typical structure determination project consists of repeated ARIA runs followed by manual correction of the input data.

provided in the form of a simple XML *conversion file*. Most important, the user has to specify the mapping between nuclei and frequency dimensions; for convenience, a preformatted project file is autogenerated. The conversion file should be filled in carefully to avoid the introduction of unnecessary errors at the beginning of a project. A common source for inconsistencies is, for example, mismatches in the residue numbering in the molecule definition and in the other data files. The command

```
aria2 --convert -t conversion.xml
```

creates an empty conversion file template, conversion.xml, that must be completed. To start the conversion, invoke the command

```
aria2 --convert conversion.xml
```

A general conversion issue is atom names. Most software packages rely on their own nomenclature and also use wild cards or pseudoatom names. These have to be translated into their IUPAC counterparts. The conversion module is mostly automated. However, in dubious cases, the user is either asked to resolve the ambiguity or is notified on the program's decision. After conversion, the user should nevertheless check whether the data were translated in the intended way. **Figure 4** shows an excerpt of a converted chemical shift list stemming from the HRDC-domain data *(35)*. Further examples as well as a complete conversion file can be found in the directory */examples/werner/ready-to-use* of the distribution. It is important to bear in mind how chemical shift assignments are converted, as this determines the way they are translated into distance restraints.

```
                                        <shift_assignment method="FLOATING">
                                          <spin_system averaging_method="NONE">
                                            <atom segid="    " residue="10" name="HB2"/>
                                            <chemical_shift value="1.97" error="0.001"/>
167   2.051 0.001 HB2   10                 <chemical_shift value="2.051" error="0.001"/>
168   1.970 0.001 HB3   10                </spin_system>
169 999.000 0.000 QB    10                <spin_system averaging_method="NONE">
                                            <atom segid="    " residue="10" name="HB3"/>
                                            <chemical_shift value="1.97" error="0.001"/>
                                            <chemical_shift value="2.051" error="0.001"/>
                                          </spin_system>
                                        </shift_assignment>
```

Fig. 4. Example for the conversion of a chemical shift list. The *left hand side* shows chemical shift assignments in XEasy format. The *right hand side* displays the corresponding assignments in ARIA XML format. Because protons HB2 and HB3 belong to a methylene group, they make up a floating chirality assignment. Pseudo atom names are superfluous (as QB in this case).

5.2. Nonstereospecific Chemical Shift Assignments

ARIA uses r^{-6} averaging (**Eq. 2**) to treat protons with degenerate chemical shift assignments. Methyl protons usually exhibit degenerate chemical shifts because they are in fast exchange. Consequently, only the whole group can match the proton shift dimension of an NOE peak. The same holds for N-terminal amide protons. Protons in an aromatic ring are also equivalent when assigned to the same chemical shift. In case both frequencies of the two substituents of a prochiral group are identical, the substituents make up a group of equivalent spins handled by r^{-6} averaging. This applies to methylene groups and the isopropyl groups in valine and leucine. Averaging might not always be physically justified. The user should verify whether it is indeed correct. Thymine is the only nucleotide that has a group of fast exchanging methyl protons at the C7 carbon.

Prochiral groups often have ambiguous chemical shift assignments. The generic case is a methylene group giving rise to two distinct resonances that cannot be assigned stereospecifically. Consider, for example, the case where HB2 and HB3 were assigned to different chemical shifts, δ_1 and δ_2. Floating chirality assignment takes both possibilities, HB2 $\to \delta_1$, HB3 $\to \delta_2$ and HB2 $\to \delta_2$, HB3 $\to \delta_1$, into account. The same applies when only one shift is given or in the case of an assignment of a wild card corresponding to a methylene group (e.g., HG# in Glu). The prochiral proline N-terminus is handled like a methylene group. The two methyl groups in valine and leucine may have nonstereospecific assignments in which case the floating chirality method is applied. Floating chirality applies also if the HD and HE protons of aromatic rings are assigned to different shifts. In nucleotides methylene groups are located at the 5′ carbon and, in the case of DNA, also at C2′.

Parameter	XML element / attribute	Location in GUI	Default value
Project environment	infrastructure	Project	
Project nickname	run		1
File root	file_root		
Working directory	working_directory		
Temporary directory	temp_root		
Data specification	data	Data	
Frequency windows	freq_window_proton1	→ Spectra	0.02
Frequency windows	freq_window_hetero1	→ Spectra	0.5
Trust assignments	trust_assigned_peaks	→ Spectra	no
CNS topology file	topology_definition	→ Sequence	topallhdg5.3.pro
CNS linkage file	linkage_definition	→ Sequence	topallhdg5.3.pep
CNS parameter file	parameter_definition	→ Sequence	parallhdg5.3.pro
Initial ensemble	template_structure	→ Template structures	
Protocol parameters	protocol	Protocol	
Number of structures	n_structures	→ Iterations	20
Distance cutoff d_c	distance_cutoff	→ Iterations	6.0
Violation tolerance t	violation_tolerance	→ Iterations	1000.0 – 0.1
Violation threshold	violation_threshold	→ Iterations	0.5
Ambiguity cutoff w_c	weight_threshold	→ Iterations	1.0 – 0.8
Maximum number of contributions, max_n	max_contributions	→ Iterations	20
Number of lowest energy structures S	n_best_structures	→ Iterations	7
Solvent for refinement	solvent	→ Water refinement	water
Structure calculation	job_manager	Structure Generation	
Local CNS executable	default_executable	→ CNS	
Command to start calculation on remote machine	command	→ Job Manager	
CNS executable on remote machine	executable	→ Job Manager	
High temperature steps	steps_high	→ CNS → Dynamics	10000

Fig. 5. Important protocol parameters.

5.3. Specification of Protocol Parameters

Data conversion generates a project file that the user has to complete. Alternatively, he can choose the "New" command in the GUI menu "Project" or execute

```
aria2 --project_template project.xml
```

on the command line to create a new project file. Important program and protocol parameters are compiled in **Fig. 5**. A minimal set of input data consists of the description of the molecule and a NOESY spectrum with a corresponding chemical shift list. To supply further data, such as additional spectra, J couplings, or RDCs, choose the menu "Add" in the user interface. For each spectrum, the default frequency window sizes should be adjusted. Other mandatory parameters concern the project's infrastructure and shall be explicitly listed here:

1. *Working directory:* directory for storing results of an ARIA run.
2. *File root:* used throughout the project as code for the molecular system, for example, when writing PDB files.
3. *Temporary path:* temporary directory used by CNS during structure calculation.
4. *Local cns executable:* path of the CNS executable on the same machine where ARIA is running.

For all other parameters default settings are provided.

5.4. Project Setup

Before calculations can start, the project must be set up:

```
aria2 --setup project.xml
```

This command reads and validates the project file project.xml. If both steps are successful, ARIA will create the directory tree shown in **Fig. 6**. Structure calculation results are deposited in *structures/* where each iteration has its own subdirectory *structures/it*/*. All data needed by CNS reside in a separate subdirectory *cns/*. These include protocols (*cns/protocols/*), topology, parameter and linkage files (*cns/toppar/*), and data files in CNS format (*cns/data/*/*). CNS specific files usually do not need to be modified. Only if the users want to change a protocol or the molecular topology in a nonstandard way, will they need to edit them.

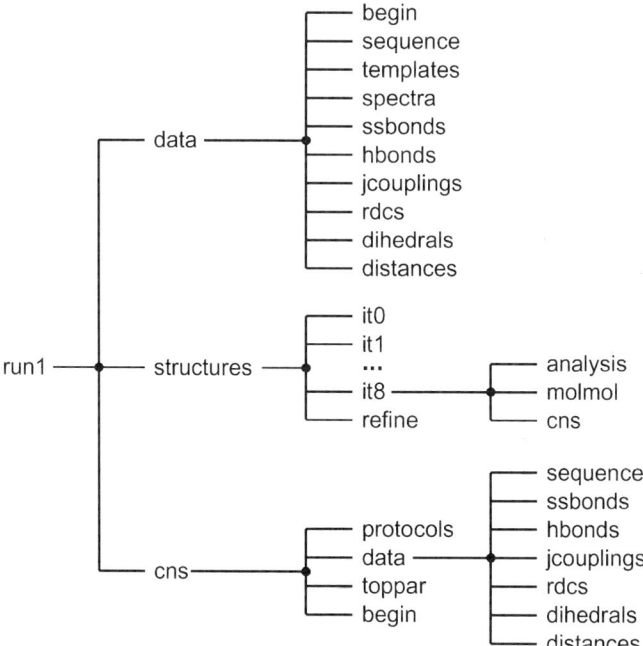

Fig. 6. Directory tree of an ARIA project.

Copies of the experimental data files as well as report files of the data filtering procedures are stored in *data/*. Only local copies are used for structure calculation. Thus, changes in the original files will become active in the next project setup.

5.5. Running ARIA

The setup of the project directory completes the necessary arrangements for starting ARIA. Without further specification, calculations will be performed on the same processor from which ARIA has been started.

5.5.1. Parallel Setup

To accelerate ARIA, we recommend distributing the calculations to multiple processors. We use a temporary shell script to launch the calculations. Hence, the user must provide ARIA with a command (project file, attribute `command`) that first establishes the connection to the remote machine and then starts the script—for example, ssh *machine_name* csh. The parallel setup is easily prepared by using the GUI.

For the first round of calibration, the user can specify an initial ensemble of *template structures* (attribute `template_structure`, *see* **Fig. 5**). In this case, the default violation tolerance (1000Å) must be reduced to a smaller value (e.g., 5Å).

To start ARIA, use the following command:

```
aria2 project.xml
```

The following sections give some advice on how to inspect the calculation results.

5.6. Convergence Checks

The performance of the minimization protocol depends on both the particular molecular system and on the quality and completeness of the data. Protocol parameters, in particular the number of dynamic steps required for convergence, are primarily determined by the system size. Default values work well for systems up to about 150 residues. However, for larger systems it might become necessary to increase the number of steps. To evaluate whether structure calculation has converged, one first examines the average and the variation of the total energy in the ensemble. As a rule of thumb, well-converged ensembles for systems of about 100 residues show average energies of the order of 1000 kcal mol^{-1} or less. An energy variation of 10% can be considered normal. The average energy scales approximately linear with the system size. Energy distributions not meeting these requirements can be categorized as follows:

- *Large average energy or large energy variation* indicates that the calculation has not converged. This can be verified by inspecting the conformational variance of the ensemble, for instance in terms of the RMSD (in *it8/analysis/rmsdave.disp*).

In most cases, poor convergence is owing to erroneous protocol settings, in particular the number of dynamics steps or the size of frequency windows, as well as poor data quality.

* *Low average energy, large energy variation.* This case indicates a broad and unpronounced energy minimum that typically occurs when using a small number of restraints. Check the `report` file for information on the number of restraints used for structure calculation. A tight violation tolerance might also cause high restraint rejection rates leading to data too sparse to achieve convergence. In particular, highly ambiguous, incomplete, or noisy data sets require a more generous violation threshold in the first iterations.

5.7. Checking Automatic Resonance Assignments

Figure 7 shows an excerpt of the analysis file `noe_restraints.violations`. It provides information on all restraints rejected by violation analysis. Restraints being violated in the majority of the structures (≥85%) in conjunction with large upper-bound violations (≥5Å) are usually inconsistent or stem from wrong assignments and should not be used in a subsequent calculation. To be on the safe side, restraints showing systematic violations of more than 0.1 Å ought to be inspected manually. One should also check frequency window sizes because narrow windows affect the completeness of a cross-peak assignment. It may therefore be advisable to increase the individual window size by 10% in the subsequent run and check the cross-peak again. For automatically generated assignments, one proceeds in the same way. Correct automatic assignments should be treated like manual assignments in a subsequent run.

5.8. Setting Up a New Run

Based on the analyses mentioned in the previous sections, wrong or inconsistent assignments as well as noise peaks have been corrected or marked. The user may also want to tell the program which cross-peak assignments are definitely correct. ARIA therefore distinguishes between NOEs with reliable and

ref_spec[1]	ref_no[1]	lower[2]	upper[2]	d_avg[3]	u_viol	%_viol	a_type[4]	n_c[5]
...								
13C NOESY	1568	2.03	5.00	14.73	9.73	100.0	S	2
13C NOESY	1612	1.76	3.30	12.93	9.63	100.0	M	1
13C NOESY	2996	1.89	3.90	13.38	9.49	100.0	A	3
...								

Fig. 7. An excerpt of the file `noe_restraints.violations`. The restraints have been violated in all structures (`%_viol`) and show upper-bound violations (`u_viol`) of more than 5Å. [1]Reference cross-peak; [2]lower and upper distance bounds; [3]ensemble average distance; [4]assignment type: (A)utomatic, (S)emi-automatic, (M)anual; [5]number of contributions.

NOEs with questionable assignments. Reliable assignments always enter structure calculation, that is, some care must be taken not to declare erroneous peak assignments as reliable. To declare a cross-peak as reliable, set its attribute `reliable` (spectrum file) to "yes". To declare all fully assigned cross-peaks of a spectrum as reliable, set the spectrum-specific attribute `trust_ assigned_ peaks` to "yes" (*see* **Fig. 5**).

Changes in protocol parameters, such as those controlling simulated annealing or the initial NOE assignment, will become operative when rerunning the program and do not require further intervention. Previous results, however, in particular PDB files and analysis files, are overwritten. For better bookkeeping, we recommend to use the same project but to create a new run (cf. **Subheading. 5.4.**). Of course, the new run must be set up. To exploit the information generated in the previous calculation the input data must be modified:

- Cross-peak assignments can be augmented by automatically created assignments.
- Erroneous cross-peak assignments can be removed from the respective spectrum.
- Low-energy structures could serve as templates or starting structures.
- Reliable distance restraints can serve to improve convergence.
- Noise peaks should be deleted.

After the new run has successfully been set up, the user proceeds as normal by running ARIA, analyzing the results, and so forth.

5.9. Inspecting the Final Result

5.9.1. Software

When all cross-peaks have been assigned, the quality of the final structure ensemble must be evaluated. However, no cogent definition of a "good" NMR structure exists. Various programs attempt to assess the quality of structures with respect to general properties of folded proteins, such as packing, dihedral angle, or hydrogen-bond configuration. Most methods compare the target structure on the basis of statistics gathered from high-resolution X-ray structures. ARIA uses the programs WHAT IF, PROCHECK, and PROSA II to calculate several quality indices; a summary can be found in the report file. Most indexes depend on both the system under investigation and on data quality, hence no exact values can be given here. However, the following scores should be investigated further:

1. *PROCHECK Ramachandran coverage.* For typical X-ray structures, about 90% of all dihedral angles lie within the preferred region of the Ramachandran plot. For high-quality NMR structures a comparable value is desirable. Unstructured loop regions, however, tend to decrease that value. Typical NMR structures deposited in the PDB show Ramachandran coverages of about 80% (cf. **Fig. 3** in **ref. *36***).

2. *PROSA II Z-score.* Prosa uses a knowledge-based potential, derived from high-resolution crystal structures, to assess how well a structure fits a sequence. The Prosa Z-score quantifies the match between target structure and target sequence compared to alternative folds compiled from the PDB. Z-scores of native proteins are sequence dependent; native structures with 150 residues typically have Z-scores less than -5.

3. *WHAT IF packing quality and bump-score.* The packing quality measures the agreement between the distribution of local atom configurations and the equivalent distributions derived from a set of high-resolution X-ray structures *(37)*. The bump-score reports van der Waals clashes *(32)*.

5.9.2. Completeness

To further judge the credibility of the structures, it is advisable to reexamine the data. The *completeness (38)* of the restraint set, in particular, provides an insight into the local reliability of each structure. An atom-wise completeness is defined by the ratio of numbers of observed to expected restraints involving the atom. The completeness is zero, if the atom is not restrained at all, and 1 if it is included in all restraints that are structurally possible. Atoms with completeness factors >1 are involved in too many restraints; this might be owing to a wrong resonance assignment. A completeness factor significantly less than 1 indicates an unreliable atom position. The most likely reasons for low completeness values are:

1. Cross-peaks are missing in the data set. They might have either not been picked or could not be assigned because of undersized frequency windows.
2. If the chemical shift list is incomplete some resonances can never be correctly assigned at all.
3. The corresponding cross-peaks were rejected by violation analysis and have thus not been used during calculation (cf. **Subheading 5.7.**).
4. If neighboring atoms also show a low completeness, the structure might be locally misfolded.
5. If the local fold of the structure is correct, missing cross-peaks could also provide an indication of local motion or multiple conformations. This hypothesis, however, cannot be verified without further experimental investigation.

6. Conclusions

ARIA is a software package for automated structure determination from NOESY spectra. In principle, the user only has to supply data (chemical shift assignments and NOESY peak lists) stored in one of the standard NMR formats. ARIA then proceeds with data conversion, derivation of ADR lists, and structure calculation. An automated approach has several advantages. First, the data are interpreted in an objective way thus reducing human bias. The method does not require manual intervention, and the results can be judged by external quality criteria. Second, although the program cannot always be expected to find cross-peak assignments, it is of great use in assisting the experimenter in analyzing his or her data. An otherwise tedious and time-consuming examination of the data can be highly facilitated.

At the time we write this chapter, more than 60 structures have been deposited in the PDR referring to ARIA.

Acknowledgments

M. Habeck and W. Rieping were supported by the European Community (5th Framework program NMRQUAL, contract number QLG2-CT-2000-01313). J. P. Linge thanks the Pasteur Institute and the SPINE network (EU 5th Framework program, contract number QLG2-CT-2002-00988) for financial support.

References

1. Nilges, M. (1995) Calculation of protein structures with ambiguous distance restraints. Automated assignment of ambiguous NOE crosspeaks and disulphide connectivities. *J. Mol. Biol.* **245,** 645–660.
2. Nilges, M. and O'Donoghue, S. I. (1998) Ambiguous NOEs and automated NOESY assignment. *Prog. NMR Spec.* **32,** 107–139.
3. Linge, J. P., O'Donoghue, S. I., and Nilges, M. (2001) Automated assignment of ambiguous nuclear Overhauser effects with ARIA. *Methods Enzymol.* **339,** 71–90.
4. Linge, J. P., Habeck, M., Rieping, W., and Nilges, M. (2003) ARIA: automated NOE assignment and NMR structure calculation. *Bioinformatics* **19,** 315, 316.
5. Görler, A. and Kalbitzer, H. R. (1997) Relax, a flexible program for the back calculation of NOESY spectra based on complete relaxation matrix formalism. *J. Magn. Reson.* **124,** 177–188.
6. Yip, P. F. and Case, D. A. (1991) Incorporation of internal motions in NMR refinements based on NOESY data. In *Computational Aspects of the Study of Biological Macromolecules by Nuclear Magnetic Resonance Spectroscopy* (Hoch, J. C., Poulsen, F. M., and Redfield, C., eds.). Plenum, New York, pp. 317–330.
7. van Rossum, G. and de Boer, J. (1991) Linking a stub generator (AIL) to a prototyping language (Python). In *EurOpen: UNIX Distributed Open Systems in Perspective: Proceedings of the Spring 1991 EurOpen Conference, Tromsø, Norway, May 20–24, 1991* (EurOpen, ed.), EurOpen, Buntingford, Herts, UK, pp. 229–247.
8. Brünger, A. T., Adams, P. D., Clore, G. M. Delano, W. L., Gros, P., Grosse-Kunstleve, R. W., et al. (1998) Crystallography and NMR system (CNS): a new software suite for macromolecular structure determination. *Acta Crystallogr. D* **54,** 905–921.
9. Nilges, M., Macias, M. J., O'Donoghue, S. I., and Oschkinat, H. (1997) Automated NOESY interpretation with ambiguous distance restraints: the refined NMR solution structure of the pleckstrin homology domain from spectrin. *J. Mol. Biol.* **269,** 408–422.

10. Fogh, R. H., Ionides, J., Ulrich, E., Boucher, W., Vranken, W., Linge, J. P., et al. (2002) The CCPN project: an interim report on a data model for the NMR community. *Nat. Struct. Biol.* **9,** 416–418.
11. The World Wide Web Consortium (1999) *Extensible Markup Language (XML) 1.0, W3C recommendation.* Available at http://www.w3.org/TR/REC-xml. Accessed 03/04/04.
12. Kraulis, P., Domaille, P. J., Campbell-Burk, S. L., van Aken, T., and Laue, E. D. (1994) Solution structure and dynamics of ras p21.GDP determined by heteronuclear three- and four-dimensional NMR spectroscopy. *Biochemistry* **33,** 3515–3531.
13. Delaglio, F., Grzesiek, S., Vuister, G. W., Zhu, G., Pfeifer, J., and Bax, A. (1995) NMRPipe: a multidimensional spectral processing system based on UNIX pipes. *J. Biomol. NMR* **6,** 277–293.
14. Johnson, B. A. and Blevins, R. A. (1994) NMRView: a computer program for the visualization and analysis of NMR data. *J. Biomol. NMR* **4,** 603–614.
15. Garrett, D., Powers, R., Gronenborn, A., and Clore, G. (1991) A common sense approach to peak picking two-, three- and four-dimensional spectra using automatic computer analysis of contour diagrams. *J. Magn. Reson.* **95,** 214–220.
16. Kjær, M., Andersen, K. V., and Poulsen, F. M. (1994) Automated and semiautomated analysis of homo- and heteronuclear multidimensional nuclear magnetic resonance spectra of proteins: the program PRONTO. *Meth. Enzymol.* **239,** 288–308.
17. Güntert, P., Mumenthaler, C., and Wüthrich, K. (1997) Torsion angle dynamics for NMR structure calculation with the new program DYANA. *J. Mol. Biol.* **273,** 283–298.
18. Bartels, C., Xia, T.-H., Billeter, M., Güntert, P., and Wüthrich, K. (1995) The program XEASY for computer-supported NMR spectral analysis of biological macromolecules. *J. Biomol. NMR* **5,** 1–10.
19. Güntert, P., Braun, W., and Wüthrich, K. (1991) Efficient computation of three-dimensional protein structures in solution from nuclear magnetic resonance data using the program DIANA and the supporting programs CALIBA, HABAS and GLOMSA. *J. Mol. Biol.* **217,** 517–530.
20. Hall, S. R. and Spadaccini, N. (1994) The STAR file: Detailed specifications. *J. Chem. Inf. Comput. Sci.* **34,** 505–508.
21. Tjandra, N., Garrett, D. S., Gronenborn, A. M., Bax, A., and Clore, G. M. (1997) Defining long range order in NMR structure determination from the dependence of heteronuclear relaxation times on rotational diffusion anisotropy. *Nat. Struct. Biol.* **4,** 443–449.
22. Meiler, J., Blomberg, N., Nilges, M., and Griesinger, C. (2000) A new approach for applying residual dipolar couplings as restraints in structure calculations. *J. Biomol. NMR* **16,** 245–252.
23. Stein, E. G., Rice, L. M., and Brünger, A. T. (1997) Torsion-angle molecular dynamics as a new efficient tool for NMR structure calculation. *J. Magn. Reson.* **124,** 154–164.

24. Cornilescu, G., Delaglio, F., and Bax, A. (1999) Protein backbone angle restraints from searching a database for chemical shift and sequence homology. *J. Biomol. NMR* **13**, 289–302.

25. Wishart, D. S. and Sykes, B. D. (1994) The 13C chemical-shift index: a simple method for the identification of protein secondary structure using 13C chemical-shift data. *J. Biomol. NMR* **4**, 171–180.

26. Markley, J. L., Bax, A., Arata, Y., Hilbers, C. W., Kaptein, R., Sykes, B. D., et al. (1998) Recommendations for the presentation of NMR structures of proteins and nucleic acids. *J. Mol. Biol.* **280**, 933–952.

27. Folmer, R. H., Hilbers, C. W., Konings, R. N., and Nilges, M. (1997) Floating stereospecific assignment revisited: application to an 18 kDa protein and comparison with J-coupling data. *J. Biomol. NMR* **9**, 245–258.

28. Linge, J. P. (2001) *New Methods for Automated NOE Assignment and NMR Structure Calculation.* Books on Demand, Norderstedt, Germany.

29. Mumenthaler, C. and Braun, W. (1995) Automated assignment of simulated and experimental NOESY spectra of proteins by feedback filtering and self-correcting distance geometry. *J. Mol. Biol.* **254**, 465–480.

30. Linge, J. P., Williams, M. A., Spronk, C. A., Bonvin, A. M., and Nilges, M. (2003) Refinement of protein structures in explicit solvent. *Proteins* **20**, 496–506.

31. Koradi, R., Billeter, M., and Wüthrich, K. (1996) MOLMOL: a program for display and analysis of macromolecular structures. *J. Mol. Graph.* **14**, 51–55.

32. Vriend, G. (1990) WHAT IF: a molecular modeling and drug design program. *J. Mol. Graph.* **8**, 52–56.

33. Laskowski, R. A., MacArthur, M. W., Moss, D. S., and Thornton, J. M. (1993) PROCHECK: a program to check the stereochemical quality of protein structures. *J. Appl. Cryst.* **26**, 283–291.

34. Sippl, M. J. (1993) Recognition of errors in three-dimensional structures of proteins. *Proteins* **17**, 355–362.

35. Liu, Z., Macias, M. J., Bottomley, M. J., Stier, G., Linge, J. P., Nilges, M., et al. (1999) The three-dimensional structure of the HRDC domain and implications for the Werner and Bloom syndrome proteins. *Fold. Des.* **7**, 1557–1566.

36. Doreleijers, J. F., Rullmann, J. A., and Kaptein, R. (1998) Quality assessment of NMR structures: a statistical survey. *J. Mol. Biol.* **281**, 149–164.

37. Vriend, G. and Sander, C. (1993) Quality control of protein models: directional atomic contact analysis. *J. Appl. Cryst.* **26**, 47–60.

38. Doreleijers, J. F., Raves, M. L., Rullmann, T., and Kaptein, R. (1999) Completeness of NOEs in protein structure: a statistical analysis of NMR data. *J. Biomol. NMR* **14**, 123–132.

19

Membrane Protein Structure Determination Using Solid-State NMR

Anthony Watts, Suzana K. Straus, Stephan L. Grage, Miya Kamihira, Yuen Han Lam, and Xin Zhao

Summary

Solid-state NMR is emerging as a method for resolving structural information for large biomolecular complexes, such as membrane-embedded proteins. In principle, there is no molecular weight limit to the use of the approach, although the complexity and volume of data is still outside complete assignment and structural determinations for any large (M_r > approx 30,000) complex unless specific methods to reduce the information content to a manageable amount are employed. Such methods include specific residue-type labeling, labeling of putative segments of a protein, or examination of complexes made up of smaller, manageable units, such as oligomeric ion channels. Labeling possibilities are usually limited to recombinant or synthesized proteins, and labeling strategies often follow models from a bioinformatics approach. In all cases, and in common with most membrane studies, sample preparation is vital, and this activity alone can take considerable effort before NMR can be applied—peptide or protein production (synthesis or expression) followed by reconstitution into bilayers and resolution of suitable sample geometry is still technically challenging. As experience is gained in the field, this development time should decrease. Here, the practical aspects of the use of solid-state NMR for membrane protein structural determinations are presented, as well as how the methodology can be applied. Some successes to date are discussed, with an indication of how the area might develop.

Key Words: Proteins; membranes; anisotropic interactions; solid-state NMR; GPCR; receptors; ion channels; antimicrobial peptides.

1. Introduction

Of the approx 20,000 total structures in the protein database, some 6000 have unique folds, and only about 40 are of membrane proteins (*1*). Such a dearth of information about membrane proteins would not be so important if prediction methods had some basis on which to predict structures with minimal direct

From: *Methods in Molecular Biology, vol. 278: Protein NMR Techniques*
Edited by: A. K. Downing © Humana Press Inc., Totowa, NJ

data. However, even this direct data is still not available in the quantity desired. It is often suggested that membrane proteins present the last remaining major challenge in structural biology, and in view of the new high-throughput screening (HTS) methods being adopted in determining soluble protein structures by crystallography and by NMR (with varying degrees of success), it is clear that transferring this technology and experience to membrane proteins will bring benefit in helping to address this challenge at some point in the future. As a general rule, however, membrane protein structure determinations have been excluded so far from most major HTS/structural genomics programs as a direct result of their complexity and challenging nature.

The lack of structural data does not parallel the intense interest in this class of proteins. Some 85% of all cellular signaling occurs through the plasma membrane, and membranes feature at some point in every described operon to date. Coupled with this pivotal role in cell function is the major drive in the pharmaceutical industry to exploit membrane proteins as new drug targets *(2)*—about 70% of all known drugs target to approx 5% of all membrane proteins, leaving a large number of undiscovered targets still to be exploited. A structural biology approach to membrane proteins as drug targets is the basis of a new major initiative in the drug industry for the next decade as an alternative to combinatorial chemistry screening programs, and thus new methods for their structural resolution are urgently needed. What is still more telling about the difficulty of resolving membrane protein structures using direct methods are the armies of people using indirect bioinformatics approaches to model targets and simulate drug interactions, in the absence of much, if any, structural data. A classic example here is the use of bacteriorhodopsin (bR) as a structural paradigm for 7 transmembrane domain (7TMD) G protein-coupled receptors (GPCRs) for approx 15 yr based on low (>0.35 nm) resolution two-dimensional (2D) diffraction data for the TMDs. When the higher resolution three-dimensional (3D) crystallographic structures of bR *(3)* and bovine rhodopsin *(4)* were resolved (in 1997 and 2000, respectively), with some major surprises about helix disposition being described for both proteins, the new information produced dictated some modification of modeling constraints. Even now, rhodopsin is not a good paradigm for GPCRs because it is not a ligand-activated receptor in the more usual sense assumed for most (modeled and putative) GPCRs. Thus, there is a clear need for more structural data, either for whole proteins, or at least for functionally significant regions (ligand-binding domains, conformational changes on activation, etc.), based perhaps on gross scaffolds determined from modeling.

Membrane proteins vary in size from small, 20–30mers, which can either be surface associated or span the membrane, such as melittin, gramicidin A, and some fusion peptides, to the largest multicomplex proteins. Families usually are described in terms of the number of TMDs they (often putatively) possess in integral proteins *(1)*. Single TM peptides can then associate to form ion channels and larger complexes, or they may even combine with other proteins (such as phospholamban and the Ca^{++}-adenosine triphosphatase [ATPase]). Interestingly, *crystallographic* structure determinations have been made mainly for very large (M_r > approx 300,000), multi-subunit complexes (respiratory complexes, bacterial antennae complexes), with smaller proteins (7TMDs and less; M_r approx 40,000) being very much the exception in the database—bR, rhodopsin, the potassium channel KcsA, and the mechanosensitive channels MscL and MscS being the only examples at present. Conversely, it is this lower molecular weight range that is more accessible to solid- and solution-state NMR, in view of the complexity of the spectra for larger proteins—currently five membrane peptide structures exist in the Protein Data Bank (PDB) (gramicidin A [1MAG, 1GRM, 1AV2], the ion channel-lining segment M2 from the nicotinic acetylcholine receptor [1CEK], Fd bacteriophage Pviii coat protein [1MZT], and the H^+ channel protein M2 from influenza virus [1MP6, 1NYJ]). Approaches, such as transverse relaxation optimized spectroscopy, which have been applied to (deuterated) β-barrel proteins in solution, have potential for proteins rich in α-helices, and developments are awaited eagerly *(5)*.

Finally, it could be argued that it is the detail of sidechains, residue locations, and local structure within a membrane protein that is central to understanding function, detail which can be resolved rather well using solid-state NMR. Ligand binding and functional descriptions, such as we have them, invariably involve groups of residues from several helices or sheets, implying that how these vital residues are located with respect to each other is of major interest.

2. Materials

2.1. General Comments

Solid-state NMR opens up ways for resolving both specific distance and torsion angle constraints between sites of interest within a protein, or for overall structural determinations. Specific distance measurements depend on predetermined labeling sites, and these are introduced chemically (say between residues in a peptide, or moieties in a ligand), and can be at isolated or multiple locations. The distance range depends on the strength of the dipolar coupling, with ^{13}C pairs potentially giving a distance range of up to 0.7 nm at good precision ($\pm$ < 0.03 nm) in ideal cases—that is, rigid systems.

Because membranes are complexes that are anisotropic both functionally and structurally, orientational constraints can be extracted from macroscopically oriented membranes by exploiting the anisotropic interactions using solid-state NMR. Although the angular information obtained often cannot be used to resolve a complete structure, the orientational constraints resolved can nevertheless be very useful and are perhaps the only direct information available. In this case, these parameters can be incorporated into models or used with other data to refine still further structural models.

Numerous solid-state NMR methods have been specifically tailored to determine complete structures of membrane proteins. Such structures have so far been confined to smaller proteins (usually small TM peptides) or larger proteins in which specific labeling has been possible, although new labeling strategies based on intein technology do open up the possibility of structural determinations of defined domains of larger proteins.

Both oriented and random dispersions have been used in static- and magic-angle spinning (MAS) techniques, respectively *(6–9)*. The choice of solid-state NMR methods (**Subheading 3.**) that can be used on a sample may be constrained by the nature or state of the sample (**Subheading 2.6.**). One approach relies on the use of either mechanically or magnetically oriented samples. Macroscopically oriented membranes give NMR spectra from which the spectral anisotropy inherent in the chemical shifts, quadrupolar interactions, and dipolar couplings is exploited to yield molecular orientations. Another methodology is to use MAS on an unoriented, random dispersion. In MAS, the sample container (rotor) is set spinning (0.5–12 kHz) about an axis that subtends an angle of 54.74°, the magic angle, with respect to the static field B_0. Now, the broad static line shape, resulting from anisotropy of the spin interactions, breaks up into a center band at the isotropic position and a set of spinning sidebands separated by the spinning frequency (approx < 4 kHz) (**Subheading 3.2.1.**). As the spinning frequency increases (approx > 7–10 kHz), the sideband intensities decrease, and an increase in the center band intensity is observed, giving much enhanced spectral intensities and resolution over static samples. Finally, the combined approach of magic-angle-oriented sample spinning (MAOSS) first proposed by C. Glaubitz and A. Watts *(10)*, where the sample is oriented and spun slowly (0.5–4 kHz) at the magic angle, can be used to obtain orientational information and also benefits from the higher spectral resolution of MAS.

The various sample geometries used and then the applicable solid-state NMR methods generally available are described in greater detail in **Subheadings 2.4.–2.6.** and **Subheading 3.**, respectively, with specific examples given in **Subheading 3.3.** New developments on every front, from protein production and labeling, sample geometry, pulse methods, spectral representation and modeling, are still ongoing, and the number of proteins available is increasing, with little shortage of membrane proteins for study.

2.2. Solvents and Buffers

Small hydrophobic peptides generally dissolve well in organic solvents (chloroform, methanol, trifluoroacetic acid) and are usually kept dry or in stock solution at concentrations of 1–10 mg/mL. This ensures that the sample volume is kept low for reconstitution with lipids. Solvents need to be dry, and if solubility in apolar, organic solvents is a problem, methanol can be added.

The range of detergents used for preventing aggregation of membrane proteins and peptides, and permitting purification and reconstitution, is as wide ranging as the number of available proteins. High critical micelle concentration (cmc) detergents often are used for reconstitutions because of their ease of removal in the presence of lipid, whereas ionic, low-cmc detergents are used for protein removal from its natural environment. In most cases, NMR is carried out after preparation of samples through established isolation, purification, and reconstitution methods.

Buffers for membrane studies usually are isotonic with physiological conditions, namely, approx 150 m*M* NaCl, and approx pH 6.0–8.0 (*see* **Subheading 4.1.**). If salt tolerance is a problem, then water can be used, although any ionic strength effects on protein folding and structure may need to be taken into account.

2.3. Membrane-Associating Peptides

Membrane-associating peptides can be produced by solid-phase synthesis or expression, and the method chosen is dictated by the type of structural information required.

Solid-phase peptide synthesis (SPPS) using Fmoc chemistry is used when specific sites within a peptide are to carry an NMR label. Most residues are available commercially (CDN Isotopes, Quebec, Canada; Cambridge Isotopes Labs, Andover, MA; Isotec, Miamisburg, OH) with labels at most positions within the peptide backbone (^{13}Cα, ^{13}C = O, ^{15}N), and some simpler residues are available uniformly labeled (including ^{2}H). A major difficulty is in protecting functional side groups for those amino acids that require them. Some NMR-labeled residues are available Fmoc protected, although they are expensive. However, most are not, and so some form of side-group protection is required, with concomitant losses of labeled material. It is always worth trying a synthesis with unlabeled residues and unprotected side groups first, just to check if yields of required peptide (after purification) can be obtained without the need to protect (e.g., with tertial butyl) the specific residues that will carry a label. Further cycling of couplings may be necessary to increase yields, especially as the peptide gets longer—most single-pass membranc peptides are 25–35 residues long—or for hydrophobic residues (e.g., Val, Ile, Trp, Ala, etc.). In designing labeling schemes, it is often more economical to make as long a portion of the desired peptide unlabeled as possible, and then add labeled residues toward the end. In this way, the start of the sequence can be purified first before continuing with the labeled part, and thus the labeled material will be used more efficiently. Solvents useful for SPPS are di-isopropylethylamine, dimethyl formamide (DMF), and *N*-methylpyrrolidine (NMP), and coupling reagents include *N*-hydroxybenzotriazole, *O*-benzotriazole–*N,N,N′,N′*,-tetramethyl-uronium-hexafluorophosphate, benzotriazole-1-yl-oxy-tris-pyrrolidino-phosphonium-hexafluorophosphate, and *O*–(7-azabenzotriazol-1-yl)-*N,N,N′,N′*-tetramethyl-uronium-hexafluorophosphate. Deprotecting solvents used include 20% piperidine in DMF/NMP, and the choice of cleaving scavenger will depend on the peptide sequence.

Uniform labeling of a peptide is best accomplished by expression in a bacterial vector (*Escherichia coli*) with fusion proteins (often glutathione *S*-transferease). For ^{15}N labeling, ^{15}N ammonium chloride is used, and for ^{13}C, ^{13}C sodium acetate metabolites *(11)* or ^{13}C-glucose is incorporated in a minimal growth medium, respectively. Although not exploited so far for membrane peptides, back regulation in bacterial metabolism provides the opportunity for specific labeling for all residues of one type, if sufficient amounts (usually 0.1–1 g per 1–10 L culture) of a labeled amino acid is available. Only certain residues can be incorporated in this way, because some (cysteine, glutamate, e.g.) scramble during metabolism leading to loss of label and nonincorporation. Although peptide release from the fusion protein can be achieved with peptidases, the preferable way is with cyanogen bromide at an inserted methione site (with removal of all other methionines in the required peptide) under standard conditions *(12)*.

2.4. Model and Reconstituted Membranes

Model membranes containing proteins or peptides can be made using conventional reconstitution approaches (e.g., dialysis, centrifugation–dilution, cosolubilization, followed by solvent removal), which vary depending on the protein and lipid. Often, a fairly rigid environment is required in structural studies (motion defeats spectral anisotropy), and saturated lipids are preferred (dimyristoyl [$C_{14:0}$], dipalmitoyl [$C_{16:0}$], and distearoyl [$C_{18:0}$], and phosphatidylcholines are popular). However, it is sometimes necessary to have a mobile lipid environment and to work in the liquid crystalline phase to achieve well-resolved spectra, in particular in oriented samples *(13)*. For proteins with functional characteristics, it may be found that these are lost in such bilayers, and a judgment needs to be made about whether this is tolerable for NMR studies. Additional restriction of molecular motion can be achieved at low hydration (<100 wt % water), which can be obtained by air-drying or incubation in a controlled atmosphere. Because limiting hydration for most lipids is approx 30 wt%, levels below this are difficult to achieve and maintain (lipids are very hygroscopic). For orientational studies, bilayers are sealed in Parafilm or plastic to maintain hydration levels, and then the NMR coil is wound around the sample.

2.5. Natural Membranes

For sensitivity reasons, few natural membranes contain sufficient protein of one type to be of use in direct structural studies at the moment. However, some cell membranes, such as mitochondrial inner membranes, ATPase, or receptor-enriched membranes and the purple membrane of *Haloarchaea,* are composed of around 25–75% of one particular membrane protein as a protein–lipid complex. Isolated and purified membranes can be used directly for structural and functional investigations, although global structural determinations are not yet possible for larger proteins. In the rather unusual case of purple membranes in which bR is the major (>95%) protein component, the purple membrane is obtained by harvesting *Halobacterium* culture, lysing the cells, and purifying using a sucrose gradient *(14)*. The isolated purple membrane suspension is concentrated to make a pellet by centrifugation, which is introduced into an NMR rotor for MAS measurements. In this ideal case, for structural studies, the protein has been labeled directly with, for example, all residues of one type *(15)*. Another exceptional natural membrane is the disk membrane from retinal rods that contain approx 95% rhodopsin, but here, incorporation of labels can only be achieved in a recombinant protein (using bacculovirus or human embryonic kidney [HEK] cells) followed by purification and then reconstitution *(16,17)*.

However, in many cases, the proteins under study may not be sufficiently abundant. Consequently, they may need to be expressed or synthesized, which is in itself a major hurdle for the amounts ($\geq$1–10 mg for approx 30 kDa) required in structural biology. Once sufficient quantities are obtained, proteins may need to be purified by chromatographic methods, such as high-performance liquid chromatography, ion-exchange chromatography, and so forth. Subsequently, the pure protein is reconstituted into lipid bilayers using lipids isolated from the cellular membrane, or using synthetic lipids, mainly phosphatidylcholines. The choice of detergent and reconstituted lipid is vital to retain the activity and thus the functionally relevant structure of the protein. In addition, it is occasionally necessary to add saline buffers to the protein preparation to reproduce physiological conditions. This requires the use of salt-tolerant probes for the NMR measurements (*see* **Subheading 4.1.**), which are equipped for variable temperature experiments (110–310 K) and are designed with higher filling factors for more efficient use of the rf powers available. Finally, for the NMR measurements, the volume of the reconstituted protein preparation is usually reduced as much as possible by centrifugation or solvent evaporation to achieve the required sensitivity.

2.6. Sample Geometry and Preparation

2.6.1. Random Dispersions

Membrane protein samples studied by solid-state NMR can be prepared in various forms to yield random dispersions: lyophilized powders, microcrystalline samples, centrifuged pellets, frozen solutions, and even hydrated suspensions. In this case, structural information typically is obtained by using MAS experiments. For membrane proteins, to ensure retention of activity and the correct intercalation into a membrane, either the natural membrane or a reconstituted membrane are the ideal systems for structural studies. Pelleting (10–100,000*g*) and transfer to an NMR rotor, or pelleting directly into the rotor, is readily achieved.

2.6.2. Oriented Samples

Because membrane proteins can be embedded in lipid bilayers or bicelles (magnetically oriented lipid disks; **Subheading 2.6.2.2.**), it is possible to prepare samples that have a high degree of local orientational order. The relative order present in the membrane imposes a certain degree of order on the peptide or protein, allowing the measurement of NMR parameters as a function of the orientation of the biomolecule. To relate the measured parameters to the arrangement of the biomolecule in the sample, the bilayer orientation should be fixed with respect to the external magnetic field. Numerous ways exist to achieve this, namely mechanical or magnetic orientation.

2.6.2.1. MECHANICALLY ORIENTED SAMPLES

Preparative methods for mechanically oriented membranes have been pioneered by Seelig and others in the 1970s *(18–22)*. Several techniques to obtain highly aligned membranes with low mosaic spread have been devised, such as deposition of single bilayers on a solid support by Langmuir–Blodgett techniques *(23,24)*. However, to achieve the necessary amount of material and sample robustness needed for NMR experiments, orientation with the aid of glass slides or alternative polymer materials such as mica has become the standard technique in static solid-state NMR applications *(25)*. Two approaches can be used for the formation of oriented samples using glass slides. One involves the hydration-induced self-assembly of lipids deposited initially on a planar surface, yielding bilayers that adopt the orientation of the surface. In the other approach, preformed vesicular structures of bilayers can be forced mechanically to anneal and orient on a planar surface.

The procedure of sample preparation using glass slides is outlined in **Fig. 1A**. Square or rectangular slides (1–2cm^2; 0.06- to 0.08-mm thick; Marienfeld GmbH, Germany) are used for static NMR experiments, round ones for

A

1- Spread sample disolved in organic solvent or as vesicle dispersion

2- Dry to leave a thin film of sample

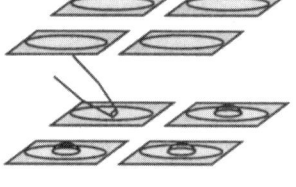

3- Hydrate directly, or via atmosphere after stacking

4- Stack glass slides and leave in atmosphere with controlled humidity (over saturated salt solution)

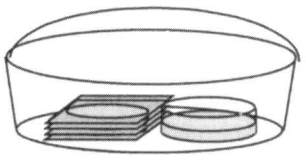

5 -Seal sample

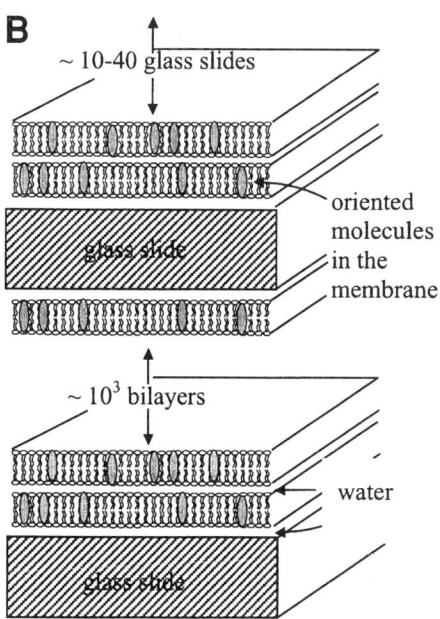

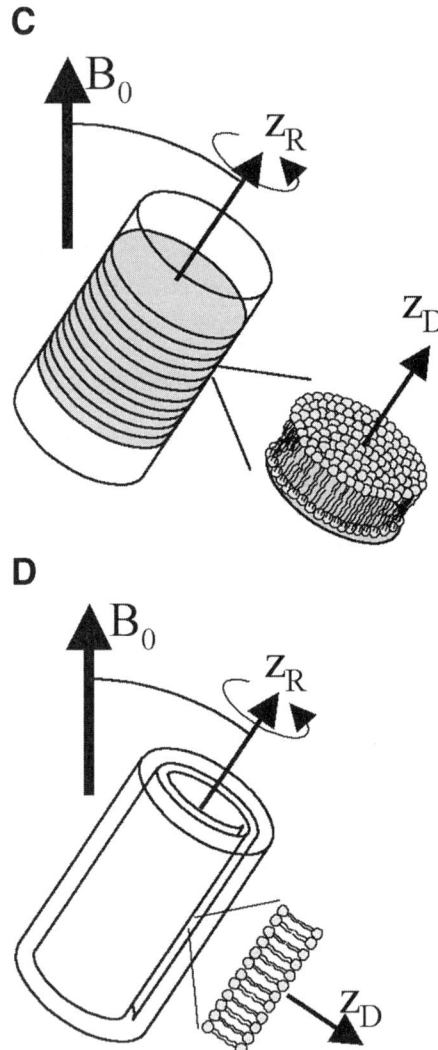

Fig. 1. Steps for the preparation of mechanically oriented samples using glass slides (**A**). Stacked glass plates can be wrapped in Parafilm or plastic to preserve hydration levels, and the r.f. probe coil can be wound directly around them. Alternatively, circular plates can be mounted inside a rotor for MAOSS (**C**) (courtesy of C. Glaubitz). Resulting sample form showing how many membrane bilayers are stacked between plates (**B**). Bilayers can also be spread on polymer sheets and rolled up with their central axis colinear with the spinning axis at the magic angle (**D**).

MAOSS applications, and more recently, polymer films rather than glass slides have also been explored as membrane supports *(10,26–28)*. The protein–lipid mixtures can be prepared by reconstituting the purified protein or peptide either in an organic solvent (e.g., trifluoroethanol, chloroform, methanol) or from detergent solutions in water followed by solvent or detergent removal. For natural membrane fragments, a concentrated solution can be used directly. The protein–lipid mixtures are deposited on the glass slides to form a thin film of sample material, and the solvent is subsequently evaporated. Typically 10–40 slides are then stacked on top of each other to achieve the required amounts needed for NMR experiments. On hydration and annealing under an atmosphere of controlled humidity, typically several thousands of oriented bilayers form between each pair of slides (**Fig. 1B**). Modifications to this approach have also been used successfully to prepare protein samples that cannot be dissolved in solvent, or proteins that are not reconstituted but left in their native membrane. These include spreading under pressure exerted either by squeezing the glass plates, or by isopotential centrifugation *(29)* (**Fig. 2**). Finally, prior to performing the solid-state NMR experiments, the sample is sealed to maintain the hydration levels of the lipid bilayers during experimentation.

To be able to apply oriented solid-state NMR methods, it is crucial that the samples are well aligned. The formation of well-oriented bilayers is governed by the hydration level and can be promoted by the addition of certain components to the lipid–protein mixture. For example, in the hydration and annealing step, water can be added either directly or via a humid atmosphere, or a combination of both. The amount of water has to be controlled to achieve a high degree of alignment, where a lower (<100 wt %) rather than a high hydration level tends to yield better orientation with less mosaic spread. The limiting hydration for lipids, which are hygroscopic, is approx 30 wt %. Recently, however, a relatively low water content of approx 2–10 molecules per lipid has been used, yielding "hydration-optimized samples" with sufficiently good alignment for solid-state NMR studies *(30)*. Another recent innovation is the addition of sublimable solids, such as naphthalene, to the sample in the early stages of the preparation. The alignment resulting from these oriented samples could be improved significantly, and being sublimable, this auxiliary compound was nonperturbing for the NMR experiments *(31)*.

The quality of alignment in the samples can be determined by the use of [31]P- (or ^{2}H if the lipids are labeled) NMR of the phosphorus in the lipid head group (or ^{2}H in the chains or head group) *(29,32)*. Here, the orientational dependence of the resonance resulting from the ^{31}P CSA (or ^{2}H quadrupolar anisotropy) can be used to assess mosaic spread and contributions from non- (or poorly) oriented membranes. A sample ^{31}P NMR spectrum is shown in **Fig. 3B**, where the signal at 25 ppm originates from oriented lipids.

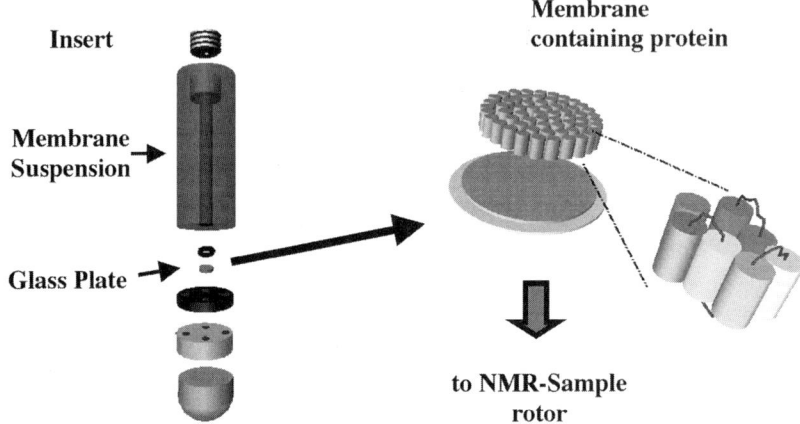

Insert

Membrane Suspension

Glass Plate

Membrane containing protein

to NMR-Sample rotor

Fig. 2. Device for producing oriented natural and model membranes by isopotential spin drying ultracentrifugation (ISDU) (courtesy of G. Gröbner). The cell has been built for use in a SW 28 rotor (Beckman). The cell reservoir body, support plug, and fill insert for the bottom of the tube are made of polycarbonate, which, in tests, showed higher resistance to deformation at high rotor speeds. Stainless steel is used for the three-dimensional isopotential surface plate. The evaporation rate is controlled by screw caps with different pinhole sizes (50–1000 mm). For the preparation of membrane samples without dehydration, sealed caps are used. Oriented samples are prepared on grease-free Melinex (Agar Scientific, Stansted, UK) or 60–80 μm-thin glass plates (Marienfeld, Germany) previously cleaned in hot nitric acid and rinsed extensively in ddH$_2$O. The isopotential cell is filled with membrane suspension and centrifuged (40,000–90,000g at 47°C for up to 18 h) with and without simultaneous evaporation of water. The pelleted membrane samples are kept at controlled humidity and temperature. For NMR experiments, typically two plates are stacked together with spacers, each containing up to 4 mg of total membrane. The samples are then placed in 6- to 10-mm sealed NMR tubes with 30 μL of appropriate salt solution for controlled humidity.

2.6.2.2. MAGNETICALLY ORIENTED SAMPLES

Exploiting the anisotropic diamagnetic susceptibility of lipids opens another possibility for bilayer orientation in the magnetic field. This susceptibility gives rise to a weak alignment bilayer with the lipid long axis perpendicular to the magnetic field *(33)*. This phenomenon has been exploited in disk-shaped bicelles made from a combination of detergent micelles and bilayers, and it was originally intended to provide a more membrane-compatible environment for micelle-like studies of membrane proteins *(34–37)*. The properties of these systems have been found to be useful in solution-state NMR to impose weak alignment on proteins in solution, but they have also been used in solid-state

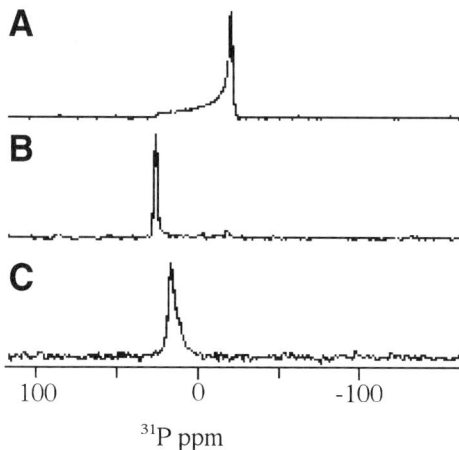

Fig. 3. ^{31}P-NMR spectra of a random dispersion sample of 1,2-dioleoyl-sn-glycero-3-phosphocholine (DOPC) bilayers **(A)**, an oriented sample of DOPC bilayers using glass slides at full level of hydration (approx 10 H$_2$O/lipid) with the field parallel with the membrane normal for checking orientation and mosaic spread **(B)** and at hydration-optimized conditions (approx 2 H$_2$O/lipid). (Adapted with permission from **ref. *30*.**)

NMR studies of peptides and small membrane proteins, and in the interaction of membrane-associated proteins with lipid bilayers *(36,38–40)*. Typically, bicelles are prepared from mixtures of a long-chain lipid, such as dilauryl phosphatidylcholine, diC$_{14:0}$, dimyrisotyl phosphatidylcholine, diC$_{16:0}$ (DMPC), or dipalmitoyl phosphatidylcholine, diC$_{18:0}$, with a short-chain lipid, such as dihexanoyl phosphatidylcholine, diC$_{6:0}$, at a molar ratio of 2.5:1 to 5:1. The long-chain lipid forms a disk-like bilayer, surrounded at the edges by the short-chain lipid. In the magnetic field, they align with their membrane normal perpendicular to the magnetic field so that embedded membrane peptides or proteins give rise to 90°-oriented NMR spectra. Of more use is a bicelle orientation, with the membrane normal aligned along the magnetic field, because the NMR parameters for a TM peptide or protein are more sensitive to small angular deviations than at the other orientation. This orientation can be achieved by the use of lanthanide ions, which, when chelated by modified lipid head groups, or bound by negatively charged lipids such as phosphatidylglycerols, confer their anisotropic susceptibility to the lipid and cause the bicelles to flip orientation *(33)*. Alternatively, lipids with aromatic moieties in their chains have been used to achieve alignment of the membrane normal along the magnetic field *(41,42)*. The morphology of the mixed membrane systems, which can be reoriented with the aid of lanthanide ions, were characterized recently by neutron scattering and other techniques, revealing a larger variety of structures than the original bicelle model *(43–45)*. Depending on the temperature and lipid concentration, these systems not only adopt nematic bicellar shapes but can form other phases such as perforated lamellar phases as well. The short-chain lipids are then accommodated in the form of pore-like defects in a lamellar bilayer rather than at the rim of a bicellar disk. Possessing smectic order, this lamellar phase aligns in the magnetic field in the same orientation as the bicelle/lanthanide systems.

A recent application of bicelles is their combination with sample spinning techniques. Thus, it is possible to benefit from both the orientational information provided by the alignment and the increased resolution from the averaging process of sample spinning, as well as influence the alignment of the bicelles by the choice of the rotation axis orientation *(46–48)*.

3. Solid-State NMR Methods

As already mentioned, numerous solid-state NMR methods can be used to obtain local and global structural information. The choice of solid-state NMR experiment depends on the nature and state of the sample. The methods that can be applied to randomly oriented samples mostly use MAS and are outlined in **Subheadings 3.2.1.–3.2.4.** Finally, static NMR methods that can be applied to oriented samples will be outlined in **Subheading 3.2.5.** In addition to the techniques presented here, there are additional solid-state NMR experiments that are required to assign the spectra, but their application to large membrane proteins for total structures is limited, except as discussed later. For additional information, the reader is referred to recent reviews *(6,7,49,50)*.

3.1. Theoretical Description

3.1.1. Spin Hamiltonians and Reference Frames in Solid-State NMR

Generally, the spin interactions for a multiple spin system in solids involve the interactions of the spins with the external fields, such as the static field B_0, and the radio-frequency (rf) field B_1, denoted as H_{ext}, and the interactions with the internal fields, such as the chemical-shielding field and dipolar fields, denoted as H_{int}. The total spin Hamiltonian may be expressed as:

$$H_{tot} = H_{ext} + H_{int} \tag{1}$$

where

$$H_{ext} = H_z + H_{rf} \tag{2}$$

and

$$H_{int} = H^{CS} + H^{DD} + H^J + H^Q \tag{3}$$

H_z is the Zeeman interaction with the static field; H_{rf} is the interaction with the rf field. H^{CS} are the chemical shifts (including isotropic chemical shifts, and chemical shift anisotropies); H^{DD}, H^J, and H^Q are the dipolar, J-couplings, and quadrupolar interactions, respectively. J-couplings are typically too small (approx 100 Hz) to be detected, except under special conditions and narrow lines, but it is all the anisotropic spin interactions, such as dipolar couplings (0.1–50 kHz), CSA (approx 10 kHz), and quadrupolar couplings (approx 200 kHz for ^{2}H) that are exploited in solid-state NMR in biology, and these require a knowledge of the system geometry.

Dealing with spin interactions in solid-state NMR involves a set of reference frames and the transformations between them that depend on the spin systems being studied (**Fig. 4**). The spin Hamiltonian of any spin interaction Λ may be expressed as a scalar product of two irreducible spherical tensors because it is very convenient to use the transformation properties of irreducible spherical tensors to describe 3D rotations. A general rotation $R(\Omega_{AB})$ of an irreducible spherical tensor of rank k from the frame A to frame B may be

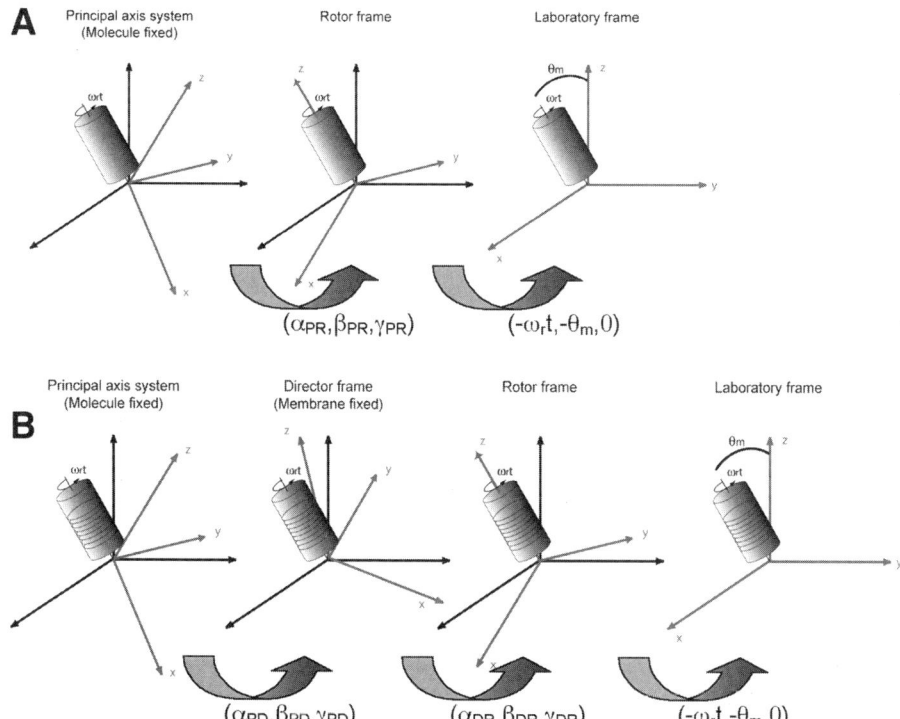

Fig. 4. Rotations necessary to relate the principal spin axis frame to the laboratory frame for MAS (**A**) and for MAOSS (**B**).

expressed by a set of three Euler angles $\Omega_{AB} = \{\alpha_{AB}, \beta_{AB}, \gamma_{AB}\}$ leading to the transformation

$$T^B_{kq} = R(\Omega_{AB})T^A_{kq}R(\Omega_{AB})^\dagger = \sum_{p=-k}^{+k} T^A_{kp}D^k_{pq}(\Omega_{AB}) \tag{4}$$

where $D^k_{pq}(\Omega_{AB})$ is an element of the $(2k + 1)$ dimensional Wigner matrix.

The *principal axis frame* P^Λ is unique for the spin interaction Λ. The second-rank spherical tensors are diagonalized and have defined forms for each spin interaction Λ in this frame.

The *molecular frame M* is a common frame for all spin interactions and is fixed with respect to the molecular structure. The principal axis frame P^Λ is related to the molecular frame M by a set of three Euler angles, denoted $\Omega_{PM} = \{\hat{\alpha}_{PM}, \hat{\beta}_{PM}, \hat{\gamma}_{PM}\}$. This set of Euler angles describes the relative orientation of the spin interaction Λ with respect to the molecular structure. Depending on the complexity of the system, $P^\Lambda = M$ is often assumed for many practical situations to study a single spin interaction.

The *rotor frame R* is defined such that its z-axis is along the rotor-spinning axis. The relative orientation of the molecular frame and the rotor frame is related by the set of Euler angles, $\Omega_{MR} = \{\alpha_{MR}, \beta_{MR}, \gamma_{MR}\}$. This set of Euler angles is randomly distributed in powdered samples. In some cases, one or more reference frames are included in between M and R. For example, in the MAOSS experiment *(10)*, the *director frame D*, is introduced in between M and R with the two sets of linked Euler angles $\Omega_{MD} = \{\alpha_{MD}, \beta_{MD}, \gamma_{MD}\}$ and $\Omega_{DR} = \{\alpha_{DR}, \beta_{DR}, \gamma_{DR}\}$.

The *laboratory frame L* is defined so that its z-axis is along the external magnetic field, and the rotor frame is related to the laboratory frame by the Euler angles $\Omega_{RL} = \{\alpha_{RL}, \beta_{RL}, \gamma_{RL}\}$. Suppose that the sample is spun at an angular frequency ω_r along the magic-angle axis, then, at a certain time-point t, the relevant Euler angles are given by $\Omega_{RL} = \{\alpha^0_{RL} - \omega_r t, -\theta_m, 0\}$ where α^0_{RL} defines the rotor position at the time-point $t = 0$, and is often considered to be 0, and $\theta_m = \tan^{-1}\sqrt{2}$ is the magic angle.

Therefore, the complete transformation of spin interaction Λ from its PAS (P^Λ) to the laboratory frame (L), may be expressed as:

$$D^l(\hat{\Omega}_{PL}) = D^l(\hat{\Omega}_{PM})D^l(\Omega_{MR})D^l(\Omega_{RL}) \tag{5}$$

3.1.2. Spin Interactions

3.1.2.1. ZEEMAN INTERACTION

The Zeeman interaction H_z describes the interaction of the S-spin with the external static field B_0,

$$H^S_z = -\gamma_S B_0 S_z = \omega_0 S_z \tag{6}$$

where $\omega_0 = -\gamma_S B_0$ is the Larmor frequency of the S-spin species. In most cases, the Zeeman interaction is the strongest spin interaction and determines the spectral line position.

3.1.2.2. CHEMICAL SHIFT INTERACTION

The chemical shift spin interaction H_j^{CS} represents an interaction of spin S_j with the magnetic field induced by the external field in the local electronic environment. It can be described by the chemical shift tensor $\boldsymbol{\delta}_j$:

$$H_j^{CS} = -\gamma_j \mathbf{S}^T \cdot \boldsymbol{\delta}_j \cdot B_0 \tag{7}$$

The Cartesian tensor $\boldsymbol{\delta}_j$ may be decomposed into an irreducible spherical tensor of rank 0, 1, and 2 parts. The rank 0 part (δ_0) is the *isotropic chemical shift*. The antisymmetric chemical shift is represented by the rank 1 part (δ_1) and ignored in the high-field approximation. The CSA, which is so useful in oriented membrane studies, is represented by the rank 2 part (δ_2). The sum of δ_0 and δ_2 is diagonal in its own principal axis frame. The diagonal elements δ_{xx}^P, δ_{yy}^P, and δ_{zz}^P are called the principal values of the tensor, and

$$\delta_{iso}^j = \frac{1}{3}\left(\delta_{xx}^P + \delta_{yy}^P + \delta_{zz}^P\right) \tag{8}$$

$$\left|\delta_{zz}^P - \delta_{iso}^j\right| \geq \left|\delta_{xx}^P - \delta_{iso}^j\right| \geq \left|\delta_{yy}^P - \delta_{iso}^j\right| \tag{9}$$

The CSA can be expressed in terms of an anisotropic chemical shift frequency, ω_{aniso}^j:

$$\omega_{aniso}^j = \omega_0\left(\delta_{zz}^P - \delta_{iso}^P\right) \tag{10}$$

and an asymmetry parameter, η:

$$\eta = \frac{\delta_{yy}^P - \delta_{xx}^P}{\delta_{zz}^P - \delta_{iso}^j} \tag{11}$$

3.1.2.3. DIRECT DIPOLE–DIPOLE COUPLING

The direct dipole–dipole coupling between two spins S_j and S_k can be expressed by a traceless symmetric Cartesian tensor $\mathbb{D}_{jk}$:

$$H_{jk}^{DL} = S_j \cdot \mathbb{D}_{jk} \cdot S_k \tag{12}$$

In the high-field approximation, the homonuclear dipolar Hamiltonian, for a rotating sample, may be described as

$$H_{jk}^{DD} = \omega_{jk}^{DD}\left[2S_{jz}S_{kz} - \frac{1}{2}\left(S_j^- S_k^+ + S_j^+ S_k^-\right)\right] \tag{13}$$

$\omega_{jk}^{DD}(t)$ is the dipolar frequency, and can be expressed in a Fourier series:

$$\omega_{jk}^{DD}(t) = \sum_{m=-2}^{+2} \omega_{jk}^{(m)} \exp\{im\omega_r t\} \tag{14}$$

where $\omega_{jk}^{(m)}$ is the *m*th component of $\omega_{jk}^{DD}(t)$ in the Fourier expansion, and ω_r is the sample spinning frequency. Because the dipolar tensor is axially symmetrical for a good approximation, the $\omega_{jk}^{(m)}$ may be expressed as:

$$\omega_{jk}^{(m)} = \sum_{m'=-2}^{+2} b_{jk} D_{0m'}^2(\Omega_{PM}^{\wedge}) D_{m'm}^2(\Omega_{MR}) d_{m0}^2(\beta_{RL}) \tag{15}$$

where $d_{m0}^2(\beta_{RL})$ is the reduced Wigner matrix element, and b_{jk} is the dipolar coupling constant, and defined as:

$$b_{jk} = -\frac{\mu_0}{4\pi} \frac{\gamma_j \gamma_k \hbar}{r_{jk}^3} \tag{16}$$

r_{jk} is the interhomonuclear distance and γ_j is the gyromagnetic ratio of the spin S_j. If we take the transformation directly from the PAS to the laboratory frame for the dipolar tensor, then the dipolar frequency, under all approximations as mentioned previously, can be simply expressed as:

$$\omega_{jk}^{DD}(t) = b_{jk} \frac{1}{2}(3\cos^2 \beta_{PL} - 1) \tag{17}$$

In the case of heteronuclear spin pair *IS*, the dipolar Hamiltonian can be expressed as:

$$H_{IS}^{DD}(t) = \omega_{IS}^{DD}(t) \cdot 2I_z S_z \tag{18}$$

$\omega_{IS}^{DD}(t)$ can be defined analogously to $\omega_{jk}^{DD}(t)$ with the dipolar coupling constant now given by

$$b_{IS} = -\frac{\mu_0}{4\pi} \frac{\gamma_I \gamma_S \hbar}{r_{IS}^3} \tag{19}$$

where r_{IS} is the interheteronuclear distance.

3.1.2.4. QUADRUPOLAR INTERACTIONS

Deuterium ($\delta = 1$) is an informative nucleus in both orientational and dynamic studies of membranes. The quadrupolar interaction for a spin S_j may be expressed in the form:

$$H_j^Q = \mathbf{S}_j^T \cdot Q \cdot \mathbf{S}_j \tag{20}$$

where Q is the quadrupolar tensor. The truncated Hamiltonian under the high-field approximation can be expressed as:

$$H_j^Q = \omega_j^Q(3S_{jz}^2 - \mathbf{S}_j \bullet \mathbf{S}_j) \tag{21}$$

where ω_j^Q is the quadrupolar frequency and has the form:

$$\omega_j^Q(\Theta) = \frac{3eQ_j}{4S_j(2S_j - 1)} V_{zz}^j(\Theta) \tag{22}$$

where $V_{zz}^j(\Theta)$ is the component of the electric field gradient at the nucleus S_j. In solids, the quadrupolar frequency depends on the sample orientation (Θ), and changes if the sample is rotated with respect to the magnetic field. The full spectral anisotropy of approx 127 kHz for a $-C^2H$ is reduced through rotation for the more usual $-C^2H_3$ reporter on, for example, Ala or Val residues. In addition, anisotropy averaging can give valuable dynamic information in the μs–ms time range.

3.2. Experimental Techniques

3.2.1. Magic-Angle Spinning (MAS)

MAS NMR produces high resolution spectra in solids *(51,52)*. The sample container (rotor) is set spinning about an axis that subtends an angle of 54.74°, the magic angle, with respect to the static field B_0. The spatial rotation of the sample introduces time dependence to the anisotropic spin interactions (ω_{int}), such as CSA, quadrupolar interactions, homonuclear dipole–dipole couplings, and heteronuclear dipole–dipole couplings, all of which are averaged out more efficiently as the sample spinning frequency increases $\omega_r \gg \omega^{DD,CSA}$. Because of the periodic time dependence of the spin interactions, the broad static line shape breaks up into a center band at the isotropic position (ω_{iso}) and a set of spinning sidebands separated by the spinning frequency (ω_r). As the spinning frequency increases, the time averaging is more effective, leading to a decrease in the sideband intensities and an increase in the center band intensity. The advantage of MAS is that both resolution and sensitivity are greatly increased. The disadvantage is that the spectrum loses all the molecular geometry information. However, full spectral assignments have been achieved using correlation spectroscopy in MAS experiments of extensively or uniformly ^{13}C, ^{15}N-labeled peptide/proteins *(50,53–55)*.

3.2.2. High-Power Heteronuclear Decoupling and Cross-Polarization

Proton coupling to rare spins complicates the observation of the rare spins, such as ^{13}C and ^{15}N. Usually, it is necessary to remove the heteronuclear dipolar couplings to protons and the observed rare spins by strong rf irradiation at the proton resonance frequency, the so-called high-power proton decoupling. Continuous-wave decoupling is a routine scheme for decoupling of the heteronuclear dipolar spin interactions. The rf field induces a fast rotation of the proton spin states averaging out their interaction with the rare spins *(56)*. For fast MAS, more sophisticated pulse sequences, such as TPPM, C12, R24, and XiX achieve a better decoupling efficiency *(57–60)*.

The sensitivity of rare spin species (*S*-spins) with low γ, such as ^{13}C and ^{15}N, can be enhanced by transferring magnetization from abundant spin species (*I*-spins) with high γ, such as ^{1}H, to the S-spins. Cross-polarization (CP) is the most widely used method to transfer polarization between unlike spin species through the heteronuclear dipolar couplings (*61*). This is achieved by (a) preparing the *I*-spin transverse magnetization by irradiating the *I*-spins with a 90° rf pulse; (b) simultaneous irradiation of both *I*-spins and *S*-spins with two rf fields, B_{1I} and B_{1S}. Magnetization transfer is achieved when the strengths of the two fields match the Hartmann–Hahn (HH) condition $|\gamma_I B_{1I}| = |\gamma_S B_{1S}|$, where γ_I and γ_S are the magnetogyric ratios of the *I*-spins and *S*-spins, respectively (*62*). The enhancement of the *S*-spin magnetization is proportional to the ratio of the two magnetogyric ratios $|\gamma_I/\gamma_S|$. Methods have also been developedwhere one of the rf fields is ramped in a linear fashion (*63–65*) or adiabatically (*65*). These methods can improve the stability and reproducibility of HH–CP, especially under fast MAS conditions.

3.2.3. Recoupling in MAS

Fast MAS leads to high resolution and sensitivity in solid-state NMR, which is the basic requirement for the sequential assignment of protein spectra. However, all anisotropic spin interactions that can be used to extract molecular geometry information are, in principle, averaged out by fast MAS. Therefore, to recover the anisotropic spin interactions in the presence of MAS, a range of recoupling techniques have been developed. Generally, there are two approaches to reintroduce anisotropic dipolar interactions, either mechanically, in which the recoupling is achieved through sample rotation, or by rf pulse-driven methods in which the recoupling is achieved by applying rf pulse trains. Thus, MAS NMR recoupling techniques have been extensively used to give a selective restoration of informative anisotropic nuclear spin interactions that, in turn, can yield structural information about distance and orientational constraints (*66,67*).

3.2.4. MAS NMR Experiments

3.2.4.1. DISTANCE MEASUREMENTS USING HOMONUCLEAR
RECOUPLING TECHNIQUES

Rotational resonance (R^2) occurs when the sample spinning frequency ω_r matches the isotropic chemical shift difference ($\Delta\omega$) of a pair of coupled homonuclear spins S_j and S_k under the condition

$$\Delta\omega = \left|\omega_j^{iso} - \omega_k^{iso}\right| = n\omega_r \tag{23}$$

where *n* is a small integer (*63,68,69*).

Under this condition, the homonuclear dipole–dipole interaction is recoupled, leading to spectral broadening and splitting as well as an exchange of longitudinal magnetization between the two spins in a 2D experiment *(68,69,70)*. The system behaves homogeneously because of truncation of the large isotropic chemical shift difference and results in both abundant and rare spin magnetization transfer rates, being considerably enhanced at the R^2 condition *(69–72)*. Magnetization transfer rates are much higher for doubly labeled samples containing closely coupled spin pairs compared to those with distant spins.

In the R^2 exchange experiment, after initial signal enhancement by cross-polarization, the signal is restored to the z direction by a 90° pulse (**Fig. 5A**). One of the two labeled-site resonances is then selectively inverted (**Fig. 5B**), and the exchange of longitudinal magnetization between nuclei as a function of mixing time is monitored with the aid of an observation pulse. From the change in intensity of the two labeled sites, the dipolar coupling constant (b_{jk}) can be extracted by numerical simulation for the best fit of the experimental data with zero quantum T_2^{ZQ} (**Fig. 5C**) *(68,73)* (*see* **Eq. 19**). T_2^{ZQ} is the time-constant zero-quantum magnetization transfer from the energy and similar to T_2 or the spin–spin relaxation time constant. Because the total Zeeman quantum numbers must be conserved, the populations of the two central states only mix with each other and with zero-quantum coherences. As a result, the relaxation of zero-quantum coherence strongly influences the time-course of magnetization transfer.

In the earlier versions of the R^2 experiment, the accuracy relied on the sample spinning speed. At $n > 1$, the mean square deviation (χ^2) between the calculated decay in magnetization transfer as a function of the dipolar coupling and T_2^{ZQ} and the measured decay is more dependent on the shielding anisotropies of the two spin species and their relative orientation *(68)*. If the internuclear distance is small, the longitudinal magnetization exchange trajectory is dominated by the direct dipole–dipole coupling, but only minor effects are the result of homogeneous zero-quantum relaxation. However, for large internuclear distances, the influence of the homogeneous zero-quantum relaxation is more significant *(74)*. Also the use of a long weak rf pulse, or a long train of short pulses, synchronously with the sample spinning *(75,76)*, can cause a disruption of the coherent averaging and result in dipolar dephasing of spin polarizations or coherences *(77)*. These methods lead to losses through relaxation and are susceptible to interference with the CSA *(75,76)*. These additional attenuation mechanisms of the time-dependent signals complicate data analysis and are associated with zero-quantum relaxation $(R = (T_2^{ZQ})^{-1})$ *(68,72,74)* and other types of relaxation processes *(78)*. Thus, some difficulties in measuring the internuclear distance between weakly coupled homonuclear spins in systems such as membrane-bound proteins can occur, although only approximate information about nuclei proximity may be helpful in

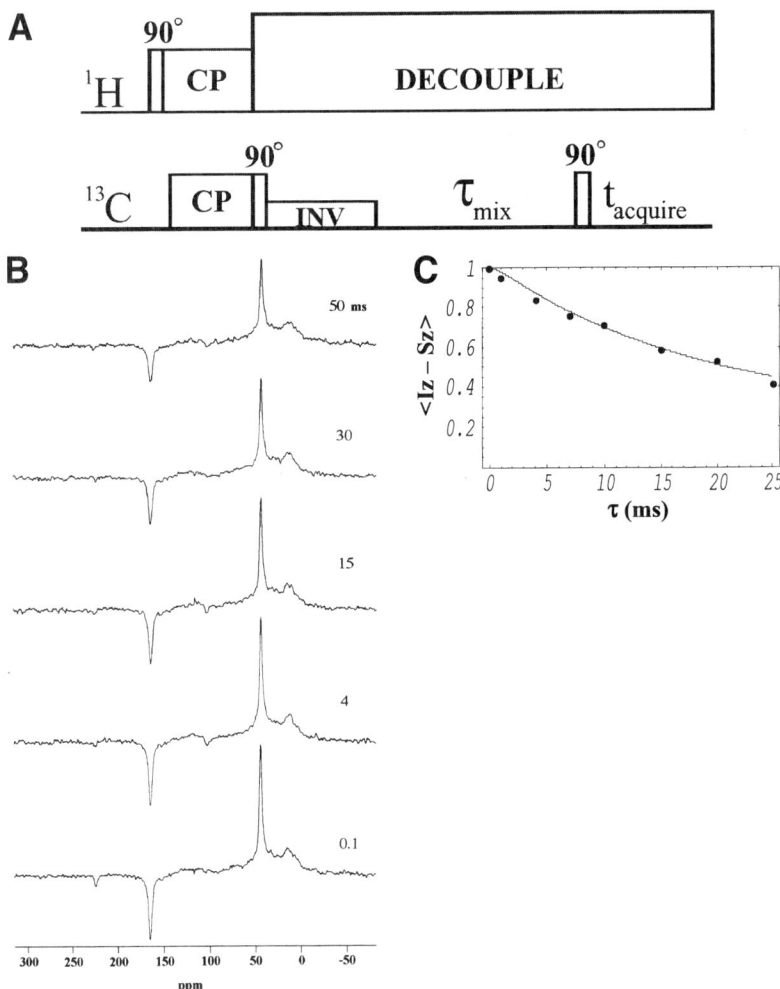

Fig. 5. Rotational resonance (R^2) pulse sequence (A), and R^2 spectra (B) of labeled melittin lyophilized from methanol with representative mixing times: 0.1, 4, 15, 30, and 50 ms The increase in mixing time resulted in a decrease in signal for both the labeled $C\alpha$ and carbonyl of melittin (B). The magnetization exchange curve determined from the R^2 spectra for labeled melittin (C). The symbols on the graph are the data points obtained from experiment and the fitted curve was obtained by nonlinear least squares fitting of T_{2ZQ} and the dipolar coupling. (Figure courtesy of Y. H. Lam.)

modeling. Reduction in intensity occurs because of both ^{13}C T_1 decay during the longer mixing times, as well as from rotor-driven magnetization exchange at the rotational resonance (R^2) condition *(79)*.

Experiments that minimize the relaxation effects also enhance the accuracy of distance measurements. The idea that the spinning frequency dependence can serve as an alternative for time dependence of the recoupling was suggested *(78)*. Different approaches characterized by constant time experiments have also been introduced. The rotational resonance tickling (R^2T) technique *(79)* employs rf-field variations sweeping through the resonance criteria. R^2T uses a ramped rf field to induce fast passage through the dipolar resonance condition, thereby greatly reducing the dependence of the exchange dynamics on T_2^{ZQ} and CSA parameters. R^2T involves spinning slightly faster than the standard $n = 1$ R^2 condition and simultaneously applying a relatively weak rf field to the homonuclear spin pair to induce dipola-driven dynamics. Under the appropriate conditions, these dynamics show markedly reduced dependence on both homogeneous and nonhomogeneous zero-quantum line shape parameters, with the result that distance information of increased accuracy can be extracted. An additional feature of the technique is that it does not demand the very high ^1H decoupling fields typically required in other recoupling experiments to limit signal loss during mixing *(79)*.

The constant-time homonuclear dipolar recoupling method uses two radio frequency driven recoupling (RFDR) periods that are varied to maintain an overall constant-mixing time period *(80)*. The first period recouples the dipolar interaction and the second partially refocuses the dipolar evolution in a way analogous to the transverse echo simple π-pulse excitation for the dephasing of rotational-echo amplitudes (t-SEDRA) experiment *(81)*. Constant-time R^2 experiments eliminate the need for control spectra to correct for effects from variable rf heating, which is critical for long-distance measurements. This reduces the total number of experiments needed by as much as a factor of 2 because the phase cycling scheme chosen for the constant-time R^2-pulse sequence succeeds in canceling out all residual magnetization in the xy plane arising from imperfect and/or mis-set $\pi/2$ pulses and missettings of the inversion delay. Other improvements include achieving selective inversion with a delay rather than a weak pulse *(79)*, which results in the elimination of oscillations in peak intensities for short mixing time-points. This is achieved by introducing an additional variable delay before spin inversion. The sum of the constant-time delay (τ_d) and the mixing time (τ_m) is kept the same for all mixing time-points to ensure a constant overall high-power decoupling time *(82)*.

Recently, a pulse sequence that achieves double quantum excitation via direct dipolar coupling under the $n = 1$ R^2 condition has been introduced *(83)*. Here, the double quantum coherence was excited by exploiting the mechanical excitation of zero-quantum coherence at R^2. This R^2-double-quantum filter (DQF) pulse sequence features a series of simple three-pulse subsequences and does not require rotor synchronization. For spin systems where the CSA is only a small fraction of the isotropic chemical shift difference, high efficiencies are found for both large and small dipolar coupling interactions. In the presence of significant CSA, the overall efficiencies decrease and become strongly dependent on the duration of the excitation period. This enables the application of the experiment on spin systems with more than two spins *(66)*. In 2D R^2-double-quantum (DQ) spectrum the absorption peaks originating from the J coupling and the dipolar interaction are separated from each other, allowing for the determination of dipolar couplings from their intensity ratio, provided that the values of the J couplings are known from solution spectra *(84)*. This method is not restricted to isolated spin pairs and can be applied to multispin systems *(66,79,82–84)*.

R^2 and modified R^2 methods have been applied extensively to structural determination of biological membranes *(28,63,85–92)*, by measuring inter- and intramolecular distances of membrane-bound proteins and peptide-inhibitor complexes (**Subheading 3.3.**).

3.2.4.2. DISTANCE MEASUREMENTS USING HETERONUCLEAR RECOUPLING TECHNIQUES

The rotational echo double resonance (REDOR) experiment was introduced by Gullion and Schaefer in 1989 *(93–95)* as a method for heteronuclear distance determination. Since then, it has proved its robustness and applicability to biological systems in numerous applications for up to 1 MDa protein complexes *(96)*. The aim of the REDOR experiment is to prevent the heteronuclear dipolar coupling from being averaged by MAS by rotor synchronized rf pulses.

As mentioned previously, MAS renders the dipolar coupling constant time dependent such that **Eq. 17** becomes:

$$\omega_{IS}^{DD}(t) = b_{IS} \frac{1}{4} \left(\sin^2 \beta_{PR} \cos 2 \left(\gamma_{PR} + 2\pi t / T_R \right) - \sqrt{2} \sin 2\beta_{PR} \cos \left(\gamma_{PR} + 2\pi t / T_R \right) \right) \quad (24)$$

where β_{PR} and γ_{PR} are the Euler angles relating the heteronuclear dipolar vector from its PAS to the rotor frame.

After a full rotor period T_R, the average dipolar coupling Hamiltonian is given by

$$\overline{H_{IS}} = \frac{\hbar}{T_R} \int_0^{T_R} \omega_{IS}^{DD}(t) \cdot 2I_z S_z dt = 0 \quad (25)$$

Because the Hamiltonian determines the evolution of the spin system, this means that, on average, the dipolar coupling does not contribute to the evolution of the spin system and hence the NMR signal. This averaging results from the time dependence of the *spatial* part $\omega_{IS}^{DD}(t)$ of the Hamiltonian. However, it

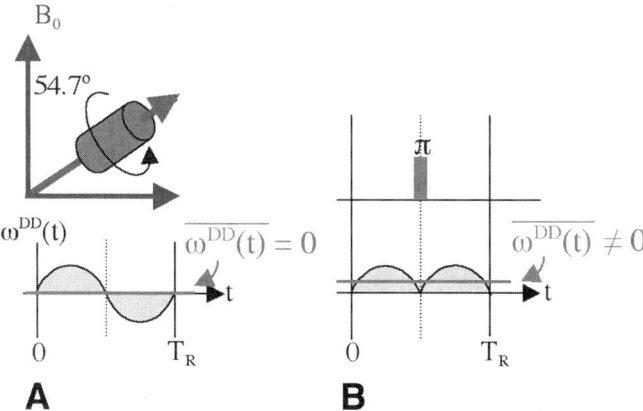

Fig. 6. Principle of the REDOR experiment. *Left*: MAS renders the dipolar coupling frequency ω(t) time dependent (curve shown is schematic to illustrate the principle). Without pulses, the average of the evolution frequency over a rotor period T_R is zero (indicated by the *shaded areas*). *Right*: A radiofrequency π-pulse on the dephase channel can "flip" the spin and invert the sense of time evolution. Now, the evolution frequency of the heteronuclear dipolar coupling does not average to zero; hence the time development of the spin system and the observed signal will reflect the coupling.

can be prevented by influencing the *spin* part (*see* **Fig. 6**). In the REDOR sequence, therefore, π-pulses on the indirect channel (S) in the center of a rotor period cause a change in sign of the Hamiltonian, and a nonzero value remains as the average over a rotor period:

$$\overline{H}_{IS} = \frac{\hbar}{T_R}\left(\int_0^{T_R/2} \omega_{IS}^{DD}(t)2I_zS_zdt - \int_{T_R/2}^{T_R} \omega_{IS}^{DD}(t)2I_zS_zdt\right) = \hbar\overline{\omega}_{IS}^{DD}2I_zS_z \qquad (26)$$

with

$$\overline{\omega}_{IS}^{DD} = b_{IS}\sqrt{2}\sin 2\beta_{PR}\sin\gamma_{PR} \qquad (27)$$

Time evolution of the spin system now occurs under an average Hamiltonian with a nonzero contribution from the heteronuclear dipolar coupling, and the measured signal $S(T_R)$ therefore depends on the coupling strength b_{kl}, providing a measure for the internuclear distance:

$$S(T_R) = \frac{1}{4\pi}\int_0^\pi \sin\beta_{PR}d\beta_{PR}\int_0^{2\pi} d\gamma_{PR}\cos\left(T_Rb_{IS}2\sqrt{2}\sin 2\beta_{PR}\sin\gamma_{PR}\right) \qquad (28)$$

In practice, two experiments are performed, one with and the other without the "dephasing" pulse on the S-channel giving rise to the dephased signal S and the full signal S_0, respectively. Because both S and S_0 are subject to the same loss by relaxation, the quantity $\Delta S/S_0 = (S_0-S)/S_0$ is free of relaxation effects

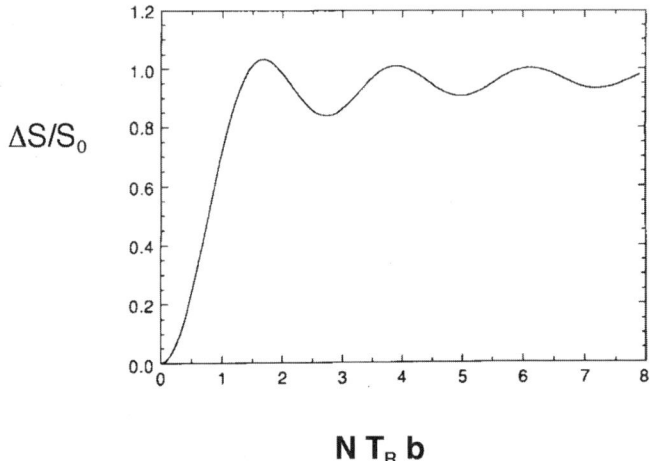

N T_R b

Fig. 7. Universal REDOR-curve. The loss in signal $\Delta S/S_0$ as a function of the product of dephasing time $N\,T_R$ and dipolar coupling strength b_{IS}. By fitting experimental values of $\Delta S/S_0$, measured for several values of the dephasing time $N\,T_R$, with the REDOR curve, the dipolar coupling constant b_{IS}, and hence the internuclear distance can be determined. (Adapted with permission from **ref. 95**.)

and displays the reduction in signal only because of the dipolar coupling. To obtain the dipolar coupling constant, a set of experiments with different evolution times $N\,T_R$ is conducted. The signal $\Delta S/S_0$ as a function of dephasing time $N\,T_R$ is then given by

$$\frac{\Delta S}{S_0} = 1 - \frac{1}{4\pi}\int_0^\pi \sin\beta_{PR}\,d\beta_{PR}\int_0^{2\pi} d\gamma_{PR}\,\cos\!\left(N T_R b_{IS}\,2\sqrt{2}\,\sin 2\beta_{PR}\,\sin\gamma_{PR}\right) \qquad (29)$$

By fitting the data with calculated "REDOR-curves," a dipolar coupling and hence a distance can be determined. As $\Delta S/S_0$ depends on the dipolar coupling in the form of the product $N\,T_R\,b_{IS}$, the dephasing follows a "universal" curve (**Fig. 7**), representing an identical behavior that depends on the dipolar coupling only through scaling along the abscissa.

Numerous experimental improvements have been introduced to the basic REDOR experiment described previously and have lead to various REDOR derivatives. A typical example of the REDOR sequence is shown in **Fig. 8**. Important for the performance of the experiment is refocusing the isotropic chemical shifts, which is achieved by a π-pulse on the observe channel. To avoid the buildup of signal loss owing to pulse imperfections, compensating phase cycles such as xy8 have been employed for the dephasing pulses *(97)*. Also, the use of composite pulses has been shown to increase the performance of dephasing *(69,98)*. Furthermore, the dephasing pulses are not restricted to the S channel, but can be applied on both channels in an alternating fashion, or even almost entirely on the observed channel *(99,100)*. Thus, it is possible to adapt the pulse sequence to sample- or hardware-specific circumstances and, for example, apply most of the pulses on the channel with the better inversion efficiency.

Adaptation of the experiment and data evaluation is also necessary if the dephasing nucleus is not spin-1/2, such as ^{2}H *(100,101)*. Numerous sequences have been developed for couplings between spin-1/2 and quadrupolar nuclei: for example, rotational echo adiabatic passage double resonance *(102)* or transfer of populations in double resonance, *(103)*. The REDOR experiment has also inspired a range of related experiments. For example, continuous irradiation with simultaneous frequency and amplitude modulation (SFAM) can be used in place of discrete pulses *(104)*, or coherence transferred by transferred echo double resonance *(105)*.

Often the spin system deviates from the ideal situation of dilute spin pairs with identical distances. Several multispin situations have been analyzed theoretically *(106)*. In the case of diluted spin pairs, but a distribution of dipolar couplings, the net dephasing $\Delta S/S_0$ is the arithmetic average of the individual, pair-wise dephasings weighted according to the dipolar coupling distribution. In this situation, a "spectrum" of dipolar couplings can be obtained directly from the REDOR-curve using a "REDOR-transform" *(107)*. A more complicated situation arises if the dephasing of the observed spin is owing to more than one spin.

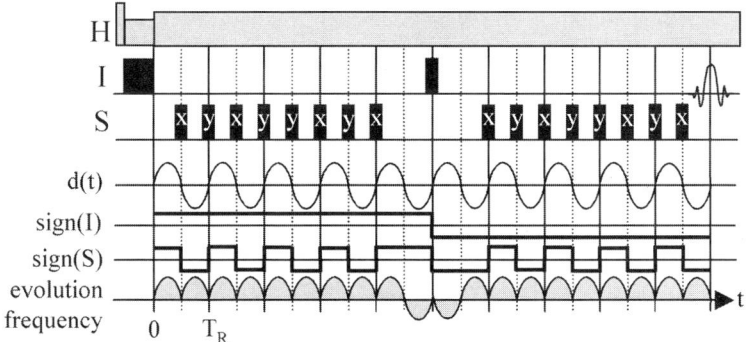

Fig. 8. Typical example of a REDOR sequence. After an initial cross polarization (CP) from protons (H) to the observed spin (I), rotor synchronized dephasing pulses on the S channel recouple the dipolar coupling between I and S spins. An xy8 phase-cycling scheme applied to the dephasing pulses avoids the build up of pulse errors, and a π-pulse on the observe channel in the center of the dephasing period refocuses the isotropic chemical shift. The dipolar coupling frequency $\omega^{DD}(t)$ follows an oscillatory behavior with an average of zero. The radiofrequency pulses of the sequence invert the sign of the evolution frequency and prevent averaging of the overall evolution frequency.

In this situation, the relative positions within the spin system have to be taken into account in the interpretation of the REDOR-curve. Pulse-sequence modifications such as θ–REDOR have been suggested as well, allowing the analysis of the spin system in a pair-wise fashion and thus eliminating this dependence on geometry *(108)*. Furthermore, in multispin situations, homonuclear couplings or J-couplings might also influence the dephasing coherences. This issue has been addressed in studies of weak heteronuclear couplings, and experimental schemes on the basis of selective pulses have been suggested *(109)*.

Typically, the REDOR experiment is used for distance measurements. However, it can also provide orientational information. If the MAS frequency chosen is sufficiently slow, then REDOR dephasing will affect the individual sidebands differently. Although the integral intensity of the difference spectrum S_0-S depends only on the dipolar coupling strength, the sideband pattern of the difference spectrum then also reflects the relative orientation of the CSA tensor (or quadrupolar interaction tensor) of the observed spin with respect to the internuclear vector. The analysis of the spinning sidebands thus provides relative orientation in addition to the internuclear distance *(110)* in membranes *(28)*.

3.2.4.3. Torsion Angle Measurements

Torsion angle constraints may be obtained by estimating the relative orientation of two anisotropic spin interactions. Generally, there are two approaches used to determine torsion angles. The first approach is called 2D exchange NMR spectroscopy. Magnetization of the spin S_1 is allowed to evolve under the first anisotropic spin interaction, which may be through CSA or dipole–dipole coupling for a certain time interval. Then, the magnetization is transferred from spin S_1 to the spin S_2, followed by detection of the signal from S_1, which evolves under the second anisotropic spin interaction. The correlation of S_1 and S_2 is revealed in the 2D exchange spectrum, which depends on the relative orientations of the two anisotropic spin interactions, allowing the torsion angle information to be extracted by simulation. Experiments of this type were first demonstrated in static solids *(111,112)*, but the low sensitivity of the static 2D spectrum often makes such measurements impractical for biological applications. Two-dimensional MAS exchange experiments were introduced later to increase the sensitivity of the measurements *(113–117)*. The second approach is based on multiple-quantum (MQ) coherence excitation. Here, a DQ/MQ correlated spin state is excited and then allowed to evolve under the influence of anisotropic spin interactions, such as CSA or heteronuclear dipolar couplings. The evolution of the DQ/MQ coherence is not influenced by the dipolar couplings between the different spin sites that are involved in the excitation of the DQ/MQ coherence. The modulation of the DQ/MQ coherence by the anisotropic spin interactions is sensitive to their relative orientation *(118–129)*.

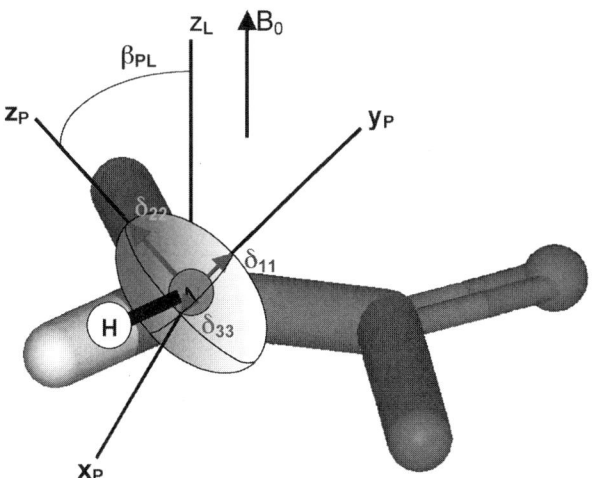

Fig. 9. Angles relating the CSA and NH internuclear vector orientation from the peptide plane to the laboratory frame. z_L is the z-axis of the peptide plane and z_P is the z-axis of the principal axis frame.

In either way, estimates of the torsion angles ϕ and ψ for adjacent peptide bonds or neighboring nuclei in ligands can be made to within $\pm$ approx 10°.

3.2.5. Static NMR Experiments

Under static conditions, the Hamiltonians in **Eqs. 1–3** and **12** preserve their orientational dependence. By aligning the sample on a solid support that has itself a specific direction with respect to the external magnetic field, orientationally dependent parameters, such as the chemical shift, quadrupolar interactions, or the dipolar coupling, can be measured for each labeled position, such as a peptide plane in the membrane protein. The chemical shift and dipolar coupling can be written in terms of the Euler angles α_{PL}, β_{PL} (for chemical shift or dipolar coupling):

$$\delta_S\left(\alpha_{PL}, \beta_{PL}\right) = \delta_{11} \cos^2 \alpha_{PL} \sin^2 \beta_{PL} + \delta_{22} \sin^2 \alpha_{PL} \sin^2 \beta_{PL} + \delta_{33} \cos^2 \beta_{PL} \tag{30}$$

and

$$\omega_{IS}^{DD}\left(\alpha_{PL}, \beta_{PL}\right) = b_{IS} \frac{1}{2}\left(3 \cos^2 \beta_{PL} - 1\right) \tag{31}$$

where δ_{11}, δ_{22}, δ_{33} are the principal CSA tensor components, and b_{IS} is the dipolar coupling constant (**Fig. 9**). Thus, solid-state NMR experiments of oriented systems can be used to resolve both local and global structures from an observation of the spectral anisotropy.

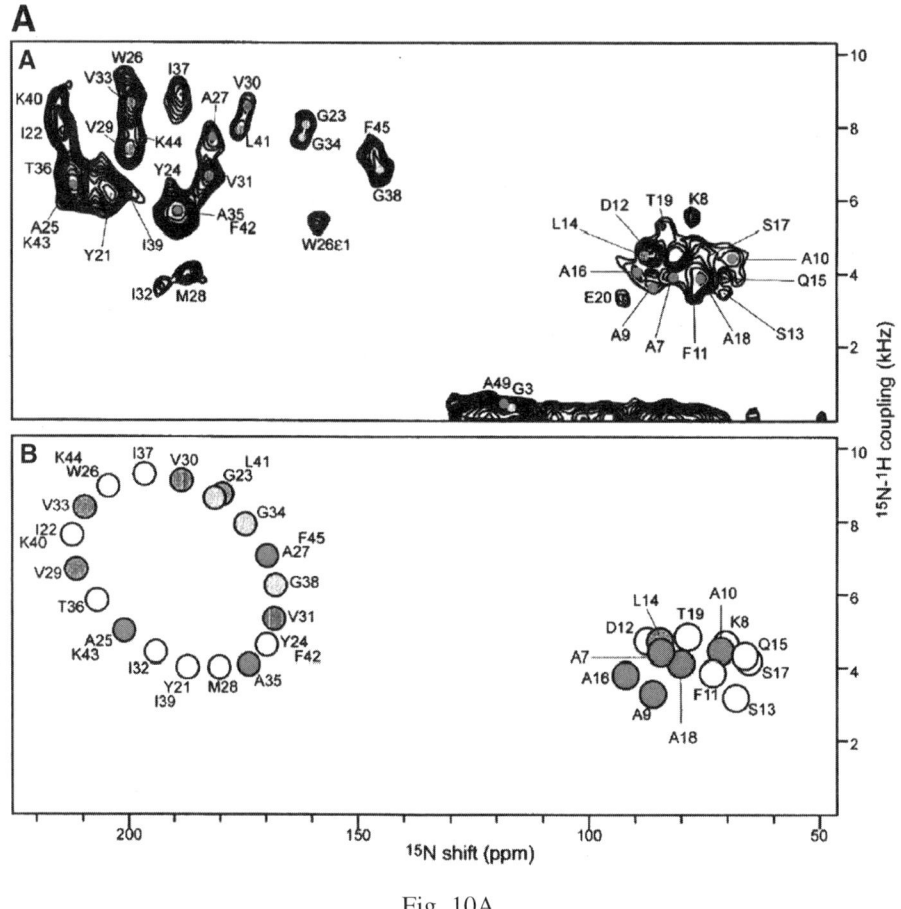

Fig. 10A.

3.2.5.1. POLARIZATION INVERSION SPIN EXCHANGE AT THE MAGIC ANGLE (PISEMA)

The polarization inversion spin exchange at the magic angle (PISEMA) *(130)* experiment yields a 2D correlation of the chemical shift of spin S ($S = {}^{13}C$, ${}^{15}N$) *with the heteronuclear* IS ($I = {}^1H$) dipolar coupling. Structural information is obtained by relating experimentally determined chemical shift/dipolar splitting pairs for a given peptide plane to four possible solutions of (α_{PL}, β_{PL}). These four solutions arise because of the symmetry of the interactions. By finding pairs of angles for all peptide planes it was possible, for instance, to resolve a family of structures for gramicidin A *(131)* and M2 from the nicotinic acetylcholine receptor *(132)* (**Subheading 3.3.1.**).

The global arrangement of helices can be estimated by fitting experimentally determined chemical shift/dipolar splitting pairs with calculated parameters for given (α_{PL}, β_{PL}) angles for a set of peptide planes. To relate (α_{PL}, β_{PL}) for each peptide plane to a global orientation of the helix relative to the external magnetic field, a specific geometry must be defined. One possibility is to assume that the helix is ideal, that is, that the (Φ,Ψ) torsion angles are identical throughout *(133,134)*. The implications of this choice of torsion angles and other parameters, including motion, used to obtain the calculated chemical shift/dipolar splittings have been recently investigated in great detail *(135,136)*. For the regular α-helix structure in uniformly labeled single-pass peptides with ^{15}N in the backbone, the PISEMA experiment leads to wheel-like patterns called polarization index slant angle (PISA) wheels (**Fig. 10**) *(133,134)*. From such wheels, it is possible to determine the secondary structure, the tilt and rotational angles of α-helices or β-sheets with respect to the magnetic field, and in

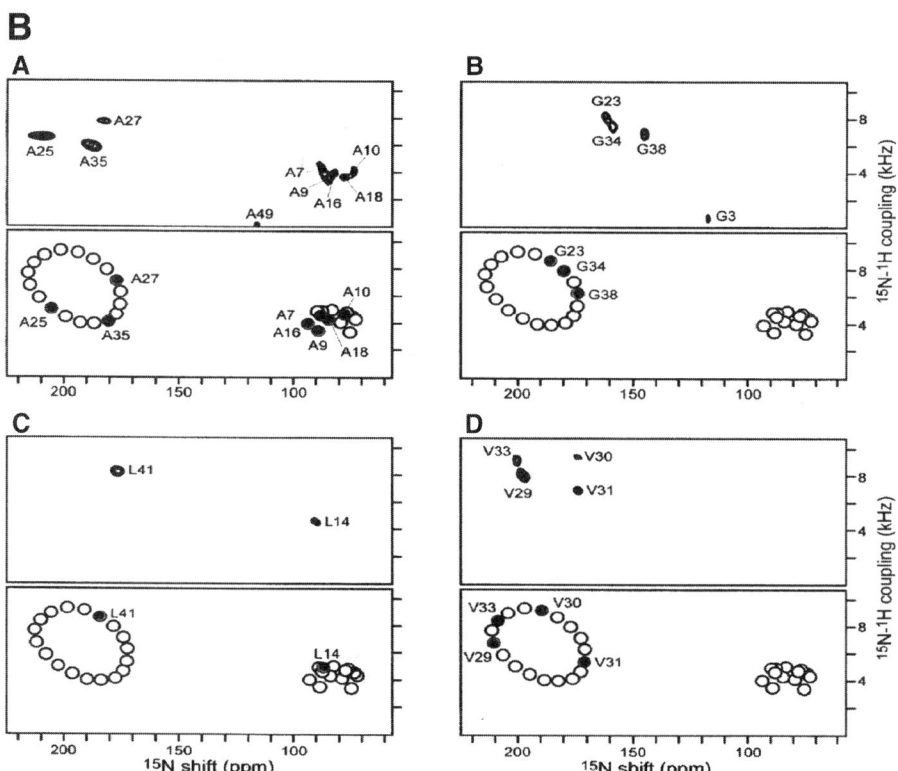

Fig. 10. A "shotgun" approach was used to assign PISEMA spectra from uniformally ^{15}N-labeled samples of the membrane-bound fd bacteriophage coat protein, a single membrane pass peptide (**A**). Labeling of a single type of amino acids (alanine, glycine, leucine, or valine) gave the clues to relate the PISA wheel resonances to the respective helical wheels (**B**). (Adapted with permission from **ref. *159*.**)

favorable cases sequential assignment of all resonances on the basis of a single unique resonance assignment *(137,138)*.

For an NMR experiment to be useful when applied to membrane peptides and proteins, it is important that the method yields well-resolved spectra, and the PISEMA experiment offers comparatively high resolution relative to the separated local field experiment (SLF) *(139–141)*. Line narrowing is achieved by incorporating the frequency-switched Lee–Goldburg (FSLG) *(142,143)* proton homonuclear decoupling sequence during the *IS* dipolar evolution in t_1 (**Fig. 11**). This eliminates the contribution of the homonuclear dipolar interaction given in **Eq. 12**. By repeatedly inverting the effective field of the *I* spins every time τ

$$\tau = \frac{2\pi}{\sqrt{\omega_{1I}^2 + \Delta\omega_I^2}} \tag{32}$$

where ω_{1I} is the rf field on the I-spin channel and $\Delta\omega_I$ is the I-spin offset, as defined in **Fig. 11**, then the homonuclear decoupling is achieved more efficiently, whereas the heteronuclear interaction remains (during t_1). The dipolar Hamiltonian for the PISEMA experiment, for a heteronuclear spin pair, can now be written as

$$\overline{H}_{IS} = \hbar \sin \theta_m \omega_{I_k S}^{DD} \frac{1}{4} \left(S^+ I_k^- + S^- I_k^+ \right) \tag{33}$$

where a single *S* spin is considered to be coupled to a number of *I* spins and $\omega_{I_k S}^{DD}$ as defined in **Eqs. 18** and **24** *(144)*. The scaling factor $\sin\theta_m = 0.82$ compares favorably to the one obtained for the SLF experiment, where

$$\overline{H}_{IS} = \hbar \sum_k \cos \theta_m \omega_{I_k S}^{DD} S_z I_{kz} \tag{34}$$

3.2.5.2. HETERONUCLEAR CORRELATION EXPERIMENTS (HETCOR)

In the same way that the PISEMA experiment can be used to obtain orientational information in oriented membrane peptides and proteins, static heteronuclear correlation experiments, for example, between ^1H and ^{15}N (**Fig. 12**) *(130,145)* (and references therein), or between ^{13}C and ^{15}N *(146)*, can also be used. In this case, two pairs of angles are obtained, one from each chemical shift, such as for $(\alpha^N_{PL}, \beta^N_{PL})$ and for $(\alpha^H_{PL}, \beta^H_{PL})$. These can be related to each other to get four unique combinations of $(\alpha_{PL}, \beta_{PL})$, which can then be related to global orientation, in much the same way as was outlined in **Subheading 3.2.5.1.** *(147)*. For ^1H–^{15}N heteronuclear correlations, similar factors need to be considered when setting up the experiment (*see* **Subheading 4.6.**).

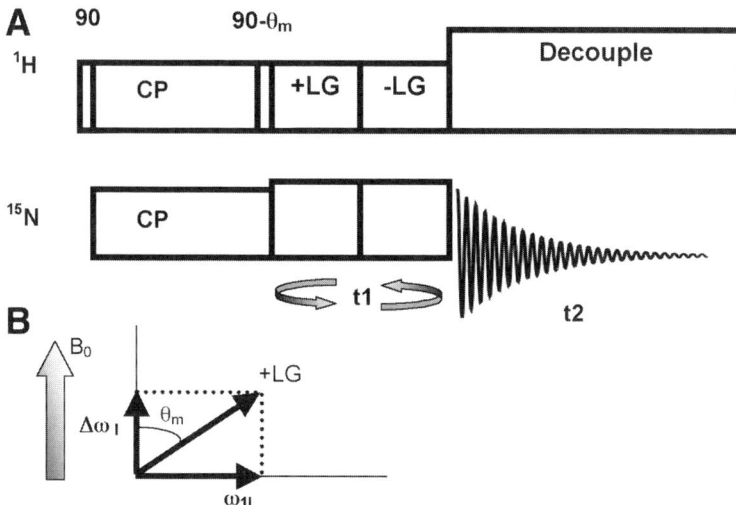

Fig. 11. Pulse sequence for the PISEMA experiment. During t_1, the homonuclear II dipolar interaction is suppressed by using FSLG, leaving the heteronuclear IS dipolar interreaction (*see* text) (**A**). Proton effective field during the Lee–Goldburg decoupling sequence (**B**). –LG would be at 180 degrees in the opposite direction.

3.2.6. Magic-Angle-Oriented Sample Spinning (MAOSS)

In the MAOSS approach, stacked membranes (**Fig. 2**) are set spinning at the magic angle with the membrane normal parallel to the rotor axis. Here, the powder spectral envelope is resolved into a series of spinning sidebands, from which orientational and mosaic spread information can be gained (**Fig. 13A**). In the fast spinning case, narrow isotropic line shapes are achieved, as shown for a pure lipid membrane in which even ^{1}H line widths are sufficient for assignment (**Fig. 13B**) (*10*).

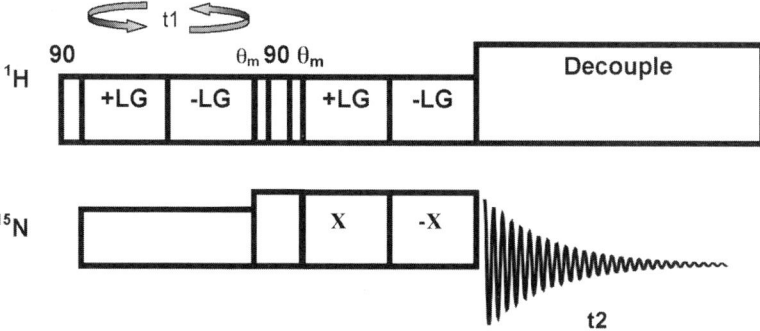

Fig. 12. Pulse sequence for static ^{1}H–^{15}N heteronuclear chemical shifts correlation.

3.3. Membrane Protein Structures Determined by Solid-State NMR

3.3.1. Single-Pass Peptides

Single TM peptides are, as a class of protein, the most abundant in number in many life-forms, as predicted from all known genomes. Although many will be leader sequences used in processing and targeting to a specific location, single TM peptides do form an important class of proteins functioning alone or through interactions with similar or identical units to form complexes such as ion channels, or with larger proteins and acting as regulators of function. It is generally assumed that single TM peptides adopt an α-helical structure in the membrane. Even if this is the general rule, kinks and helical tilts do have a significant bearing on function and mechanism—solid-state NMR methods that exploit spectral anisotropy and give distance measurements can make significant contributions to this knowledge base, helping functional descriptions.

Complete structures by solid-state NMR of numerous TM peptides, small single-pass proteins, or TM helices of larger membrane proteins have recently been elucidated. Examples deposited in the PDB include gramicidin A, the M2 segment of the nicotinic acetylcholine receptor and of the proton channel of influenza, and the fd bacteriophage coat protein (**Fig. 14**). Most of the structural information for these systems was obtained from orientational constraints from oriented samples of uniformly ^{15}N-labeled peptides or proteins.

The first membrane protein structure elucidated by solid-state NMR and deposited in the protein database is the antimicrobial peptide gramicidin A *(148)*. The structural knowledge has since then been used in numerous methodological, bioinformatics, and biophysical studies, and gramicidin A has become a prototype for solid-state NMR on membrane proteins *(149)*. It was determined entirely on the basis of orientational constraints deduced for the protein embedded in oriented bilayers, piecing together consecutive peptide plane alignments from a set of selectively labeled peptides. Gramicidin A spans the bilayer as a head-to-head dimer, each monomer formed by a 15-residue right-handed β-helix. Because of alternating L and D amino acids, all sidechains point outside the β-helix, leaving an ion channel in the center of the peptide, which is responsible for its antimicrobial activity. Important to the function are four tryptophan residues on each end of the channel, anchoring the peptide in the membrane, which have been the focus of ^{2}H–NMR refinement as well as functional studies *(150–154)*.

Viral ion channels have been another focus of solid-state NMR structure determinations. Viral genomes are often relatively small in size, thus encoding for moderately sized oligomeric viral proteins, which are amenable to solid-state NMR studies. The structure of the M2 TM segment of the influenza H$^+$ channel has been obtained this way *(155)*. The complete channel plays an important role

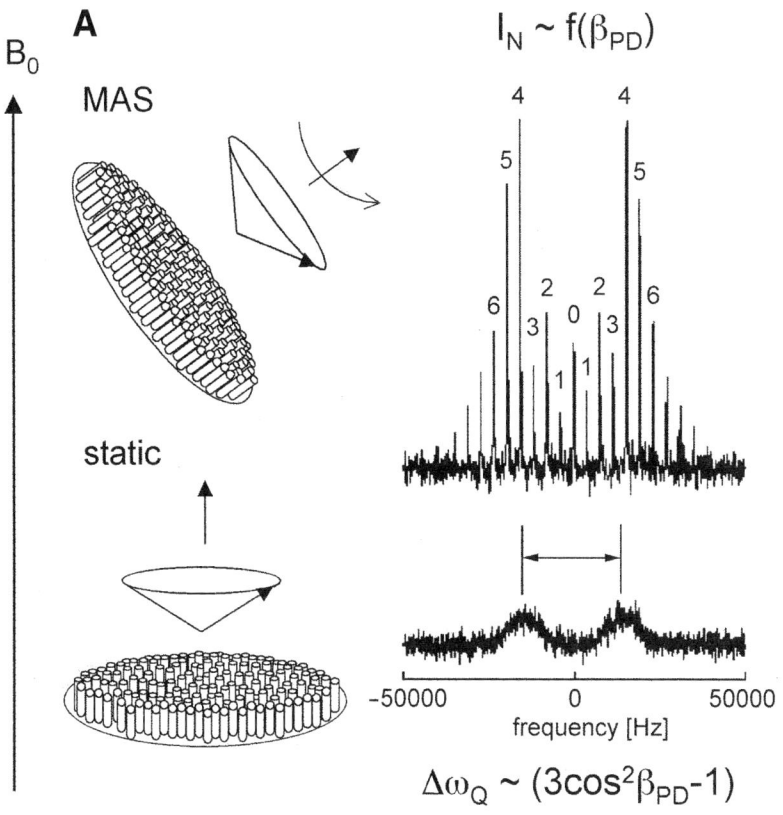

Fig. 13A.

in the viral life cycle and facilitates the uncoating of the virion in the endosome. It is formed by a cysteine crosslinked tetramer, each monomer possessing a TM helix (M2) and a cytosolic domain. Important for the channel activity is the M2 segment, which was found to be formed by four almost ideal α-helices, tilted at approx 25° with respect to the membrane normal *(156)*. A histidine residue pointing toward the center of the M2 bundle plays a crucial role in the pH-dependent gating of the proton channel. Extensive use was made of the PISEMA experiment (**Subheading 3.2.2.1.**) to provide the necessary orientational constraints. However, to identify intermolecular connectivities and, hence, the oligomeric state of the bundle, and to determine the position of functionally important sidechains, these orientational studies were combined with REDOR distance measurements *(157)* (**Subheading 3.2.1.2.2.**).

A similar strategy was adopted in the study of the nicotinic acetylcholine receptor, where a TM segment (again M2) was investigated individually *(158)*.

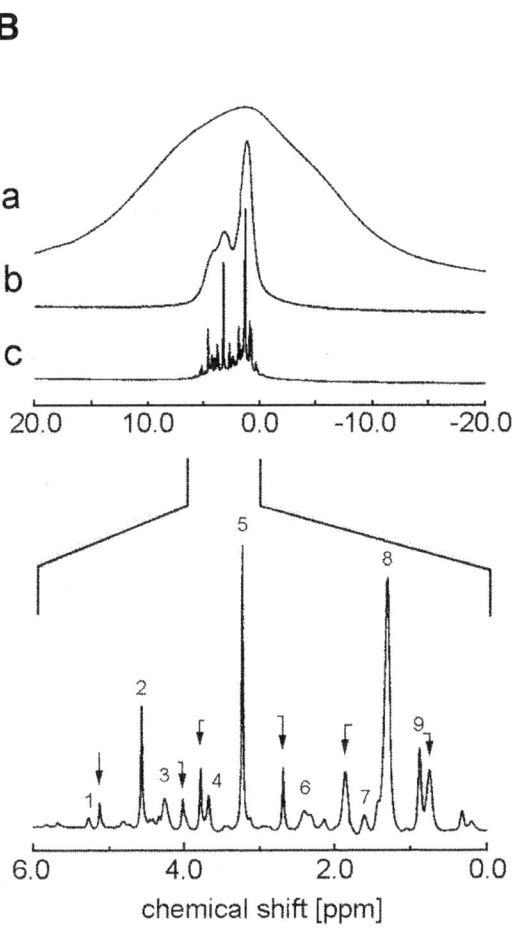

Fig. 13. Demonstration of sensitivity and resolution improvements when using MAOSS compared with static, wide-line solid-state NMR on membrane bilayers. A deuterium wide-line NMR spectrum (from a —C^2H_3 labeled retinal in membrane-bound rhodopsin) (**A** lower spectrum) and the equivalent MAOSS spectrum (spinning frequency approx 500 Hz) (**A** upper spectrum) showing how the spectral maxima (from which the orientationally informative quadrupole splitting [$\Delta\omega_Q$] is measured) are better resolved (**A**). Proton NMR spectra of fluid lipid bilayers static (**B**); static, random distributed sample (**B.a**); static, oriented sample with bilayer normal at the magic angle (**B.b**); and at a MAS speed of 22 Hz (**B.c**). (Courtesy of C. Glaubitz, Frankfurt, Germany.)

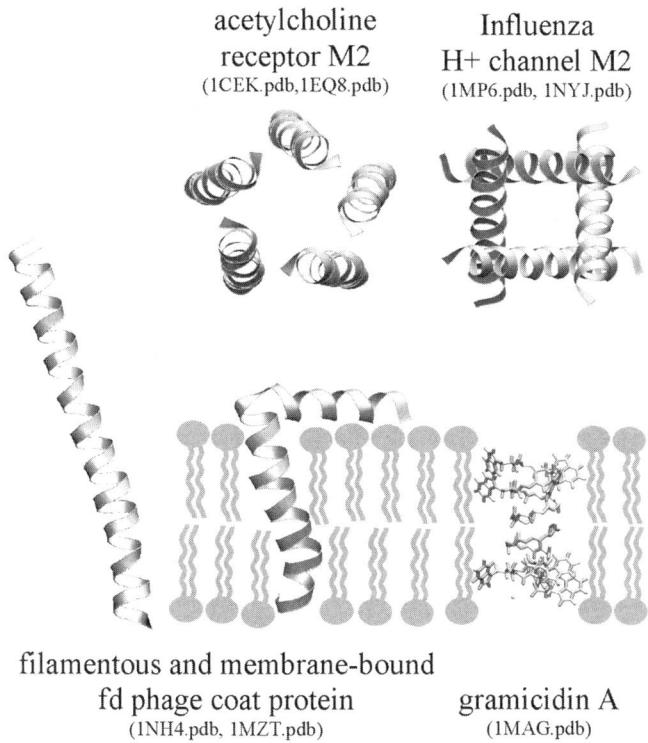

acetylcholine
receptor M2
(1CEK.pdb,1EQ8.pdb)

Influenza
H+ channel M2
(1MP6.pdb, 1NYJ.pdb)

filamentous and membrane-bound
fd phage coat protein
(1NH4.pdb, 1MZT.pdb)

gramicidin A
(1MAG.pdb)

Fig. 14. Structures of membrane proteins or peptides solved by solid-state NMR and deposited in the Protein Data Bank (PDB).

It was found to form pentamers, even without the context of the entire protein, thus enabling insight into the structure on the basis of a smaller peptide. A funnel-like structure, formed by a bundle of straight helices tilted at 12° with respect to the membrane normal was determined this way using, again, oriented samples and the PISEMA approach.

A further example of the use of oriented samples in solid-state NMR structure elucidation is demonstrated by the fd bacteriophage major coat protein *(159–161)*. This 50-residue protein was investigated in the two forms relevant for its function. Inserted in the bacterial membrane, it assists the release of phage particles from the host cell. In its filamentous form it then coats the phage DNA and constitutes a structural part of the phage assembly. Both structures were found to be distinctly different (**Fig. 14**). In its membrane-bound form the coat protein possesses two distinct domains, a TM helix as well as an amphiphilic helix lying on the membrane surface. In contrast, when embedded in the phage particle, the protein forms a single, almost straight α-helix. Two strategies have been instrumental in solving these structures by solid-state NMR. To solve the assignment problem inherent to

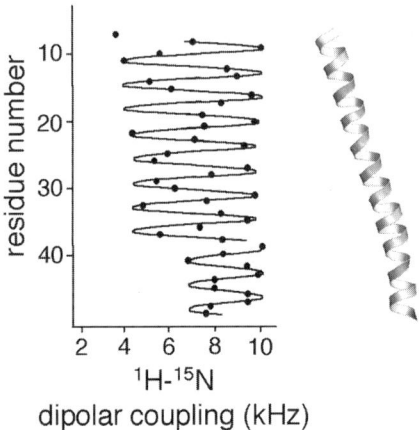

Fig. 15. Dipolar waves. The ^{1}H–^{15}N dipolar coupling as a function of residue position presents a spectroscopic representation of the helix structure, indicating structural details, such as changes in helix tilt and helix kinks in the filamentous fd bacteriophage coat protein. (Adapted with permission from **ref. *261*.**)

the uniformly ^{15}N-labeled samples used for this study, a "shotgun" approach was employed. By including a low number of spectra from protein samples, which were sparsely ^{15}N-labeled in only one type of amino acid, the complete assignment of PISEMA spectra from uniformly ^{15}N-labeled samples (**Fig. 10**) became accessible. This is possible because secondary structure elements, such as α-helices, are mapped to characteristic patterns in the PISEMA spectrum, and a few connectivities between these two representations of the protein structure are sufficient to correlate both. In the case of the filamentous coat protein, the concept of *dipolar waves* was used to characterize the details in the helix structure (**Fig. 15**). By plotting the ^{15}N–^{1}H dipolar coupling as a function of residue number, an oscillatory pattern becomes apparent. The amplitude, pitch and offset of these dipolar waves characterize the local structure of helix segments, and regions with different orientation as well as a small helix kink could be identified.

3.3.2. Polytopic Proteins—Seven TMD Proteins

Polytopic membrane proteins have more than one TMD connected by loops which themselves may have structural elements. Examples of single-chain and multicomplex proteins exist, although the tractability of total structures for polytopic proteins is hampered by assignment and spectral complexity for ^{15}N-labeled proteins, as shown from simulations of PISEMA spectra for rhodopsin and porin (**Fig. 16**). Thus, most approaches are designed to simplify the information content or focus on one part, such as a prosthetic group structure and its binding site, or through the use of selective labeling.

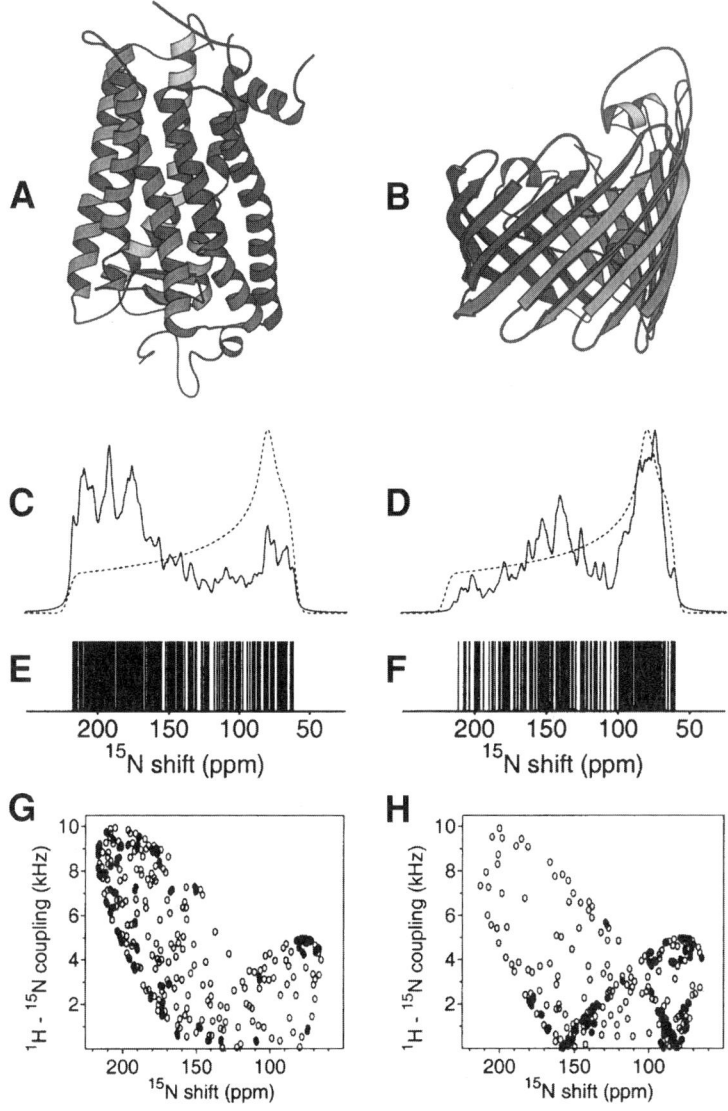

Fig. 16. SIMPSON simulations of 1D ^{15}N spectra and ^{15}N chemical shift/^{15}N–^{1}H dipolar PISEMA spectra for two membrane proteins under ideal conditions in macroscopically oriented membranes with their average α-helix and β-strand axes parallel to the external field to show the expected complexity of the spectra. Molecular structures of bovine rhodopsin (1FB8) (**A**) and the porin OmpA (2POR) (**B**) and their simulated 1D spectra (*solid line*) and powder spectra (*dashed line*). Stick plots of the resonance spectral line positions (**E, F**) and from the simulated spectra (**C, D**), and the simulated PISEMA spectra (**G, H**) with open and closed ellipses representing resolved and unresolved resonances, respectively. (Adapted with permission from **ref. 257**.)

Seven TMDs connected by three extracellular loops and three intracellular loops are classified as one of the largest superfamilies of membrane proteins—they make up 1% of the proteins encoded by open reading frames from the human genome. In higher life forms, the family includes the group GPCRs that are involved in sensory (visual, taste) and neurological responses (for example, dopamine, substance P, serotonin-level regulation). As such, they are important therapeutic targets, and structural and functional understanding of GPCRs and the interaction with their ligands is a route to aid the design of novel drugs. bR is an archetypical 7TMD protein and differs from later evolved members of the family by having short loops connecting the helices.

Although X-ray diffraction has been used to resolve the 3D structures of some 7TMD proteins, including the ground state of bovine rhodopsin and several photointermediates of bR, solid-state NMR spectroscopy gives complementary information without the need for crystallization, and especially it has been used to provide specific details about dynamics, electronic changes, and retinal isomerization that occur during activation.

3.3.2.1. RHODOPSIN

Rhodopsin is one of the most studied GPCRs and is used as a structural paradigm for the family, although it has a highly specialized structure composed of opsin and retinal, a chromophore. Over 300 7TMD proteins are known, all of which are photochemically reactive *(162)*. The retinylidene chromophore (retinal) is covalently bound to a lysine in the seventh helix of the protein by forming a Schiff base (SB). From the primary structure and function, rhodopsin is classified as type II in the retinylidine protein family and bR is classified as type I, together with halorhodopsin, sensory rhodopsins. They function as light-driven ion transporters. Light induces a change in conformation in retinal, which in turn affects the retinal-binding pocket and the remainder of the protein. Spectrally distinct species called *photointermediates* are observed during the photocycle. Being able to characterize these vitally important conformational changes is a major challenge for any structural method. Thus, the elucidation of the 3D structures of not only the type II retinylidene family, but also of the type I proteins, has been the focus of intensive activity in efforts to elucidate their structure–function relationship.

Rhodopsin is composed of a protein (opsin) with M_r approx 39,000, linked covalently through a SB to an *11-Z*-retinylidene choromophore to a lysine, which is Lys296 in the case of bovine rhodopsin. It is activated by photons in the rod-shaped photoreceptor cell of mammals and initiates the visual transduction cascade. Recently the ground-state structure of bovine rhodopsin was determined at 2.6 and 2.8 Å by X-ray crystallography *(4,163)*. Solid-state NMR has been used to determine the chromophore conformation at greater resolution than in the crystal studies, using selectively or uniformly ^{13}C-labeled retinal. In the ground state, the retinal exists in the *11-Z* conformation, which photoisomerizes to the all-*trans* isomer by the photointermediate, metarhodopsin I (Meta I), which is in a dynamic equilibrium under physiological conditions with metarhodopsin II (Meta II). Meta II is characterized by having an unprotonated SB imine and corresponds to the active receptor conformation *(164)*. Because the state can be trapped in membranes at relatively high temperature, Meta I has been studied by solid-state NMR spectroscopy.

^{2}H NMR spectra of bovine rhodopsin in which the chromophore is deuterated at the methyl group position C19 or C20 in aligned DMPC bilayers showed that the C–CD$_3$ bond vectors of the C19 and C20 segment orient at $42.5 \pm 5°$ and $30 \pm 5°$ with respect to the membrane normal from the deuterium quadrupolar coupling tensor *(165)*. From ^{2}H MAOSS NMR measurement for rhodopsin containing a C5–C18D$_3$ group on the cyclohexene ring of the retinal, the C–CD$_3$ bond vector orientation was changed markedly from $21 \pm 5°$, with respect to the membrane normal at the ground state, to $62 \pm 7°$ at Meta I state trapped below 213 K (**Fig. 17**) *(27)*.

^{13}C CP–MAS NMR has also been employed to study the conformation of retinal in bovine rhodopsin in detail. Early studies *(166)* comparing the chemical shifts with model retinoid compounds determined that the C6–C7 single bond has the unperturbed *cis* conformation. R^2 was used to determine the relative orientation of the β-ionone ring and the polyene chain of retinal in bovine rhodopsin. The existence of two conformations of retinal in the ground state was observed showing two distinct ^{13}C NMR signals, and the resulting distance constraints were obtained from C8 to C16 and C17 (4.05 ± 0.25 Å) and from C8 to C18 (2.95 ± 0.15 Å) for the major conformation (approx 74%) with a twisted 6-s-*cis* conformation *(167)*. R^2 was also used to determine the distances between C10 and C20, and between C11 and C20, which were found to be 3.04 ± 0.15 Å and 2.93 ± 0.15 Å, respectively, for the ground state, whereas in the Meta I state the distances changed significantly to ≥ 4.35 Å and to 2.83 ± 0.15 Å, respectively. This finding agrees with a model of a relaxed all-*E* structure *(168)*. Retinal that was extensively labeled in the chain, or uniformly ^{13}C–labeled throughout, was used to provide the complete ^{13}C and ^{1}H chemical shift assignment in 2D homonuclear (^{13}C–^{13}C) and heteronuclear (^{1}H–^{13}C) dipolar correlation spectra to analyze the electronic structure of the ground-state retinal *(53,169)*. The ^{1}H and ^{13}C chemical shifts of the ionone ring methyl groups were found to be strongly perturbed in the protein, suggesting nonbonding interactions and close contact between the ring moiety of retinal and the protein.

Torsion angles in the retinal chromophore in rhodopsin in its ground state and Meta I state at the C10–C11 bond have been determined by the HCCH-double quantum heteronuclear local field spectroscopy (HCCH-DQ-HLF) method

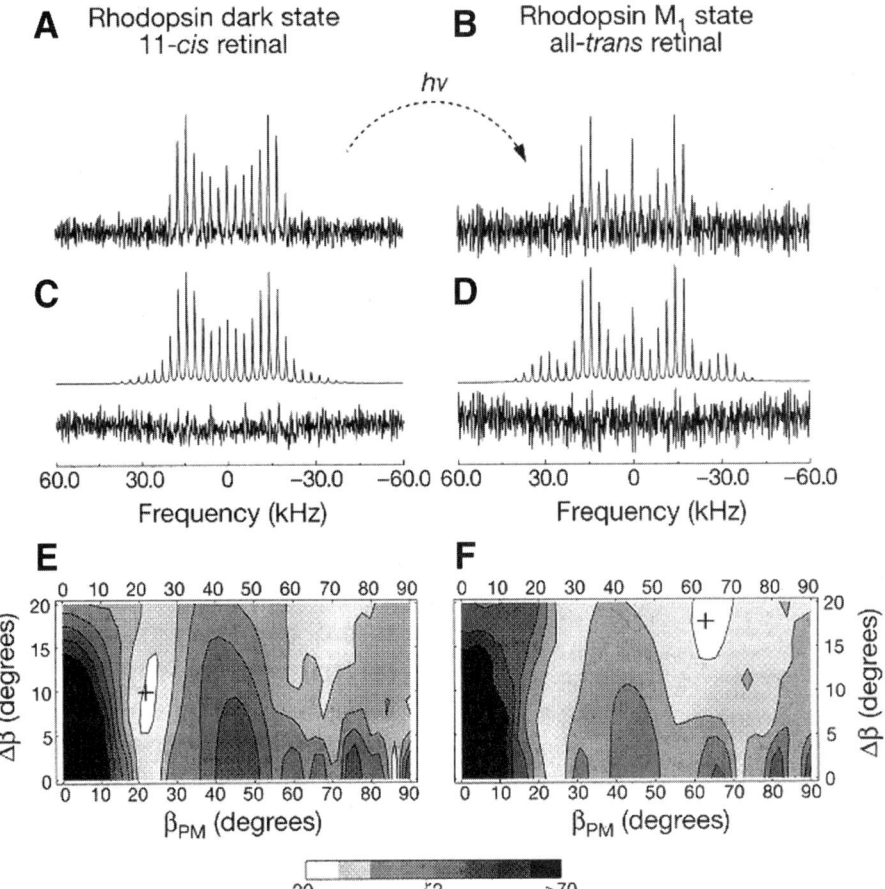

Fig. 17. ^{2}H MAOSS NMR spectra of [18-CD$_3$] retinal in dark-adapted rhodopsin and **(A)** on photoactivation in the Meta I state **(B)** (illuminated below 273 K, at 2860 MAS frequency and T = 213 K), and the respective simulations **(C, D)**, and the residuals (below). The spectra were analyzed by minimizing the RMS deviation of the MAS sideband intensities in the [β$_{PM}$, Δβ]-parameter space as shown by the ζ^2 contour plots **(E)** and **(F)**. The C-CD$_3$ tilt angle is determined to be β$_{PM}$ = 21 ± 5° in the dark-adapted state and 62 ± 7° in the Meta I state indicating a significant change on photoactivation. (Reproduced with permission from **ref. 27**.)

(**Subheading 3.2.4.3.**). The results show that the retinal chromophore adopts a nonplanar 11-*cis* conformation in the ground state and becomes an all-*trans* conformation after photoisomerization to the Meta I state *(120,127)*. More specifically, the H–C10–C11–H torsional angle of retinal was estimated to be 160 ± 10° at the ground state, whereas an angle of 180 ± 25° was determined for the trapped Meta I state. These results indicate that the strain of retinal chromophore is probably released in the Meta I intermediate, when compared to the ground state. Similar experiments have been applied to bR on the C14–C15 bond *(170)*. Using additional chemical shift measurements, the following mechanism is proposed for the conformational changes in retinal: (a) The electrostatic interactions with the protonated SB lead to a distorted retinal conformation, and (b) deprotonation of the SB relaxes the retinal and changes its connectivity from the cytoplasmic to the extracellular side as needed for the proton pumping cycle *(171)*.

Rhodopsin has been expressed in baculovirus/Sf9 cells and in the HEK stable cell line to label with highly ^{15}N-enriched, all [α,ε-^{15}N]-L-Lysines *(16)*. The chemical shift of the SB ^{15}N in [α,ε-^{15}N]-L-Lysine-rhodopsin showed it to be protonated and stabilized by a complex counterion *(16,17)*. This work demonstrates the feasibility for future studies on selectively or uniformly residue-labeled membrane proteins from eukaryotes.

3.3.2.2. Bacteriorhodopsin (bR)

bR is one of the most comprehensively studied proteins largely because of its stability, ease of production with, specifically in NMR studies, selective isotope labeling. It works as a light-driven proton pump of *Haloarchaea* that converts light energy into an electrochemical proton gradient from the cytoplasmic to the extracellular side. This process is driven by photoisomerization of chromophore retinal from all-*trans* to the 13-*cis,* 15-*anti* configuration. This proton transport process is cyclic, and some residues in the proton pathway are protonated/deprotonated as the counter ions. A number of high-resolution X-ray crystal structures of bR have been determined at up to 1.43 Å resolution for the ground state, and of some functional photointermediates *(172–178)*. However, solid-state NMR studies have given complementary information, such as accurate conformational determinations of local active sites, not only for the ground state but also for some other photointermediates trapped by adjusting the temperature, pH, or by using mutants *(179)*.

Because bR forms a 2D hexagonal crystalline lattice arranged in trimeric units in the purple membrane, it orients well on glass plates *(180,181)* (**Subheading 2.6.2.1.**). bR oriented on glass plates preserves the directional quality of membrane protein structure and also maintains the structural and functional integrity of the protein under a wide range of pH, temperature, humidity, or chemical environments *(182)*. Wide-line ^{2}H NMR on oriented purple membrane has been used to determine the orientation of a specifically labeled molecular segment relative to the magnetic field (B_0), which is parallel to the membrane normal when the glass plates are oriented perpendicular to B_0. From the splittings of the quadrupole couplings in the ^2H NMR spectra, the $C–CD_3$ angle with respect to the magnetic field was measured. Specifically, deuterium labeled at C16 and 17 methyls on the cyclohexene ring incorporated into the oriented protein was found to be oriented at angles of $94 \pm 2°$ and $75 \pm 2°$ with respect the membrane normal in the ground state *(183)*. Furthermore, results from deuterated methyl groups at C18, C19, and C20 showed that the $C–CD_3$ bond vectors were oriented at $37 \pm 1°$, $40 \pm 1°$, and $32 \pm 1°$, respectively *(184)*. The M intermediate trapped using guanidium hydrochloride at $-60°C$ showed the angle between the $C–CD_3$ bond and the magnetic field at the C19 methyl group changed to $44 \pm 2°$, which is consistent with a slight upward tilting of the polyene chain. Additional line broadening compared with that of the ground state suggested some two-state heterogeneity *(185)*. ^2H NMR also revealed that the $C9–CD_3$ bond changes by $7°$ from the ground to the M state *(186)*.

MAS has also been successfully applied to the study of purple membrane pellet to determine local conformation around the labeled positions in bR. R^2 was used to determine the distances of $4.1 \pm <0.4$ Å from the C8–C18 positions and $3.3–3.5 \pm <0.4$ Å for the average C8–C16/C8–C17 positions for selectively labeled retinal in the ground state, showing that retinal is in a 6-s-*trans* conformation *(187)*. The same approach was also used to determine the distance between $^{13}C14$ in retinal and the $^{13}C_\varepsilon$-Lys216, which was found to be 4.1 ± 0.3 Å in bR_{568} and 3.9 ± 0.1 Å in the thermally trapped M_{412} state. This demonstrated that the $C = N$ bond is *anti* in these photostates. The distance of 3.0 ± 0.2 Å between these same two groups in bR_{555} showed a change to a $C = N$ *syn* bond in this state *(188,189)*.

For structural studies of bR, one type of amino acid can be labeled selectively. However, because bR has 29 alanines, 21 valines, and 9 methionine residues, and many are located in helices, the MAS NMR signals may be overlapped because of similar environments and conformations. In the case of ^{13}C, the chemical shift is very sensitive to local conformations, and chemical shifts can be assigned in spectral envelopes, rather than individual assignments, to local secondary structures of α-helices, loops, or β-sheets in most cases *(190–192)*. Further assignment for each residue is determined by selective resonance knockout using site-directed mutagenesis in favorable cases and proteolysis to cleave C-terminus residues *(193,194)*, or by comparing with model compounds *(179)*.

^{15}N spectra are also sensitive to conformational changes, although the isotropic chemical shifts are weakly dependent on local secondary structure because the amide nitrogen chemical shift is affected by the neighboring amino acids and existence or strength of hydrogen bonds *(195)*. Recently, assignments have been reported of ^{15}N CP–MAS spectra of [^{15}N]Met-bR. This was achieved by resonance knockouts using mutants and by substitution of easily exchangeable protons to deuterium located in residues from loops and at the surface of the protein *(15)*. Further examination of 2D HETCOR MAS spectra showed some further signal separation in the 1D spectrum *(15)*, whereas PISEMA and ^{1}H–^{15}N HETCOR and MAOSS methods have been applied to uniaxially oriented PMs to determine structural and orientational constraints of some of the Met residues *(15,196)*. The spectra allowed clear distinction of helix and loop resonances (**Fig. 18A,D**), and using SIMPSON *(197)* and SIMMOL *(198)* simulation software the helix signal was deconvoluted to provide information about helix tilt angles and structural constraints. Using a combination of experimental and simulated spectra, it was estimated that the extracellular section of helix B has a tilt of less than 5° from the membrane normal, in agreement with most crystal structures. Furthermore, the experimental solid-state NMR spectra were directly compared with recent X-ray and electron crystal structures, 1C3W *(173)* (**Fig. 18B,E**) and 1FBB *(199)* (**Fig. 18C,F**), respectively, showing that the NMR spectral methods are extremely sensitive to local conformations and crystal differences *(196)*.

Local protein dynamics around the residues in helices or loops can be exquisitely probed using solid-state NMR from measurements of relaxation times (T_1, T_2, T_{C-H} and $T_{1\rho}{}^{H}$), or from suppressed signals caused by the motions that are modulated by proton decoupling or by the MAS frequency *(192)*. Local conformational changes can also be examined under various conditions, such as ionic strength and pH *(193)*, under a range of temperatures *(200)*, and before and after reconstitution of retinal in bacterio-opsin *(201)*. Dynamics can be determined for loop structures that are sometimes missing in crystal structures because of mobility. Using this approach, the location of cation-binding sites in bR were proposed to be at the F–G loop *(192,202)*.

[ζ-^{15}N]Lys-labeled bR is a good system to study the conformation of bR in the photocycle and detect photointermediates because retinal is covalently attached to Lys216. The ^{15}N signals from the SB nitrogen in the L, M_{412}, M_{405}, and N intermediates can be distinguished, as well as the thermal equilibrium of ground states, bR_{568} and bR_{555} (the subscript indicates the wavelength of maximum visible absorbance in nm) *(203,204)*. The strongly shielded chemical shift in the ground state suggests that the SB counterion is extraordinarily delocalized, whereas in the L state, the SB interacts more strongly with its counterion as shown by a downfield shift (approx 130 ppm) of the signal. Finally, the deprotonation of the SB nitrogen in the M state clearly caused a far downfield shift *(179)*. Some aspartic acids in bR play critical roles during the pumping. ^{13}C CP–MAS spectra of the ground state and M intermediate of [4-^{13}C]Asp-bR yielded changes in protonated/deprotonated aspartic acids from the shift positions *(205–207)* and the distance between [4-^{14}C]Asp212 and [14-^{13}C]retinal revealed a displacement by 0.4

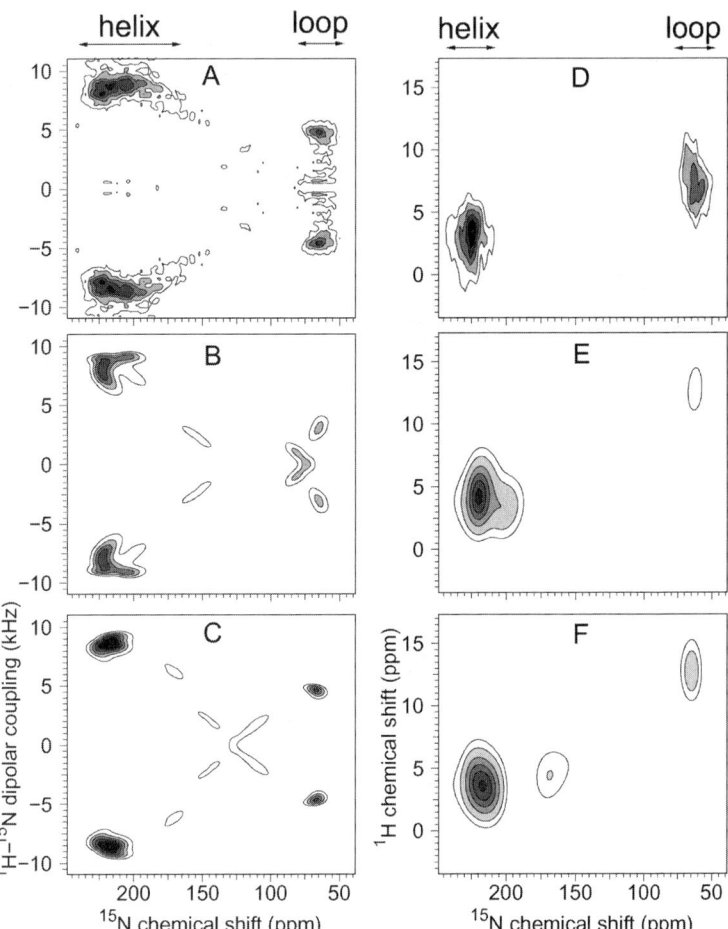

Fig. 18. Experimental PISEMA (**A**) and ^{15}N–^{1}H HETCOR (**D**) spectra of [^{15}N-Met]-bR. Simulated PISEMA (**B**) and simulated HETCOR (**E**) spectra, using SIMP-SON/SIMMOL and structural parameters from the X-ray structure 1C3W.pdb. Simulated PISEMA (**C**) and simulated HETCOR (**F**) spectra using the structural parameters for the membrane-bound protein in 1FBB.pdb. Resonances from helix-and loop-positioned labeled residues are well resolved and show that the NMR spectra are rather sensitive to the local protein structure.

Å at M_{412} state from RFDR experiments, suggesting the existence of bound water between them *(208)*. Use of D85N-bR mutant suggested that the photocycle is accompanied with a conformational change at the C-terminal α-helix *(209)*. The existence of long-range interactions between Asp96 and the extracellular surface through some residues was shown at the resting state from displacement or suppression of the [1–^{13}C]Val signals using site-directed mutants *(210,211)*. Some Trp residues are also present in the retinal-binding pocket. The distances between ^{13}C at the C14 position of retinal and [indole-^{15}N]Trp86 was determined to be 4.2 Å in the 13-*cis*-15-*syn*-retinal form and 3.9 Å in the all-*trans*-15-*anti* conformer as determined by simultaneous frequency and amplitude–modulation rotational echo double resonance (SFAM REDOR) methods (**Subheading 3.2.4.2.**) *(212)*. The dipolar interactions in [20-^{13}C]retinal, [indole-^{15}N]Trp-bR measured by the REDOR method yielded a distance of 3.36 ± 0.2 Å between the C20 retinal and the indole nitrogen of Trp182, which is its closest residue, in the light-adapted state. In the early M state, this distance changes only slightly to 3.16 ± 0.4 Å *(213)*. All X-Pro bonds in bR were shown to be *trans* in the resting state *(214)*, and those showed only slight changes in the early and late M states from the view of distances measured between [1-^{13}C]Val and [^{15}N]Pro *(179,215)*.

3.3.2.3. Sensory Rhodopsin

Recently, selectively ^{13}C-labeled *pharonois* phoborhodopsin (*p*pR) was studied by solid-state NMR spectroscopy *(216)*. Phoborhodopsin (sensory rhodopsin II) belongs to the same family as bR—in the retinylidine proteins—and functions via a common photochemical reaction. The ^{13}C NMR spectra of [3-^{13}C]Ala-, and [1-^{13}C]Val-labeled *p*pR incorporated into egg phosphatidylcholine bilayer showed well-resolved signals from α-helices and loops, with a comparable spectral profile to that of bR *(216)*. Complex formation of *p*pR with a truncated cognate transducer *p*Htr II (1-159), which occurs following light incidence was also investigated. The results show that the complexation leads to changes in conformation and dynamics of *p*pR, in particular at the C-terminal α-helix protruding from the membrane surface *(216)*.

3.3.3. Amyloid Proteins and Membrane Interactions

So far, some 16 proteins are known to form amyloid fibrils in various pathological conditions (neurodegenerative and system amyloidosis) *(217,218)*, including Alzheimer's disease. Although these proteins vary considerably in their primary structure, function, size, and tertiary structure, they all appear to form amyloid fibrils that show very few structural differences. Electron microscopy has shown *(219,220)* that amyloid fibrils are straight, unbranching fibrils of 70–120 Å in diameter and of indeterminate length. Being ordered, relatively immobile, and noncrystalline macromolecular complexes, and because the initial events in fibril formation are membrane mediated, they are ideal systems for solid-state NMR studies.

Alzheimer's disease is characterized by the deposition of insoluble amyloid fibrils as amyloid plaques in the neurophil, as well as the communication of neurofibrillary tangles in cell bodies of neurons *(221)*. Neurofibrillary tangles consist primarily of hyperphosphorylated tau, whereas the major fibrillar component of senile plaques is the amyloid-β peptide (Aβ), a 40–42-amino acid fragment of the Alzheimer precursor protein (APP) *(222)*. The sequence of $A\beta_{1-42}$ is: Asp-Ala-Glu-Phe-Arg-His-Asp-Ser-Gly-Tyr-Glu-Val-His-His-Gln-Lys-Leu-Val-Phe-Phe-Ala-Glu-Asp-Val-Gly-Ser-Asn-Lys-Gly-Ala-Ile-Ile-Gly-Leu-Met-Val-Gly-Gly-Val-Val-Ile-Ala. X-ray diffraction analysis, circular dichroism spectroscopy, and Fourier transform infrared (FTIR) spectroscopy have revealed that amyloid has a generic "cross-β" conformation, composed of multiple copies of the constituent polypeptide arranged into stacked β-sheets running perpendicular to the fibril axis *(223–225)*. These observations suggest an explanation for the ability of different proteins with unique native conformations to adopt an essentially identical structure: the formation of amyloid may not be a peculiarity of the proteins that form fibrils in vivo, but rather a generic property of all polypeptide chains, albeit one which evolution has presumably not favored *(226,227)*.

The proposed mechanism of amyloid fibril formation involves initial membrane-mediated polymerization of the soluble Aβ peptide into intermediates leading finally to amyloid deposition in vivo *(226,228)*. This is supported by studies of Aβ peptide in various membrane-mimicking solvents using solution NMR. The soluble Aβ peptide was shown to have an α-helical conformation *(229–240)*, whereas fibrillogenesis of Aβ has been shown to be linked to a conversion to β-sheet structure. Residues 11–16 are thought to be important in forming electrostatic interactions, and residues 17–21 are important in hydrophobic interactions. A short peptide composed of residues 14–23 is capable of forming amyloid fibrils in vitro, and the region 11–24 is thought to play an important role in α to β conversion. Residues 25–35 have been implicated in neurotoxicity, and residues 34–42 are very hydrophobic and are thought to be situated in the TM region in APP. Peptides consisting of residues in this region are extremely insoluble and have been implicated in nucleating amyloid fibril formation.

Two types of structural information are of interest, namely, the molecular conformations of peptides and proteins in amyloid fibrils, as well as the supramolecular organization of the fibrils, that is, intermolecular interactions and packing. Understanding amyloid fibril formation, mediated by membrane interactions, will increase our knowledge about protein structure, folding, and stability. However, the insoluble fibrils that are finally formed by several other proteins limit the capabilities for structural studies using several biophysical techniques, such as X-ray diffraction and solution NMR. Therefore, the use of solid-state NMR and FTIR is becoming important in examining amyloid fibril structure and could yield considerable information about the morphology of amyloid fibrils and their internal structural conformation. Thus, several solid-state NMR approaches have been applied for structural elucidation of the fibrils, including dipolar recoupling with a windowless sequence *(241)*, REDOR *(93,106,242)*, R² *(69,243)*, spin echo *(244)*, and MQ ¹³C NMR *(245–248)*. These methods can be used to measure interactions over relatively short ranges (approx 6 Å) and therefore provide information about the local interactions within the β-sheets. Regions that are of interest include the extracellular domain *(249)* and the TMD *(114,250–253)*.

4. Appendix

4.1. Salt-Tolerant Probes

Most conventional solid-state NMR is carried out on dry materials, where filling factor and susceptibility are not a problem. However, biological material requires water and, in certain cases, salt buffers to be functionally competent, and even crystalline material has a high trapped water content. Commercial probes therefore need to be ordered so that they not only tune but also can take the required power levels with wet material (most of a sample will be water, even though it is centrifuged in a rotor). No commercial probe should be accepted until fully checked under "real" conditions.

4.2. Rotor Packing

An important practical concern in biological MAS NMR is the transfer of fragile and often expensive samples in the NMR rotor. This can be done by centrifuging into the rotor itself, usually repeatedly. For membranes that have a high protein content and saturated lipid, bench centrifugation may be sufficient. In samples with lower protein content or with unsaturated lipids, higher speed centrifugation may be necessary (the relative partial specific volume of the complex to the buffer density is crucial in determining the conditions). Rotors can, of course, withstand centrifugation, and samples can be removed by centrifuging the rotor upside down in a centrifuge tube for sample recovery or for manipulation.

4.3. Local Molecular Dynamics

CP–MAS NMR methods are widely used for protein structural studies. However, because of the differences in "rigidity" (mobility) in different motifs of the protein, cross-polarization efficiency is not always the same for the residues in, for example, a TM helix or in a loop. Furthermore, the frequency of the molecular motion can interfere with proton decoupling frequency (around 5×10^4 Hz *(254,255)* or the MAS frequency (approx $10^3–10^4$ Hz *(255)*, thereby suppressing signal intensities. In many cases it is difficult to observe signals in loops compared with helices in the CP–MAS spectra.

Some cross-polarization methods, such as ramped CP (**Subheading 3.2.2.**), can be applied to achieve a wider range of cross-polarization efficiency than normal CP. Recently, the $^1H–^{15}N$ HETCOR MAS approach was used to permit observation of signals from residues in loops as well as in helices (15). Temperature variation is another way of manipulating molecular motion, although conformational heterogeneity by sample freezing sometimes causes signal broadening. On the other hand, the suppressed signals sometimes can be exploited to give information about the differential molecular motion of the protein, by comparing spectra measured at different temperatures or by comparing the spectra obtained with direct excitation and a decoupling during the acquisition *(192)*.

In static samples, signals from both helices and loops can be observed at reasonable signal intensities *(15,196)*. From the 1D CP spectrum, local motions can be analyzed separately for the loops and helices, although some signals may overlap. However, in the 2D PISEMA and HETCOR spectra, signals from different structural elements can be resolved, better permitting some analysis of structures and differential motions.

4.4. REDOR Settings

Achieving full dephasing is of practical importance for a reliable interpretation of the REDOR results in terms of distance. Experimentally, issues such as accurate adjustment of the spinning speed or concentrating the sample in the center of the rotor to minimize the effect of rf inhomogeneity are important to avoid incomplete dephasing and an overestimation of the determined distance *(95,256)*.

4.5. Which Static NMR Method to Use—PISEMA or HETCOR?

Numerical simulations for labeled membrane proteins in oriented bilayers *(138,257)* show that PISEMA provides good signal separation for TM helices with modest tilt angles relative to the bilayer normal, whereas HETCOR provides good signal separation for tilt angles in the 60–90° regime being typical for in-plane structure or loop regions. In the case of the oriented membrane sample on the glass plates (*see* **Subheading 2.6.2.1.**), the sample is normally placed in a flat-coil NMR probe perpendicular to the external magnetic field (B_0), such that the bilayer normal is parallel to B_0. Parallel application of the two types of 2D static NMR methods potentially gives more complete structural and orientational information of helices and loops of membrane proteins in the oriented samples.

4.6. Sensitivity and Resolution in PISEMA Experiments

When performing a PISEMA experiment on membrane peptides and proteins, numerous factors need to be taken into account to achieve a better sensitivity and resolution and to minimize artifacts. The first is rapid phase switching between the +LG and −LG steps *(130)*. If switching is too slow, then phase transients will arise, which in turn will result in additional dipolar splitting frequencies appearing in the spectrum. This can be remedied by either modifying the length of the Lee–Goldburg pulses such that a $[(2\pi-\pi)_{x+LG}-(2\pi)_{-x-LG}]_n$ is used *(130)*, where $n = 0, 1, 2, \ldots t_{1max}$. Alternatively, the pulses can be cycled such that a block consists of the pulses *(258)* $[(2\pi)_{x+LG}(2\pi)_{-x-LG}(2\pi)_{-x-LG}(2\pi)_{x+LG}]_n$, where n is as defined above. Another important experimental factor is the exact matching of the rf fields used for +LG and −LG. If a mismatch occurs between the two pulses, then homonuclear decoupling becomes inefficient. This manifests itself experimentally as a broadening of the lines and a less than optimal scaling factor. In addition, a good match of the HH condition at the magic angle should be achieved. This can be performed by ramping the amplitude of the *S* spin lock, as recently described by R. Fu and coworkers *(259)*, and results in removal of the center artifact. Finally, as was demonstrated by Z. Gan *(144)*, the resonance offset should be set carefully (**Eq. 32**) to permit the measured dipolar splitting to be accurately determined. In HETCOR, as with PISEMA, the ^{1}H chemical shift evolves under FSLG. Other types of correlations and analyses, in terms of sensitivity and the global structural information that can be extracted, have been discussed elsewhere from simulations *(257)*. Recently, a new pulse sequence has been developed that eliminates the need for the FSLG component of the PISEMA pulse sequence, thus making it easier to implement on spectrometers *(260)*.

4.7. Peptide-Labeling Schemes

Using R^2 determinations of spin pairs within peptides it is possible to decide on various secondary structural motifs, assuming no interhelical interactions. Computer modeling studies of the TM peptide sequences identify suitable spin-pair labeling sites in a peptide that could be able to distinguish between the different secondary structure conformations adopted by a peptide using R^2 distance measurements (**Subheading 3.2.4.1.**). Distance measurements have been obtained on Silicon Graphics workstations using the *Insight® II* software program (from Biosym/MSI of San Diego, CA) to illustrate this principle. The distances measured between two carbon labels (**Fig. 19**) placed i and $i + n$

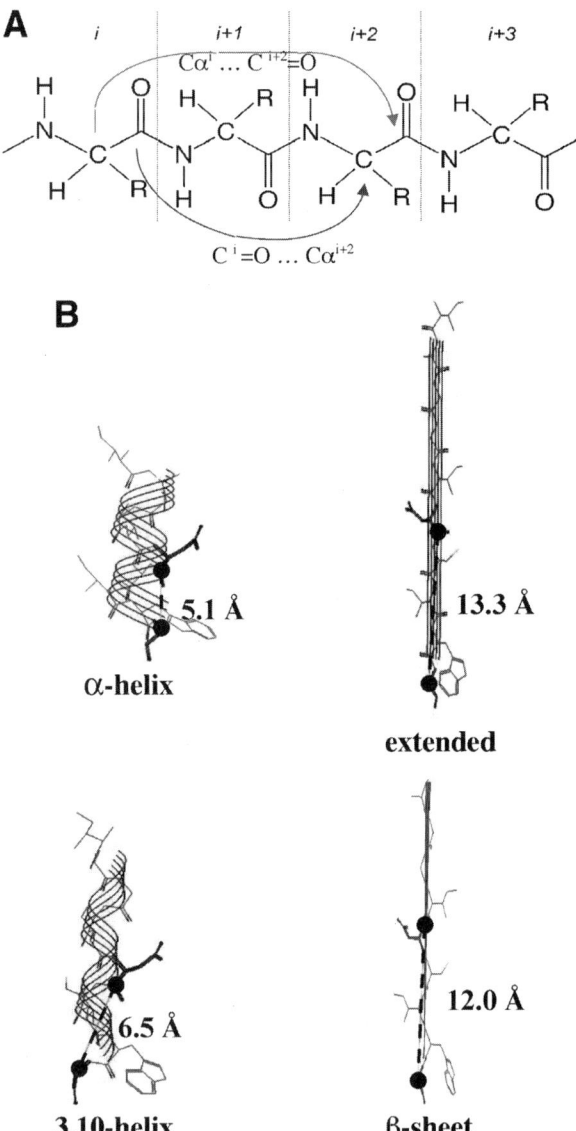

Fig. 19. Possible ^{13}C-isotope labeling schemes for a polypeptide chain from Cα to C = O, and *vice versa* from residue *i* to residue *i* + *n* (*n* = 2 here) to show convention used (**A**). As an example, distances from C = O to Cα, shown here from residue *i* to (*i* + 4) (*see* **Table 1** for other possibilities) for ideal rigid secondary structures in a generic protein for various motifs are shown (**B**). With suitable labeling schemes, differentiation between various secondary structures can be made by determining distances from the dipolar coupling range in an R^2 experiment (up to approx 6.5 Å) for ^{13}C when residues are close enough. If magnetization transfer does not occur, then extended structures would be assumed. Similar arguments can be made for REDOR approaches with ^{13}C and ^{15}N but with reduced range.

Table 1
Distances Calculated for Two Carbon Backbone Atoms Placed *i* and *i* + *n* Residues Apart in a Peptide, and for Different Secondary Structures

Secondary structure		Distance between residues *i* and *i*+1/Å	Distance between residues *i* and *i*+2/Å	Distance between residues *i* and *i*+3/Å	Distance between residues *i* and *i*+4/Å	Distance between residues *i* and *i*+5/Å
Extended	CO to Cα	*2.4*	*6.1*	9.7	13.3	17.0
	Cα to CO	*4.4*	8.5	12.1	15.7	19.3
β-sheet	CO to Cα	*2.4*	*5.4*	8.8	12.0	15.3
	Cα to CO	*4.4*	7.9	11.2	14.5	17.8
α-helix	CO to Cα	*2.4*	*4.5*	*4.5*	*5.1*	7.4
	Cα to CO	*4.7*	*5.5*	*5.8*	7.5	9.6
3,10-helix	CO to Cα	*2.4*	*4.3*	*4.6*	*6.3*	8.6
	Cα to CO	*4.8*	*5.6*	*6.5*	9.8	10.6

Distances in bold italics show values accessible by R^2 NMR methods.

apart in the peptide backbone (from the N-terminus of the peptide to the C-terminus), and which adopt different secondary structure conformations, are shown in **Table 1**.

It is an advantage to be able to characterize the local secondary structure of a longer peptide sequence rather than a shorter sequence, although the internuclear distances between the labels must be within the upper limits of R^2 (< 6.5 Å *(73)*). Short internuclear distances between CO to Cα (*i* and *i* + 1) and (*i* and *i* + 2), have internuclear distances that are very similar and are within the limits of R^2 for all secondary structure motifs. This makes it difficult to distinguish between the different secondary structure types. For CO to Cα atoms that are (*i* and *i* + 3) residues apart, the α-helix and 3,10-helix conformations are distinguishable from the extended and β-sheet conformations, although the distances are very similar for the α-helix and 3,10-helix secondary structure types. Increasing the CO to Cα distance for (*i* and *i* + 4)-labeled peptides provides a means of differentiating the helix secondary structure types from the extended and β-sheet conformations, and the internuclear distances are sufficiently different for the α-helix and 3,10-helix. Increasing the labels further to (*i* and *i* + 5) results in internuclear distances that are too long to be determined by R^2 experiments. It can be seen that the (*i* and *i* + 4)-labeling scheme is most discriminating because all the different secondary structure conformations are within the limit of R^2 (<6.5 Å).

Acknowledgments

The Medical Research Council (UK), BBSRC, CLRC, and EPSRC are thanked for their support for grants to A. Watts, and HEFCE, Magnex, and

Varian for equipment support. S. L. Grage is supported by an Emmy-Noether Fellowship from the DFG. S. K. Straus thanks the Royal Society of London for a Dorothy Hodgkin Research Fellowship.

References

1. Arkin, I. T., Brünger, A. T., and Engelman, D. M. (1997) Are there dominant membrane protein families with a given number of helices? *Proteins* **28,** 465–466.
2. Terstappen, G. C. and Reggiani, A. (2001) *In silico* research in drug discovery. *Trends Pharmacol. Sci.* **22,** 23–26.
3. Pebay-Peyroula, E., Rummel, G., Rosenbusch, J. P., and Landau, E. M. (1997) X-ray structure of bacteriorhodopsin at 2.5 angstroms from microcrystals grown in lipidic cubic phases. *Science* **277,** 1676–1681.
4. Palczewski, K., Kumasaka, T., Hori, T., Behnke, C. A., Motoshima, H., Fox, B. A., et al. (2000) Crystal structure of rhodopsin: a G protein-coupled receptor. *Science* **277,** 687–690.
5. Wüthrich, K. (1998) The second decade—into the third millenium. *Nat. Struct. Biol.* **5,** 492–495.
6. Davis, J. H., and Auger, M. (1999) Static and magic angle spinning NMR of membrane peptides and proteins. *Prog. NMR Spectrosc.* **35,** 1–84.
7. Griffin, R. G. (1998) Dipolar recoupling in MAS spectra of biological solids. *Nat. Struct. Biol.* **5(Suppl.),** 508–512.
8. De Groot, H. J. M. (2000) Solid-state NMR spectroscopy applied to membrane proteins. *Curr. Opin. Biotechnol.* **10,** 593–600.
9. Smith, S. O., Aschheim, K., and Groesbeek, M. (1996) Magic angle spinning NMR spectroscopy of membrane proteins. *Q. Rev. Biophys.* **29,** 395–449.
10. Glaubitz, C., and Watts, A. (1998) Magic angle-oriented sample spinning (MAOSS): a new approach toward biomembrane studies. *J. Magn. Reson.* **130,** 305–316.
11. Goto, N. K., and Kay, L. E. (2000) New developments in isotope labeling strategies for protein solution NMR spectroscopy. *Curr. Opin. Struct. Biol.* **10,** 585–592.
12. Opella, S. J., Ma, C., and Marassi, F. M. (2001) Nuclear magnetic resonance of membrane-associated peptides and proteins. *Methods Enzymol.* **339,** 285–313.
13. Marassi, F. M., Ramamoorthy, A., and Opella, S. J. (1997) Complete resolution of the solid-state NMR spectrum of a uniformly ^{15}N-labeled membrane protein in phospholipid bilayers. *Proc. Natl. Acad. Sci. USA* **94,** 8551–8556.
14. Oesterhelt, D., and Stoeckenius, W. (1974) Isolation of the cell membrane of *Halobacterium halobium* and its fractionation into red and purple membrane. *Methods Enzymol.* **31,** 667–678.
15. Mason, A. J., Grage, S. L., Glaubitz, C., Strauss, S. K., and Watts, A. (2004) Identifying anisotropic constraints in multiply labeled membrane proteins by ^{15}N MAS NMR. *Biophys. J.* **86,** 1610–1617.
16. Creemers, A. F. L., Klaassen, C. H. W., Bovee-Geurts, P. H. M., Kelle, R., Kragl, U., Raap, J., et al. (1999) Solid state ^{15}N NMR evidence for a complex Schiff base counterion in the visual G-protein-coupled receptor rhodopsin. *Biochemistry* **38,** 7195–7199.

17. Eilers, M., Reeves, P. J., Ying, W., Gobind Khorana, H. G., and Smith, S. O. (1999) Magic angle spinning NMR of the protonated retinylidene Schiff base nitrogen in rhodopsin: expression of ^{15}N-lysine- and ^{13}C-glycine-labeled opsin in a stable cell line. *Proc. Natl. Acad. Sci. USA* **93,** 487–492.

18. Seelig, J. (1970) Spin label studies of oriented smectic liquid crystals: a model system for bilayer membranes. *J. Am. Chem. Soc.* **92,** 3881–3887.

19. Smith, I. C. P. (1971) A spin label study of the organization and fluidity of hydrated phospholipid multilayers: a model membrane system. *Chimia* **25,** 349–380.

20. Powers, L. and Clark, N. A. (1975) Preparation of large monodomain phospholipid bilayer smectic liquid crystals. *Proc. Natl. Acad. Sci. USA* **72,** 840–843.

21. Clark, N. A., Rothschild, K. J., Luippold, D. A., and Simon, B. A. (1980) Surface-induced lamellar orientation of multilayer membrane arrays: theoretical analysis and a new method with application to purple membrane fragments. *Biophys. J.* **31,** 65–96.

22. Rothgeb, T. M., and Oldfield, E. (1981) Nitrogen-14 nuclear magnetic resonance spectroscopy as a probe of lipid bilayer headgroup structure. *J. Biol. Chem.* **256,** 6004–6009.

23. Macnaughtan, W., Snook, K. A., Capsi, E., and Franks, N. P. (1985) An X-ray diffraction analysis of oriented lipid multilayers containing basic proteins. *Biochim. Biophys. Acta* **818,** 132–148.

24. Tamm, L. K., and Mcconnell, H. M. (1985) Supported phospholipid bilayers. *Biophys. J.* **47,** 105–113.

25. Marassi, F. M. (2002) NMR of peptides and proteins in oriented membranes. *Conc. Magn. Reson.* **14,** 212–224.

26. Sizun, C., and Bechinger, B. (2002) Bilayer sample for fast or slow magic angle oriented sample spinning solid state NMR spectroscopy. *J. Am. Chem. Soc.* **124,** 1146–1147.

27. Gröbner, G., Burnett, I. J., Glaubitz, C., Chol, G., Mason, A. J., and Watts, A. (2000) Observation of light-induced structural changes of retinal within rhodopsin. *Nature* **405,** 810–813.

28. Glaubitz, C., Gröbner, G., and Watts, A. (2000) Structural and orientational information of the membrane embedded M13 coat protein by 13C-MAS NMR spectroscopy. *Biochim. Biophys. Acta* **1463,** 151–161.

29. Grobner, G., Taylor, A., Williamson, P. T., Choi, G., Glaubitz, C., Watts, J. A., et al. (1997) Macroscopic orientation of natural and model membranes for structural studies. *Anal. Biochem.* **254,** 132–138.

30. Marassi, F. M., and Crowell, K. J. (2003) Hydration-optimized oriented phospholipid bilayer samples for solid-state NMR structural studies of membrane proteins. *J. Magn. Reson.* **161,** 64–69.

31. Hallock, K. J., Henzler-Wildman, K., Lee, D. K., and Ramamoorthy, A. (2002) An innovative procedure using a sublimable solid to align lipid bilayers for solid state NMR studies. *Biophys. J.* **82,** 2499–2503.

32. Moll, F., and Cross, T. A. (1990) Optimizing and characterizing alignment of oriented lipid bilayers containing gramicidin D. *Biophys. J.* **57,** 351–362.

33. Prosser, R. S., and Shiyanovskaya, I. V. (2001) Lanthanide ion assisted magnetic alignment of model membranes and macromolecules. *Conc. Magn. Reson.* **13,** 19–31.

34. Sanders, C. R., and Prosser, R. S. (1998) Bicelles: a model membrane system for all seasons? *Structure* **6,** 1227–1234.

35. Sanders, C. R., Hare, B. J., Howard, K. P., and Prestegard, J. H. (1994) Magnetically-oriented phospholipid micelles as a tool for the study of membrane-associated molecules. *Prog. NMR Spectrosc.* **26,** 421–444.

36. Sanders, C. R., and Landis, G. C. (1995) Reconstitution of membrane-proteins into lipid-rich bilayered mixed micelles for NMR-studies. *Biochemistry* **34,** 4030–4040.

37. Ram, P., and Prestegard, J. H. (1988) Magnetic-field induced ordering of bile-salt phospholipid micelles—new media for NMR structure investigations. *Biochim. Biophys. Acta* **940,** 289–294.

38. Marcotte, I., Wegener, K. L., Lam, Y. H., Chia, B. C. S., Planque, M. R. R. D., Bowie, J.H., et al. (2003) Interaction of antimicrobial peptides from Australian amphibians with lipid membranes. *Chem. Phys. Lett.* **122,** 107–120.

39. Whiles, J. A., Glover, K. J., Vold, R. R., and Komives, E. A. (2002) Methods for studying transmembrane peptides in bicelles: consequences of hydrophobic mismatch and peptide sequence. *J. Magn. Reson.* **158,** 149–156.

40. Whiles, J. A., Deems, R., Vold, R. R., and Dennis, E. A. (2002) Bicelles in structure-function studies of membrane-associated proteins. *Bioorg. Chem.* **30,** 431–442.

41. Prosser, R. S., Volkov, V. B., and Shiyanovskaya, I. V. (1998) Novel chelate-induced magnetic alignment of biological membranes. *Biophys. J.* **75,** 2163–2169.

42. Prosser, R. S., Bryant, H., Bryant, R. G., and Vold, R. R. (1999) Lanthanide chelates as bilayer alignment tools in NMR studies of membrane-associated peptides. *J. Magn. Reson.* **141,** 256–260.

43. Prosser, R. S., Hwang, J. S., and Vold, R. R. (1998) Magnetically aligned phospholipid bilayers with positive ordering: a new model membrane system. *Biophys. J.* **74,** 2405–2418.

44. Nieh, M.-P., Glinka, C.-J., Krueger, S., Prosser, R. S., and Katsaras, J. (2002) SANS study on the effect of lanthanide ions and charged lipids on the morphology of phospholipid mixtures. *Biophys. J.* **82,** 2487–2498.

45. Nieh, M. P., Glinka, C. J., Krueger, S., Prosser, R. S., and Katsaras, J. (2001) SANS study of the structural phases of magnetically alignable lanthanide-doped phospholipid mixtures. *Langmuir* **17,** 2629–2638.

46. Zandomeneghi, G., Tomaselli, M., Williamson, P. T. F., and Meier, B. H. (2003) NMR of bicelles: orientation and mosaic spread of the liquid-crystal director under sample rotation. *J. Biomol. NMR* **25,** 113–123.

47. Carlotti, C., Aussenac, F., and Dufourc, E. J. (2002) Towards high-resolution 1H-NMR in biological membranes: magic angle spinning of bicelles. *Biochim. Biophys. Acta* **1564,** 156–164.

48. Zandomeneghi, G., Williamson, P. T. F., Hunkeler, A., and Meier, B. H. (2003) Switched-angle spinning applied to bicelles containing phospholipid-associated peptides. *J. Biomol. NMR* **25,** 125–132.

49. Baldus, M. (2002) Correlation experiments for assignment and structure elucidation of immobilized polypeptides under magic angle spinning. *Prog. NMR Spectrosc.* **41,** 1–47.

50. Straus, S. K., Bremi, T., and Ernst, R. R. (1998) Experiments and strategies for the assignment of fully $^{13}C/^{15}N$-labelled polypeptides by solid state NMR. *J. Biomol. NMR* **12,** 39–40.

51. Andrew, E. R., Bradbury, A., and Eades, R. G. (1959) Removal of dipolar broadening of nuclear magnetic resonance spectra of solids by specimen rotation. *Nature* **183,** 1802–1803.

52. Lowe, I. J. (1959) Free induction decay of rotating solids. *Phys. Rev. Lett.* **2,** 285–287.

53. Creemers, A. F., Kiihne, S., Bovee-Geurts, P. H., Degrip, W. J., Lugtenburg, J., and De Groot, H. J. (2002) (1)H and (13)C MAS NMR evidence for pronounced ligand-protein interactions involving the ionone ring of the retinylidene chromophore in rhodopsin. *Proc. Natl. Acad. Sci. USA* **99,** 9101–9106.

54. Yao, X. L., and Hong, M. (2001) Dipolar filtered ^1H-^{13}C heteronuclear correlation spectroscopy for resonance assignment of proteins. *J. Biomol. NMR* **20,** 263–274.

55. Yao, X. L., Schmidt-Rohr, K., and Hong, M. (2001) Medium- and long-distance ^1H-^{13}C heteronuclear correlation NMR in solids. *J. Magn. Reson.* **149,** 139–143.

56. Mehring, M. (1983) Principles of high resolution NMR in solids. In *NMR: Basic Principles and Progress, vol. II* (Fluck, E., Diehl, P., and Kosfeld, R., eds.). Springer, New York.

57. Bennett, A. E., Rienstra, C. M., Auger, M., Lakshmi, K. V., and Griffin, R. G. (1995) Heteronuclear decoupling in rotating solids. *J. Chem. Phys.* **103,** 6951–6958.

58. Detken, A., Hardy, E. H., Ernst, M., and Mcier, B. H. (2002) Simple and efficient decoupling in magic-angle spinning solid-state NMR: the XiX scheme. *Chem. Phys. Lett.* **356,** 298–304.

59. Carravetta, M., Eden, M., Zhao, X., Brinkmann, A., and Levitt, M. H. (1000) Symmetry principles for the design of radiofrequency pulse sequences in the nuclear magnetic resonance of rotating solids. *Chem. Phys. Lett.* **321,** 205–215.

60. Eden, M., and Levitt, M. H. (1999) Pulse sequence symmetries in the nuclear magnetic resonance of spinning solids: application to heteronuclear decoupling. *J. Chem. Phys.* **111,** 1511–1519.

61. Pines, A., Gibby, M. G., and Waugh, J. S. (1973) Protein-enhanced NMR of dilute spins in solids. *J. Chem. Phys.* **59,** 569–590.

62. Hartmann, S., and Hahn, E. L. (1962) Nuclear double resonance in the rotating frame. *Phys. Rev.* **128**, 2042–2053.
63. Peersen, O. B., and Smith, S. O. (1993) Rotational resonance NMR of biological membranes. *Conc. Magn. Reson.* **5**, 303–317.
64. Metz, G., Wu, X., and Smith, S. O. (1994) Ramped-amplitude cross polarization in magic-angle-spinning NMR. *J. Magn. Reson.* **110**, 219–227.
65. Hediger, S., Meier, B. H., Kurur, N. D., Bodenhausen, G., and Ernst, R. R. (1994) NMR cross polarization by adiabatic passage through the Hartmann-Hahn condition (APHH). *Chem. Phys. Lett.* **223**, 283–288.
66. Dusold, S., and Sebald, A. (2000) Double-quantum filtration under rotational-resonance conditions: numerical simulations and experimental results. *J. Magn. Reson.* **145**, 340–356.
67. Bennett, A. E., Becerra, L. R., and Griffin, R. G. (1994) Frequency-selective heteronuclear recoupling in rotating solids. *J. Chem. Phys.* **100**, 812–814.
68. Levitt, M. H., Raleigh, D. P., Creuzet, F., and Griffin, R. G. (1990) Theory and simulations of homonuclear spin pair systems in rotating solids. *J. Chem. Phys.* **92**, 6347–6364.
69. Raleigh, D. P., Levitt, M. H., and Griffin, R. G. (1988) Rotational resonance in solid state NMR. *Chem. Phys. Lett.* **146**, 71–76.
70. Colombo, M. G., Meier, B. H., and Ernst, R. R. (1988) Rotor-driven spin diffusion in natural abundance 13C spin systems. *Chem. Phys. Lett.* **146**, 189–196.
71. Maas, W. E. J. R., and Veeman, W. S. (1988) Natural abundance ^{13}C spin diffusion enhanced by magic angle spinning. *Chem. Phys. Lipids* **149**, 170–174.
72. Kubo, Y., Miyashita, T., and Murata, Y. (1998) Structural basis for a Ca^{2+}-sensing function of the metabotropic glutamate receptors. *Science* **279**, 1722–1725.
73. Peersen, O. B., Groesbeek, M., Aimoto, S., and Smith, S. (1995) Analysis of rotational resonance magnetization exchange curves from crystalline peptides. *J. Am. Chem. Soc.* **117**, 7228–7237.
74. Karlsson, T., and Levitt, M. H. (1998) Longitudinal rotational resonance echoes in solid state nuclear magnetic resonance: investigation of zero quantum spin dynamics. *J. Chem. Phys.* **109**, 5493–5507.
75. Caravatti, P., Bodenhausen, G., and Ernst, R. R. (1983) Selective pulse experiments in high-resolution solid state NMR. *J. Magn. Reson.* **55**, 88–103.
76. Bodenhausen, G., Freeman, R., and Morros, G. A. (1976) A simple pulse sequence for selective excitation in Fourier transform NMR. *J. Magn. Reson.* **23**, 171–175.
77. Goobes, G., and Vega, S. (2002) MAS NMR structure refinement of uniformly ^{13}C enriched chlorophyll a/water aggregates with 2D dipolar correlation spectroscopy. *J. Magn. Reson.* **154**, 236–251.
78. Goobes, G., Boender, G. J., and Vega, S. (2000) Spinning-frequency-dependent narrowband rf-driven dipolar recoupling. *J. Magn. Reson.* **146**, 204–219.
79. Costa, P. R., Sun, B., and Griffin, R. G. (1997) Rotational resonance tickling: accurate internuclear distance measurements in solids. *J. Am. Chem. Soc.* **119**, 10,821–10,830.

80. Bennett, A. E., Weliky, D. P., and Tycko, R. (1998) Quantitative conformational measurements in solid state NMR by constant-time homonuclear dipolar recoupling. *J. Am. Chem. Soc.* **120,** 4897–4898.

81. Weintraub, O., Vega, S., Hoelger, C., and Limbach, H. H. (1994) Distance measurements between homonuclear spins in rotating solids. *J. Magn. Reson.* **109,** 14–25.

82. Balazs, Y. S., and Thompson, L. K. (1999) Practical methods for solid-state NMR distance measurements on large biomolecules constant-time rotational resonance. *J. Magn. Reson.* **139,** 371–376.

83. Karlsson, T., Edén, M., Luthman, H., and Levitt, M. H. (2000) Efficient double-quantum excitation in rotational resonance NMR. *J. Magn. Reson.* **145,** 95–107.

84. Nielsen, N. C., Creuzet, F., Griffin, R. G., and Levitt, M. H. (1992) Enhanced double-quantum nuclear magnetic resonance in spinning solids at rotational resonance. *J. Chem. Phys.* **96,** 5668–5677.

85. Feng, X., Verdegem, P. J. E., Lee, Y. K., Helmle, M., Shekar, S.C., De Groot, H. J. M., et al. (1999) Rotational resonance NMR of ^{13}C2-labelled retinal quantitative internuclear distance determination. *Solid State Nucl. Magn. Reson.* **14,** 81–90.

86. Nomura, K., Takegoshi, K., Terao, T., Uchida, K., and Kainosho, M. (1999) Determination of the complete structure of a uniformly labeled molecule by rotational resonance solid-state NMR in the tilted rotating frame. *J. Am. Chem. Soc.* **121,** 4064, 4065.

87. Lam, Y.-H., Wassall, S. R., Morton, C. J., Smith, R., and Separovic, F. (2001) Solid-state NMR structure determination of melittin in a lipid environment. *Biophys. J.* **81,** 2752–2761.

88. Lam, Y.-H., Morton, C. J., and Separovic, F. (2002) Solid-state NMR conformational studies of a melittin-inhibitor complex. *Eur. Biophys. J.* **31,** 383–388.

89. Ahmed, Z., Reid, D. G., Watts, A., and Middleton, D. A. (2000) A solid-state NMR study of the phospholamban transmembrane domain: local structure and interactions with Ca(2+)-ATPase. *Biochim. Biophys. Acta* **1468,** 187–198.

90. Middleton, D. A., Robins, R., Feng, X., Levitt, M., Spiers, I. D., Schwalbe, C., et al. (1997) The conformation of an inhibitor bound to the gastric proton pump. *FEBS Lett.* **410,** 269–274.

91. Middleton, D. A., Rankin, S., Esmann, M., and Watts, A. (2000) Structural insights into the binding of cardiac glycosides to the digitalis receptor revealed by solid-state NMR. *Proc. Natl. Acad. Sci. USA* **97,** 13,602–13,607.

92. Smith, S. O., Peersen, O. B., Yoshimura, S., and Aimoto, S. (1995) Determination of peptide structure in membranes by rotational resonance NMR. *Pept. Chem.* **32,** 109–112.

93. Gullion, T., and Schaefer, J. (1989) Rotational-echo double resonance NMR. *J. Magn. Reson.* **81,** 196–200.

94. Gullion, T., and Schaefer, J. (1989) Detection of weak heteronuclear dipolar coupling by rotational-echo double-resonance nuclear magnetic resonance. *Adv. Nucl. Magn. Reson.* **13,** 57–83.

95. Gullion, T. (1998) Introduction to rotational-echo, double-resonance NMR. *Conc. Magn. Reson.* **10,** 277–289.

96. Goetz, J., Poliks, B., Studelska, D., Fischer, M., Kugelbrey, K., Bacher, A., et al. (1999) Investigation of the binding of fluoroluminazes to the 1-MDa capsid of luminaze synthase by ^{15}N{^{19}F} REDOR NMR. *J. Am. Chem. Soc.* **121,** 7500–7508.

97. Gullion, T., Baker, D. B., and Conradi, M. S. (1990) New, compensated Carr-Purcell sequences. *J. Magn. Reson.* **89,** 479–484.

98. Sack, I., Goldbourt, A., Vega, S., and Buntkowsky, G. (1999) Deuterium REDOR: principles and applications for distance measurements. *J. Magn. Reson.* **138,** 54–65.

99. Merritt, M. E., Goetz, J. M., Whitney, D., Chang, C.P., Heux, L., Halary, J. L., et al. (1998) Location of the antiplasticizer in cross-linked epoxy resins by ^{2}H, ^{15}N and ^{13}C REDOR NMR. *Macromolecules* **31,** 1214–1220.

100. Gullion, T. (2000) Measuring ^{13}C-^{2}D dipolar couplings with a universal REDOR dephasing curve. *J. Magn. Reson.* **146,** 220–222.

101. Schmidt, A., Mckay, R. A., and Schaefer, J. (1992) Internuclear distance measurement between deuterium (I = 1) and a spin-1/2 nucleus in rotating solids. *J. Magn. Reson.* **96,** 644.

102. Gullion, T. (1995) Measurement of dipolar interactions between spin-1/2 and quadrupolar nuclei by rotational-echo, adiabatic-passage, double-resonance NMR. *Chem. Phys. Lett.* **246,** 325–330.

103. Grey, C. P., Veeman, W. S., and Vega, A. J. (1993) Rotational echo ^{14}N/^{13}C/^{1}H triple resonance solid state magnetic resonance: a probe of ^{13}C-^{14}N internuclear distances. *J. Chem. Phys.* **98,** 7711–7724.

104. Fu, R., Smith, S. A., and Bodenhausen, G. (1997) Recoupling of heteronuclear dipolar interactions in solid state magic-angle spinning NMR by simultaneous frequency and amplitude modulation. *Chem. Phys. Lett.* **272,** 361–369.

105. Hing, A. W., Vega, S., and Schaefer, J. (1992) Transferred-echo double-resonance NMR. *J. Magn. Reson.* **96,** 205–209.

106. Goetz, J. M., and Schaefer, J. (1997) REDOR dephasing by multiple spins in the presence of molecular motion. *J. Magn. Reson.* **127,** 147–154.

107. Mueller, K. T., Jarvie, T. P., Aurentz, D. J., and Roberts, B. W. (1995) The REDOR transform: direct calculation of internuclear couplings from dipolar-dephasing NMR data. *Chem. Phys. Lett.* **242,** 535–542.

108. Gullion, T., and Pennington, C. H. (1998) Theta-REDOR: an MAS NMR method to simplify coupled heteronuclear spin systems. *Chem. Phys. Lett.* **290,** 88–93.

109. Jaroniec, C. P., Tounge, B. A., Rienstra, C. M., Herzfeld, J., and Griffin, R. G. (1999) Measurement of ^{13}C-^{15}N distances in uniformly ^{13}C labeled biomolecules: J-decoupled REDOR. *J. Am. Chem. Soc.* **121,** 10,237–10,238.

110. Goetz, J. M., and Schaefer, J. (1997) Orientational information in solids from REDOR sidebands. *J. Magn. Reson.* **129,** 222–223.

111. Weliky, D. P., Dabbagh, G., and Tycko, R. (1993) Correlation of chemical-bond directions and functional-group orientations in solids by 2-dimensional NMR. *J. Magn. Reson.* **104,** 10–16.

112. Dabbagh, G., Weliky, D. P., and Tycko, R. (1994) Determination of monomer conformations in noncrystalline solid polymers by 2-dimensional NMR exchange spectroscopy. *Macromolecules* **27,** 6183–6191.

113. Weliky, D. P., and Tycko, R. (1996) Determination of peptide conformations by two-dimensional magic angle spinning NMR exchange spectroscopy with rotor synchronization. *J. Am. Chem. Soc.* **118,** 8487, 8488.

114. Tycko, R. (1999) Selective rules for multiple quantum NMR excitation in solids: derivation from time-reversal symmetry and comparison with simulations and C-13 NMR experiments. *J. Magn. Reson.* **139,** 302–307.

115. Ishii, T., Terao, T., and Kainosho, M. (1996) Relayed anisotropy correlation NMR: determination of dihedral angles in solids. *Chem. Phys. Lipids* **265,** 133–140.

116. Ishii, Y., Hirao, K., Terao, T., Terauchi, T., Oba, M., Nishiyama, K., et al. (1998) Determination of peptide γ angles in solids by relayed anisotropy correlation NMR. *Solid State Nucl. Magn. Reson.* **11,** 169–175.

117. Takegoshi, K., Imaizumi, T., and Terao, T. (2000) One- and two-dimensional C-13-H-1/N-15-H-1 dipolar correlation experiments under fast magic-angle spinning for determining the peptide dihedral angle phi. *Solid State Nucl. Magn. Reson* **16,** 271–278.

118. Schmidt-Rohr, K. (1996) Double-quantum solid-state NMR technique for determining torsion angles in polymers. *Macromolecules* **29,** 3975–3981.

119. Feng, X., Lee, Y. K., Sandstroem, D., Eden, M., Maisel, H., Sebald, A., et al. (1996) Direct determination of a molecular torsional angle by solid-state NMR. *Chem. Phys. Lett.* **257,** 314–320.

120. Feng, X., Eden, M., Brinkmann, A., Luthman, H., Ericksson, L., Graslund, A., et al. (1997) Direct determination of a peptide torsion angle psi by double-quantum solid-state NMR. *J. Am. Chem. Soc.* **119,** 12,006–12,007.

121. Feng, Y., and Gregor, P. (1997) Cloning of a novel member of the G protein-coupled receptor family related to peptide receptors. *Biochem. Biophys. Res. Commun.* **231,** 651–654.

122. Costa, P. R., Gross, J. D., Hong, M., and Griffin, R. G. (1997) Solid-state NMR measurement of Psi in peptides: a NCCN 2Q-heteronuclear local field experiment. *Chem. Phys. Lett.* **280,** 95–103.

123. Hong, M., Gross, J. D., and Griffin, R. G. (1997) Site-resolved determination of peptide torsion angle γ from the relative orientations of backbone N-H and C-H bonds by solid-state NMR. *J. Phys. Chem. B* **101,** 5869–5874.

124. Hong, M., Gross, J. D., Rienstra, C. M., Griffin, R. G., Kumashiro, K. K., and Schmidt-Rohr, K. (1997) Coupling amplification in 2D MAS NMR and its application to torsion angle determination in peptides. *J. Magn. Reson.* **129,** 85–92.

125. Gregory, D. M., Mehta, M. A., Shiels, J. C., and Drobny, G. P. (1997) Determination of local structure in solid nucleic acids using double quantum nuclear magnetic resonance spectroscopy. *J. Chem. Phys.* **107**, 28–42.

126. Bower, P. V., Oyler, N., Mehta, M. A., Long, J. R., Stayton, P. S., and Drobny, G. P. (1999) Determination of torsion angles in proteins and peptides using solid state NMR. *J. Am. Chem. Soc.* **121**, 8373–8375.

127. Feng, X., Verdegem, P. J., Eden, M., Sandstrom, D., Lee, Y. K., Bovee, G., et al. (2000) Determination of a molecular torsional angle in the metarhodopsin-I photointermediate of rhodopsin by double-quantum solid-state NMR. *J. Biomol. NMR* **16**, 1–8.

128. Huster, D., Arnold, K., and Gawrisch, K. (2000) Strength of Ca(2+) binding to retinal lipid membranes: consequences for lipid organization. *Biophys. J.* **78**, 3011–3018.

129. Huster, D., Yamaguchi, S., and Hong, M. (2000) Efficient β-sheet identification in proteins by solid-state NMR spectroscopy. *J. Am. Chem. Soc.* **122**, 11,320–11,327.

130. Ramamoorthy, A., Wu, C. H., and Opella, S. J. (1999) Experimental aspects of multidimensional solid-state NMR correlation spectroscopy. *J. Magn. Reson.* **140**, 131–140.

131. Kim, S., Quine, J., and Cross, R. (2001) Complete cross-validation and R-factor calculation of a solid-state NMR derived structure. *J. Am. Chem. Soc.* **123**, 7292–7298.

132. Opella, S. J., Marassi, F. M., Gesell, J. J., Valente, A.P., Kim, Y., Oblatt-Montal, M., et al. (1999) Structures of the M2 channel-lining segments from nicotinic acetylcholine and NMDA receptors by NMR spectroscopy. *Nat. Struct. Biol.* **6**, 374–379.

133. Marassi, F. M., and Opella, S. J. (2000) A solid-state NMR index of helical membrane protein structure and topology. *J. Magn. Reson.* **144**, 150–155.

134. Wang, J., Denny, J., Tian, C., Kim, S., Mo, Y., Kovacs, F., et al. (2000) Imaging membrane protein helical wheels. *J. Magn. Reson.* **144**, 162–167.

135. Fares, C., Sharom, F. J., and Davis, J. H. (2002) ^{15}N, ^{1}H heteronuclear correlation NMR of gramicidin A in DMPC-d_{67}. *J. Am. Chem. Soc.* **124**, 11, 232–11,233.

136. Straus, S. K., Scott, W., and Watts, A. (2003) Assessing the effects of time- and spatial-averaging in ^{15}N chemical shift/^{15}N-^{1}H dipolar correlation solid state NMR experiments. *J. Biomol. NMR* **26**, 283–295.

137. Denny, J. K., Wang, J., Cross, T. A., and Quine, J. R. (2001) PISEMA powder patterns and PISA wheels. *J. Magn. Reson.* **152**, 217–226.

138. Marassi, F. M. (2001) A simple approach to membrane protein secondary structure and topology based on NMR spectroscopy. *Biophys. J.* **80**, 994–1003.

139. Waugh, J. S. (1976) Uncoupling of local field spectra in nuclear magnetic resonance: determination of atomic positions in solids. *Proc. Natl. Acad. Sci. USA* **73**, 1394.

140. Hester, R. K., Ackerman, J. L., Cross, V. R., and Waugh, J. S. (1975) Resolved dipolar coupling spectra of dilute nuclear spins in solids. *Phys. Rev. Lett.* **34,** 993.

141. Opella, S. J., and Waugh, J. S. (1977) Two-dimensional ^{13}C NMR of highly oriented polyethylene. *J. Chem. Phys.* **66,** 4919.

142. Lee, M., and Goldburg, W. (1965) Nuclear-magnetic-resonance line narrowing by a rotating rf field. *Phys. Rev.* **140,** 1261–1271.

143. Bielecki, A., Kolbert, A. C., and Levitt, M. (1989) Frequency-switched pulse sequences: homonuclear decoupling and dilute spin NMR in solids. *Chem. Phys. Lett.* **155,** 341–346.

144. Gan, Z. (2000) Spin dynamics of polarization inversion spin exchange at the magic angle in multiple spin systems. *J. Magn. Reson.* **143,** 136–143.

145. Ramamoorthy, A., Wu, C. H., and Opella, S. J. (1995) Three-dimensional solid-state NMR experiment that correlates the chemical shift and dipolar coupling frequencies of two heteronuclei. *J. Magn. Reson. B* **107,** 88–90.

146. Gu, Z. T. and Opella, S. J. (1999) Three dimensional ^{13}C shift/^{1}H-^{15}N coupling/^{15}N shift solid state NMR correlation spectroscopy. *J. Magn. Reson.* **138,** 193–198.

147. Ramamoorthy, A., Wu, C. H., and Opella, S. J. (1997) Magnitudes and orientations of the principal elements of the ^{1}H chemical shift, ^{1}H-^{15}N dipolar coupling and ^{15}N chemical shift interaction tensors in ^{15}Nε-1-tryptophan and ^{15}Nε-histidine sidechains determined by three dimensional solid state NMR spectroscopy of polycrystalline samples. *J. Am. Chem. Soc.* **119,** 10,479–10,486.

148. Ketchem, R. R., Hu, W., and Cross, T. A. (1993) High-resolution conformation of gramicidin A in a lipid bilayer by solid-state NMR. *Science* **261,** 1457–1460.

149. Cross, T. A., Tian, F., Cotten, M., Wang, J., Kovacs, F., and Fu, R. (1999) Correlation of structure, dynamics and function in the gramicidin channel by solid-state NMR spectroscopy. In *Gramicidin and Related Ion Channel-Forming Peptides. Novartis Foundation Symposium 225.* Wiley, Chichester, England, pp. 4–22.

150. Hu, W., Lazo, N. D., and Cross, T. A. (1995) Tryptophan dynamics and structural refinement in a lipid bilayer environment: solid state NMR of the gramicidin channel. *Biochemistry* **34,** 14,138–14,146.

151. Hu, W. and Cross, T. A. (1995) Tryptophan hydrogen bonding and electric dipole moments: functional roles in the gramicidin channel and implications for membrane proteins. *Biochemistry* **34,** 14,147–14,155.

152. Separovic, F., Ashida, J., Woolf, T., Smith, R., and Terao, T. (1999) Determination of chemical shielding tensor of an indole carbon and application to tryptophan orientation of a membrane peptide. *Chem. Phys. Lett.* **303,** 493–498.

153. Koeppe, R. E., II, Killian, J. A., and Greathouse, D. V. (1994) Orientations of the tryptophan 9 and 11 sidechains of the gramicidin channel based on deuterium nuclear magnetic resonance spectroscopy. *Biophys. J.* **66,** 14–24.

154. Cotten, M., Tian, C., Busath, D. D., Shirts, R. B., and Cross, T. A. (1999) Modulating dipoles for structure-function correlations in the gramicidin A channel. *Biochemistry* **38,** 9185–9197.

155. Wang, J. F., Kovacs, F., and Cross, T. A. (2001) Structure of the transmembrane region of the M2 protein H⁺ channel. *Protein Sci.* **10,** 2241–2250.

156. Tian, C., Tobler, K., Lamb, R. A., Pinto, L. H., and Cross, T. A. (2002) Expression and initial structural insights from solid state NMR of the M2 proton channel from influenza A virus. *Biochemistry* **41,** 11,294–11,300.

157. Nishimura, K., Kim, S. G., Zhang, L., and Cross, T. A. (2002) The closed state of a H⁺ channel helical bundle combining precise orientational and distance restraints from solid state NMR—1. *Biochemistry* **41,** 13,170–13,177.

158. Montal, M. and Opella, S. J. (2002) The structure of the M2 channel-lining segment from the nicotinic acetylcholine receptor. *Biochim. Biophys. Acta* **1565,** 287–293.

159. Marassi, F. M. and Opella, S. J. (2003) Simultaneous assignment and structure determination of a membrane protein from NMR orientational restraints. *Protein Sci.* **12,** 403–411.

160. Shon, K.-J., Kim, Y., Colnago, L. A., and Opella, S. L. (1991) NMR studies of the structure and dynamics of membrane-bound bacteriophage Pf1 coat protein. *Science* **252,** 1303–1305.

161. Nambudripad, R., Stark, W., Opella, S. J., and Makowski, L. (1991) Membrane-mediated assembly of filamentous bacteriophage Pf1 coat protein. *Science* **252,** 1305–1308.

162. Spudich, J. L., Yang, C. S., Jung, K. H., and Spudich, E. N. (2000) Retinylidene proteins: structures and functions from archaea to humans. *Annu. Rev. Cell Dev. Biol.* **16,** 365–392.

163. Teller, D. C., Okada, T., Behnke, C. A., Palczewski, K., and Stenkamp, R. E. (2001) Advances in determination of a high-resolution three-dimensional structure of rhodopsin, a model of G-protein-coupled receptors (GPCRs). *Biochemistry* **40,** 7761–7772.

164. Menon, S. T., Han, M., and Sakmar, T. P. (2001) Rhodopsin: structural basis of molecular physiology. *Physiol. Rev.* **81,** 1659–1688.

165. Grobner, G., Choi, G., Burnett, I. J., Glaubitz, C., Verdegem, P.J., Lugtenburg, J., et al. (1998) Photoreceptor rhodopsin: structural and conformational study of its chromophore 11-*cis* retinal in oriented membranes by deuterium solid state NMR. *FEBS Lett.* **422,** 201–204.

166. Mollevanger, L. C. P. J., Kentgens, A. P. M., Pardoen, J. A., Courtin, J. M. L., Veeman, W. S., Lugtenburg, J., et al. (1987) High-resolution solid-state ¹³C-NMR study of carbons C-5 and C-12 of the chromophore of bovine rhodopsin. *Eur. J. Biochem.* **163,** 9–14.

167. Spooner, P. J. R., Sharples, J. M., Verhoeven, M. A., Lugtenburg, J., Glaubitz, C., and Watts, A. (2002) Relative orientation between the beta-ionone ring and the polyene chain for the chromophore of rhodopsin in native membranes. *Biochemistry* **41,** 7549–7555.

168. Verdegem, P. J., Bovee-Geurts, G., De Grip, W. J., Lugtenburg, J., and De Groot, H. J. (1999) Retinylidene ligand structure in bovine rhodopsin, metarhodopsin-I, and 10-methylrhodopsin from internuclear distance measurements using ¹³C-labeling and 1-D rotational resonance MAS NMR. *Biochemistry* **38,** 11,316–11,324.

169. Verhoeven, M. A., Creemers, A. F., Bovee-Geurts, P. H., De Grip, W. J., Lugtenburg, J., and De Groot, H. J. (2001) Ultra-high-field MAS NMR assay of a multispin labeled ligand bound to its G-protein receptor target in the natural membrane environment: electronic structure of the retinylidene chromophore in rhodopsin. *Biochemistry* **40,** 3282–3288.

170. Lansing, J. C., Hohwy, M., Jaroniec, C. P., Creemers, A.F., Lugtenburg, J., Herzfeld, J., et al. (2002) Chromophore distortions in the bacteriorhodopsin photocycle: evolution of the H-C14-C15-H dihedral angle measured by solid-state NMR. *Biochemistry* **41,** 431–438.

171. Hatcher, M. E., Hu, J. G., Belenky, M., Verdegem, P., Lugtenburg, J., Griffin, R. G., et al. (2002) Control of the pump cycle in bacteriorhodopsin: mechanisms elucidated by solid-state NMR of the D85N mutant. *Biophys. J.* **82,** 1017–1029.

172. Edman, K., Nollert, P., Royant, A., Belrhali, H., Pebay-Peyroula, E., Hajdu, J., et al. (1999) High-resolution x-ray structure of an early intermediate in the bacteriorhodopsin photocycle. *Nature* **401,** 822–826.

173. Luecke, H., Schobert, B., Richter, H. T., Cartailler, J. P., and Lanyi, J. K. (1999) Structure of bacteriorhodopsin at 1.55Å resolution. *J. Mol. Biol.* **291,** 899–911.

174. Sass, H., Büldt, G., Gessenich, R., Hehn, D., Neff, D., Schlesinger, J., et al. (2000) Structural alterations for proton translocation in the M state of wild-type bacteriorhodopsin. *Nature* **40,** 649–653.

175. Faham, S. and Bowie, J. U. (2002) Bicelle crystallization: a new method for crystallizing membrane proteins yields a monomeric bacteriorhodopsin structure. *J. Mol. Biol.* **316,** 1–6.

176. Royant, A., Edman, K., Ursby, T., Pebay-Peyroula, E., Landau, E. M., and Neutze, R. (2001) Spectroscopic characterization of bacteriorhodopsin's L-intermediate in 3D crystals cooled to 170K. *Photochem. Photobiol.* **74,** 794–804.

177. Schobert, B., Cupp-Vickery, J., Hornak, V., Smith, S., and Lanyi, J. (2002) Crystallographic structure of the K intermediate of bacteriorhodopsin: conservation of free energy after photoisomerization of the retinal. *J. Mol. Biol.* **321,** 715–726.

178. Lanyi, J. and Schobert, B. (2002) Crystallographic structure of the retinal and the protein after deprotonation of the Schiff base: the switch in the bacteriorhodopsin photocycle. *J. Mol. Biol.* **321,** 727–737.

179. Herzfeld, J. and Lansing, J. C. (2002) Magnetic resonance studies of the bacteriorhodopsin pump cycle. *Annu. Rev. Biophys. Biomol. Struct.* **31,** 73–95.

180. Seiff, F., Wallat, I., Ermann, P., and Heyn, M. P. (1985) A neutron diffraction study on the location of the polyene chain of retinal in bacteriorhodopsin. *Proc. Natl. Acad. Sci. USA* **82,** 3227–3231.

181. Fitter, J., Lechner, R. E., and Dencher, N. A. (1999) Interactions of hydration water and biological membranes studied by neutron scattering. *J. Phys. Chem.* **103,** 8036–8050.

182. Oesterhelt, D., Brauchle, C., and Hampp, N. (1991) Bacteriorhodopsin: a biological material for information processing. *Q. Rev. Biophys.* **24,** 425–478.

183. Ulrich, A. S., Heyn, M. P., and Watts, A. (1992) Structure determination of the cyclohexene ring of retinal in bacteriorhodopsin by solid-state deuterium NMR. *Biochemistry* **31,** 10,390–10,399.

184. Ulrich, A. S., Watts, A., Wallat, I., and Heyn, M. P. (1994) Distorted structure of the retinal chromophore in bacteriorhodopsin resolved by ^{2}H-NMR. *Biochemistry* **33,** 5370–5375.

185. Ulrich, A. S., Wallat, I., Heyn, M. P., and Watts, A. (1995) Re-orientation of retinal in the M-photointermediate of bacteriorhodopsin. *Nat. Struct. Biol.* **2,** 190–192.

186. Moltke, S., Wallat, I., Sakai, N., Nakanishi, K., Brown, M. F., and Heyn, M. P. (1999) The angles between the C(1)-, C(5)-, and C(9)-methyl bonds of the retinylidene chromophore and the membrane normal increase in the M intermediate of bacteriorhodopsin: direct determination with solid-state (2)H NMR. *Biochemistry* **38,** 11,762–11,772.

187. Mcdermott, A. E., Creuzet, F., Gebhard, R., Van Der Hoef, K., Levitt, M. H., Herzfeld, J., et al. (1994) Determination of internuclear distances and the orientation of functional groups by solid-state NMR: rotational resonance study of the conformation of retinal in bacteriorhodopsin. *Biochemistry* **33,** 6129–6136.

188. Thompson, L. K., Mcdermott, A. E., Raap, J., Van Der Wielen, C. M., Lugtenburg, J., Herzfeld, J., et al. (1992) Rotational resonance NMR study of the active site structure in bacteriorhodopsin: conformation of the Schiff base linkage. *Biochemistry* **31,** 7931–7938.

189. Lakshimi, K. V., Auger, M., Raap, J., Lugtenburg, J., Griffin, R. G., and Herzfeld, J. (1993) Internuclear distance measurement in a reaction intermediate: solid-state ^{13}C NMR rotational resonance determination of the Shiff base configuration in the M photointermediate of bacteriorhodopsin. *J. Am. Chem. Soc.* **115,** 8515, 8516.

190. Saito, H., Tuzi, S., Yamaguchi, S., Tanio, M., and Naito, A. (2000) Conformation and backbone dynamics of bacteriorhodopsin revealed by ^{13}C-NMR. *Biochim. Biophys. Acta* **1460,** 39–48.

191. Saito, H., Tuzi, S., and Naito, A. (1998) Empirical vs nonempirical evaluation of secondary structure of fibrous and membrane proteins. *Annu. Rep. NMR Specrosc.* **36,** 79–121.

192. Saito, H., Tuzi, S., Tanio, M., and Naito, A. (2002) Dynamic aspect of membrane proteins and membrane associated peptides as revealed by ^{13}C NMR: lessons from bacteriorhodopsin as an intact protein. *Annu. Rep. NMR Spectrosc.* **47,** 39–108.

193. Yamaguchi, S., Yonebayashi, K., Konishi, H., Tuzi, S., Naito, A., Lanyi, J. K., et al. (2001) Cytoplasmic surface structure of bacteriorhodopsin consisting of interhelical loops and C-terminal alpha helix, modified by a variety of environmental factors as studied by (13)C-NMR. *Eur. J. Biochem.* **268,** 2218–2228.

194. Tuzi, S., Hasegawa, J., Kawaminami, R., Naito, A., and Saito, H. (2001) Regioselective detection of dynamic structure of transmembrane alpha-helices as

revealed from (13)C NMR spectra of [3-13C]Ala-labeled bacteriorhodopsin in the presence of Mn^{2+} ion. *Biophys. J.* **81,** 425–434.

195. Shoji, A., Ozaki, T., Fujito, T., Deguchi, K., Ando, S., and Ando, I. (1990) ^{15}N chemical shift tensors and conformation of solid polypeptides containing ^{15}N-labeled L-alanine residues by ^{15}N NMR, 2: secondary structure reflected in sigma22. *J. Am. Chem. Soc.* **112,** 4693–4697.

196. Kamihira, M., Vosegaard, T., Mason, A. J., Straus, S., Nielsen, N. C., and Watts, A. (2004) Structural and orientational constraints on bacteriorhodopsin in purple membranes determined by oriented-sample solid-state NMR spectroscopy. *J. Mol. Biol.*, submitted.

197. Bak, M., Rasmussen, J. T., and Nielsen, N. C. (2000) SIMPSON: a general simulation program for solid-state NMR spectroscopy. *J. Magn. Reson.* **147,** 296–330.

198. Bak, M., Schultz, R., Vosegaard, T., and Nielsen, N. C. (2002) Specification and visualization of anisotropic interaction tensors in polypeptides and numerical simulations in biological solid-state NMR. *J. Magn. Reson.* **154,** 28–45.

199. Subramaniam, S. and Henderson, R. (2000) Molecular mechanism of vectorial proton translocation by bacteriorhodopsin. *Nature* **406,** 653–657.

200. Tuzi, S., Naito, A., and Saito, H. (1996) Temperature-dependent conformational change of bacteriorhodopsin as studied by solid-state ^{13}C NMR. *Eur. J. Biochem.* **239,** 294–301.

201. Yamaguchi, S., Tuzi, S., Tanio, M., Naito, A., Lanyi, J. K., Needleman, R., et al. (2000) Irreversible conformational change of bacterio-opsin induced by binding of retinal during its reconstitution to bacteriorhodopsin, as studied by (13)C NMR. *J. Biochem. (Tokyo)* **127,** 861–869.

202. Tuzi, S., Yamaguchi, S., Tanio, M., Konishi, H., Inoue, S., Naito, A., et al. (1999) Location of a cation-binding site in the loop between helices F and G of bacteriorhodopsin as studied by ^{13}C NMR. *Biophys. J.* **76,** 1523–1531.

203. Hu, J. G., Sun, B. Q., Petkova, A. T., Griffin, R. G., and Herzfeld, J. (1997) The predischarge chromophore in bacteriorhodopsin: a ^{15}N solid-state NMR study of the L photointermediate. *Biochemistry* **36,** 9316–9322.

204. Hu, J. G., Sun, B. Q., Bizounok, M., Hatcher, M.E., Lansing, J. C., Raap, J., et al. (1998) Early and late M intermediates in the bacteriorhodopsin photocycle: a solid-state NMR study. *Biochemistry* **37,** 8088–8096.

205. Engelhard, M., Hess, B., Erneis, D., Metz, G., Kreutz, W., and Siebert, F. (1989) Magic angle sample spinning ^{13}C nuclear magnetic resonance of isotopically labeled bacteriorhodopsin. *Biochemistry* **28,** 3967–3975.

206. Metz, G., Engelhard, M., and Siebert, F. (1992) High-resolution solid state ^{13}C NMR of bacteriorhodopsin: characterization of [4-14C]Asp resonances. *Biochemistry* **31,** 455–462.

207. Metz, G., Siebert, F., and Engelhard, M. (1992) Asp85 is the only internal aspartic acid that gets protonated in the M intermediate and the purple-to-blue transition of bacteriorhodopsin: a solid-state ^{13}C CP-MAS NMR investigation. *FEBS Lett.* **303,** 237–241.

208. Griffiths, J. M., Bennett, A. E., Engelhard, M., Siebert, F., Raap, J., Lugtenburg, J., et al. (2000) Structural investigation of the active site in bacteriorhodopsin: geometric constraints on the roles of Asp-85 and Asp-212 in the proton-pumping mechanism from solid state NMR. *Biochemistry* **39**, 362–371.

209. Kawase, Y., Tanio, M., Kira, A., Yamaguchi, S., Tuzi, S., Naito, A., et al. (2000) Alteration of conformation and dynamics of bacteriorhodopsin induced by protonation of Asp 85 and deprotonation of Schiff base as studied by ^{13}C NMR. *Biochemistry* **39**, 14,472–14,480.

210. Tanio, M., Inoue, S., Yokota, K., Seki, T., Tuzi, S., Needleman, R., et al. (1999) Long-distance effects of site-directed mutations on backbone conformation in bacteriorhodopsin from solid state NMR of [1-13C]Val-labeled proteins. *Biophys. J.* **77**, 431–442.

211. Tanio, M., Tuzi, S., Yamaguchi, S., Kawaminami, R., Naito, A., Needleman, R., et al. (1999) Conformational changes of bacteriorhodopsin along the proton-conduction chain as studied with (13)C NMR of [3-(13)C]Ala-labeled protein: arg(82) may function as an information mediator. *Biophys. J.* **77**, 1577–1584.

212. Helmle, M., Patzelt, H., Ockenfels, A., Gartner, W., Oesterhelt, D., and Bechinger, B. (2000) Refinement of the geometry of the retinal binding pocket in dark-adapted bacteriorhodopsin by heteronuclear solid-state NMR distance measurements. *Biochemistry* **39**, 10,066–10,071.

213. Petkova, A. T., Hatanaka, M., Jaroniec, C. P., Hu, J. G., Belenky, M., Verhoeven, M., et al. (2002) Tryptophan interactions in bacteriorhodopsin: a heteronuclear solid-state NMR study. *Biochemistry* **41**, 2429–2437.

214. Engelhard, M., Finkler, S., Metz, G., and Siebert, F. (1996) Solid-state C-13-NMR of [(3-C-13)Pro]bacteriorhodopsin and [(4-C-13)Pro]bacterhiorhodopsin-evidence for a flexible segment of the C-terminal tail. *Eur. J. Biochem.* **235**, 526–533.

215. Lansing, J. C., Hu, J. G., Belenky, M., Griffin, R. G., and Herzfeld, J. (2003) Solid-state NMR investigation of the buried X-Proline bonds of bacteriorhodopsin. *Biochemistry* **42**, 3586–3593.

216. Arakawa, T., Shimono, K., Yamaguchi, S., Tuzi, S., Sudo, Y., Kamo, N., et al. (2002) Dynamic structure of pharaonis phoborhodopsin (sensory rhodopsin II) and complex with a cognate truncated transducer as revealed by site-directed ^{13}C solid-state NMR. *FEBS Lett.* **536**, 237–240.

217. Sundle, M. and Blake, C. (1998) From the globular to the fibrous state: protein structure and structural conversion in amyloid formation. *Q. Rev. Biophys.* **31**, 1–39.

218. Sundle, M., Serpell, L., Bartlam, M., Fraser, P., Pepys, M., and Blake, C. (1997) Common core structure of amyloid fibrils by synchrotron X-ray diffraction. *J. Mol. Biol.* **33**, 729–739.

219. Shirahama, T. and Cohen, A. S. (1967) High resolution electron microscopic analysis of the amyloid fibril. *J. Cell Biol.* **33**, 679–706.

220. Cohen, A. S., Shirahama, T., and Skinner, M. (1982) Electron microscopy of amyloid. In *Electron Microscopy of Protein* (Harris, I. R., ed.). Academic Press, London, pp. 165–205.

221. Selkoe, D. J. (1991) The molecular pathology of Alzheimer's disease. *Neuron* **6,** 487–498.

222. Cottingham, M., Hollinshead, M., and Djt, V. (2002) Amyloid fibril formation by a synthetic peptide from a region of human acetylcholinesterase that is homologous to the Alzheimer's amyloid-β peptide. *Biochemistry* **41,** 13,539–13,547.

223. Perutz, M. F., Finch, J. T., Berriman, J., and Lesk, A. (2002) Amyloid fibers are water-filled nanotubes. *Proc. Natl. Acad. Sci. USA* **99,** 5591–5595.

224. Sundle, M. and Blake, C. (1997) The structure of amyloid fibrils by electron microscopy and X-ray diffraction. *Adv. Protein Chem.* **50,** 123–159.

225. Astbury, W. T., Dickinson, S., and Bailey, K. (1935) The X-ray interpretation of denaturation and the structure of seed globulins. *Biochem. J.* **29,** 2351–2360.

226. Lansbury, P. (1999) Evolution of amyloid: what normal protein folding may tell us about fibrillogenesis and disease. *Proc. Natl. Acad. Sci. USA* **96,** 3342–3344.

227. Bucciantini, M., Giannoni, E., Chiti, F., Baroni, F., Formigli, L., Zurdo, J., et al. (2002) Inherent toxicity of aggregates implies a common mechanism for protein misfolding diseases. *Nature* **416,** 507–511.

228. Roher, A. E., O'Chaney, M., Kuo, Y., Webster, S. D., Stine, W. B., Haverhams, L. J., et al. (1996) Morphology and toxicity of Aβ-(1-42) dimer derived from neuritic and vascular amyloid deposits of Alzheimer's disease. *J. Biol. Chem.* **271,** 20,631–30,635.

229. Barrow, C. J. and Zagorski, M. G. (1991) Solution structures of β-peptide and its constituent fragments relation to amyloid deposition. *Science* **253,** 179–182.

230. Zagorski, M. G. and Barrow, C. J. (1992) NMR studies of amyloid β-peptide: proton assignments, secondary structure and mechanism of an α-helix-β-sheet conversion for a homologous 28 residue N-terminal fragment. *Biochemistry* **31,** 5621–5631.

231. Talafous, K., Marcinowski, K., Klopman, G., and Zagorski, M. G. (1994) Solution structure of residues 1-28 of the amyloid β-peptide. *Biochemistry* **33,** 7788–7796.

232. Sorimachi, K. and Craik, D. (1994) Structural determination of extracellular fragments of amyloid proteins involved in Alzheimer's disease and Dutch-type hereditary cerebral haemorrhage with amyloidosis. *Eur. J. Biochem.* **219,** 237–251.

233. Fletcher, F. and Keire, D. (1997) The interaction of β-amyloid protein fragment (12-28) with lipid environments. *Protein Sci.* **6,** 666–675.

234. Elagnaf, O., Guthrie, D., Walsh, D., and Irvine, G. (1998) The influence of the central region containing residues 19-25 on the aggregation properties and secondary structure of Alzheimer's β-amyloid peptide. *Eur. J. Biochem.* **256,** 560–569.

235. Kohno, T., Kobayashi, K., Maeda, T., Sato, K., and Takashima, A. (1996) Three-dimensional structures of amyloid β peptide (25-35) in membrane-mimicking environment. *Biochemistry* **35,** 16,094–16,104.

236. Soto, C., Castano, E., Frangione, B., and Inestrosa, N. (1995) The α-helical to β-sheet transition in the amino-terminal fragment of the amyloid β-peptide modulates amyloid formation. *J. Biol. Chem.* **270,** 3063–3067.

237. Coles, M., Bicknell, W., Watson, A., Fairlie, D., and Craik, D. (1998) Solution structure amyloid β-peptide (1-40) in a water-micelle environment: is the membrane-spanning domain where we think it is? *Biochemistry* **37,** 11,064–11,077.

238. Watson, A., Fairlie, D., and Craik, D. (1998) Does oxidation affect conformation switching? *Biochemistry* **37,** 12,700–12,706.

239. Sticht, H., Bayer, P., Willbold, D., Dames, S., Hilbich, K., Beyreuther, K., et al. (1995) Structure of amyloid Aβ(1-40)-peptide of Alzheimer's disease. *Eur. J. Biochem.* **233,** 293–298.

240. Shao, H., Jao, S., Ma, J., and Zagorski, M. G. (1999) Solution structures of micelle-bound amyloid β-(1-40) and β-(1-42) peptides of Alzheimer's disease. *J. Mol. Biol.* **285,** 755–773.

241. Gregory, D., Mitchell, D., Stringer, J., Kihne, S., Shiels, J. C., Callahan, J., et al. (1995) Windowless dipolar recoupling the detection of weak dipolar couplings between spin 1/2 nuclei with large chemical shift anisotropies. *Chem. Phys. Lett.* **246,** 654–663.

242. Pan, Y., Guillon, T., and Schaefer, J. (1990) Determination of C-N internuclear distances by rotational-echo double resonance NMR of solids. *J. Magn. Reson.* **90,** 330–340.

243. Peersen, O. B., Wu, X. L., Kustanovich, I., and Smith, S. O. (1993) Variable-amplitude cross-polarization MAS NMR. *J. Magn. Reson. Ser. A* **104,** 334–339.

244. Ireland, P. S., Olson, L. W., and Brown, T. L. (1975) Spin echo double resonance detection of deuterium quadrupole resonance transitions in pentacarbonylmanganese-d. *J. Am. Chem. Soc.* **97,** 3548, 3549.

245. Warren, W. S., Weitekamp, D. P., and Pines, A. (1980) Theory excitation of multiple-quantum transitions. *J. Chem. Phys.* **73,** 2084–2099.

246. Yen, Y.-S. and Pines, A. (1983) Multiple quantum NMR in solids. *J. Chem. Phys.* **78,** 3579–3582.

247. Baum, J., Munowitz, M., Garroway, A. N., and Pines, A. (1985) Multiple quantum dynamics in solid state NMR. *J. Chem. Phys.* **83,** 2015–2025.

248. Suter, D., Liu, S. B., Baum, J., and Pines, A. (1987) Multiple quantum NMR excitation with a one-quantum Hamiltonian. *Chem. Phys. Lett.* **114,** 103–109.

249. Benzinger, T. L. S., Gregory, D. M., Burkoth, T. S., Miller-Auer, H., Lynn, D. G., Botto, R. E., et al. (2000) Two-dimensional structure of beta-amyloid (10-35) fibrils. *Biochemistry* **39,** 3491–3499.

250. Antzutkin, O. and Tycko, R. (1999) High-order multiple quantum excitation in C-13 nuclear magnetic resonance spectroscopy of organic solids. *J. Chem. Phys.* **110,** 2749–2752.

251. Antzutkin, O. N., Balbach, J. J., Leapman, R. D., Rizzo, N. W., Reed, J., and Tycko, R. (2000) Multiple quantum solid state NMR indicates a parallel, not antiparallel, organization of the beta-sheets in Alzheimer's beta amyloid fibrils. *Proc. Natl. Acad. Sci. USA* **97,** 13,045–13,050.

252. Balbach, J. J., Petkova, A. T., Oyle, N. A., Antzutkin, O. N., Gordon, D. J., Meredith, S.C., et al. (2002) Supramolecular structure in full-length Alzheimer's

β-amyloid fibrils evidence for a parallel β-sheet organization from solid state NMR. *Biophys. J.* **83,** 1205–1216.

253. Grobner, G., Glaubitz, C., Williamson, P. T. F., Hadingham, T., and Watts, A. (2001) Structural insight into the interaction of amyloid-β peptide with biological membranes by solid state NMR. *Focus Struct. Biol.* **1,** 203–214.

254. Suwelack, D., Rothwell, W. P., and Waugh, J. S. (1980) Slow molecular motion detected in the NMR spectra of rotating solids. *J. Chem. Phys.* **73,** 2559–2569.

255. Rothwell, W. P. and Waugh, J. S. (1981) Transverse relaxation of dipolar coupled spin system under rf irradiation: detecting motions in solid. *J. Chem. Phys.* **74,** 2721–2732.

256. Nishimura, K., Fu, R., and Cross, T. A. (2001) The effect of rf inhomogeneity on heteronuclear dipolar recoupling in solid state NMR: practical performance of SFAM and REDOR. *J. Magn. Reson.* **152,** 227–233.

257. Vosegaard, T. and Nielsen, N. C. (2002) Towards high-resolution solid-state NMR on large uniformly ^{15}N- and [^{13}C, ^{15}N]-labeled membrane proteins in oriented lipid bilayers. *J. Biomol. NMR* **22,** 225–247.

258. Fu, R., Tian, C., and Cross, T. A. (2002) NMR spin locking of proton magnetization under a frequency-switched Lee-Goldburg pulse sequence. *J. Magn. Reson.* **154,** 130–135.

259. Fu, R., Tian, C., Kim, H., Smith, S. A., and Cross, T. A. (2002) The effect of Hartmann-Hahn mismatching on polarization inversion spin exchange at the magic angle. *J. Magn. Reson.* **159,** 167–174.

260. Nevzorov, A. A. and Opella, S. J. (2003) Structural fitting of PISEMA spectra of aligned proteins. *J. Magn. Reson.* **160,** 33–39.

Index